Fifth International Conference on Materials Science and Manufacturing Technology (ICMSMT-2023)

AA007495

IOP Conference Series: Materials Science and Engineering
Volume 1291

Tamil Nadu, India
12-14 April 2023

Editors:

Ramya Muthusamy
Selvaraju M

ISBN: 978-1-7138-7970-1
ISSN: 1757-8981

Printed from e-media with permission by:

Curran Associates, Inc.
57 Morehouse Lane
Red Hook, NY 12571

Some format issues inherent in the e-media version may also appear in this print version.

Institute of Physics
Dirac House, Temple Back
Bristol BS1 6BE UK

Phone: 44 1 17 929 7481
Fax: 44 1 17 920 0979

techtracking@iop.org

Additional copies of this publication are available from:

Curran Associates, Inc.
57 Morehouse Lane
Red Hook, NY 12571 USA
Phone: 845-758-0400
Fax: 845-758-2633
Email: curran@proceedings.com
Web: www.proceedings.com

TABLE OF CONTENTS

ADVANCES IN MATERIALS SCIENCE AND ENGINEERING

IOP Conf. Series: Materials Science and Engineering 1291 (2023) 011001 doi:10.1088/1757-899X/1291/1/011001

Preface

Materials and technologies are significant elements for all kinds of high-tech industries that pave the road for advancements in the manufacturing area. With the rapid development of computer technology, communications technology, and network technology, the traditional manufacturing process has evolved to intelligent manufacturing that is more technologically flexible and efficient.

The Fifth International Conference on Materials Science and Manufacturing Technology 2023 (ICMSMT 2023) was a premier interdisciplinary platform for researchers, academicians, and practitioners worldwide to present recent developments in materials sciences and manufacturing technology. The fifth edition of the ICMSMT conference series (ICMSMT 2023) has been organized with partnered by Akshaya College of Engineering and Technology, Coimbatore, and Diligentec Solutions, Coimbatore, India. The conference received submissions from 15 countries, and the editorial board of ICMSMT has accepted the papers after the stringent screening and review process. ICMSMT is a scientific event and aimed to be conducted in the second week of April every year.

ICMSMT 2023 was organized in an hybrid model with both online and physical participations. Each presenter was given ten minutes for the presentation and three minutes for queries and discussion. The online meetings have been conducted using the Google Meet platform.

Dr. Ramya Muthusamy

Editor / Chair – TPC

ICMSMT 2023

IOP Conf. Series: Materials Science and Engineering 1291 (2023) 011001 doi:10.1088/1757-899X/1291/1/011001

Committees

Conference Chair(s)
Dr. Thangaprakash Sengodan, Diligentec Solutions, Coimbatore, India
Dr. Nagaraja S, Akshaya College of Engineering and Technology, Coimbatore, India

General Chair(s)
Dr. Baharudin Bin Ismail, Universiti Malaysia Perlis, Malaysia
Dr. Kapilan S, Akshaya College of Engineering and Technology, India

Organizing Chair(s)
Dr. Senthilkumar S, Diligentec Solutions, Coimbatore, India
Dr. Karthik A, Akshaya College of Engineering and Technology, India

Editorial Committee

Editors
Dr. Ramya Muthusamy, Diligentec Solutions, Coimbatore, India
Prof. Selvaraju M, Akshaya College of Engineering and Technology, India

Review Committee

Chair
Dr. Markas Law Lee Lam, National University of Singapore, Singapore

Members
Prof. Sevugan Rajesh, Akshaya College of Engineering and Technology, India
Dr. Sumathi K, Akshaya College of Engineering and Technology, India

Information Contact
Dr. Rajkumar Ramanujam, Akshaya College of Engineering and Technology, India

Publicity and Promotions
Mr. Sivakumar K, Diligentec Solutions, Coimbatore

Technical Program Committee Chairs
Dr. Siva Sankari K, Vice-Principal, Akshaya College of Engineering and Technology, India
Dr. Sabareesaan K J, Mechanical and Industrial Section, Nizwa College of Technology, Oman
Dr. Moorthi, S R Engineering College, Warangal, India
Dr. Rajamani R, PSG College of Technology, Coimbatore, India

Technical Program Committee
Dr. Jaroslaw Drelich, Michigan Technological University, USA
Dr. Chong Wen Tong, University of Malaya, Malaysia
Dr. Mohamed Arezki Mellal, M'Hamed Bougara University, Algeria
Dr. Khairel Rafezi Ahmad, University Malaysia Perlis, Malaysia
Dr. Mahanthi S D, Michigan State University, USA
Dr. Mohammed Henini, University of Nottingham, UK
Dr. Sharifah Shahnaz Syed Bakar, University Malaysia Perlis, Malaysia
Dr. Alexandru Bogdan Andreescu, Transilvania University of Brasov, Romania
Dr. Norinsan Kamil Othman, University Science Malaysia, Malaysia
Dr. Mitsuru Itoh, Tokiyo Institute of Technology, Japan
Dr. Sanjeev Kumar, Aalborg University, Denmark

Dr. Ramani Kannan, University Technology Petronas, Malaysia
Dr. Satyabrata Das, Indian Institute of Technology, Kanpur
Dr. Singh S P, Indian Institute of Technology, Delhi
Dr. Rajiv O. Dusane, Indian Institute of Technology, Bombay
Dr. Kamal K Kar, Indian Institute of Technology, Kanpur
Dr. Ashish Kumar Mishra, Indian Institute of Technology, Varanasi
Dr. Anand K, Indian Institute of Technology Madras, India
Dr. Sushil Mishra, Indian Institute of Technology, Bombay
Dr. Ajit Panigrahi, CSIR-Institute of Minerals and Materials Technology (IMMT), Bhubaneswar
Dr. Somnath Bhowmick, Indian Institute of Technology, Kanpur
Dr. Patil P, Indian Institute of Technology, Varanasi
Dr. Amrita Bhattacharya, Indian Institute of Technology, Bombay
Dr. Akhilesh Kumar Singh, Indian Institute of Technology, Varanasi
Dr. Krishnakant, Indian Institute of Technology, Delhi
Dr. Alankar Alankar, Indian Institute of Technology, Bombay
Dr. Vanitha C, National Institute of Technology, Warangal
Dr. Tanmoy Maiti, Indian Institute of Technology, Kanpur
Dr. Jitendra Kumar, Indian Institute of Technology, Varanasi
Dr. Rer. Nat. Ravikumar, Indian Institute of Technology, Madras
Dr. Bikash Kumar, CSIR-IMMT, Bhubaneswar
Dr. Ramkumar J, Indian Institute of Technology, Kanpur
Dr. Jyoti Prasad Borah, National Institute of Technology, Nagaland
Dr. Chandana Rath, Indian Institute of Technology, Varanasi
Dr. Mani V N, C-MET, DIETY, Government of India, India
Dr. Bimal Prasad Singh, CSIR-IMMT, Bhubaneswar
Dr. Vivek Verma, Indian Institute of Technology, Kanpur
Dr. Parag Tandaiya, Indian Institute of Technology, Bombay
Dr. Honey John, Cochin University of Science and Technology, Cochin
Dr. Manas K Dalai, CSIR-IMMT, Bhubaneswar
Dr. Nagaraja S, Akshaya College of Engineering Technology, India
Dr. Mathan Kumar N, Akshaya College of Engineering and Technology, India
Prof. Selvaraju M, Akshaya College of Engineering and Technology, India
Dr. Barkunan S R, Akshaya College of Engineering and Technology, India
Dr. Balamurugan, Amrita School of Engineering, Coimbatore, India
Dr. Josephine R L, National Institute of Technology Trichy (NITT), India
Dr. Kulkarni S M, National Institute of Technology (NITK), Surathkal, India
Dr. Gururaj Punekar, National Institute of Technology (NITK), Surathkal, India

Peer Review Statement

All papers published in this volume have been reviewed through processes administered by the Editors. Reviews were conducted by expert referees to the professional and scientific standards expected of a proceedings journal published by IOP Publishing.

- **Type of peer review:** Single Anonymous
- **Conference submission management system:** Morressier
- **Number of submissions received:** 93
- **Number of submissions sent for review:** 64
- **Number of submissions accepted:** 58
- **Acceptance Rate (Submissions Accepted / Submissions Received × 100):** 62.4
- **Average number of reviews per paper:** 2.35
- **Total number of reviewers involved:** 86
- **Contact person for queries:**
  Name: Dr. Ramya Muthusamy
  Email: ramsthangam@gmail.com
  Affiliation: Diligentec Solutions, Coimbatore, Tamil Nadu, India

IOP Conf. Series: Materials Science and Engineering 1291 (2023) 012003 doi:10.1088/1757-899X/1291/1/012003

Organic-inorganic perovskite structure-based (OIPSB) solar cells: A progress from invention

Mehtab Ur Rehman and Wang Qun

Faculty of Materials and Manufacturing, Beijing University of technology, No.100, Pingleyuan, Chaoyang District, Beijing, China
Emails: mehtaburrehman7@gmail.com, wanq@bjut.edu.cn

Abstract. The organic-inorganic perovskite structure-based (OIPSB) materials are key material for optoelectronic industry. In this review paper, we have discussed some physical properties of OIPSB compounds (lead-based) which we got from previous research findings. The main compounds of OIPSB have been given with their electronic and optical properties. Furthermore, their applications are discussed in solar cells with device structures. A full mini review of organic-inorganic perovskite solar cells (OIPSCs) is given from its initial Power conversion efficiency (PCE) <4% to highest PCE of > 24%.

Keywords: Solar cells, Organic-inorganic Perovskite, Power conversion efficiency (PCE), Green energy

1. Introduction

With the passage of time, the energy demands rapidly increases with world population. Presently, the electrical energy is a life basics, which is used in charging mobile phones, lamps, electrical cars, machines and in every gadget, which reduces the human efforts and work smartly. With increase in population, the number of electrical appliances increases but the energy stock is limited. To get electrical energy we have two ways, to go for non-renewable energy and renewable energy resources. The non-renewable energy resources are burning fuels like petrol, diesel, kerosene oil, wood and coal for generating energy. Firstly, these resources are limited in stock and even cannot be used twice. Secondly, they are expensive, noisy to produce energy, not eco-friendly and most dangerous. Alternatively, the renewable energy resources include air-turbines, building dams and solar cells for producing electrical energy. The air-turbines and building dams requires more space and installation cost as compared to solar cells. The sun provides continuously energy to earth almost 8 to 14 hours per day. So, we can use this Sun energy directly by using solar cells. The traditional solar cells are silicon solar cells, which are quite expensive that reduces its availability to local market. The third-generation solar cells are organic-inorganic perovskite solar cells (OIPSCs) which uses organic-inorganic compounds as an absorber in solar cells architecture to produce cost effective, clean, ecofriendly and efficient electrical energy. The PCE of these solar cells increased from 3.7% to 24% within a very short period of time. The PCE is dependent on two main components, firstly organic-inorganic material and secondly device structure. We have provided literature regarding materials and device architecture of OIPSB solar cells [1-10].

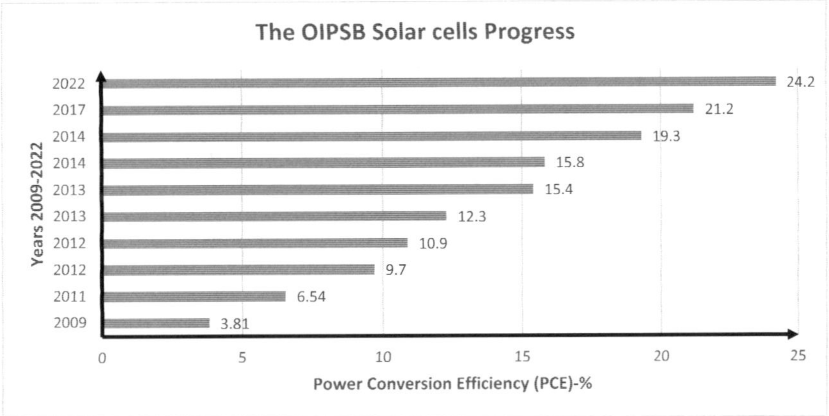

Figure. 1. Progress of OIPSB solar cells from its invention (2009-2022) [1-10].

2. Organic-inorganic perovskite structure-based (OIPSB) compounds

The number of OIPSB compounds has been developed, the organic-inorganic lead halides ($CH_3NH_3PbX_3$) where (X=Cl, Br, I) and organic-inorganic tin halides ($CH_3NH_3SnX_3$) where (X=Cl, Br, I) with their blend structures using different techniques. Among all these compounds the OIPSB $CH_3NH_3PbI_3$ Shows exceptional character in solar cells as radiation absorber which crossed PCE of 24% [10]. The verified efficiency is available at National Renewable Energy Laboratory (NREL) efficiency chart [11]. The easy processing, cost effectiveness and natural abundancy make this material beyond solar cells devices such as water splitting [12], light emitting diodes [13, 14] and lasers [15-19].

Recently we used DFT approach to investigate electronic and optical properties of all lead- based OIPSB compounds ($CH_3NH_3PbX_3$) where (X=Cl, Br, I) and found that $CH_3NH_3PbI_3$ has narrow band gap as compared to $CH_3NH_3PbCl_3$ and $CH_3NH_3PbBr_3$ compounds. The direct band gap of $CH_3NH_3PbI_3$ is found to be 1.9 eV and optical absorption α (ω) is 4283.20 at 3.18 eV which is quite impressive compared with silicon [20]. The absorption spectra increase in visible range and reduces after 3.2 eV and make some ups and down due to multi-elements in of $CH_3NH_3PbI_3$ compound. The Optical conductivity σ (ω) is 69.071 cm^{-1} at 372 nm (3.34 eV) determined by GGA. The valence band and conduction bands are mostly occupied by halides and lead orbitals respectively. The overall $CH_3NH_3PbI_3$ is most suitable material for organic solar cells device [21].

In this review we have given the details of mainly $CH_3NH_3PbI_3$-based solar cells progress based on PCE and device configuration. The review is taken from 2009 since first organic-inorganic hybrid solar cells discovered.

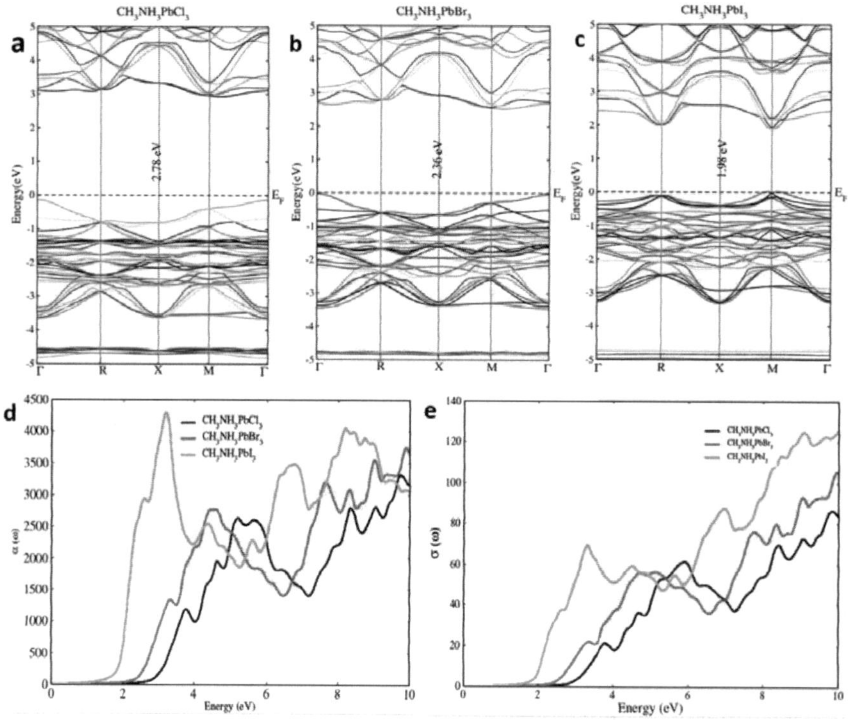

Figure. 2. Band structures of (a) $CH_3NH_3PbCl_3$, (b) $CH_3NH_3PbCl_3$, (c) $CH_3NH_3PbI_3$, (d) Absorption coefficient and (e) optical conductivity of $CH_3NH_3PbX_3$.

3. OIPSB solar cells progress

The OIPSB solar cells Perovskite solar cells are a type of solar cell that use a material called perovskite as the light-absorbing layer. The OIPSB solar cells have attracted significant attention due to their high efficiency, low-cost, and easy-processing features. In recent years, significant progress has been made in the development of OIPSB solar cells. The efficiency of these cells has increased rapidly, with some research groups reporting efficiencies above 24%. This is a significant improvement over the initial efficiencies of around 4% that were reported when OIPSB solar cells were first developed in 2009 [1].

Actually, in OPISB solar cells the organic-inorganic compound is sandwich between hole transport material (HTL) and electron transport material (ETL). The device configuration has been changing time to time by adding transparent electrodes and more elements layers for better efficiency. The general structure of OIPSB solar cell structure is given in fig 3 in which organic-inorganic perovskite is sandwich between ETL and HTL on substrate usually glass.

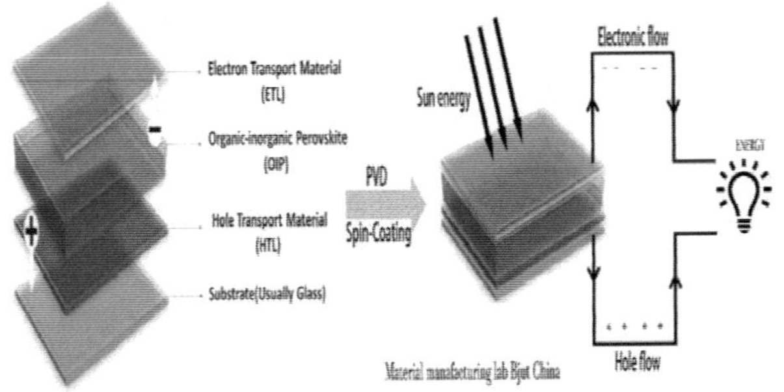

Figure. 3. General configuration of OIPSB solar cells.

In this paper we will full review the efficiency progress starts from 3.8 % to 24.42 % of lead iodide-based solar cells with device configuration, open-circuit voltage V_{oc}, fil factor ff, short-circuit photocurrent density J_{sc} (mA/cm^2) etc.

In 2009, the first OIPSB solar cells was introduced by kojima and workers. They used $CH_3NH_3PbI_3$ and $CH_3NH_3PbBr_3$ as sensitizer in dye-sensitized solar cell (DSSC) which shown PCE of 3.81 % and 3.13 % respectively. The $CH_3NH_3PbI_3$-based solar cells shown high PCE while high voltage 0.96 V was obtained using $CH_3NH_3PbBr_3$-based solar cells. Structurally, they coated N-type semiconductor TiO_2 nanocrystalline paste on fluorine-doped SnO_2 transparent glass (FTO, Nippon Sheet Glass 10 Ω/sq). The thickness of TiO_2 was kept 8-12 μm using screen printers and sintering at temperature of 480 °C for 1 hour in air. The nanocrystalline particles of $CH_3NH_3PbX_3$ (X = Br, I) were deposited by self-organization process which started with precursor solution having stoichiometric amounts of PbX_2 and CH_3NH_3X. The film formation of $CH_3NH_3PbX_3$ on TiO_2 was done by spin-coating and X-ray diffraction (XRD) analysis also confirmed that films formation on TiO_2 of both OIPSB compounds are crystalline in nature as well as perovskite structure formation. Finally, the common device configuration was as FTO/TiO_2/$CH_3NH_3PbX_3$, (X = Br, I) for first OIPSB solar cells. The short-circuit photocurrent densities J_{sc} (mA/cm^2) and fil factor are 11.0, 5.57 and 0.59, 0.57 for $CH_3NH_3PbI_3$ and $CH_3NH_3PbBr_3$ respectively [1].

After two years, in 2011 Jeong Hyeok Im *et .al* prepared efficient quantum dots sensitized solar cell with 2-3 nm $CH_3NH_3PbI_3$ nanocrystal as light absorber layer. The achieved PCE of 6.54% with V_{oc} of 0.605, ff = 0.513 and J_{sc} = 13.31. The device architecture is as follows; FTO/TiO_2/ $CH_3NH_3PbI_3$. The FTO was washed with ethanol using ultrasonic bath for 20 mints and heated at 500 °C continuously for 15 mints. After that, TiO_2 particles having size of 20 nm are deposited on FTO substrate and again heated to 1 h at 550 °C. After that, the perovskite solution was spread on TiO_2 film and keep it for 1 min so that it penetrates in film and then spun it for 40 s at 2000 rpm (round per minute) in ambient atmosphere. Finally, the last perovskite layer was dried for 30 mints at temperature ranges from 40-160 °C [2].

After that in 2012, Hui Seon Kim et .al report solid state mesoscopic heterojunction organic-inorganic solar cells having $CH_3NH_3PbI_3$ nanoparticles. The achieved PCE of 9.7 % with V_{oc} = 0.88 V, ff = 0.62 and J_{sc} = 17.6 mA/cm^2 when exposed to sun light. The device configuration is as FTO/TiO_2+spiro-MeOTAD/$(CH_3NH_3)PbI_3$. The pores of submicron TiO_2 film was filled by spiro-MeOTAD which acts as hole conductor. So, introducing hole conductor in OIP solar cells configuration increased the PCE as well as device stability as compared to solution based- cells [3].

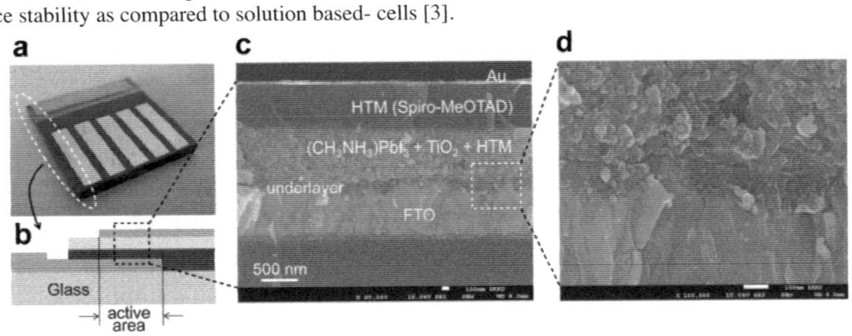

Figure. 4. (a) Solid-state device structure, (b) cross-sectional device structure, (c) cross-sectional SEM image of device and (d) active layer-underlayer FTO interfacial structure [3].

At the same year, Michael M. Lee with coworkers fabricated cost-effective solution-processable highly efficient solar cells using highly crystalline $CH_3NH_3PbI_3$ material as light absorber. They achieved PCE of

10.9 % when exposed to light with Voc = 0.98 V, J_{sc} = 17.8 mA/cm^2 and ff = 0.63. the device configuration was FTO/TiO_2+Al_2O_3/$CH_3NH_3PbI_3$/spiro-MeOTAD/ and with outer Ag layer as electrode. This meso-superstructured solar cells containing additional mesoporous alumina contains few energy losses as compared with previous fabricated OIPSB solar cells. By using mesoporous alumina in TiO_2 layer enhanced the electron transportation through $CH_3NH_3PbI_3$ absorber towards electrode [4].

Just after one year, Jun Hong Noh et .al in 2013 fabricated more efficient, low cost and more stable nano-structures solar cells by using $CH_3NH_3Pb(I_{1-x}Br_x)_3$ as light absorber. They used bi-layers (bl) and additional polymer polytriarylamine (PTAA) to enhance the efficiency and stability of OIPSB solar cells. The device PCE approached to 12.3 % with J_{sc} = 18 mA/cm^2, V_{oc} = 1.13 V and ff = 0.74. The device architecture was as FTO/bl-TiO_2/TiO_2+ $CH_3NH_3Pb(I_{1-x}Br_x)_3$/PTAA with electrode Au for charge transportation at outer end. The TiO_2 and PTAA in device fabrication enhanced the hole transferred ratio towards electrode when it was illuminated in sun light, hence acts as hole transport material (HTM) [5].

In the same year, Mingzhen Liu and his team introduced a simpler planer hetrojunction solar cells with improved PCE and demonstrated that nanostructering is not necessary for getting high efficiency. They used physical vapor deposition method (PVD) for coating organic-inorganic layer and achieved PCE of 15.4 % with J_{sc} = 21.5 mA/cm^2, V_{oc} = 1.07 V and ff = 0.67. They used the simplest architecture as FTO/compact TiO_2/$CH_3NH_3PbI_{3-x}Cl_x$/ spiro-MeOTAD with Ag electrode at the end. The formation of TIO_2 layers was done by spin coating and organic-inorganic materials are coated by PVD method with organic CH3NH3I and inorganic $PbCl_2$ separate targets on compact TiO_2 layer at 1025 mbar. Finally, the device was capped by silver Ag metal electrods using thermal evaporation method at 1026 mbar [6].

Just after one year, Nam Joong Jeon et .al and his team-mates developed efficient OIPSB solar by solution-based spin coating method and achieved 15.8 % PCE. The measured optical parameters J_{sc}, V_{oc} and ff were measured as 19.9 mA/cm^2, 1.06 V and 0.75 respectively. The device architecture follows the pattern as FTO/bl-TiO_2/Perovskite+TiO_2/Perovskite/PTAA and Au as electrode capped on device. The poly-triarylamine (PTAA) acts as HTM which improved hole conductivity of OIPSB solar cells. The absorbers layers consist of two parts, the mixture of TiO_2 and $CH_3NH_3PbI_3$ and upper layer of $CH_3NH_3PbI_3$ without TiO_2. Finally, a better solar cell with improved photovoltaic properties were fabricated with simplest processing techniques. The figure (4) shows the architecture and properties of OIPSB solar cells architecture with XRD pattern of annealed perovskite film (at 100 °C) and formation procedures [7].

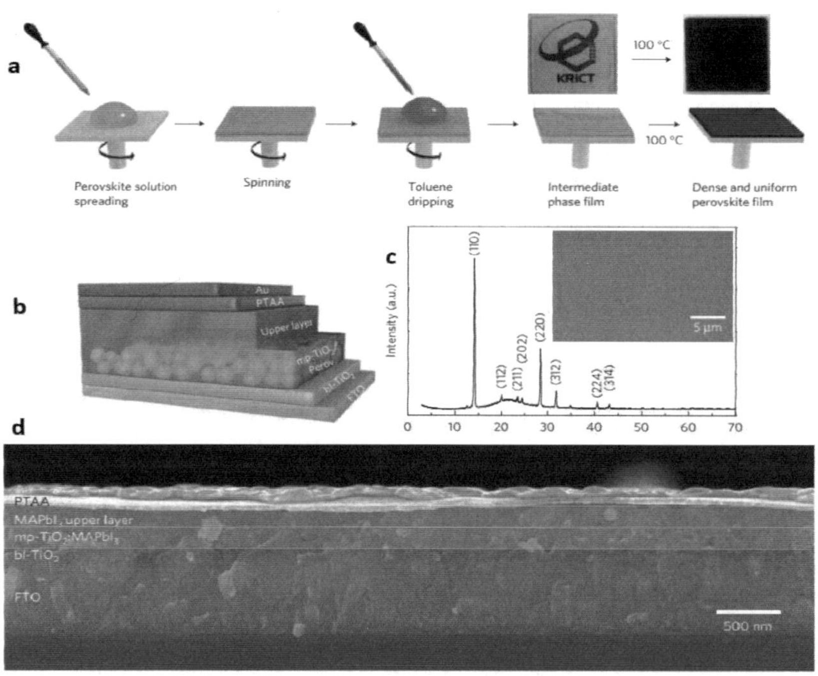

Figure. 5. The OIPSB solar cells (a) formation process, (b) layers 3-D design, (c) X-ray diffraction peaks and (d) SEM image from side for better understanding of device architecture.

After that, Zhou, H et al manufactured OIPSB solar cells with much improved PCE of 19.3 % with solution method at low temperature of less than 150 C. They used $CH_3NH_3PbI_{3-x}Cl_x$ as light absorber with spiro-MeOTAD as light absorbing material and indium Titanium oxide (ITO) with polyethyleneimine ethoxylated (PEIE) as electrode instead of FTO that reduced work function. Additionally, Yttrium-doped TiO_2 (Y-TiO_2) was used as electron transport material to enhance charge extraction from absorber. The device architecture follows as ITO/PEIE/Y-TiO_2/$CH_3NH_3PbI_{3-x}Cl_x$ with Au electrode for charge transfer. The device formation was conducted in air with low temperature and achieved higher efficiency. Finally, when exposed to sun light the optical parameters measured as; J_{sc} = 22.75 mA/cm2, V_{oc} = 1.13 V and ff = 0.75 [8].

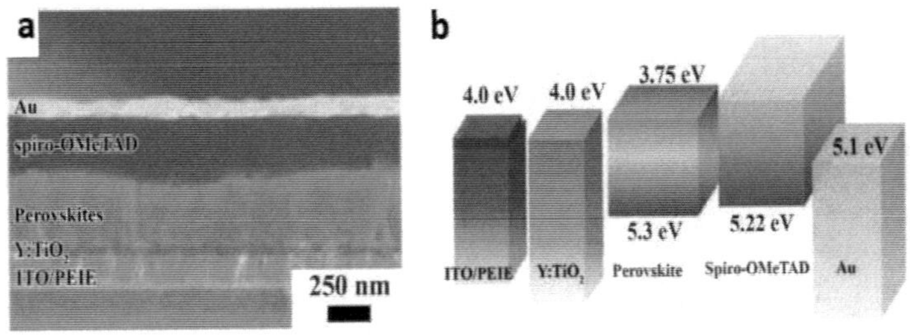

Figure 6. (a) SEM image from side and (b) different layers general designs with electronic band gapes.

Moreover, in 2017 Seong Sik Shin et .al fabricated the OIPSB solar cells with much improved PCE of 21.2 %. They used lanthanum-doped $BaSnO_3$ (LBTO) layer on FTO to enhance electronic mobility and stability of OIPSB solar cells. The device architecture was FTO/LBSO/$CH_3NH_3PbI_3$/PTAA with outer Au electrode encamped. The value of J_{sc}, ff, and V_{oc} was 23.4 mA/cm2, 81.3% and 1.12 respectively when exposed to light. Finally, the device stability was checked and up to 42 days (1000 h) the OIPSB solar cells retain 93 % of its initial performance when fabricated [22, 9].

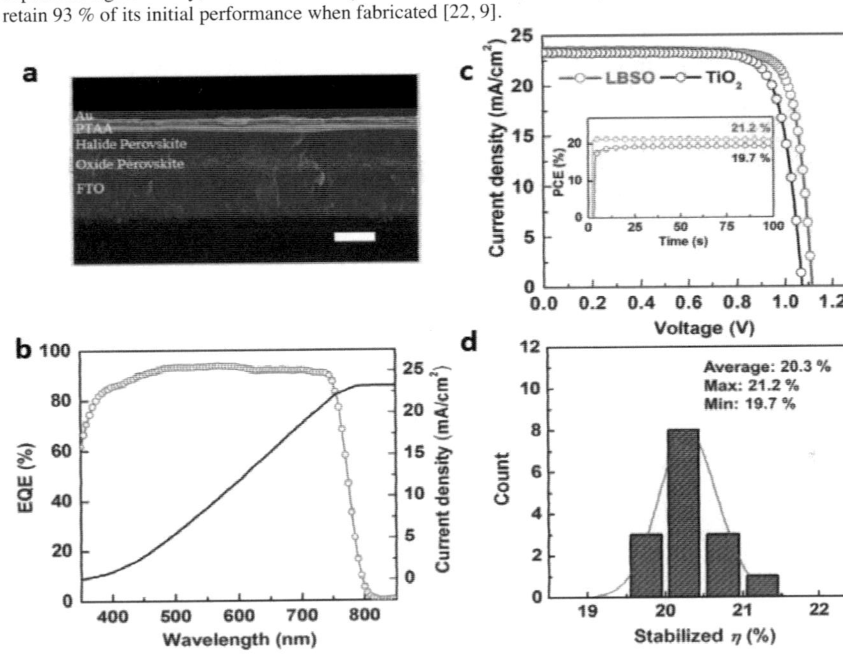

Figure 7. (a) SEM cross-sectional image of OIPSB solar cell, (b) J-V curves, (c) external quantum efficiency (EQE) spectrum and J_{sc} of OIPSB solar cells and (d) the Histograms of PCEs taken from a photocurrent density stabilized at the maximum power point during 100 s [9].

Finally, in 2022 Hang Li and coworker reported 24.2 % PCE using formamidinium iodide (FAI), cesium iodide (CsI), lead iodide (PbI_2) and lead chloride ($PbCl_2$). They used evaporation deposition method to fabricate solar cells with improved efficiency. The device architecture as FTO/FAI/$Cs_{0.05}PbI_{2.05-x}Cl_x$ with J_{sc} = 25.91 mA/cm^2, ff = 80.99 and V_{oc} = 1.145 V. The FAI film first deposited on FTO substrate and then individual $PbCl_2$, PbI_2 and CsI targets are deposited using vacuum evaporation method. These solar cells shown outstanding performance and highest stability when exposed to light. The solar cells were tested for 4000 h in air and shown negligible decline in performance, hence very effective solar cells for future commercialization [10].

Table 1. Perovskite solar cells (2009-2022) architecture and photovoltaic properties measurements.

Year	Device architecture	J_{sc} (mA/cm^2)	V_{oc} (V)	Fil-factor (%)	PCE (%)	Ref.
2009	FTO/TiO_2/$CH_3NH_3PbI_3$	11.0	0.61	0.59	3.8	[1]
2011	FTO/TiO_2/ $CH_3NH_3PbI_3$	13.31	0.60	0.513	6.54	[2]
2012	FTO/TiO_2+spiro-MeOTAD/(CH_3NH_3)PbI_3	17.6	0.88	0.62	9.7	[3]
2012	FTO/TiO_2+Al_2O_3/$CH_3NH_3PbI_3$/spiro-MeOTAD/	17.8	0.98	0.63	10.9	[4]
2013	FTO/bl-TiO_2/TiO_2+ $CH_3NH_3Pb(I_{1-x}Br_x)_3$/PTAA	18.0	1.13	0.74	12.3	[5]
2013	FTO/Cm-TiO_2/$CH_3NH_3PbI_{3-x}Cl_x$/spiro-MeOTAD	21.5	1.07	0.67	15.4	[6]
2014	FTO/bl-TiO_2/Perovskite+TiO_2/Perovskite/PTAA	19.9	1.06	0.75	15.8	[7]
2014	ITO/PEIE/Y-TiO_2/$CH_3NH_3PbI_{3-x}Cl_x$	22.75	1.13	0.75	19.3	[8]
2017	FTO/LBSO/CH33NH3PbI3/PTAA	23.4	1.12	0.81	21.2	[9]
2022	FTO/FAI/$Cs_{0.05}PbI_{2.05-x}Cl_x$	25.91	1.14	0.81	24.4	[10]

4. Conclusion of work

The OIPSB solar cells attracted the photovoltaic community because of excellent power conversion efficiency (PCE). The PCE is exceeded from 20% in a single decade with device structure and manufacturing method. The OIPSB solar cells can be manufacturing by spin-coating, solution method or by physical vapor deposition. In all methods, solution method is most easy method for manufacturing OIPSB solar cells. Furthermore, it is clear from the review that PCE depends on both device structure and other transport material like, ETL, HTL and electrodes.

References

[1] Kojima, A., Teshima, K., Shirai, Y. & Miyasaka, T. (2009) Organometal halide perovskites as visible-light sensitizers for photovoltaic cells. *Journal of the american chemical society*, *131*(17), 6050-6051.

[2] Im, J. H., Lee, C. R., Lee, J. W., Park, S. W. & Park, N. G. (2011) 6.5% efficient perovskite quantum-dot-sensitized solar cell. *Nanoscale*, *3*(10), 4088-4093.

[3] Kim, H. S., Lee, C. R., Im, J. H., Lee, K. B., Moehl, T., Marchioro, A. & Park, N. G. (2012) Lead iodide perovskite sensitized all-solid-state submicron thin film mesoscopic solar cell with efficiency exceeding 9%. *Scientific reports*, *2*(1), 591.

[4] Lee, M. M., Teuscher, J., Miyasaka, T., Murakami, T. N. & Snaith, H. J. (2012) Efficient hybrid solar cells based on meso-superstructured organometal halide perovskites. *Science*, *338*(6107), 643-647.

[5] Noh, J. Hπ., Im, S. H., Heo, J. H., Mandal, T. N. & Seok, S. I. (2013) Chemical management for colorful, efficient, and stable inorganic–organic hybrid nanostructured solar cells. *Nano letters*, *13*(4), 1764-1769.

[6] Liu, M., Johnston, M. B. & Snaith, H. J. (2013) Efficient planar heterojunction perovskite solar cells by vapπour deposition. *Nature*, *501*(7467), 395-398.

[7] Jeon, N. J., Noh, J. H., Kim, Y. C., Yang, W. S., Ryu, S. & Seok, S. I. (2014) Solvent engineering for high-performance inorganic–organic hybrid perovskite solar cells. *Nature materials*, *13*(9), 897-903.

[8] Zhou, H., Chen, Q., Li, G., Luo, S., Song, T. B., Duan, H. S. & Yang, Y. (2014) Interface engineering of highly efficient perovskite solar cells. Science, *345*(6196), 542-546.

[9] Shin, S. S., Yeom, E. J., Yang, W. S., Hur, S., Kim, M. G., Im, J. & Seok, S. I. (2017) Colloidally prepared La-doped $BaSnO_3$ electrodes for efficient, photostable perovskite solar cells. *Science*, *356*(6334), 167-171.

[10] Li, H., Zhou, J., Tan, L., Li, M., Jiang, C., Wang, S. & Yi, C. (2022) Sequential vacuum-evaporated perovskite solar cells with more than 24% efficiency. *Science Advances*, *8*(28), 7422.

[11] https://www.nrel.gov/pv/cell-efficiency.html

[12] Luo, J., Im, J. H., Mayer, M. T., Schreier, M., Nazeeruddin, M. K., Park, N. G., & Grätzel, M. (2014) Water photolysis at 12.3% efficiency via perovskite photovoltaics and Earth-abundant catalysts. *Science*, *345*(6204), 1593-1596.

[13] Tan, Z. K., Moghaddam, R. S., Lai, M. L., Docampo, P., Higler, R., Deschler, F., & Friend, R. H. (2014) Bright light-emitting diodes based on organometal halide perovskite. *Nature nanotechnology*, *9*(9), 687-692.

[14] Gil-Escrig, L., Longo, G., Pertegás, A., Roldán-Carmona, C., Soriano, A., Sessolo, M. & Bolink, H. J. (2015) Efficient photovoltaic and electroluminescent perovskite devices. *Chemical Communications*, *51*(3), 569-571.

[15] Hoye, R. L., Chua, M. R., Musselman, K. P., Li, G., Lai, M. L., Tan, Z. K. & Credgington, D. (2015) Enhanced performance in fluorene-free organometal halide perovskite light-emitting diodes using tunable, low electron affinity oxide electron injectors. *Advanced Materials*, *27*(8), 1414-1419.
[16] D'Innocenzo, V., Srimath Kandada, A. R., De Bastiani, M., Gandini, M. & Petrozza, A. (2014) Tuning the light emission properties by band gap engineering in hybrid lead halide perovskite. *Journal of the American Chemical Society*, *136*(51), 17730-17733.
[17] Sutherland, B. R., Hoogland, S., Adachi, M. M., Wong, C. T. & Sargent, E. H. (2014) Conformal organohalide perovskites enable lasing on spherical resonators. *ACS nano*, *8*(10), 10947-10952.
[18] Deschler, F., Price, M., Pathak, S., Klintberg, L. E., Jarausch, D. D., Higler, R. & Friend, R. H. (2014) High photoluminescence efficiency and optically pumped lasing in solution-processed mixed halide perovskite semiconductors. *The journal of physical chemistry letters*, *5*(8), 1421-1426.
[19] Xing, G., Mathews, N., Lim, S. S., Yantara, N., Liu, X., Sabba, D. & Sum, T. C. (2014) Low-temperature solution-processed wavelength-tunable perovskites for lasing. *Nature materials*, *13*(5), 476-480.
[20] Geist, J., Schaefer, A. R., Song, J. F., Wang, Y. H. & Zalewski, E. F. (1990) An accurate value for the absorption coefficient of silicon at 633 nm. *Journal of research of the National Institute of Standards and Technology*, *95*(5), 549.
[21] Jin, X., Wang, Q. & Jadoon, A. M. (2020) Opto-Electronic Properties of Methyl-Ammonium Lead Halide: A First Principle Approach. In *Journal of Physics: Conference Series* (Vol. 1622, No. 1, p. 012105). IOP Publishing
[22] Scanlon, D. O. (2013) Defect engineering of $BaSnO_3$ for high-performance transparent conducting oxide applications. *Physical Review B*, *87*(16)s 161201.

IOP Conf. Series: Materials Science and Engineering 1291 (2023) 012004 doi:10.1088/1757-899X/1291/1/012004

An Experimental Investigation of Surface Quality in Single Point Incremental Forming of Cu-Al Composite Sheets

Ankitkumar C Pambhar[1], **Janak B Valaki**[1,2]

[1]Gujarat Technological University, Ahmedabad, India
[2]Government Engineering College, Bhavnagar; Gujarat Technological University, Ahmedabad, India

Email: pambharankit@gmail.com
Email: janakvalaki@gmail.com (Corresponding Author)

Abstract. Single-point incremental forming (SPIF) has been developed as a customized substitute of traditional sheet metal forming techniques. SPIF is a great method for rapidly prototyping sheet metal objects, although it faces challenges in achieving surface quality and thickness distribution. In this investigation, sheets of Cu-Al bimetallic composite were formed utilizing a hemispherical-headed tool with a variety of step depths, spindle speeds, forming speeds, layer arrangements, and lubrication types. Surface roughness and minimum thickness were evaluated as response parameters in relation to the significance of process variables. The best surface quality was attained with 0.4 mm step depth, 750 mm/min forming speed, and 1500 rpm spindle speed with liquid lubrication support for both types of layer arrangements. Layer arrangement and lubricant type were the decisive factors for the minimum thickness response parameters.

1. Introduction:

The single point incremental forming (SPIF) technique is thought to be an inexpensive method for producing intricate and specialized components with a rapid turnaround time in low volume. This advantage has positioned SPIF apart from traditional sheet metal forming procedures. In SPIF, a hemispherical-headed tool deforms the metal sheet locally along a predetermined tool path to get the required shape. [1][2].

Surface quality and thickness variation were regarded as the key considerations for the acceptability of the formed component. There have been several studies done to improve surface quality by varying input process parameters and using lubrication. The mixture of white petroleum jelly and MoS_2 powder combined in the right proportion can be employed as a lubricant to improve the surface finish when forming CP Ti blanks incrementally [3]. The surface finish has also been altered by the application method and the lubrication type. The ceramic coating of MoS2 or solid graphite, and the Nano-K2Ti4O9 whisker-improved solid lubricating layer were added during the formation of the AZ31 sheet. It was claimed that solid graphite or ceramic coating of MoS2 had an amazing greasing and self-lubrication effect. [4].

One of the forming flaws that restrict the process's widespread deployment has always been sheet thinning. According to a study on multiple passes of forming, an increase in the overall plastic deformation zone leads to improved thickness uniformity during the process [5]. The incremental forming of DC04 sheet demonstrates that, when employed with a conventional tool trajectory, the minimum thickness is strongly correlated with tool diameter, whereas the region of the minimum thickness is substantially determined by step depth [6]. The tool and sheet blank contact conditions have an impact on surface irregularity. Forming AA7075T0 sheet reveals that, in comparison to a non-spinning tool, a rotating forming tool provided a greater surface quality[7]. The surface roughness of Al 5052 initially increases with increasing incremental depth up to a particular angle, and subsequently reduces during incremental forming [8].
The arrangement of the layers is essential for describing surface quality in terms of surface roughness. The surface quality of the bimetal blank was impacted by friction conditions between its surface and the tool throughout the forming process. For low-carbon steel/CP-titanium bimetal blank, when a tool made contact with a layer of titanium in a bimetal sheet, there was sticky friction that caused the metal blank to adhere to the tool's tip and increase the surface irregularity. However, the St-Ti arrangement has produced improved surface quality[2]. In comparison to SUS/Al, it has been found that Al/SUS arrangements exhibit early fracture incidence. However, after the occurrence of fracture, the SUS/Al layer arrangement experiences delamination. The forming wall angle and the surface irregularity are both positively impacted by the low values of the step size. The SUS/Al layer arrangement has been demonstrated to have a greater forming wall angle and excellent surface consistency [9][10]. It was clear that sheet thinning occurred in the pyramid's wall section during the formation of St/CP-Ti sheets. It was discovered that the corners of the pyramid form showed the greatest rate of thickness variation [11].
Numerous investigations have been carried out to look into bimetallic composite forming and the process variables involved [12][13]. However, there is relatively little research available on incremental bimetallic sheet forming. The present experimental investigation concentrated on the influence of step depth, forming speed, and spindle speed for two lubrication conditions and layer arrangements. Surface roughness and minimum thickness were two important responses taken into account for the present experimental investigation.

2. Materials and Methodology:

2.1 Material:

The Cu-Al bimetallic composite sheet that was employed for the studies has layers of Copper 11000 and Aluminum 1060 that are each 0.20 mm thick and 0.80 mm thick, respectively. table 1 demonstrates the composition of both layers. The size of the test samples is 150 mm x 150 mm x 1 mm. The review of the literature demonstrates that lubrication has a greater overall influence on the surface irregularity of the manufactured component [4][14]. For the current investigation, two lubricants—SAE-10W-30 lubricating oil and Molybdenum disulphide lithium base grease were employed as lubricants.

Table.1: Chemical composition (%wt) of Cu-Al bimetallic sheet.

Aluminum 1060 alloy	Element	Si	Fe	Cu	Mn	Zn	Ti	Ni	Zr	Al
	%wt	0.056	0.22	0.002	0.001	0.001	0.011	0.003	0.001	Balance
Copper 11000 alloy	Element	Zn	Fe	Pb	Mn	Ni	Si	Bi	Sb	Cu
	%wt	0.082	0.01	0.017	0.001	0.014	0.007	0.003	0.004	Balance

2.2 Experimental Equipment:

(a) (b)

Figure 1. (a) Set-up arrangement on the machine; (b) Formed component.

For the SPIF experiments on the Cu-Al bimetallic sheets, a test setup was constructed. The clamping plate, support column, and hemispherical-headed shaping tool comprised the setup for the experiments. A CNC milling center was equipped with the developed setup, and a hemispherical-headed tool was fixed on the spindle. The hemispherical-headed tool is made of tungsten carbide. Figure 1 (a) depicts the arrangement of the setup and the holding of a Cu-Al bimetallic blank on the setup. Figure 2 (b) depicts the formed component for the Cu-Al and Al-Cu layer arrangements using truncated geometry. The Cu-Al layer arrangement refers to the forming tool coming into contact with the Cu layer as it is being formed, and vice versa.

2.3 Process parameters and response measurement

The continuous process variables used in the investigation are step depth, forming speed, and spindle speed. Layer arrangement and type of lubrication were considered categorical process parameters. In order to establish the range of the input variables prior to the core experimental work, a number of trial experiments were conducted. For the experiments, three levels of continuous parameters and two levels of categorical parameters were considered. The levels of the input variables taken into consideration are displayed in Table 2.

Table.2: Levels of input variables for the experiments.

Input Variables	**Level-1**	**Level-2**	**Level-3**
Step depth (mm)	0.4	0.6	0.8
Spindle Speed (rpm)	1000	1500	2000
Forming speed (mm/min)	500	750	1000
Input Variables	**Level-1**	**Level-2**	
Layer Arrangement	Al-Cu	Cu-Al	
Lubrication type	Molybdenum disulfide lithium base grease (Solid)	SAE 30 lubrication oil (Liquid)	

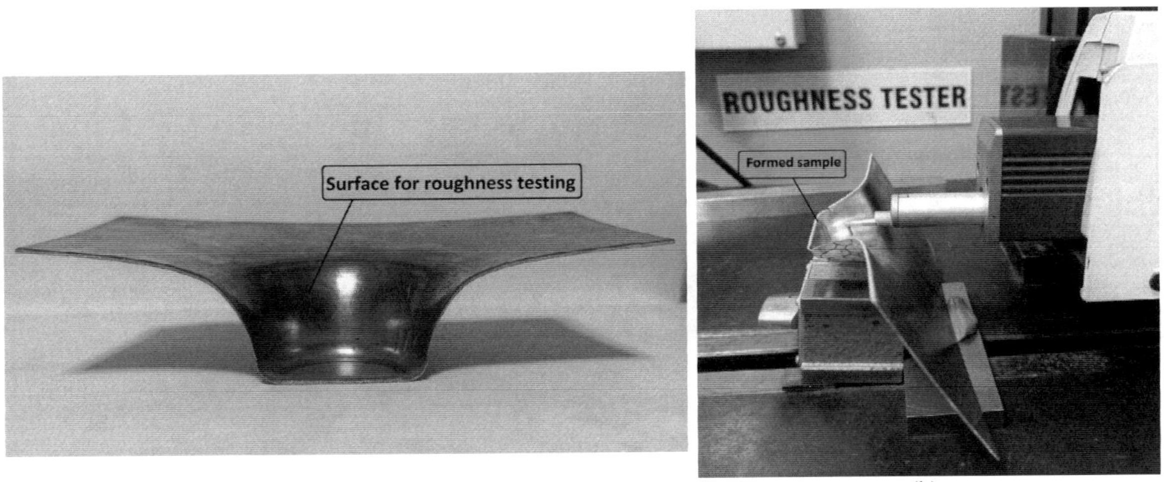

(a) (b)

Figure 2. (a) Formed samples cut from the center; (b) Measurement on roughness tester.

Formed samples were cut from the center to measure the minimum thickness and surface roughness. Figure 2 (a) depicts the formed sample that has been cut out of the center and the surface where roughness will be assessed. Readings close to the fracture location were acquired to determine the minimum thickness value. In order to examine the surface finish for the current investigation, the average roughness value (Ra) was assessed. Measurements of the roughness were made on the inside of the produced workpiece. Mitutoyo SJ-400 was used to utilize to gauge the roughness of the surface at the predetermined region. The setup for the measurement of the formed sample on the surface roughness tester is shown in Figure 2(b). In order to ensure precision, three measurements were obtained at the same location. The sample length for surface roughness measurement was 0.80 mm, while the evaluation length was 4.00 mm.

3. Result and Discussion:

Several tests have been carried out to evaluate the surface irregularity and thickness of the formed components in order to examine the influence of input variables. The following process variables are taken into account for establishing the quality of the formed workpiece in terms of surface roughness and minimum thickness value: step depth, spindle speed, forming speed, lubricants, and sheet layer arrangements. Figure 1 (b) illustrates the generated component employing the truncated cone-type shape from the Cu and Al sides. Surface roughness (Ra) and minimum thickness values were measured on samples cut from the center. The graphs are constructed using data points that represent the average of five measurements, and the error bar represents the standard deviation among these measurements. It was evident that minimum thickness and surface roughness were crucial factors for successfully manufactured components. The results were discussed in this section while taking into consideration the effects of the lubrication condition and layer arrangement.

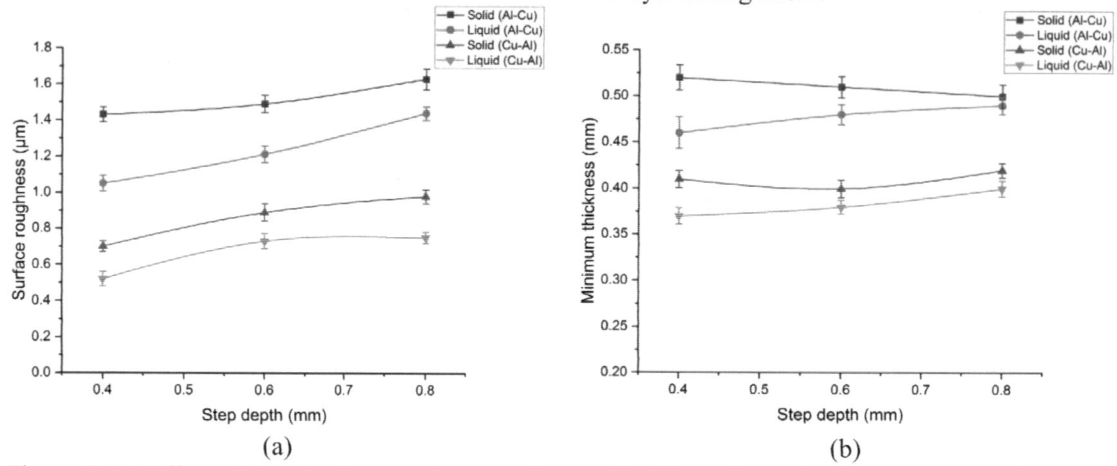

Figure 3. (a) Effect of step depth on surface roughness; (b) Effect of step depth on minimum thickness.

To study the impact of step-depth on surface roughness and minimum thickness, spindle speed and forming speed were maintained constant at mid-range values. Figure 3(a), illustrates that the value of surface roughness increases as the step depth increases. Lower roughness values have been achieved for both the Al-Cu and Cu-Al layer arrangements using liquid lubricants. To achieve good surface quality, a Cu-Al layer with liquid lubrication assistance was used, as depicted in figure 3(a). Figure 3(b) demonstrates that for both types of lubrication conditions, lower minimum thickness values were obtained for the Cu-Al layer arrangement. The minimum thicknesses have also increased as the step depth increased. However, for the Al-Cu arrangement, the minimum thickness has somewhat lowered with an increase in step depth with solid lubrication. The error bar plot depicted on the graphs reveals a significant decrease in measurement error, indicating improved accuracy.

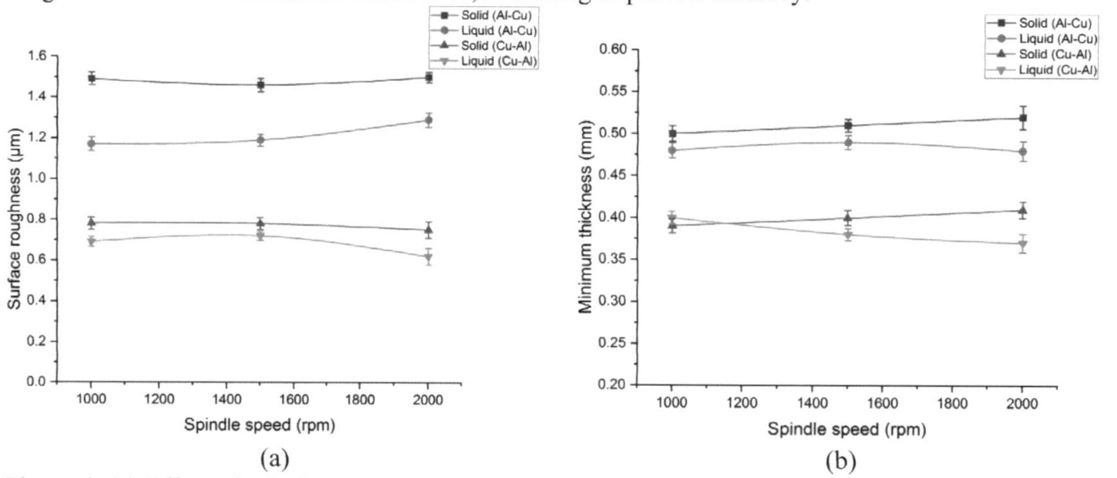

Figure 4. (a) Effect of spindle speed on surface roughness; (b) Effect of spindle speed on minimum thickness.

Step depth and forming speed were held constant at mid-range levels to evaluate the effects of spindle speed on surface roughness and minimum thickness. As seen in Figure 4(a), under both lubrication conditions, the value of surface roughness rises with spindle speed for the Al-Cu layer configuration. As the spindle speed rises, the surface roughness for the Cu-Al configuration slightly reduces. Liquid lubrication enhances the surface quality of both types of layer arrangements. Figure 4(b) illustrates how the minimum thickness value rises as spindle speed rises. In comparison to a Cu-Al layer arrangement, there is a significant variation in the minimum thickness value for an Al-Cu layer arrangement. The analysis of the error bars suggests that there is no significant variation in the minimum thickness for the Cu-Al arrangement at 1000 rpm. The liquid lubrication has obtained greater minimum thickness values for both types of layer arrangements.

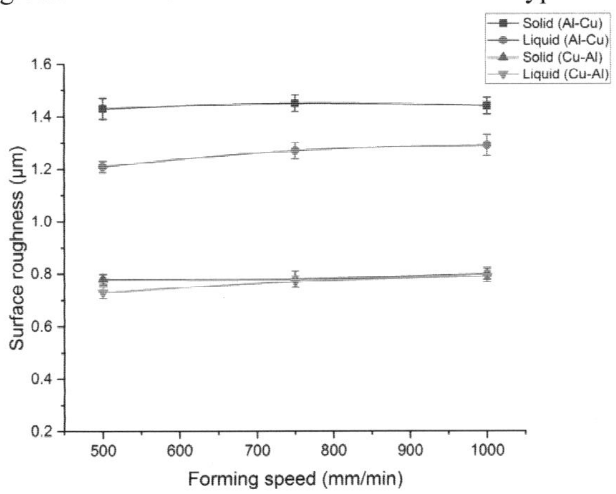

(a)

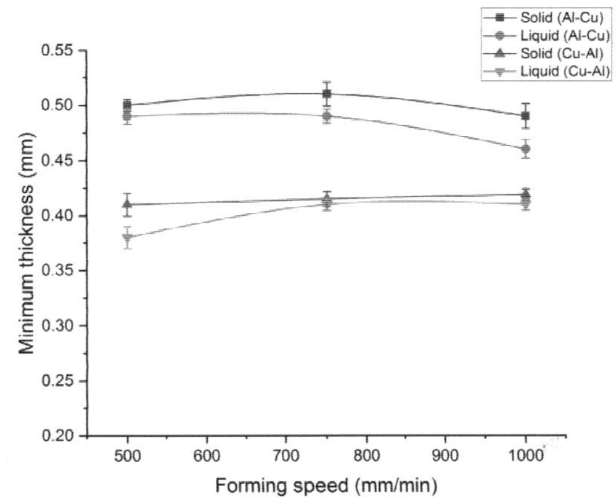

(b)

Figure 5. (a) Effect of forming speed on surface roughness; (b) Effect of forming speed on minimum thickness.

To assess the impacts of forming speed on surface roughness and minimum thickness, spindle speed and step depth were maintained at mid-range levels. Figure 5(a) illustrates that surface roughness appears to be somewhat unaffected by forming speed. However, the Al-Cu layer arrangement shows significant variations with various lubrication conditions. Cu-Al configuration with solid and liquid lubrication support has demonstrated greater surface quality. The minimum thickness values have grown up to the middle level with an increase in forming speed before decreasing, as illustrated in figure 5. (b). In comparison to Cu-Al arrangement, Al-Cu arrangement delivers superior minimum thickness values for both lubrication conditions. The error bars overlapping each other for surface roughness and minimum thickness indicate that the impact of lubrication on the Al-Cu arrangement was not found to be significant.

The outcomes of the experiment highlight the significance of layer arrangement and lubrication during the forming of Cu-Al bimetallic sheets. As compared to the lubrication type, layer arrangement is crucial, with response factors including minimum thickness and surface roughness being taken into account.

4. Conclusion:

On a Cu-Al bimetallic blank, the SPIF is performed, with the layer arrangement and lubrication type acting as categorical process parameters whereas step depth, forming speed, and spindle speed are considered continuous process parameters. Surface roughness and minimum thickness were studied on samples that had been formed while taking into consideration a combination of several process factors. The experimental investigation yielded the following findings.

- The minimum surface roughness values of 0.52 μm for Cu-Al arrangement and 1.05 μm for Al-Cu arrangement have been achieved using 0.4 mm step depth, 750 mm/min forming speed, and 1500 rpm spindle speed with liquid lubrication support.
- The average surface quality in terms of surface roughness is better in the Cu-Al arrangement for liquid lubrication assistance than in the Al-Cu arrangement. The average minimum thickness value of an Al-Cu arrangement with solid lubricant is higher than that of a Cu-Al arrangement.
- Layer configuration, step depth, lubricant type, and spindle speed have been identified as the process variables that have a significant effect on surface roughness. However, the impact of the forming speed was very minimal.
- The most important factors that influenced the minimum thickness response parameter included layer arrangement, lubricant type, forming speed, spindle speed, and step depth, respectively.

References:

[1] Vahdani M, Mirnia M J, Gorji H and Bakhshi-Jooybari M 2019 Experimental Investigation of Formability and Surface Finish into Resistance Single-Point Incremental Forming of Ti–6Al–4V Titanium Alloy Using Taguchi Design *Trans. Indian Inst. Met.*

[2] Sakhtemanian M R, Honarpisheh M and Amini S 2018 Numerical and experimental study on the layer arrangement in the incremental forming process of explosive-welded low-carbon steel/CP-titanium bimetal sheet *Int. J. Adv. Manuf. Technol.* **95** 3781–96

[3] Hussain G, Gao L, Hayat N, Cui Z, Pang Y C and Dar N U 2008 Tool and lubrication for negative incremental forming of a commercially pure titanium sheet *J. Mater. Process. Technol.* **203** 193–201

[4] Zhang Q, Xiao F, Guo H, Li C, Gao L, Guo X, Han W and Bondarev A B 2010 Warm negative incremental forming of magnesium alloy AZ31 Sheet: New lubricating method *J. Mater. Process. Technol.* **210** 323–9

[5] Li J, Hu J, Pan J and Geng P 2012 Thickness distribution and design of a multi-stage process for sheet metal incremental forming *Int. J. Adv. Manuf. Technol.* **62** 981–8

[6] Li J C, Li C and Zhou T G 2012 Thickness distribution and mechanical property of sheet metal incremental forming based on numerical simulation *Trans. Nonferrous Met. Soc. China (English Ed.* **22** s54–60

[7] Durante M, Formisano A and Langella A 2011 Observations on the influence of tool-sheet contact conditions on an incremental forming process *J. Mater. Eng. Perform.* **20** 941–6

[8] Bhattacharya A, Singh S, Maneesh K, Reddy N V and Cao J 2011 Formability and Surface Finish Studies in Single Point Incremental Forming *ASME 2011 Int. Manuf. Sci. Eng. Conf. Vol. 1* **133** 621–7

[9] Abd Ali R, Chen W, Jin K, Bao Y and Hussein A W 2019 Formability and failure analyses of Al/SUS bilayer sheet in single point incremental forming *Int. J. Adv. Manuf. Technol.* **105** 2785–98

[10] Raneen Abd Ali , Wenliang Chen , M.S.H. Al-Furjan , Xia Jin Z 2019 Experimental Investigation and Optimal Prediction

[11] Ben W and Abdelkader B 2020 ScienceDirect ScienceDirect Numerical Investigation of the Influence of some Parameters in SPIF Process on the Forming Forces and Thickness Distributions of a Bimetallic Sheet CP-Titanium / Low-carbon Steel Compared to an Individual Layer *Procedia Manuf.* **47** 1319–27

[12] Kotobi M, Mansouri H and Honarpisheh M 2019 Investigation of laser bending parameters on the residual stress and bending angle of St-Ti bimetal using FEM and neural network *Opt. Laser Technol.* **116** 265–75

[13] Afshin E and Kadkhodayan M 2015 An experimental investigation into the warm deep-drawing process on laminated sheets under various grain sizes *Mater. Des.* **87** 25–35

[14] Li Y, Chen X, Liu Z, Sun J, Li F, Li J and Zhao G 2017 A review on the recent development of incremental sheet-forming process *Int. J. Adv. Manuf. Technol.* **92** 2439–62

The synthesis and characterization of demulsifying poly vinylidene Fluoride (PVDF) Disc

Fida Ullah [1], Mehtab Ur Rehman Sahir [2, *], Ali Bahader [3], Haseeb Ullah[4], Abdul Hafeez[1], Yasar Ali[6], Alamgir Khan[3], Fazal dayan[6], Amir Ali Khan[3], Muhmmad Murtaza[5], Gohar Aziz[9], Kamran Khan[7], Naseer Ullah[8]

[1]Department of Environment and Life Sciences, Beijing University of technology, Beijing, China
[2]Faculty of Materials and Manufacturing, Beijing University of technology, No.100, Pingleyuan, Chaoyang District, Beijing, China
[3]Department of Chemistry, Hazara University, 21300, Mansehra, KPK, Pakistan
[4]Department of Chemistry department Quaid-i-Azam University Islamabad, Pakistan
[5]College of Metropolitan transportation, Beijing University of technology, Beijing, China
[6]MRS Institute of Physics and Mathematics, Pakistan
[7]Institute of chemical sciences, University of Peshawar Pakistan.
[8]Biotechnology Islamic international university Islamabad Pakistan
[9]School of science, Beijing university of technology, China

Correspondence: mehtaburrehman7@gmail.com, mehtaburrehman@emails.bjut.edu.cn

Abstract. Oil/water separation is still a challenging universal task. Here we fabricated an amphiphilic polymeric (APD) PVDF disc that can be used to separate free oil/water mixtures and recover the pure water from W/O emulsions. The modification of the disc includes deposition of copper oxide (CuO) was integrated on the surface via the impregnation method; the CTAB was used as a porogen to increase the roughness of the surface, Morphology of the disc was investigated by SEM and the result showed the nanoparticles were entrapped on the surface and inside the disc. FTIR results strongly predicted the correlation of CTAB, PVDF, and CuO nanoparticles of the modified disc. Separation and wettability test of the FRR% (Flux recovery ratio), permeate flux/oil rejection coefficient (R %) and the water-uptake test confirmed the antifouling properties, permeability of water, oil-rejection, and porosity of the disc. The purpose of impregnation CuO enhanced the hydrophilicity and anti-fouling properties of the PVDF disc.

Keywords: PVDF disc, hydrophilicity, anti-fouling properties, SEM

1. Introduction

Oily wastewater is one of the constituents of industrial effluent, which is drained into the freshwater stream. The oily wastewater in the presence of different salts/surfactants turned into oil/water emulsions as produced in industries including petroleum, food, textile, and leather. The emulsified water ultimately finds its way into the ecosystem and harms the general public health. Many oil-spill accidents for example at Bohai Bay in (2011) once happened, affected enormous damages by way of the leaked oil diversified

through seawater (containing a high fraction of different salts) producing emulsion, and severely affected the oceanic flora and fauna [1]. On the analytical side, petroleum crude often presents complex composition, which is difficult to analyze with high reliability and precision and ultimately affects the valuation of crude in the market. One of the main problems in such type of scenario is the stability of the emulsion, which varies from well to well and ultimately affects the analytical protocol.
Numerous approaches are in place to separate water from oil such as air flotation, adsorptive, electrochemical, and ultrasonic methods [2]. Besides, all these methods suffer from one another drawback, as electrochemical and ultrasonic methods are difficult to materialize for handling material in bulk due to safety concerns. Therefore, an alternative method/material, which is economical, efficient method and material has desirable, and has a high recyclability ratable. Membrane technology offers the aforesaid benefits and has been widely used [3, 4]. Different polymers find application in this area and were successfully used with pores in the ultra, micro, and nano dimension including polyether-sulfone (PES) [5], poly (vinylidene fluoride) (PVDF), cellulose acetate (CA) and polyacrylonitrile (PAN) to break the emulsified matrix and offer effective demulsification [6].

Among these polymers the poly (vinylidene fluoride PVDF-based membranes secured a distinctive place in the separation area due to high toughness, inertness, relative transparency, and ease of processing [7]. But with other man-made synthetic materials PVDF's membrane also suffers shortcomings of fouling [8-9], especially when employed in demulsification area [10]. Being hydrophobic, organic or oily fraction stick to PVDF's membrane surface, this is ted as fouling. Fouling resultantly appears in flux reduction due to pores clogging the need arises to continuously regenerate the surface otherwise making the process tedious and time-consuming [11-13]. However, in protocols where speed is a major concern the membrane-based separation process suffers from mechanical stability. Due to fouling the clogged pores of extremely high back pressure deteriorated the membrane surface often leading to its rupturing. One method to counter this drawback is to make the surface of VDF membrane hydrophilic [14, 15]. Mohan et al successfully modified the PVDF nano-filtration membrane with carboxymethyl chitosan (CMC) and reported that the addition of CMC not only increase the hydrophilicity but also augmented the mechanical strength of the membrane [16]. But it's still not enough mechanically strong to sustain the high flow rate, especially in automated method design. On the other hand, the disc or monolithic format of separation material minimized or even eliminated the cited problem, especially enabling the analyst to work with a high flow rate with maximum throughput [17].

In this work PVDF in Disc format was fabricated and modified its internal structure with hydrophilic copper nano-particles to counter the hydrophobicity and mechanical strength of the material [18].

2. Experimental

Material polyvinylidene fluoride (PVDF) blobs and CTAB (Cetrimonium bromide) were obtained from Sigma Aldrich, USA. Dimethylformamide (DMF) and ethanol (C_2H_5OH) were of analytical grade and achieved from Sigma Aldrich, USA. Chloroxylenol (C_8H_9ClO), Copper sulphate ($CuSO_4$), and Sodium tetrahydridoborate ($NaBH_4$) were purchased from Sigma Aldrich, Germany. Commercially available distilled water bought from Aqua Art institute of Pakistan. All chemicals used remained of analytical grade and were used as received.

2.1 Preparation and Fabrication of PVDF Disc

According to the method, the polyvinylidene fluoride solution was ready in dimethylformamide at a suitable temperature [18]. The calculated quantity of Cetrimonium bromide as porogen has been added to the desired solution and extracted from the scaffold using distilled water as a leaching agent. Polyvinylidene fluoride disc was prepared in a disposable syringe used as rot. PVDF solution was packed into a specified syringe to control the sample thickness and width. The solution gelation has achieved in more than 2 Hrs. The required PVDF solution in the syringe was dipped into cold water for coagulation. After clotting, the disc was rained with refined water up to the whole elimination of CTAB. The obtained disc became dry for 40 minutes at 60 °C.

Table 1. Recipe for preparing the modified polymeric amphiphilic PVDF disc

Sample Code	PVDF (Wt %)	Porogen (CTAB) (Wt %)
F0	100	0.0
F1	99.5	0.5
F2	99.0	1.0
F3	95.0	5.0
F4	85.0	15.0
F5	80.0	20.0
F6	75.0	25.0

2.2 Synthesis of Nanoparticles

$CuSO_4$ solution was synthesized in distilled water. A suitable amount of C_8H_9ClO as a structure-directing agent was added to the $CuSO_4$ solution [19]. The weighted amount of Sodium tetrahydridoborate solution was added to the solution at freezing temperature till the color of the mixture has been converted from blue to blackish dark, which demonstrates the formation of Copper nanoparticles (CNPs). The CNPs prepared were isolated by centrifugation at 6000 rpm, washed with distilled water followed by ethanol in triplicate, and dried in a vacuum oven at 60 °C for 3 hours.

2.3 Modification of Polymeric PVDF Disc

The pore modification of the PVDF disc was performed by the impregnation method. A nanostructure was prepared inside the pores of the disc. In a typical experiment, the disc was suspended in $CuSO_4$ containing chloroxylenol in an ultrasonic bath, and then $NaBH_4$ was added according to the method discussed earlier. Finally, the discs were sprayed with de-ionized water and dried at 60 ◦C for 24 hrs.

3. Characterization

To investigate the chemical composition of a polymeric amphiphilic PVDF disc FTIR was employed. The morphology of the disc was determined by scanning electron microscope (JSM5910, JEOL Japan). The hydrophilicity of the resulting amphiphilic polymeric disc was measured by the different filtration experiments. The filtrate sample was characterized by UV.

3.1 Oil-Water Separation Experiments

Performed different filtration tests and adopted the following methods [20-23].

3.1.1 *Antifouling test*

The disc was placed in the filtration cell. The flux of the disc J_{w1} was gauged based on the absorbed water weight at 0.4 MPa pressure for 30 minutes. The process was carried out in the following steps: In step 1 the pure PVDF disc was tested through a water-oil mixture. Pure disc flux was measured to examine the fouling level. In step 2 the modified disc was tested with water-oil emulsion for 30 min for measuring their water flux, J_{w2}. To check out the antifouling property of the prepared disc, the flux recovery ratio (FRR) was employed, high FRR showed a greater fouling-resistant ability of the disc. The equation (1) used for FRR calculation was as

$$FRR\% = (J_{w2}/J_{w1}) \times 100 \quad \text{..........} \quad (1)$$

Where Jw1 and Jw2 are the fluxes of the original PVDF-Disc and the Polymeric Amphiphilic- disc, respectively

3.1.2 *Rejection and Permeation Flux of the Amphiphilic Disc*

A calculation was made for the polymeric disc's flow. The following equation was used to calculate the permeation's volume per unit of time (2).

$$Flux=V/At \quad .... (2)$$

Where the volume of the permeate V, the valid area of the disc A (cm^2), and the retention time t, has been shown in eq (2). The filter was filled by definite volume of water-in-oil (15 mL) and every sample was measured to get the average result for each test. According to the given Equation (3) the separation efficiency was calculated and defined by oil rejection coefficient R (%).

$$R\ (\%) = (1-C_P/C_0) \times 100\% \quad .... (3)$$

The concentration in the original oil/water mixture shown by C_0 and Cp and the collected water similarly.

3.1.3 *Extractive-gravimetric method*

This method was used for the determination of water uptake by the disc, using a rounded disc sample and weighted it W_0, the discs were soaked in a beaker containing 15 mL of swelling medium (water) and shaken at 22 °C for 8 hrs. The disc was then taken out of the beaker, additional water was strained away and the disc was weighed again to get its final weight, W_f (Water uptake) was then calculated by following Equation (4).

$$W_{uptake} = (W_f - W_0) / \times 100\% \quad .... (4)$$

Where V_{sample} is the volume of wastewater sample (dm^3) and ml is the mass of a round flask with oil (in grams), m^2 is the mass of a dry and empty round flask (in grams). As shown in Figure 10.

4. Result and Discussion

4.1 Fourier Transform Infrared (FTIR) Analysis

At 833 cm^{-1} confirmed the C-F starching vibration and PVDF material is characterized by an 872 cm^{-1} peak shown to the carbon C-C-C chain asymmetric vibration [24]. Shown in Fig 1.

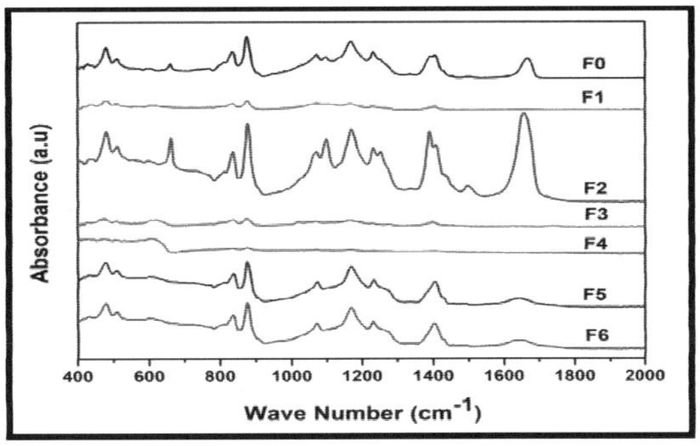

Fig. 1. FTIR spectrum of Neat PVDF and its blend with different concentrations of cetyltrimethylammonium bromide [CTAB].

The 1408 cm^{-1} absorption peak was attributed to the CH_2 wagging vibration. The peak observed at 1230 cm^{-1} and 1074 cm^{-1} was ascribed to the γ and β phase of C-F out-of-plan deformation. The peak near 514 cm^{-1} is attributed to the bending vibration of CF_2's phase [25]. It can be inferred from the data presented in Figure 1 that not only does CTAB attenuate the crystal structure of PVDF but the DMF can also have profound effects on it as can be seen in Figure 1 (F0).

As reported earlier peaks arising at 1657, 1172, 1098, 880, 662, and 469 cm $^{-1}$ different phases of PVDF were attributed. Still, the presence of 1393 cm^{-1} band has displayed the characteristic vibration of the ($N-CH_3$) group, which is a principal component of CTAB [26] observed in Figure 1 (F1). The PVDF sample containing different loading of CTAB are designated from F1 to F6. It can be observed from the data presented in Figure 1 that more CTAB facilitates the formation of different phases of PVDF [27, 28]. Although, CTAB was leached out from the PVDF disc the band at 880 cm^{-1} band suggests the presence of α phase of PVDF, and the ($N-CH_3$) stretching vibration of CTAB was also observed at 1389 cm^{-1} while the peaks at 1257, 1172 cm^{-1} indicates β phase of $C-F_2$ out-of-plan deformation. It is suggestive of the fact that CTAB has strongly interacted with the polymeric chain and caused permanent deformation of the structure and tends to remain attached to a chain. It may be possible that the coil structure of the polymer inhibits its complete removal with the help of a leaching agent like H_2O. This observation is confirmed with EDX as in the sub-set of Figure 2.3 (B), where the presence of Br confirms that some traces of CTAB are still present in the PVDF disc. The head-to-tail/head-to-head polymer chain flaw was apparent in the peak absorbed at 479 cm^{-1} [29]. From the modified PVDF disc, FTIR data shows that there is a confirmation of CuO absorption at the peak 433 cm^{-1} from this peak it is predicted that the surface of the PVDF disc is altered by CuO nanoparticles [30, 31]. The stretching vibration of CuO is predicted by the band 472 cm^{-1} in the above-mentioned Fig 1. The absorption bands at 880, 837, and 612 cm^{-1}, on the other hand, show that the alpha phase is still present. The (-C-C-) bond found in PVDF and CTAB has a similar symmetric attitude, as demonstrated by the peak absorbed at 1164 cm^{-1} once more. According to the FTIR data, which showed that the peaks assigned to the original PVDF had changed from absorption peaks at 608 and 476 cm^{-1} conforming to the CuO functional group, all of these samples were subjected to impregnation of nanoparticles inside the disc after leaching of the CTAB in good solvent (H2O). Hence, the presence of nano-sized Cu-O particles is confirmed by the highest absorbed peaks.

4.2 Morphological Analysis

To investigate the surface microstructure of the disc samples were identified under an SEM (scanning electron microscope). The morphological observation of the surface of the original PVDF sample is presented in Figure 2.

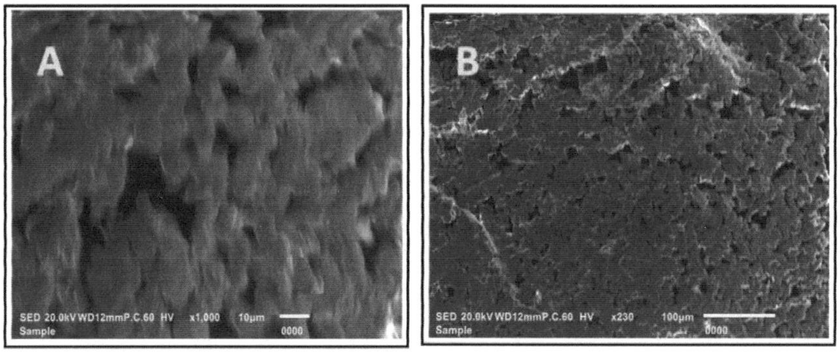

Fig. 2. SEM images (Scanning electronic microscopic) of the sample F0 at a magnification of 10 and 100μm represented as A and B.

The morphology in Figure 2 (A) is not smooth and the crystal is deformed. It means that earlier observation in the case of FTIR, that DMF being polar in nature, it can also have affected the crystalline structure of PVDF disc [32]. It can also be inferred from the micrograph that there is competition in different forms of PVDF, which crystallized from DMF during the process of de-mixing in the non-solvent-induced phase separation process. This competition leads to the generation of a small porous structure of the original PVDF disc. It may be predicted from the surface morphology that the disc when dried and DMF was evaporating may have created these nanocavities.

In Figure 3 the morphology observed had shown that CTAB is acting not only porogen but also as a nucleating agent for PVDF chains.

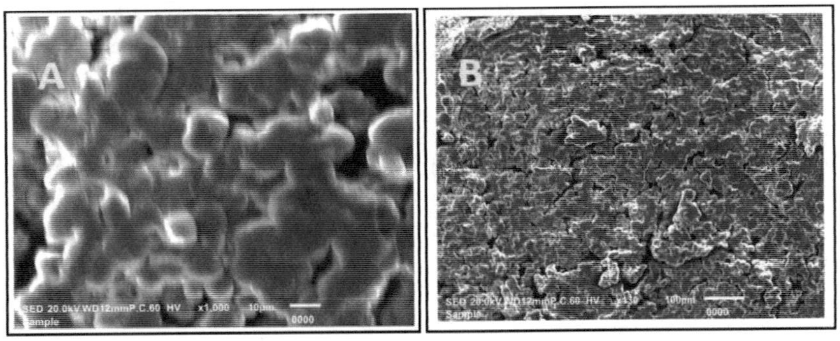

Fig. 3. SEM images of the sample F1 at 10 and 100 μm magnification.

The chains tend to attain structure morphology and this is beneficial in terms of regular pores formation. As the chain gets crystallized on a point-to-point basis, where CTAB is available nearby of the PVDF chain [21]. The void volume is extended and the disc becomes more porous. When the amount of CTAB was increased, such behavior becomes more pronounced and now spherical morphology can be seen in Figure 4.

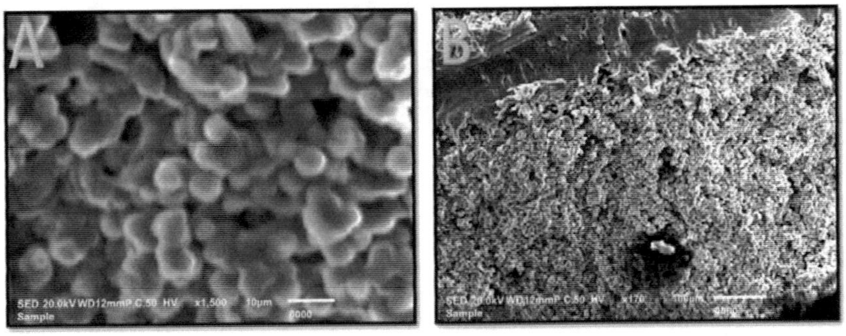

Fig. 4. Cross-sectional view of the sample F6 at 10 and 100 μm magnification.

Once these porous structures were developed the sample was subjected to in situ generation of copper (Cu) nanoparticles inside generated pores. The CTAB while self-assembled in the hydrophobic PVDF environment shows curvy nature and when it leaches out the pores remain in that shape. It was observed from the synthesis of nanoparticles in a separate experiment that shows random morphology with a curvy shape as reported earlier [33].

On the contrary, when nanoparticles are prepared inside confinement generated in PVDF with the removal of CTAB, attain the shape of the pores [26] and appear in that shape as shown in Figure 5. Scanning electronic microscopic images (Figure 5) shows that the pores of the disc were filled out by the inorganic nanoparticles (CuO). The amphiphilic properties of the PVDF were improved and the fouling properties were typically reduced by the nanoparticles [34]. The average pore size has embedded by CuO particles.

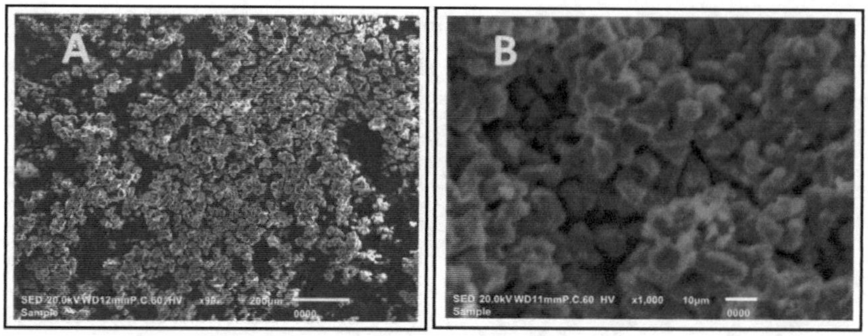

Fig. 5. Surface morphology of the amphiphilic polymeric disc.

4.3 Permeability and Separation of Amphiphilic Polymeric Disc

The inorganic nanoparticle was discovered to be an important factor in the modification of hydrophobic materials, according to several methods. The results of laboratory tests have verified that CuO nanoparticles have been deposited on the PVDF disc's surface. For the separation and permeability of the amphiphilic disc used different approaches, and many experiments were performed [19-21]. Equation (1), examined the disc's antifouling characteristics and explained the flux recovery ratio (FRR %). Greater FRR% indicates high fouling resistance. The concentration of nanoparticles increases along with the flux recovery ratio. The outcomes from equation (1) are described in Figure 6. From the FRR% graphical results, it's confirmed that the disc shows antifouling properties.

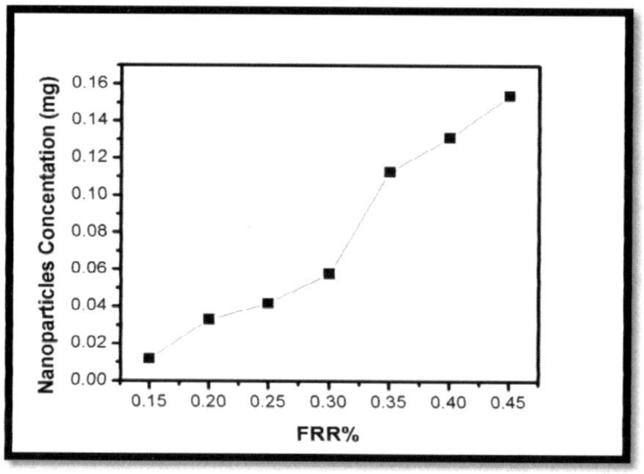

Fig. 6. Flux Recovery Ratio in percent (FRR %) shows Antifouling Properties.

In the second experiment, the mass flux rate of water was calculated using equation (2), and the flow rate per unit area in relation to the time of the disc was examined. Figure 7 illustrates how the flow rate enhanced as CuO (nanoparticles) were gradually added to the disc.

As the concentration of nanoparticles is raised, the mass flux increases, and vice versa, the water retention time in the APD (amphiphilic polymeric disc) decreases. The experiment was conducted under standard pressure, and the quantity of nanoparticles had an impact on the APD's retention period. Hence, the finding indicated that the disc's water permeability increased continuously as inorganic particles were added (Cu-O).

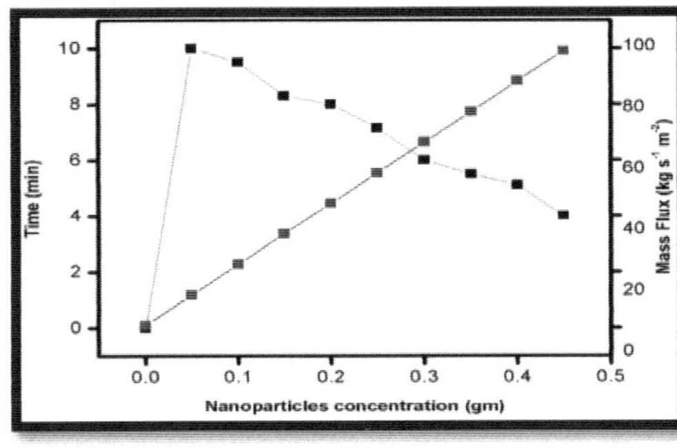

Fig. 7. Flux of water through APD

The PVDF disc's Cu-O impregnation demonstrated hydrophilicity and improved separation performance tested by the method are as shown in eq (3). The separation efficiency increased throughout the duration of the 0.6 gm of nanoparticles adding, and over 95% of the oil was rejected, allowing water to flow onto the APD alternatively. Figure 10 depicts the rejection ratio R% grows as the concentration of CuO rises.

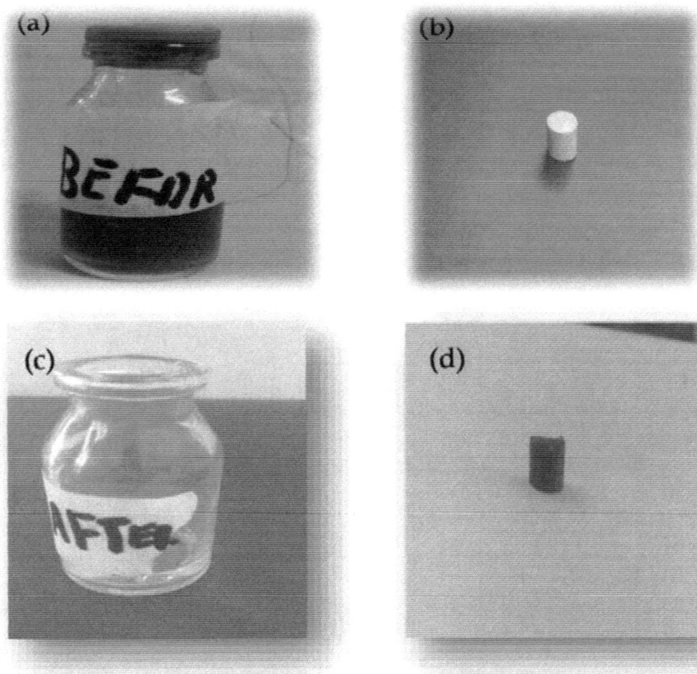

Fig. 8. (a) Oil-water Mixture, (b) Disc Before Separation, (c) Photographs of water and (d) disk after separation of the oil-water emulsion.

The oil-water combination and Amphiphilic disc were shown in Figure 8 before separation. Purified water and oil were obtained after experimentation with the APD's separation efficiency, and oil were rejected by the disc from the emulsion as shown in Figure 9 (A) below.

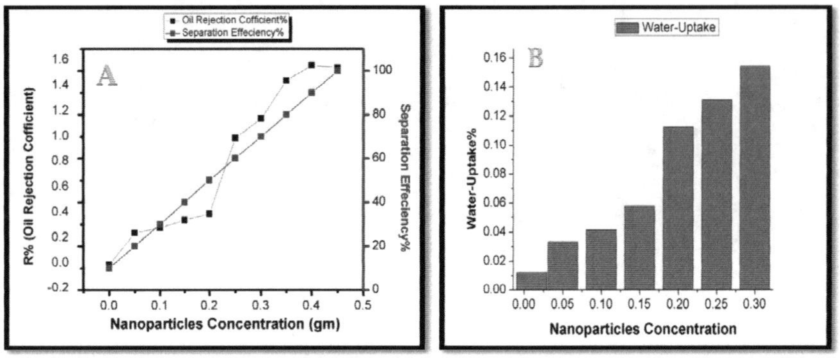

Fig. 9. (A) R % Oil rejection coefficient of the amphiphilic polymeric disc (B) Water-up taken by APD. One of the major aspects of the water uptake experiment in equation (4) that defined the hydrated disc mass ratio to that of the dry disc was the porosity of the disc. The water intake for APD was depicted in Figure 10(B). As the number of nanoparticles added to the PVDF disc increased, so did the weight of the wetted disc. The average weight of the hydrated disc revealed that its surface has tiny pores filled with CuO nanoparticles that have the ability to absorb water, increasing the disc's weight also.

4.4 UV-Visible-analysis

The differential absorbance spectra of water-in-oil emulsions before and after separation were measured by using a UV-visible spectrophotometer [35]. Several solutions with parts per million (ppm) concentrations in the range of 0.1, 0.2, 0.3, 0.4, and 0.5 were prepared, as well as standard solutions of crude oil in cyclohexane. A UV-Vis spectrophotometer was used to examine each solution in triplicate for the signal absorbance (A) versus wavelength (λ) using benzene as a reference, as illustrated in figure 10.

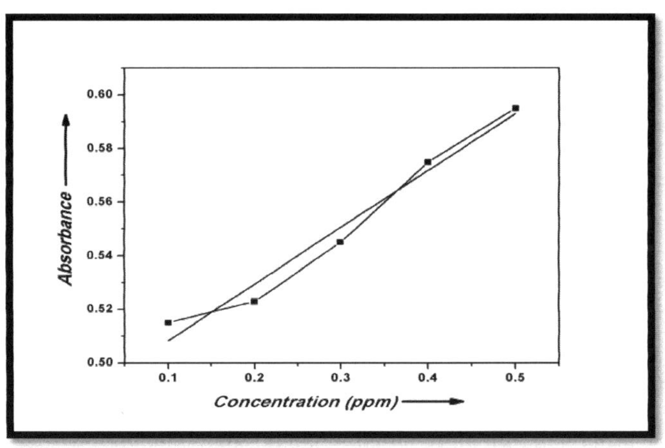

Fig. 10. Calibration curve of crude oil sample

The crude oil emulsion was passed to various disc samples and processed on a UV-visible spectrophotometer to evaluate how well the manufactured disc performed. In figure 11, the collected data were plotted.

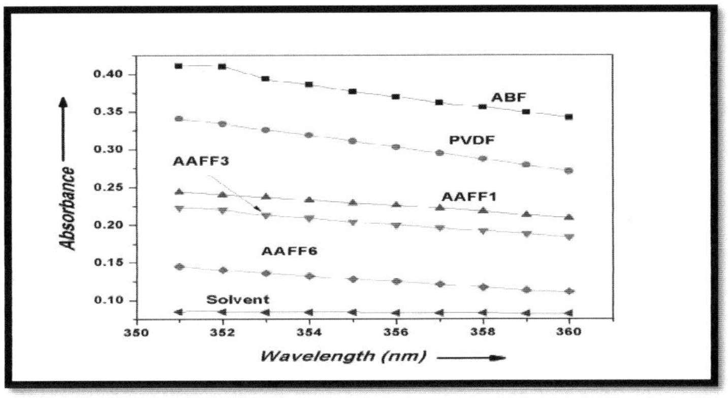

Fig. 11. Performance Evaluation by UV-Vis spectroscopy of different discs

(ABF = absorbance before filtration, AAFF1= absorbance after filtration through sample disc (F1), AAF3= absorbance after filtration through sample disc (F3), AAFF6 = absorbance after filtration through sample disc (F6).

According to the data collected, it can be shown that as the amount of nanoparticles inside the PVDF disc grows, the absorbance of the permeate oil tends to decrease. This increases the disc's hydrophilic nature and significantly reduces the amount of oil that may pass through it.

5. CONCLUSION

In order to remove oil from water samples that are high in oil numerous pore structures were developed in the amphiphilic polymeric PVDF disc (APD). CTAB was applied to the APD as a porogen to create pores. With more CTAB amount, the function of (APD) surface is raised. To create the disc more hydrophilic, nanoparticles (CuO) were applied. Cu-O was added to the disc using the impregnation technique. Nanoparticles considerably improved the disc surface's antifouling characteristics. Experimental tests on the permeation of water and oil successfully separated the water from the water-oil emulsion. The graphic analysis of the impact of different nanoparticle concentrations revealed that the flow recovery ratio (FRR %) improved with increasing nanoparticle concentration; the higher FRR% demonstrated maximal fouling resistance. Also, the mass flux rate result demonstrated that dramatically nanoparticles enhanced the flow of water over time through the disc. To support the APD's rejection of oil, experiments on the oil-rejection coefficient and water uptake were also conducted, briefly addressed in the results and conclusions. Any polymeric hydrophobic surface can readily be converted into a hydrophilic surface using the approach that is employed in polymers modification. Also, the manufactured disc has excellent potential for treating emulsion/wastewater in practical applications in industry and daily life.

References

[1] Peng, H.; Wang, H.; Wu, J.; Meng, G.; Wang, Y.; Shi, Y.; Liu, Z.; Guo, X., (2016) Preparation of superhydrophobic magnetic cellulose sponge for removing oil from water. *Industrial & Engineering Chemistry Research*, *55* (3), 832-838.

[2] Yu, L.; Han, M.; He, F., (2017) A review of treating oily wastewater. *Arabian journal of chemistry*, *10*, S1913-S1922.

[3] Ekambaram, K.; Doraisamy, M., (2017) Fouling resistant PVDF/Carboxymethyl chitosan composite nanofiltration membranes for humic acid removal. *Carbohydrate Polymers*, *173*, 431-440.

[4] Figueiredo, A. S.; Sánchez-Loredo, M. G.; Maurício, A.; Pereira, M. F. C.; Minhalma, M.; de Pinho, M. N., (2015) Tailoring of structures and permeation properties of asymmetric nanocomposite

cellulose acetate/silver membranes. *Journal of Applied Polymer Science* , *132* (21).
[5] Alenazi, N. A.; Hussein, M. A.; Alamry, K. A.; Asiri, A. M., (2017) Modified polyether-sulfone membrane:a mini review. *Designed monomers and polymers* , *20* (1), 532-546.
[6] Scharnagl, N.; Buschatz, H., (2001) Polyacrylonitrile (PAN) membranes for ultra- and microfiltration. *Desalination*, *139* (1), 191-198.
[7] Zhang, W.; Shi, Z.; Zhang, F.; Liu, X.; Jin, J.; Jiang, L., (2013) Superhydrophobic and Superoleophilic PVDF Membranes for Effective Separation of Water-in-Oil Emulsions with High Flux. *Advanced Materials*, *25* (14), 2071-2076.
[8] Jang, H.; Song, D.-H.; Kim, I.-C.; Kwon, Y.-N., (2015) Fouling control through the hydrophilic surface modification of poly(vinylidene fluoride) membranes. *Journal of Applied Polymer Science*, *132* (21).
[9] Kumar, R.; Ismail, A. F., (2015) Fouling control on microfiltration/ultrafiltration membranes: Effects of morphology, hydrophilicity, and charge. *Journal of Applied Polymer Science*, *132* (21).
[10] Otitoju, T. A.; Ahmad, A. L.; Ooi, B. S., (2016) Polyvinylidene fluoride (PVDF) membrane for oil rejection from oily wastewater: A performance review. *Journal of Water Process Engineering*, *14*, 41-59.
[11] Ahmad, A.; Majid, M.; Ooi, B., (2011) Functionalized PSf/SiO2 nanocomposite membrane for oil-in-water emulsion separation. *Desalination* 2011, *268* (1-3), 266-269.
[12] Wang, Z.; Lin, S., (2017) Membrane fouling and wetting in membrane distillation and their mitigation by novel membranes with special wettability. *Water research* , *112*, 38-47.
[13] Zhao, X.; Liu, C., (2019) Efficient preparation of a novel PVDF antifouling membrane based on the solvent- responsive cleaning properties. *Separation and Purification Technology* 2019, *210*, 100-106.
[14] Zhang, R.; Liu, Y.; He, M.; Su, Y.; Zhao, X.; Elimelech, M.; Jiang, Z., (2016) Antifouling membranes for sustainable water purification: strategies and mechanisms. *Chemical Society Reviews* 2016, *45* (21), 5888-5924.
[15] Płotka-Wasylka, J.; Szczepańska, N.; de la Guardia, M.; Namieśnik, J., (2016) Modern trends in solid phase extraction: new sorbent media. *TrAC Trends in Analytical Chemistry* 2016, *77*, 23-43.
[16] Elizalde, C. N. B.; Al-Gharabli, S.; Kujawa, J.; Mavukkandy, M.; Hasan, S. W.; Arafat, H. A., (2018) Fabrication of blend polyvinylidene fluoride/chitosan membranes for enhanced flux and fouling resistance. *Separation and Purification Technology*, *190*, 68-76.
[17] Nischang, I.; Causon, T. J., (2016) Porous polymer monoliths: from their fundamental structure to analytical engineering applications. *TrAC Trends in Analytical Chemistry*, *75*, 108-117.
[18] Nasrollahi, N.; Aber, S.; Vatanpour, V.; (2019) Mahmoodi, N. M., Development of hydrophilic microporous PES ultrafiltration membrane containing CuO nanoparticles with improved antifouling and separation performance. *Materials Chemistry and Physics*, *222*, 338-350.
[19] LIU, Q.-m.; ZHOU, D.-b.; Yamamoto, Y.; Ichino, R.; Okido, M., (2012) Preparation of Cu nanoparticles with NaBH4 by aqueous reduction method. *Transactions of Nonferrous Metals Society of China* , *22* (1), 117-123.
[20] Du, J.; Tian, Y.; Li, N.; Zhang, J.; Zuo, W., (2019) Enhanced antifouling performance of ZnS/GO/PVDF hybrid membrane by improving hydrophilicity and photocatalysis. *Polymers for Advanced Technologies*, *30* (2), 351-359.
[21] Liu, J.; Xiong, J.; Ju, X.; Gao, B.; Wang, L.; Sillanpää, M., (2018) Streaming potential for identification of foulants adsorption on PVDF membrane surface. *Journal of membrane science* , *566*, 428-434.
[22] Abdurahman, H. N.; Yunus, R. M.; Jemaat, Z., (2007) Chemical demulsification of water-in-crude oil emulsions. *Journal of Applied Sciences*, *7* (2), 196-201.
[23] Abdel-Fatah, M. A., (2018) Nanofiltration systems and applications in wastewater treatment. *Ain Shams Engineering Journal* . *9*(4), 3077-3092
[24] Bormashenko, Y.; Pogreb, R.; Stanevsky, O.; Bormashenko, E., (2004) Vibrational spectrum of PVDF and its interpretation. *Polymer testing*, *23* (7), 791-796.
[25] Gu, S.; He, G.; Wu, X.; Hu, Z.; Wang, L.; Xiao, G.; Peng, L., (2010) Preparation and characterization of poly (vinylidene fluoride)/sulfonated poly (phthalazinone ether sulfone ketone) blends for proton exchange membrane. *Journal of applied polymer science* , *116* (2), 852-860.

[26] Li, Y.; Xu, J.-Z.; Zhu, L.; Xu, H.; Pan, M.-W.; Zhong, G.-J.; Li, Z.-M., (2014) Multiple stage crystallization of gamma phase poly (vinylidene fluoride) induced by ion-dipole interaction as revealed by time-resolved FTIR and two-dimensional correlation analysis. *Polymer*, *55* (18), 4765-4775.

[27] Viana, R. B.; da Silva, A. B.; (2012) Pimentel, A. S., Infrared spectroscopy of anionic, cationic, and zwitterionic surfactants. *Advances in Physical Chemistry*, *2012*.

[28] Cai, X.; Lei, T.; Sun, D.; Lin, L., A (2017) critical analysis of the α, β and γ phases in poly (vinylidene fluoride) using FTIR. *RSC Advances*, *7* (25), 15382-15389.

[29] Zhang, Y. X.; Huang, M.; Li, F.; Wen, Z. Q., (2013) Controlled synthesis of hierarchical CuO nanostructures for electrochemical capacitor electrodes. *Int. J. Electrochem. Sci*, *8*, 8645-8661.

[30] El-Trass, A.; ElShamy, H.; El-Mehasseb, I. (2012) El-Kemary, M., CuO nanoparticles: synthesis, characterization, optical properties and interaction with amino acids. *Applied Surface Science*, *258* (7), 2997-3001.

[31] Cavallaro, G.; Lazzara, G.; Milioto, S.; Parisi, F., (2015) Hydrophobically modified halloysite nanotubes as reverse micelles for water-in-oil emulsion. *Langmuir.*, *31* (27), 7472-7478.

[32] Abdelrasoul, A.; Doan, H.; Lohi, A.; Cheng, C.-H., (2017) The influence of aggregation of latex particles on membrane fouling attachments & ultrafiltration performance in ultrafiltration of latex contaminated water and wastewater. *Journal of Environmental Sciences*, *52*, 118-129.

[33] Sundararajan PR. Small molecule self-assembly in polymer matrices. Journal of Polymer Science Part B: Polymer Physics, 15;56(6):451-78.

[34] Zhao, C.; Lv, J.; Xu, X.; Zhang, G.; Yang, Y.; Yang, F., (2017) Highly antifouling and antibacterial performance of poly (vinylidene fluoride) ultrafiltration membranes blending with copper oxide and graphene oxide nanofillers for effective wastewater treatment. Journal of colloid and interface science , 505, 341-351.

[35] Lin, X.; Chen, Y.; Liu, N.; Cao, Y.; Xu, L.; Zhang, W.; Feng, L., (2016) In situ ultrafast separation and purification of oil/water emulsions by superwetting TiO 2 nanocluster-based mesh. *Nanoscale* , *8* (16), 8525-8529.

IOP Conf. Series: Materials Science and Engineering 1291 (2023) 012006 doi:10.1088/1757-899X/1291/1/012006

Electrodeposited CZTS loaded titania nanotubes for photocatalytic water splitting

Athulya Jayakumar, Ambily Krishnan*

Govt. Victoria College, Palakkad, Kerala, India

ambilykrishnan@gvc.ac.in

Abstract. Copper Zinc Tin Sulphide (Cu_2ZnSnS_4 (CZTS)) is a promising semiconductor material with optimum direct band gap of 1.4-1.6 eV, large light absorption coefficient (> 10^4 cm^{-1}) and p-type electrical conductivity. In addition, its earth abundant and non-toxic elemental constituents make it a good candidate to replace the CdTe and CIGS absorber layers used in thin film solar cells. Also, it has properties that make it useful as a buffer layer that helps to reduce the effective band gap of TiO_2 when used as an anode for photocatalytic splitting of water. We prepared titania nano tube (TNT) arrays by anodization of titanium foil. Raman spectrum of TNT confirmed the presence of pure anatase phase of TiO_2. Field emission scanning electron microscopy images were obtained to analyse the surface morphology of TNT. CZTS sensitized titania nano tube arrays were then prepared by the electrodeposition of CZTS thin films over TNT arrays. FESEM images of the CZTS sensitized titania nano tube arrays showed that the electrodeposited CZTS layer had a fibrous structure. A photo electro chemical (PEC) cell was set up with CZTS sensitized titania nanotubes as anode. An anodic current density of 0.19 mA/cm^2 was shown by the PEC cell when the anode was irradiated with blue light of wavelength 430 nm at power 50 mW/cm^2. The incident photon to current conversion efficiency was 1.1 %.

1. Introduction

Hydrogen is an ideal fuel for the future because it is sustainable, energy-dense and has eco-friendly properties. Greenhouse gas emissions could be reduced by using hydrogen as a fuel [1]. Photocatalytic water splitting has great potential for low cost and clean hydrogen production [2]. In 1972, Honda and Fujishima discovered electrochemical water splitting over a TiO_2 electrode [3]. Titania with properties like low cost, high stability, non-toxicity, biological inertness, redox ability and environmental merits has been extensively studied. TiO_2 is an n-type semiconductor which shows optical absorption in the UV region of solar spectrum with lowest conduction band level of -0.3 eV which is more negative than the reduction potential of hydrogen (H^+/H_2 0 eV vs NHE) and a valence band maximum level of 2.9 eV which is more positive than the oxidation potential of water (O_2/H_2O 1.23 eV vs NHE [4]. Nanostructured TiO_2 such as nanoparticles, nanowires, nano tubes show enhancement in photoelectrochemical performance. One-dimensional (1D) TiO_2 nanostructures act as direct pathways of electrons and holes and cause light to be more easily scattered and trapped compared to bulk counterpart. However, the large band gap which results in inefficient utilization of solar light, fast recombination of photogenerated electron-hole pairs and photocorrosion are some disadvantages which reduce the scope of TiO_2 in this application [5]. TiO_2 can be modified using various methods like doping, co-doping, modification with inorganic acids, semiconductor coupling, constructing heterojunctions, sensitization etc.

In inorganic thin film photovoltaic technology, CZTS is one of the very popular light absorbing materials consisting entirely of non-toxic, cheap and earth abundant elemental constituents. It has an optimum band gap (1.5 eV) and has a kesterite crystal structure (KS) with I_4^- space group symmetry [6]. It is a good substitute for platinum counter electrode of dye sensitized solar cell due to its catalytic property and p-type conductivity. For photocatalytic water splitting, CZTS has been utilized to form heterojunctions with TiO_2 because of its band gap matching with the optimum value required for photocatalyst. The electronic and optical behaviour of the semiconductor have a prime contribution in this application because the redox reactions that occur on the catalyst are responsible for hydrogen generation. For this, the conduction band bottom edge must be more negative than the water reduction potential (0V versus NHE) and valence band top edge should be more positive than the water oxidation potential (1.23V versus NHE) [1]. The energy band gap of bulk CZTS is suitable for solar energy conversion but it is not sufficient for splitting water because bulk CZTS with a band gap range of 1.4-1.6 eV, has a valence band maximum slightly higher than water oxidation potential. By size reduction, the band gap is widened. CZTS when combined with the large area wide band gap material TiO_2 nano tube arrays to form heterostructure is a better photocatalyst for water splitting. During past years, CZTS thin films have been developed using several vacuum and non- vacuum techniques like sputtering, thermal and electron beam evaporation, chemical bath deposition, spin coating, successive ionic layer adsorption and reaction, nano crystal ink approaches, electrodeposition and hydrazine-based methods. Co-electro deposition technique is a simple, flexible, fast, easy and energy efficient method to deposit CZTS films using non-toxic precursors, safe solvents and cheap equipment compared to the other methods [7].

In this paper we report how we decorated titania nano tubes by electrodepositing CZTS fibres over it and investigated the possibility of photocatalytic hydrogen evolution in a water splitting reaction. We noticed that CZTS sensitized titania nano tube arrays possessed improved photoactivity properties.

2. Experimental details

2.1. Preparation of titania nanotubes

For preparing highly ordered titania nano tube (TNT) arrays, anodization of titanium foil was done in a two-electrode system using Ti foil as anode and Pt mesh as cathode. Initially, before anodization titanium foils (0.3mm thickness, 99.7% purity (from Sigma Aldrich), and 2 × 2 cm^2 area) were washed in de-ionized water, sonicated in isopropanol and acetone and dried. Anodization was carried out at 55 V for 4 hours. The electrolyte was ethylene glycol containing 0.3 wt.% NH_4F and 0.3 volume % water. To remove the organic compounds, as-grown titania nano tube (TNT) arrays were cleaned by sonicating in isopropanol for 5 minutes and dried in air. To crystallize the TNTs, it was annealed at 500 °C for 2 h in air [8].

2.2. Deposition of CZTS thin films

Using an ordinary DC power supply converted into a potentiostat in conventional three electrode mode with Ag/AgCl standard electrode as the reference electrode, Pt wire for counter electrode and TNT coated titanium foil as the working electrode, the electrodeposition of CZTS was done over the TNT layer. An aqueous solution of 0.01 M $CuSO_4$, 0.03 M $ZnSO_4.7H_2O$, 0.02M $SnCl_2.2H_2O$ and 0.01 M $Na_2S_2O_3.5H_2O$ was used as the electrolyte. 100 mM trisodium citrate was added to the bath as complexing agent. In order to maintain the pH of the electrolyte in the range 4-5, 20 mM tartaric acid was added. Deposition of CZTS thin films over TNT arrays was done at -2.5 V for 5 minutes at room temperature. The films were rinsed with de-ionized water and dried in air [7]. Thereafter the structure was annealed at 200 °C for an hour in order to form a heterojunction.

2.3. Characterization

Raman spectrum of titania nano tubes was acquired using Horiba Labram HR Evo Raman spectrometer with an excitation wavelength of 532 nm. FESEM images of TNT and CZTS sensitized titania nanotube arrays were acquired using JEOL JSM 6390.

2.4. Photoelectrochemical cell

To measure the photoactivity properties, a photo electro chemical cell (PEC) was set up using a cubical quartz vessel with titanium foil coated with CZTS sensitized titania nanotubes as anode, Pt mesh as cathode and aqueous solution of sodium sulphate (1M) as electrolyte. Photocurrents were measured using Keithley SMU 2450 set in the ammeter mode.

3. Results and discussion

FESEM image of the surface of titania nanotubes grown on titanium foils prepared by the electro deposition process is shown in figure 1(a). An average tube diameter of 86 nm can be seen on the FESEM image of cross section of the anodized Ti foil. Figure 1(b) shows the longitudinal view of the well aligned and densely formed nanotubes. Figure 1(c) shows the FESEM image of the CZTS coated titania nanotubes from the surface. We can see that the deposited CZTS has a fibrous structure. The side view in figure 1(d) shows the approximate length of the TNT as 12.6 microns. Upon anodizing Ti sheet for 4 hours, we find that length of nano tubes formed is 12.6 μm. This implies that the oxidation/etching process takes place only on the surface which is only 5 % of the thickness of the sheet. Rest of the thickness remains as Ti metal and acts as a substrate which supports the TNT and as an electrode that provides electrical contact. The length of the titania nanotubes depend on the time for which anodization is done.

Raman spectrum of TNT shows peaks corresponding to the active modes at 142.79 (E_g), 197.33(E_g), 394.21 (B_{1g}), 512.013 (A_{1g}) and 636.87(E_g) cm^{-1}, which confirms the presence of pure anatase phase of TiO_2. The active mode E_g refers to O-Ti-O symmetric stretching vibrations, B_{1g} relates to O-Ti-O symmetric bending vibrations and A_{1g} refers to anti-symmetric bending vibrations in anatase crystal.

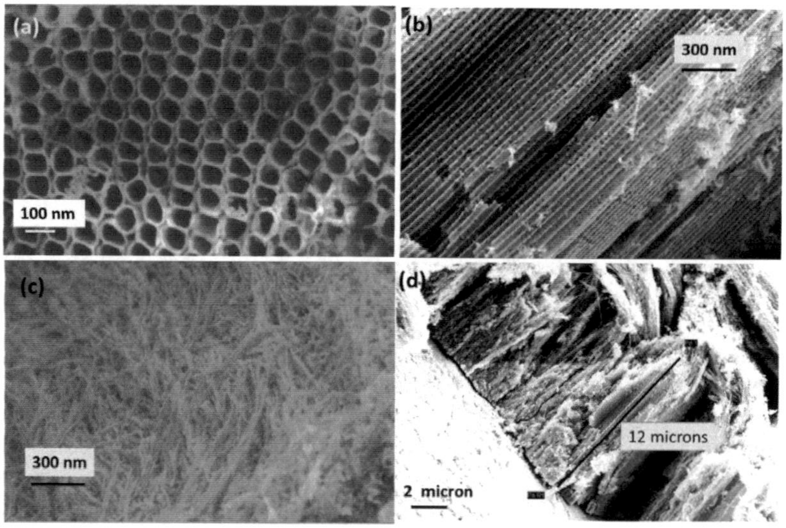

Figure 1. FESEM images of (a) bare TNT when viewed from the surface (b) longitudinal view of the titanium nanotubes showing dense and regular growth (c) surface of the CZTS layer electro deposited over TNT (d) side view of the TNT used to measure the length of the titania nano tubes. Fibrous CZTS deposited on top of the TNT can also be seen.

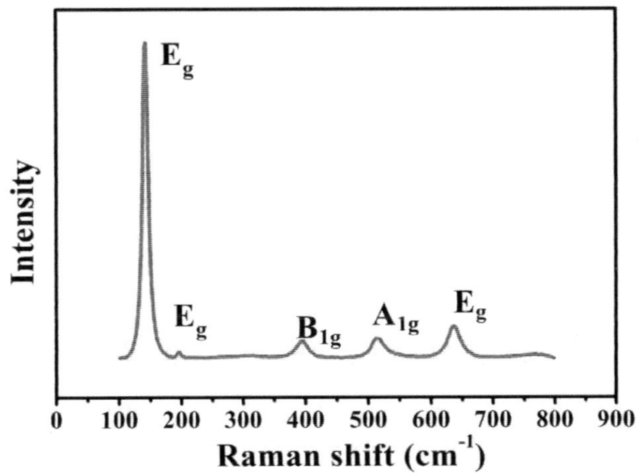

Figure 2. Raman spectrum of TiO_2 nanotubes

Though the aim was to grow a thin film of CZTS over the TNT surface, the FESEM images show that the deposited layer has a fibrous structure. Optical absorption could be higher in the fibrous structure and the electron-hole pairs generated will produce a current if they can be separated with the help of the barrier potential at the heterojunction formed at the interface of the n-type TiO_2 and p-type CZTS (Figure 3(a)). A schematic of the anode developed and the PEC set up are shown in figure 2 (b) and (c).

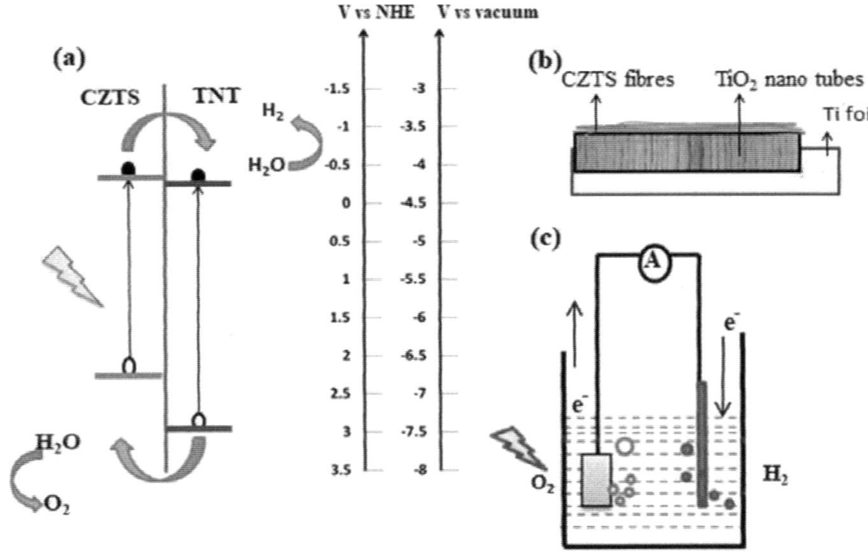

Figure 3. (a) Energy level diagram showing photo generation of electron hole pair and transport of electron and holes across the CZTS-TNT junction (b) schematic of the anode and (c) photo electro chemical cell.

Due to the large band gap of titania (> 3.2 eV) no current could be obtained in the PEC with bare titania nano tubes when the anode was irradiated with light from blue, green and red lasers at power 50 mW. The wide bandgap makes it unable to utilize visible range of the spectrum. For photocatalytic water splitting, it can utilize only UV light. But when coated with CZTS, we observed that a current flows in the external circuit when the anode was illuminated. The PEC using CZTS sensitized TNT as anode showed a current density of 0.19 mA/cm^2 when the anode was irradiated with blue light of wavelength 430 nm at power 50 mW/cm^2. The number of carriers generated per incident photon is the incident photon to current conversion efficiency (IPCE) or quantum yield. The IPCE value obtained for blue light was 1.1 %. This shows that though small, there is a distinct possibility of effecting the water splitting hydrogen evolution reaction through photocatalytic method using the CZTS/TNT combination.

In photocatalytic water splitting, when the photocatalyst is irradiated with light of energy greater than its band gap, electron-hole pairs are formed. This is due to the excitation of electron from its valence band (VB) to conduction band (CB) thereby forming holes in the VB. The photo generated electrons and holes migrate to the surface and act as reducing and oxidizing agents respectively. Water splitting is a highly endothermic process which requires a Gibbs free energy of 1.23 eV per electron. This corresponds to light with a wavelength of 1008 nm [3]. A photocatalyst of band gap > 1.23 eV is required, or else the electrons will not have enough energy to start the reaction. For water splitting the reduction and oxidation potentials of water should lie within the band gap of the photocatalyst. TiO_2 is an n-type semiconductor which shows optical absorption in UV-region of solar spectrum. Its CB level (-0.3 eV) is more negative and suitable for the redox potential of H^+/H_2 (0 eV) and its VB level (2.9 eV) is more positive than the standard redox potential of O_2/H_2O (1.23 eV) vs. normal hydrogen electrode. But due to its wide band gap, it is unable to utilize the visible light of solar spectrum [3]. We observed that no current could be obtained in the PEC with bare titania nano tubes when the anode was irradiated with light from blue, green and red lasers at power 50 mW. The wide bandgap makes it unable to utilize visible range of the spectrum. For photocatalytic water splitting, it can utilize only UV light. The formation of p-n junction of CZTS and TiO_2 reduces the effective band gap since CZTS is having an optimum band gap of 1.4-1.6 eV. When blue light of wavelength 430 nm at power 50 mW/cm^2 is irradiated over CZTS/TiO_2 junction, electron-hole pairs are formed in CZTS. We observed that a current flowed in the external circuit. The PEC using CZTS sensitized TNT as anode showed a current density of 0.19 mA/cm^2 when the anode was irradiated with blue light of wavelength 430 nm at power 50 mW/cm^2. The electric field at the p-n junction helps in separation of the two by attracting them in opposite directions. Photo generated electrons transfer from CZTS to TiO_2 and the electrons are easily drawn into the Ti metal electrode through the channel provided by the TNT nanotubes. They travel through the external circuit, reach the platinum electrode and reduce water molecules releasing hydrogen at the Pt cathode of the PEC cell. The holes are attracted in the opposite direction because of the barrier potential at the p-n junction and reach the surface of CZTS layer where they oxidise water to oxygen [3].

The current in the circuit of the PEC is negligible in the absence of light and is found to increase upon incidence of light from the blue laser. This clearly indicates that CZTS/TiO2 junction behaves as a photovoltaic p-n junction, where photons produce e-h pairs. A photo current of 0.19 mA/cm^2 indicates the possibility of water splitting process. The number of carriers generated per incident photon is the incident photon to current conversion efficiency (IPCE) or quantum yield. The IPCE value obtained for blue light was 1.1 %. This shows that though small, there is a distinct possibility of effecting the water splitting hydrogen evolution reaction through photocatalytic method using the CZTS/TNT combination.

4. Conclusion

The possibility of modifying TiO_2 in order to engineer its band gap to a value where it can absorb visible light has been a promising prospect of research. Its application in the field of photo catalytic water splitting reaction has been explored by us using the p-type semiconductor CZTS. The electrodeposition method which is simple, efficient and low cost has been employed for the fabrication of both titania nano tubes as well as the CZTS layer. The efficiency of the PEC reported is not very high, but the photocurrent clearly indicates the possibility of the process and shows that better values and more efficient hydrogen generation can be achieved by improving the interface properties and by reducing recombination of the photogenerated carriers. Several factors like length of the titania nanotubes which depends on the anodization time and presence of binary and ternary defect phases in the CZTS layer could affect the conversion efficiency of the PEC. These effects need to be investigated in future work and formation of the catalytic heterojunction for water splitting optimized accordingly.

Acknowledgements
The authors wish to thank CIF-CMFF IIT, Palakkad for Raman Spectroscopy and Department of Physics, Cochin University of Science and Technology for FESEM characterization of the samples. The authors acknowledge Department of Science and Technology, Govt. of India for providing funding to Department of Physics, Govt. Victoria College, Palakkad under DST-FIST scheme.

References
[1] Singh, Rohini, and Suman Dutta. "A review on H2 production through photocatalytic reactions using TiO2/TiO2-assisted catalysts." Fuel 220 (2018): 607-620.
[2] Liao, Chi-Hung, Chao-Wei Huang, and Jeffrey Wu. "*Hydrogen production from semiconductor-based photocatalysis via water splitting.*" Catalysts 2, no. 4 (2012): 490-516.
[3] Du, Hong, Ya-Nan Liu, Cong-Cong Shen, and An-Wu Xu. "*Nanoheterostructured photocatalysts for improving photocatalytic hydrogen production.*" Chinese Journal of Catalysis 38, no. 8 (2017): 1295-1306.
[4] Nolan, Michael, Anna Iwaszuk, Aoife K. Lucid, John J. Carey, and Marco Fronzi. "*Design of novel visible light active photocatalyst materials: surface modified TiO2.*" Advanced Materials 28, no. 27 (2016): 5425-5446.
[5] Zhu, Lumin, Jie Tao, Haijun Tao, Shanlong Chen, Yizhou Shen, Aichun Xu, JiaJia Jiang, and Lei Pan. "*In-situ growth of Cu2ZnSnS4 nanosheets on TiO2 nanowires for enhanced photoelectrochemical performance.*" Journal of Alloys and Compounds 649 (2015): 704-711.
[6] Walsh, Aron, Shiyou Chen, Su-Huai Wei, and Xin-Gao Gong. "*Kesterite thin-film solar cells: advances in materials modelling of Cu2ZnSnS4.*" Advanced Energy Materials 2, no. 4 (2012): 400-409.
[7] Krishnan, Ambily, Dharsana Vidyadharan, Sindhu Swaminathan, and Pradeesh Kannan. "*Co-electrodeposited Cu2ZnSnS4 thin films for PN junction photovoltaics and dye sensitized solar cells.*" Materials Today: Proceedings 25 (2020): 122-128.
[8] Krishnan, Ambily, and Pradeesh Kannan. "*A Non-Van der Waals two-dimensional material from microwave synthesized Cu2ZnSnS4 nanocrystal agglomerates.*" Materials Research Express 6, no. 11 (2019): 1150a3.

IOP Conf. Series: Materials Science and Engineering 1291 (2023) 012007 doi:10.1088/1757-899X/1291/1/012007

Magnetic, morphological, and photocatalytic studies of Cu^{2+} doped Mg-Zn ferrite nanoparticles

Sangita N Pund[1], Pratik A Nagwade[2], Arvind V Nagawade[1], Shankar R Thopate[3] and Arun V Bagade[1*]

[1]Department of Chemistry, BPHE Society's Ahmednagar College, Ahmednagar, 414001, Maharashtra, India.
[2]Department of Chemistry, Shri Anand College, Pathardi, Ahmednagar, 414102, Maharashtra, India.
[3]Department of Chemistry, Radhabai Kale Mahila Mahavidyalaya, Ahmednagar, 414001, Maharashtra, India.

arunbagade14@gmail.com

Abstract: Due to their distinctive characteristics, including their optical, catalytic, electrical, and magnetic properties, spinel ferrite nanoparticles attract more interest. Also, the substitution of transition metals like copper in ferrites has the potential to control their physical characteristics and could improve their catalytic and magnetic capabilities. Cu^{2+} doped Mg-Zn ferrite samples show a change in behaviour from superparamagnetic to soft ferrimagnetic. The photocatalytic studies for the $Cu_xMg_{0.5}Zn_{0.5-x}Fe_2O_4$ (x= 0.1 to 0.5, and Δx= 0.1) nano-ferrites are conducted in visible light to investigate the methylene blue photodecomposition capability. The Cu-Mg-Zn nano-ferrites displayed unique behaviour in terms of Magnetic, and photocatalytic activity. These outcomes show that the Cu-Mg-Zn ferrite samples are apply to water remediation.

1. Introduction

For around 55 years, researchers have studied and used ferrites, a large class of oxides with notable magnetic properties. The majority of ferrites have aided in the development of new applications in nanotechnology. Due to its notable modification in the physical and chemical properties compared to that of the bulk ferrites [1], nano-ferrites are used in photocatalysis [2], gas sensors [3], magnetic resonance imaging [4], photochemical hydrogen generation from water [5], and drug delivery [6]. Nano-ferrites can be produced using a variety of methods to yield their three distinct crystal systems, and the potential to produce nearly infinite solid solutions of the nano-ferrites opens up the possibility of modifying and adapting their properties for a variety of uses. These ferrite nanoparticles have evolved into a new and fascinating era in the research field of dye degradation.

Several synthetic dyes are currently employed in various sectors, including rubber, leather, pharmaceuticals, plastic, etc., but the textile business uses them the most frequently. Dyes are categorized as cationic, anionic, and non-ionic (disperse dyes). The most common synthetic dye used in the majority of the coloring industries is a cationic dye like methylene blue (MB). It is employed primarily due to its distinctive characteristics of high thermal and light-withstanding behavior [7]. It is exceedingly difficult to eliminate MB from water when it is drained out with wastewater and it has negative impacts. To remove the MB, a variety of traditional methods have been developed, such as coagulation, nanofiltration, adsorption, etc. which are slower and time-consuming, and did not show good output efficiency [7]. These days, nanoparticles are utilized to degrade dyes because they have a high surface area, good adsorption, and a quick rate of equilibrium [8].

Spinel ferrites have the potential to be used in visible-light-driven photocatalytic applications, according to certain recent publications [9–11]. To enhance the characteristics of ferrites, several substitutions have been incorporated. The size, magnetic stability, and purity of these nanoparticles further determine their applicability [12]. The substitution of magnetic ions like Cu^{2+} for Zn^{2+} has a substantial impact on the characteristics of Mg-Zn ferrites [13]. A soft magnetic ion, Cu^{2+} has a magnetic moment of 1 μB. According to M. Suleman et al. [14], adding cobalt ions to Mg-Zn ferrites increased their photocatalytic activity against benzimidazole and methylene blue. In the study by S. B. Somvanshi et al. [15], a sufficient amount of Gd^{3+} considerably increases the number of hydroxyl radicals that the ferrite produces, which in turn increases the photocatalytic activity of the material against Rhodamine B dye. M. A. Abdo et al. [16] synthesized $Co_{0.5}Cu_{0.25}Zn_{0.25}Y_xFe_{2-x}O_4$; ($0 \leq x \leq 0.1$; step 0.02) (CCZY) spinel ferrites by citrate technique and demonstrated their photocatalytic activity against MB. The nano-ferrite CCZY (0.1), which has a degradation efficiency of 95% within only 60 minutes, exhibits the best performance for the elimination of MB dye. $Zn_{0.7}Mg_{0.24}Cu_{0.06}Fe_2O_4$ (x = 0.06) nano-photocatalyst demonstrated 90% photocatalytic degradation efficiency in the photocatalytic degradation of methylene blue with a maximum rate constant of 0.0252 min^{-1} [17]. The literature [18,19] supports the use of Cu as a dopant, showing that the photocatalytic performance of the catalysts was enhanced by doping Cu into ferrites.

In this present work, magnetic and morphological characteristics of Cu^{2+} doped Mg-Zn ferrite nanoparticles were analyzed using Vibrating Sample Magnetometer (VSM) and Field Emission Scanning Electron Microscopy (FE-SEM). The photocatalytic activity on MB dye degradation was further explored with visible light irradiation.

2. Materials and methods

2.1 Materials

Copper sulfate [$CuSO_4.5H_2O$], Magnesium sulfate [$MgSO_4.7H_2O$], Zinc sulfate [$ZnSO_4.H_2O$], and Ferrous sulfate [$FeSO_4.7H_2O$] were acquired from Merck, India. Spectrochem chemicals from India supplied the potassium oxalate monohydrate [$C_2H_2K_2O_5$]. All procured chemicals were of analytical grade. The chemicals received were used as such without extra purification. Distilled water is used all over this research.

2.2 Experimental procedure

The preparation of $Cu_xMg_{0.5}Zn_{0.5-x}Fe_2O_4$ (with x ranging from 0.1 to 0.5 at an increase of 0.1) ferrites nano-particles [Cu-Mg-Zn NPs] by chemical co-precipitation technique (potassium oxalate precipitant) has been already published elsewhere [20]. The batch mode photocatalytic experiments were conducted in visible light (160 W lamp with 535 lux) for 180 minutes. Using 25 mL of a 10-ppm pollutant aqueous solution (MB) and 25 mg of catalyst, adsorption-desorption equilibrium was established in the dark. The solution was then subjected to visible light, and 2 mL aliquots were taken at specific intervals with the solid catalyst particles, which were subsequently removed with an external magnet. As seen in Figure 1, a magnet can be used to quickly separate the catalyst from the reaction mixture.

Using a UV-Vis spectrophotometer (Make: LabIndia UV 3000+), the photocatalytic degradation is determined by the subsequent equation:

$$\% \ D = \frac{C_0 - C_t}{C_0} \times 100 \quad (1)$$

Where, C_o and C_t stand for initial and after absorbance time t, respectively, and % D for percent degradation.

Figure 1. Images for MB solution (a) without photocatalyst (b) with photocatalyst, and (c) after degradation showing magnetic separation.

3. Results and discussion

The detailed structural, vibrational, and optical absorption properties of Cu-Mg-Zn NPs which were investigated by X-ray diffraction, FT-IR, and UV-DRS have been also published elsewhere [20].

3.1 Magnetization studies by VSM analysis

Figure 2 displays the M-H curves for samples of $Cu_xMg_{0.5}Zn_{0.5-x}Fe_2O_4$ (with x ranging from 0.0 to 0.5 at an increase of 0.1) at room temperature. Synthesized ferrites with Cu^{2+} doping exhibit a shift from superparamagnetic to soft ferrimagnetic behavior. The residence of diamagnetic Mg^{2+} and Zn^{2+} ions in the x= 0 sample, which have no electron spin and do not contribute to magneto-crystalline anisotropy, results in super-paramagnetic behavior [21]. The ferrimagnetic property is induced by replacing diamagnetic Zn^{2+} with magnetic Cu^{2+} ions possessing a magnetic moment of 1 μB. Samples can display soft ferrimagnetic behavior and maintain magnetic order by boosting magneto-crystalline anisotropy with the help of magnetic Cu^{2+} ions [22].

It is also evident from the figure that all of the curves furnished narrow loops accompanied by a behavior typical of soft magnetic ferrites (in the current samples, easy magnetization and demagnetization accompanied with low coercivity meant fewer hysteresis losses, which is a crucial condition for a successful electromagnet). These hysteresis curves were employed to evaluate magnetic properties such as coercivity (H_c), remanence magnetization (M_r), and saturation magnetization (M_s) (Table 1). As seen in Table 1, the M_s values rise linearly with rising copper content up to x=0.3, at which point it starts to noticeably decline. Based on the exchange interactions and the cation distribution at the A and B sites, respectively, it is possible to explain the observed fluctuation in saturation magnetization. Although A-B super-exchange interactions are more significant than those of A-A and B-B, all super-exchange interactions, including A-B, A-A, and B-B, affect ferrite magnetization. The magnetization of spinel ferrites is calculated by assuming the magnetizations at the B- and A- sites are M_B and M_A in Neel's two sub-lattice model [22]:

$$M = M_B - M_A \quad (2)$$

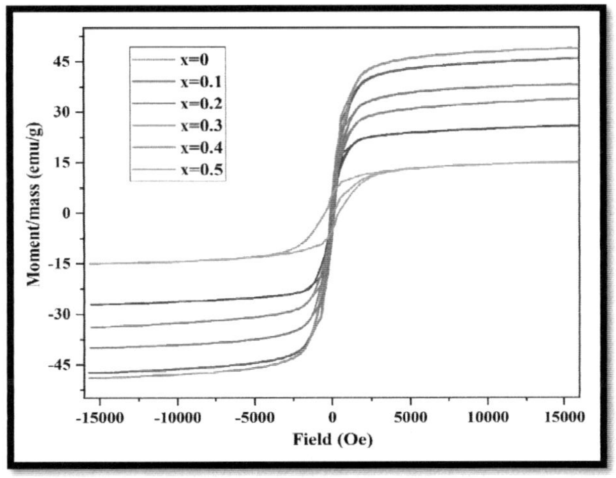

Figure 2. Magnetic Hysteresis (M-H) loops for $Cu_xMg_{0.5}Zn_{0.5-x}Fe_2O_4$ (x= 0.0 to 0.5, and Δx= 0.1) ferrites.

The first rise in the saturation magnetization value is ascribed to the preference of Cu^{2+} ions with magnetic moments of 1 μB to migrate to the B-sites over non-magnetic Zn^{2+} ions with magnetic moments of 0 μB, which exhibit high priority for the tetrahedral (A) site. As a result, Zn^{2+} moves Fe^{3+} from site A to B site, and the addition of Cu^{2+} ions to site B raises the magnetization of these sublattice sites. Because of the super exchange interaction in the ferrite lattice, which causes the magnetic spin of nearby A and B sites to be anti-ferromagnetically connected, the net effect is the rise in the magnetic moment on the B-site, and the net magnetic moment of the Cu-Mg-Zn NPs rises with Cu^{2+} ion content up to x = 0.3 as shown in Figure 2. The presence of some Fe^{2+} ions in addition to their trivalent state at B-sites will cause the magnetization of ferrites with compositions x = 0.4 and x = 0.5 to decrease. For samples with x=0.4 and x=0.5, the steady decrease in magnetization would be due to an increase in the occupation of B sites by Fe^{2+} ions. Additionally, the movement of Fe^{3+} ions to the B sub-lattice in conjunction with the movement of Mg^{2+} ions to A-sites will subsequently increase the magnetization of B-sites up to x = 0.3, increasing the M_s values of the Cu-Mg-Zn NPs. If the Cu^{2+} ion can push some Fe^{3+} ions to move from B-site to A-site during cation redistribution, the drop in the saturation magnetization for x ≥ 0.4 may also be described. In addition, the production of some Fe^{2+} with a lower magnetic moment (4 μB) will diminish the magnetization of the B sites, which will lower the overall net magnetization of synthesized samples. A similar explanation for the decrease in M_s of copper-substituted magnesium zinc ferrite has been provided by M. H. Zaki et al. [23]. One can conclude that this technique's claimed saturation magnetization of the tested ferrites is superior to that of previous ceramic techniques [24] and auto-combustion routes [25]. This indicates that the employed co-precipitation technique has the potential to enhance the ferrite with higher saturation magnetization, making this ferrite significant throughout a broad range of frequencies.

Coercivity and remanence magnetization values are shown to increase with rising Cu concentrations. The significant magneto crystalline anisotropy of the Cu^{2+} ion, carried by one unpaired electron, contributes to the rise in coercivity with Cu substitution. The coercivity and remanence magnetization values are in line with what Sharma et al. [22] and Zaki et al. [23] have previously published. As Cu^{2+} concentration increases, the M_s values increase in direct proportion to the expansion of η_B [26]. It is shown that K and H_c behave in the same way.

Table 1 displays the results of the calculations made for the other parameters, including squareness ratio (R), magnetic moment (η_B), and anisotropy constant (K), which were calculated using the following equations [26,27]. According to Table 1, the remanence ratio rises from 0.1380 for x = 0.0 to 0.3442 for x = 0.5, with the low value indicating the isotropic nature of samples [28]. Multi-magnetic domains are present when R for prepared samples is measured and is smaller than 0.5 [29].

$$R = \frac{M_r}{M_s} \quad (3)$$

$$\eta_B = \frac{M_w * M_s}{5585} \quad (4)$$

$$K = \frac{H_c \, X \, M_s}{0.96} \quad (5)$$

The Yafet-Kittel three sub-lattice model can also be used to explain the fluctuation in magnetic moment. The B sub-lattice can be split into two sub-lattices, B_1 and B_2, each of which has a triangle spin configuration (canting angle, $\alpha_{Y\text{-}K}$), according to the Yafet-Kittel model. Understanding the triangular or non-collinear arrangement of spins is provided by Yafet-Kittel's model. The following relation was employed to calculate the Yafet-Kittel (Y-K) angles [30]:

$$\eta_B = (10 - 3x) \cos \alpha_{Y-K} - 5x \quad (6)$$

The $\alpha_{Y\text{-}K}$ angles become smaller as the substitution of Cu^{2+} increases (Table 1). The rise in super-exchange interaction (A-B) and fall in the non-collinear arrangement of spins are both confirmed by a decrease in $\alpha_{Y\text{-}K}$ angles with Cu^{2+} doping [29]. The increase in wave function overlaps adjacent by surrounding magnetic ions is indicated by a decrease in $\alpha_{Y\text{-}K}$ angles, which increases A-B super-exchange interactions and decreases B-B interactions [31,32].

Table 1. Values of magnetic parameters of $Cu_xMg_{0.5}Zn_{0.5-x}Fe_2O_4$ (x= 0.0 to 0.5, and Δx= 0.1) ferrites.

Ferrite Samples	Coercivity, H_c (Oe)	Remanence Magnetization, M_r (emu/g)	Saturation Magnetization, M_s (emu/g)	Squareness Ratio, R	Magnetic Moment, η_B (μ_B)	Anisotropy Constant, K (erg/cc)	Y-K canting angles, $\alpha_{Y\text{-}K}$ (degree)
$Mg_{0.5}Zn_{0.5}Fe_2O_4$	93.66	3.6577	26.508	0.1380	1.0467	2586.19	83.99
$Cu_{0.1}Mg_{0.5}Zn_{0.4}Fe_2O_4$	64.15	2.8025	33.912	0.0826	1.3380	2266.10	79.08
$Cu_{0.2}Mg_{0.5}Zn_{0.3}Fe_2O_4$	75.89	4.3446	45.605	0.0953	1.7978	3605.17	72.68
$Cu_{0.3}Mg_{0.5}Zn_{0.2}Fe_2O_4$	105.96	5.2510	48.901	0.1074	1.9261	5397.45	67.88
$Cu_{0.4}Mg_{0.5}Zn_{0.1}Fe_2O_4$	125.16	5.9446	39.059	0.1522	1.5372	5092.32	66.30
$Cu_{0.5}Mg_{0.5}Fe_2O_4$	481.45	5.1675	15.013	0.3442	0.5903	7529.18	68.68

3.2 FE-SEM, EDS, and elemental mapping

A representative $Cu_{0.2}Mg_{0.5}Zn_{0.3}Fe_2O_4$ ferrite sample was analyzed using an FE-SEM microscope together with EDS and elemental mapping, and the results are depicted in Figure 3.

Figure 3 (A) of the FE-SEM micrograph shows the non-uniform grain growth of a representative ferrite sample. FE-SEM micrograph shows aggregated nanoparticles with almost spherical shapes and a range of sizes. Theoretically, the development of large and irregularly distributed agglomerated nanocrystals is influenced by annealing, the synthesis method, and flaws in Cu-doped Mg-Zn ferrites [33]. An energy-dispersive X-ray (EDS) spectrometer is employed to assess the purity and chemical composition of the representative sample. The EDS spectra of $Cu_{0.2}Mg_{0.5}Zn_{0.3}Fe_2O_4$ nanoparticles are shown in Figure 3 (B). Cu, Mg, Fe, O, and Zn are present in the sample, proving that these elements are present. The sample contains no additional peaks. It demonstrates the sample's purity. The inset table shows the calculated atomic percentages of Cu, Mg, Fe, O, and Zn, which are pretty close to the theoretical value. The uniform distribution of Cu, Mg, Fe, O, and Zn was confirmed with elemental mapping from a portion of the sample that was randomly chosen.

Figure 3. (A) FE-SEM micrograph and (B) EDS and elemental mapping images of representative $Cu_{0.2}Mg_{0.5}Zn_{0.3}Fe_2O_4$.

3.3 Photocatalytic activity of $Cu_xMg_{0.5}Zn_{0.5-x}Fe_2O_4$

Cu-Mg-Zn NPs are suitable for photocatalytic activity because of their low band gap (~ 1.6 eV) [20]. The efficiency of photo-catalytic degradation was determined using equation (1). To track the effects of $Cu_xMg_{0.5}Zn_{0.5-x}Fe_2O_4$ (x=0.1 to 0.5, and Δx=0.1) on the photocatalytic degradation of methylene blue (MB) dye, absorption spectra were used. As a result of the proportionality between concentration (C) and absorbance (A), the C_o/C_t can be replaced by A_o/A_t [34]. When Cu-Mg-Zn ferrites are exposed to light [240 V (160 W) 535 lux], electron-hole pairs are furnished. The production of more electron-hole pairs as a result of lengthening the exposure duration for each sample causes the degradation to occur more quickly. Due to stable electronic configuration (d^{10}) and low electron availability of Zn^{2+} ions, Cu^{2+} ions should be swapped out for them to increase electron availability. Because there are more available electrons at x=0.5, the most electron-hole pairs are produced. The basic variables that affect photodegradation are the quantity of the photocatalyst, the temperature of the reaction medium, the pH of the solution, the duration and intensity of the light source, the nature of the substrate as well as the photocatalyst, the surface area of the photocatalyst, the doping of non-metals or metals, and the structural characteristics of the photocatalyst, such as band gap, particle size, surface defects, etc. [35]. Ferrite becomes a more promising material for photodegradation applications as its band gap shrinks. The photocatalytic degradation of MB by Cu-Mg-Zn NPs under visible light irradiation was proposed and outlined below based on the literature [33].

Step-I: Cu-Mg-Zn NP valence band electron is excited by light, generating a free electron (e^-) and a hole (h^+).

$$Cu_xMg_{0.5}Zn_{0.5-x}Fe_2O_4 + h\nu \rightarrow Cu_xMg_{0.5}Zn_{0.5-x}Fe_2O_4\,(e^- + h^+) \quad (7)$$

Step-II: Valence band holes react with surface-bound water or OH^- ions to produce $OH^{\cdot}$ radicals.

$$h^+ + H_2O \rightarrow OH^{\cdot} + H^+ \quad (8)$$

$$h^+ + OH^- \rightarrow OH^{\cdot} \quad (9)$$

In addition, by interacting with the adsorbed molecule O_2, the electron from the conduction band can produce an anionic super-oxide radical:

$$e^- + O_2 \rightarrow O_2^{\cdot -} \quad (10)$$

These super-oxide radicals react with H^+ to produce $HO_2^{\cdot}$, which subsequently reacts with electrons to produce $OH^{\cdot}$ (Free radicals).

$$O_2^{\cdot-} + H^+ \rightarrow HO_2^{\cdot} \quad (11)$$
$$HO_2^{\cdot} + e^- \rightarrow OH^{\cdot} \quad (12)$$

Step-III: Degradation of MB by active species:

$$OH^{\cdot}, O_2^{\cdot-}, HO_2^{\cdot} + MB \rightarrow Oxidation\ products \quad (13)$$

Afore-discussed reaction mechanism for MB photodegradation can also be shown as a schematic given below (Figure 4):

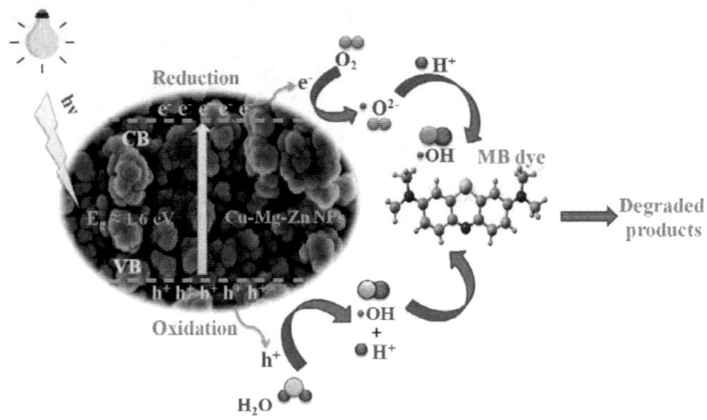

Figure 4. Schematic diagram illustrating the mechanism of MB dye photodegradation by Cu^{2+}-doped Mg-Zn ferrites.

Figure 5 shows the highest absorbance of MB measured utilizing a UV-vis spectrophotometer with a wavelength range of 200-800 nm [36].

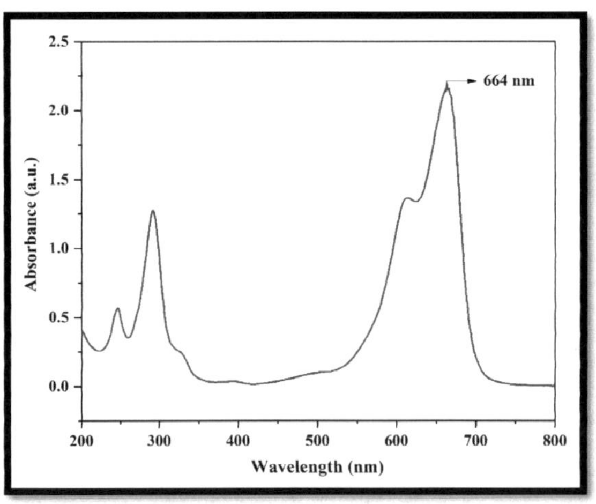

Figure 5. Lamba max (λ_{max}) of methylene blue.

The operational factors affecting the photocatalytic degradation of MB are as follows:

3.3.1 *Effect of contact time*

In the presence of $Cu_xMg_{0.5}Zn_{0.5-x}Fe_2O_4$ photocatalysts (x = 0.1 to 0.5, and Δx = 0.1), the effect of contact time (i.e., 30-180 min) on the photocatalytic degradation of MB dye was assessed. A blank experiment without any photocatalysts was used as the control in the experiment. We also conducted a photocatalytic test in the dark, which subjects MB solution with the catalyst to degradation under similar conditions but without the presence of visible light. The findings show that as the irradiation time interval is increased, the degradation rate of MB increases rapidly up to 90 min. The initial concentration of the dye molecules was higher for degradation, but it gradually decreased with time. This may be because the dye molecule is initially attacked by a large number of anionic radicals in the solution, but as their quantity slowly decreases over time after 90 min, the breakdown rate slows. The results of the degradation of MB dye by varying the contact duration were evaluated using a catalyst dosage of 25 mg in 25 mL (10 ppm) of dye solution. The degrading effectiveness of $Cu_xMg_{0.5}Zn_{0.5-x}Fe_2O_4$ (x=0.1 to 0.5, and Δx=0.1) was plotted against the contact time as shown in Figure 6. It demonstrates that the MB degradation rises considerably up to 120 minutes and gets saturated after the ideal period. N. Ali et al. observed similar findings for contact time variation [37].

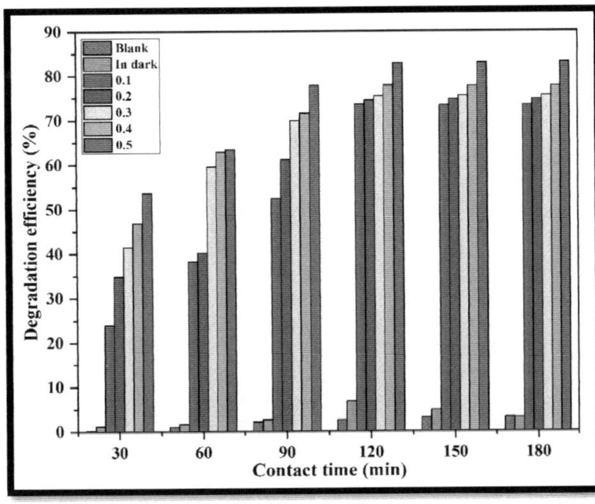

Figure 6. Degradation efficiency (%) of $Cu_xMg_{0.5}Zn_{0.5-x}Fe_2O_4$ (x= 0.1 to 0.5, and Δx= 0.1) for methylene blue.

3.3.2 Effect of Cu-Mg-Zn NPs dosing

The amount of catalyst used in the photocatalysis method is a critical factor from an economic standpoint [38]. Figure 7 displays the percentage of MB dye degradation (10 ppm) at different $Cu_xMg_{0.5}Zn_{0.5-x}Fe_2O_4$ (x = 0.1 to 0.5, and Δx = 0.1) photocatalyst dosages (10-35 mg). When the catalyst dosage is raised from 10 to 25 mg, the percentage of MB dye degradation rises linearly to 88.74 percent. The photocatalytic degradation declines after this photocatalyst dosage. The active sites on the catalyst surface grew as their quantity rose, which in turn improved superoxide and hydroxyl radical generation and accelerated dye degradation. Furthermore, over-catalyzing can lead to catalyst particle aggregation and turbidity, which reduce photon absorption and decrease the effectiveness of photocatalytic degradation [39,40]. Therefore, a modified dosage may be utilized to more efficiently to degrade the MB dye. Under visible light, the control experiment (blank) showed the least amount of MB self-degradation. The effectiveness of the catalyst dose change was evaluated over a contact time of 120 minutes with 25 mL of dye solution (10 ppm).

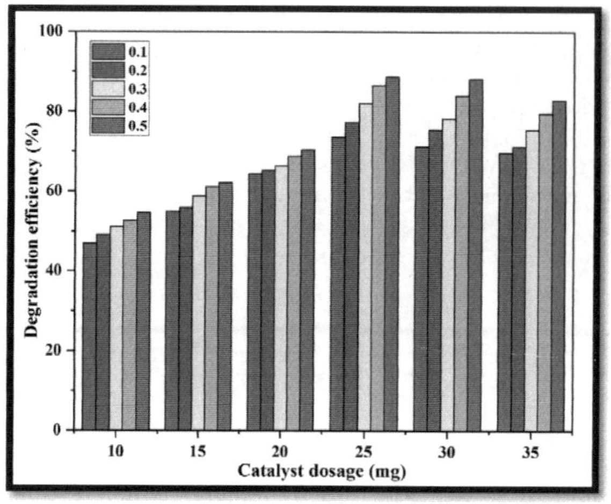

Figure 7. Photocatalytic activity of $Cu_xMg_{0.5}Zn_{0.5-x}Fe_2O_4$ (x = 0.1 to 0.5, and Δx = 0.1) for methylene blue over different catalyst dosages.

3.3.3 Effect of initial MB concentration

Figure 8 displays the degradation efficiency (%) as a function of the initial MB dye concentration. When the starting MB dye concentration is increased from 5 to 10 ppm, the degradation efficiency (%) increases to 87.58 %. The effectiveness of the degradation later decreases as the original concentration of MB dye rises.

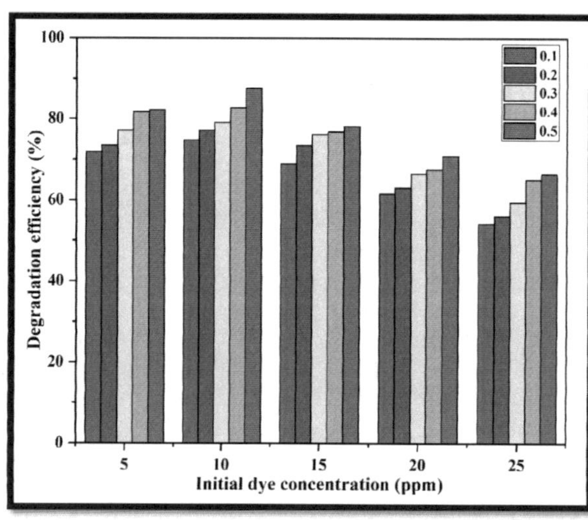

Figure 8. Degradation efficiency (%) against methylene blue as a function of initial dye concentration.

The production of hydroxyl radicals on catalyst surfaces and their interactions with dye species play a major role in controlling the photodegradation of MB [41]. The quantity of dye molecules that are first adsorbed on the catalyst surface increases as the initial dye concentration rises. As a result, the majority of the active sites are covered by more dye molecules, which reduces the generation of hydroxyl radicals. The photodegradation efficiency consequently fell. H. S. B. Naik [42] and G. Nagaraju et al. [43] claim that there is still another scenario that could be the source of the issue. The catalyst, according to the claimed explanation, was able to absorb fewer photons by providing a shielding effect for light that entered the solution at high dye concentrations. This, in turn, decreased the quantity of hydroxyl radicals that were produced on the catalyst's surface. The consequences of the initial dye concentration variation were investigated using a catalyst dosage of 25 mg in 25 mL of dye solution and a contact time of 120 min.

3.3.4 Effect of pH

The adsorption and degradation of MB dye are influenced by the surface charge of the catalyst and pH, two crucial parameters in dye degradation [44]. Figure 9 displays the degradation efficiency (%) as a function of pH after tests were conducted at three distinct pH levels (4, 7, and 10). Contact duration of 120 minutes, 25 mL of 10 ppm MB dye solution, and 25 mg of a photocatalyst were used in the experiment. Three different pH values were used to measure the breakdown of the MB (4, 7, and 10). While Figure 10 depicts the representative absorbance spectra for MB degradation at optimum pH 10. As shown in Figure 9, the rate of MB breakdown under visible light dramatically increased from pH 4 to pH 10.

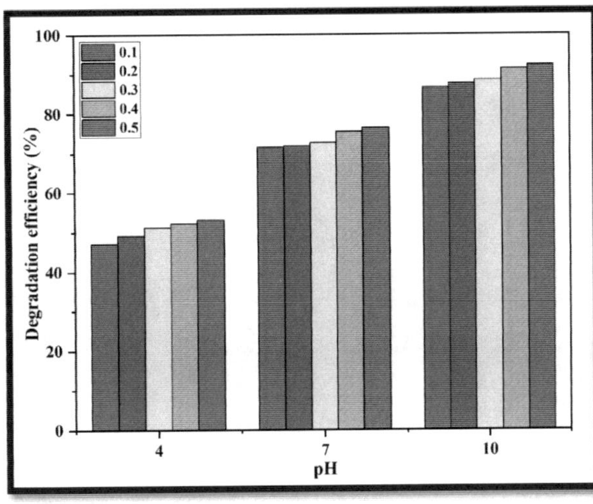

Figure 9. Effect of pH on photodegradation of methylene blue.

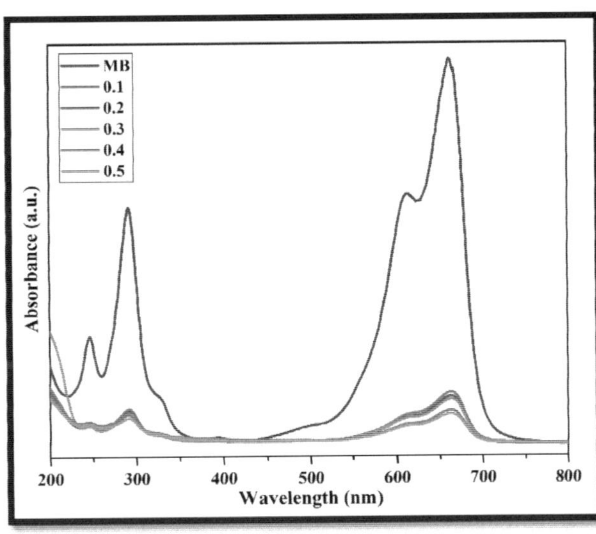

Fig. 10. Absorbance spectra for effect of pH on the methylene blue dye degradation.

Since pH 10 had the highest percentage of degradation (53.07 to 92.25%) and the shortest time is taken to degrade, it was determined in this study to be the optimal pH. $Cu_{0.2}Mg_{0.3}Zn_{0.5}Fe_2O_4$ nanoparticles have a point of zero charge (pzc) of 7.4, according to Dhiman et al. [45]. Electrostatic repulsions arise because the catalyst surface and the positively charged cationic dye molecules are at an acidic pH, which minimizes degradation. The pzc of Cu-Mg-Zn ferrite relates the pH of the solution to the photodegradation of aqueous MB dye solutions, which shows increased activity at pH 10. Above the pzc value, the surface of the Cu-Mg-Zn NPs material turns negative. Due to the cationic dye MB being drawn to the oppositely charged Cu-Mg-Zn ferrite species as result, the pollutant can be quickly destroyed.

3.3.5 *Effect of different acids*

The photocatalytic breakdown of the dye may be impacted by the acidic environment. Various acids were tested to achieve this. Figure 11 illustrates the addition of various acids to the reaction solution in varying volumes (2–8 mL) to quantify photocatalytic degradation.

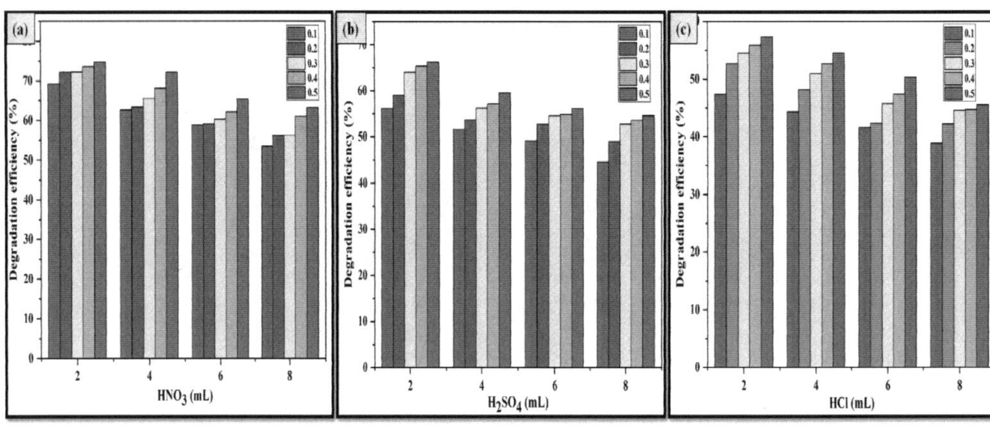

Figure 11. Influence of different acids on the methylene blue dye.

Because the MB solution contains a lot of hydrogen ions (H^+), HNO_3 (0.01 mol/L) causes the most deterioration in terms of percent. The hydrogen ions H^+ interact with the electrons in the aqueous medium to form hydrogen radicals $H^\bullet$, which then interact with the MB dye to demolish it.

$$H^+ + e_{aq}^- \rightarrow H^\cdot \quad (14)$$

Because H_2SO_4 (0.01 mol/L) produces sulfate SO_4^{2-} ions, which act as radical scavengers as shown below, MB dye degrades relatively more slowly than HNO_3.

$$SO_4^{2-} + OH^\cdot \rightarrow SO_4^{\cdot-} + OH^- \quad (15)$$

There is less photocatalytic degradation of the MB solution as a consequence of the HCl (0.01 mol/L) producing Cl^- ions in the MB solution, which more effectively scavenges $OH^\cdot$ radicles.

$$Cl^- + OH^\cdot \rightarrow ClOH^- \quad (16)$$

N. Ali et al. [37,46] observed similar results for dye deterioration in the presence of HNO_3, H_2SO_4, and HCl.

Additionally, we compared our results with those that have already been reported for the removal of MB employing ferrite-based photocatalysts (Table 2).

Table 2. Comparison of previously published on MB degradation by ferrites and their composites.

Photocatalysts	Synthesis method	Light source	Dye concentration	Catalyst dosage	Reaction time (min)	Degradation efficiency (%)	Reference
$CuFe_2O_4$	Ultrasound irradiation	Sun-light	10 mg/L	-	180	36	[47]
$CoFe_2O_4$	Hydrothermal	Visible	-	-	140	51.53	[48]
$CoFe_2O_4$-rGO	Co-precipitation	Visible	10 mg/L	0.5 g/L	180	58	[49]
$ZnFe_2O_4$	Wet chemical	Sun-light	10 mg/L	0.1 g/L	180	85	[50]
$Co_{0.1}Mg_{0.9}Fe_2O_4$	Sol-gel	Halogen lamp	10 mg/L	0.1 g/L	240	74.5	[51]
$MgFe_2O_4$	Solution combustion	Visible	-	-	60	63	[52]
$MnFe_2O_4$	Co-precipitation	UV light	10 mg/L	0.3 g/L	290	84	[53]
$Cd_{0.6}Mg_{0.4}Fe_2O_4$-rGO	Co-precipitation	Visible	5 mg/L	0.1 g/L	80	75.67	[54]
$Cu_xMg_{0.5}Zn_{0.5-x}Fe_2O_4$	Co-precipitation	Visible	10 mg/L	1 g/L	120	92.25 (MB)	Present work

3.3.6 Catalyst recycling

Two essential ideas in the photocatalysis process are catalytic efficiency and catalytic reusability. They were employed at least five times to verify the stability of materials. Figure 12 displays the results of the studies, which used 25 mg of catalyst and 10 ppm of dye at an optimum pH. The degrading efficiency of MB was practically equal, with only a 1-2% drop in efficiency after each cycle. This graph depicts the degradation of MB solution declining somewhat over 5 cycles, from 90.56 to 83.15 percent. The stable crystal structure of Cu-Mg-Zn NPs contributes to the stability of the catalytic recycling for MB degradation.

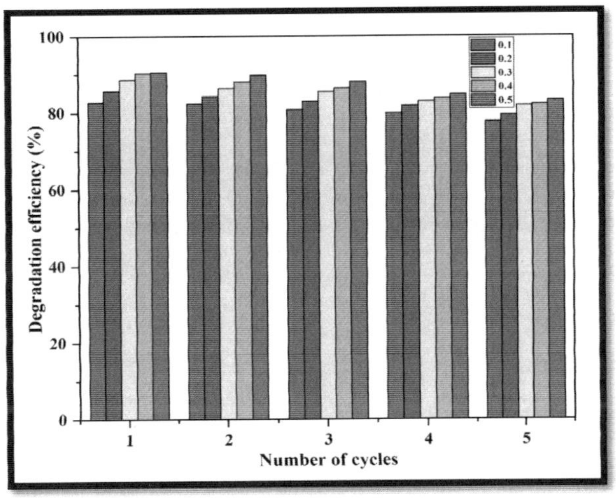

Figure 12. Recycling of $Cu_xMg_{0.5}Zn_{0.5-x}Fe_2O_4$ (x=0.1 to 0.5, and Δx= 0.1) for methylene blue dye degradation.

4. Conclusions

In $Cu_xMg_{0.5}Zn_{0.5-x}Fe_2O_4$ (x= 0.1 to 0.5, and Δx= 0.1) nano-ferrites, as the concentration of Cu doping increases, the transition from superparamagnetic to soft ferrimagnetic behavior has been observed. FE-SEM reveals the formation of spherical, non-homogeneous, agglomerated nanoparticles for $Cu_{0.2}Mg_{0.5}Zn_{0.3}Fe_2O_4$ sample. An amount of 1 g/L photocatalyst was found to provide the optimum dosage for 10 ppm MB solution, resulting in maximum degradation (92.25 %) under visible light irradiation in 120 minutes of contact time. The $Cu_{0.5}Mg_{0.5}Fe_2O_4$ photocatalyst showed MB dye degradation efficiency. The photocatalysts showed remarkable performance in alkaline medium (pH 10), it might be due to the influence of surface charge properties.

5. Declaration of competing interest:

The researchers state that they have no known competing financial interests or personal relationships that could have appeared to affect the research reported in this paper.

6. Acknowledgement:

The authors would like to show their gratitude to the CSIR and the UGC for providing fellowships for this research.

7. References:

[1] Chandra Babu Naidu K and Madhuri W 2017 Microwave processed bulk and nano NiMg ferrites: A comparative study on X-band electromagnetic interference shielding properties *Mater. Chem. Phys.* **187** 164–76

[2] Ismael M 2021 Ferrites as solar photocatalytic materials and their activities in solar energy conversion and environmental protection: A review *Sol. Energy Mater. Sol. Cells* **219** 110786

[3] Wu K, Li J and Zhang C 2019 Zinc ferrite based gas sensors: A review *Ceram. Int.* **45** 11143–57

[4] Umut E, Coşkun M, Pineider F, Berti D and Güngüneş H 2019 Nickel ferrite nanoparticles for simultaneous use in magnetic resonance imaging and magnetic fluid hyperthermia *J. Colloid Interface Sci.* **550** 199–209

[5] Han J, Lan Y, Song Q, Yan H, Kang J, Guo Y and Liu Z 2020 Zinc ferrite-based p-n homojunction with multi-effect for efficient photoelectrochemical water splitting *Chem. Commun.* **56** 13205–8

[6] Al-rawi N N, Azad B, Al-rawi N H, Tahseen A and Saad I 2020 Magnetism in drug delivery : The marvels of iron oxides and substituted ferrites nanoparticles *Saudi Pharm. J.* **28** 876–87

[7] Liu L, He D, Pan F, Huang R, Lin H and Zhang X 2020 Comparative study on treatment of methylene blue dye wastewater by different internal electrolysis systems and COD removal kinetics, thermodynamics and mechanism *Chemosphere* **238** 124671

[8] Marimuthu S, Antonisamy A J, Malayandi S, Rajendran K, Tsai P C, Pugazhendhi A and Ponnusamy V K 2020 Silver nanoparticles in dye effluent treatment: A review on synthesis, treatment methods, mechanisms, photocatalytic degradation, toxic effects and mitigation of toxicity *J. Photochem. Photobiol. B Biol.* **205** 111823

[9] Hassan Mohamed N A, Shamma R N, Elagroudy S and Adewuyi A 2022 Visible Light-Driven Photocatalytic Degradation of Ciprofloxacin, Ampicillin and Erythromycin by Zinc Ferrite Immobilized on Chitosan *Resources* **11**

[10] Sharmila M, Mani R J, Parvathiraja C, Kader S M A, Siddiqui M R, Wabaidur S M, Islam A and Lai W 2022 Visible Light Photocatalyst and Antibacterial Activity of BFO (Bismuth Ferrite) Nanoparticles from Honey *Water* **14** 1545

[11] Fei L, Ali F, Said A, Tariq N, Raziq F, Ali N, Arif U, Akhter M S, Rahdar A and Bilal M 2023 Surface-functionalized spongy zinc ferrite as a robust visible-light driven nanocatalyst for

wastewater remediation: characterization, kinetic, and mechanistic insight *Int. J. Environ. Sci. Technol.* **20** 1007–18

[12] Zhivulin V, Zaitseva O and Vinnik D 2020 Correlation between Iron Substitution Degree in the M-Type Hexaferrites Based Crystal Structures and Dopant Content in the Crystal Growth Feeding Melt *IOP Conf. Ser. Mater. Sci. Eng.* **969**

[13] Sharma R, Thakur P, Kumar M, Thakur N, Negi N S, Sharma P and Sharma V 2016 Improvement in magnetic behaviour of cobalt doped magnesium zinc nano-ferrites via co-precipitation route *J. Alloys Compd.* **684** 569–81

[14] Irshad A, Zulfiqar M, Ali H M, Shahzadi N, Abd El-Gawad H H, Chokejaroenrat C, Sakulthaew C, Anjum F and Suleman M 2022 Co-substituted Mg–Zn spinel nanocrystalline ferrites: Synthesis, characterization and evaluation of catalytic degradation efficiency for colored and colorless compounds *Ceram. Int.* **48** 29805–15

[15] Jadhav S A, Somvanshi S B, Gawali S S, Zakade K and Jadhav K M 2023 Rare earth-doped mixed Ni–Cu–Zn ferrites as an effective photocatalytic agent for active degradation of Rhodamine B dye *J. Rare Earths*

[16] Abdo M A, Mansour S F, Al-Bassami N S and Abu-Elsaad N I 2022 Yttrium substituted Co–Cu–Zn nanoferrite: A synergetic impact of Y^{3+} on enhanced physical properties and photocatalysis *Ceram. Int.* **48** 15314–26

[17] Jasrotia R, Suman, Verma A, Verma R, Ahmed J, Godara S K, Kumar G, Mehtab A, Ahmad T and Kalia S 2022 Photocatalytic dye degradation efficiency and reusability of Cu-substituted Zn-Mg spinel nanoferrites for wastewater remediation *J. Water Process Eng.* **48** 102865

[18] Bhukal S, Shivali and Singhal S 2014 Magnetically separable copper substituted cobalt-zinc nano-ferrite photocatalyst with enhanced photocatalytic activity *Mater. Sci. Semicond. Process.* **26** 467–76

[19] Alhashmialameer D, Ullah S, Irshad A, Alsafari I A, Abd El-Gawad H H, Abdelrahman Elsheikh M A, Liu X and Bashir S 2022 Copper-doped magnesium ferrite and its composite with rGO: Synthesis, characterization, and degradation of organic effluents and antibacterial study *Ceram. Int.* **48** 24100–13

[20] Pund S N, Nagwade P A, Nagawade A V. and Bagade A V. 2022 Influence of Cu^{2+} substitution on structural and optical properties of Mg-Zn ferrite nanoparticles *Mater. Today Proc.* **49** 2382–91

[21] Lodhi M Y, Mahmood K, Mahmood A, Malik H, Warsi M F, Shakir I, Asghar M and Khan M A 2014 New $Mg_{0.5}Co_xZn_{0.5-x}Fe_2O_4$ nano-ferrites: Structural elucidation and electromagnetic behavior evaluation *Curr. Appl. Phys.* **14** 716–20

[22] Sharma R, Thakur P, Sharma P and Sharma V 2017 Ferrimagnetic Ni^{2+}doped Mg-Zn spinel ferrite nanoparticles for high density information storage *J. Alloys Compd.* **704** 7–17

[23] Zaki H M, Al-Heniti S H and Elmosalami T A 2015 Structural, magnetic and dielectric studies of copper substituted nano-crystalline spinel magnesium zinc ferrite *J. Alloys Compd.* **633** 104–14

[24] Kamble S S, Jagtap V S and Pingale P C 2013 Synthesis of $Mg_{0.48}Cu_{0.12}Zn_{0.40}Fe_2O_4$ ferrite and its aptness for multilayer chip component application *Ceram. Int.* **39** 3597–601

[25] Bahiraei H, Shoushtari M Z, Gheisari K and Ong C K 2014 The effect of non-magnetic Al^{3+} ions on the structure and electromagnetic properties of MgCuZn ferrite *J. Magn. Magn. Mater.* **371** 29–34

[26] Amer M A, Meaz T M, El-Kestawy M and Ghoneim A I 2016 Magnetic and structural studies

of trivalent Co-substituted Cd-Mn ferrites *J. Magn. Magn. Mater.* **405** 137–44

[27] Kambale R C, Shaikh P A, Kamble S S and Kolekar Y D 2009 Effect of cobalt substitution on structural, magnetic and electric properties of nickel ferrite *J. Alloys Compd.* **478** 599–603

[28] Shirsath S E, Toksha B G and Jadhav K M 2009 Structural and magnetic properties of In^{3+} substituted $NiFe_2O_4$ *Mater. Chem. Phys.* **117** 163–8

[29] Bagade A, Nagwade P, Nagawade A, Thopate S, Pandit V and Pund S 2022 Impact of Mg^{2+}substitution on structural, magnetic and optical properties of Cu-Cd ferrites *Mater. Today Proc.* **53** 144–52

[30] Srinivas C, Ranjith Kumar E, Tirupanyam B V., Singh Meena S, Bhatt P, Prajapat C L, Chandrasekhar Rao T V. and Sastry D L 2020 Study of magnetic behavior in co-precipitated Ni–Zn ferrite nanoparticles and their potential use for gas sensor applications *J. Magn. Magn. Mater.* **502** 166534

[31] Pong W, Chang Y, Su M, Tseng P, Lin H, Ho G, Tsang K and Chen C 1997 Magnetic orientation of Ni in Zn-Ni ferrites studied by soft-x-ray magnetic circular dichroism *Phys. Rev. B - Condens. Matter Mater. Phys.* **55** 11409–13

[32] Topkaya R, Baykal A and Demir A 2013 Yafet-Kittel-type magnetic order in Zn-substituted cobalt ferrite nanoparticles with uniaxial anisotropy *J. Nanoparticle Res.* **15** 1–18

[33] Yuvaraj S, Manikandan N and Vinitha G 2017 Influence of copper ions on structural and non-linear optical properties in manganese ferrite nanomaterials *Opt. Mater. (Amst).* **73** 428–36

[34] Rajesh Babu B and Tatarchuk T 2018 Elastic properties and antistructural modeling for Nickel-Zinc ferrite-aluminates *Mater. Chem. Phys.* **207** 534–41

[35] Tatarchuk T R, Paliychuk N D, Bououdina M, Al-Najar B, Pacia M, Macyk W and Shyichuk A 2018 Effect of cobalt substitution on structural, elastic, magnetic and optical properties of zinc ferrite nanoparticles *J. Alloys Compd.* **731** 1256–66

[36] Kumar A 2017 A Review on the Factors Affecting the Photocatalytic Degradation of Hazardous Materials *Mater. Sci. Eng. Int. J.* **1** 106–14

[37] Ali N, Said A, Ali F, Raziq F, Ali Z, Bilal M, Reinert L, Begum T and Iqbal H M N 2020 Photocatalytic Degradation of Congo Red Dye from Aqueous Environment Using Cobalt Ferrite Nanostructures: Development, Characterization, and Photocatalytic Performance *Water. Air. Soil Pollut.* **231** 1–16

[38] Hassani A, Eghbali P, Ekicibil A and Metin Ö 2018 Monodisperse cobalt ferrite nanoparticles assembled on mesoporous graphitic carbon nitride ($CoFe_2O_4$/mpg-C_3N_4): A magnetically recoverable nanocomposite for the photocatalytic degradation of organic dyes *J. Magn. Magn. Mater.* **456** 400–12

[39] Ikramullah., Ali F, Ali Z, Humayun M and Zain Ul wahab 2018 Ac ce pte d M us *Mater. Res. Express* **5** 055509

[40] Sakthivel S, Neppolian B, Shankar M V., Arabindoo B, Palanichamy M and Murugesan V 2003 Solar photocatalytic degradation of azo dye: Comparison of photocatalytic efficiency of ZnO and TiO_2 *Sol. Energy Mater. Sol. Cells* **77** 65–82

[41] Harraz F A, Mohamed R M, Rashad M M, Wang Y C and Sigmund W 2014 Magnetic nanocomposite based on titania-silica/cobalt ferrite for photocatalytic degradation of methylene blue dye *Ceram. Int.* **40** 375–84

[42] Patil S B, Bhojya Naik H S, Nagaraju G, Viswanath R, Rashmi S K and Vijay kumar M 2018 Sugarcane juice mediated eco-friendly synthesis of visible light active zinc ferrite nanoparticles: Application to degradation of mixed dyes and antibacterial activities *Mater. Chem. Phys.* **212**

351–62

[43] Udayabhanu, Nagaraju G, Nagabhushana H, Suresh D, Anupama C, Raghu G K and Sharma S C 2017 Vitis labruska skin extract assisted green synthesis of ZnO super structures for multifunctional applications *Ceram. Int.* **43** 11656–67

[44] Mandal S S and Bhattacharyya A J 2012 Electrochemical sensing and photocatalysis using Ag-TiO_2 microwires *J. Chem. Sci.* **124** 969–78

[45] Dhiman P, Patial M, Kumar A, Alam M, Naushad M, Sharma G, Vo D V N and Kumar R 2021 Environmental friendly and robust $Mg_{0.5-x}Cu_xZn_{0.5}Fe_2O_4$ spinel nanoparticles for visible light driven degradation of Carbamazepine: Band shift driven by dopants *Mater. Lett.* **284** 129005

[46] Ali N, Ali F, Said A, Begum T, Bilal M, Rab A, Sheikh Z A, Iqbal H M N and Ahmad I 2020 Characterization and Deployment of Surface-Engineered Cobalt Ferrite Nanospheres as Photocatalyst for Highly Efficient Remediation of Alizarin Red S Dye from Aqueous Solution *J. Inorg. Organomet. Polym. Mater.* **30** 5063–73

[47] Massoud-Sharifi A, Kara G K and Rabbani M 2020 $CuFe_2O_4$@CuO: A Magnetic Composite Synthesized by Ultrasound Irradiation and Degradation of Methylene Blue on Its Surface in the Presence of Sunlight *Proceedings* **48** 17

[48] Gebreslassie G, Gebrezgiabher M, Lin B, Thomas M and Linert W 2023 Direct Z-Scheme $CoFe_2O_4$-Loaded g-C_3N_4 Photocatalyst with High Degradation Efficiency of Methylene Blue under Visible-Light Irradiation *Inorganics* **11** 119

[49] Jelokhani F, Sheibani S and Ataie A 2020 Adsorption and photocatalytic characteristics of cobalt ferrite-reduced graphene oxide and cobalt ferrite-carbon nanotube nanocomposites *J. Photochem. Photobiol. A Chem.* **403** 112867

[50] Ali N, Zada A, Zahid M, Ismail A, Rafiq M, Riaz A and Khan A 2019 Enhanced photodegradation of methylene blue with alkaline and transition-metal ferrite nanophotocatalysts under direct sun light irradiation *J. Chinese Chem. Soc.* **66** 402–8

[51] Dojcinovic M P, Vasiljevic Z Z, Pavlovic V P, Barisic D, Pajic D, Tadic N B and Nikolic M V 2021 Mixed Mg–Co spinel ferrites: Structure, morphology, magnetic and photocatalytic properties *J. Alloys Compd.* **855** 157429

[52] Heidari P and Masoudpanah S M 2021 Structural, magnetic and optical properties and photocatalytic activity of magnesium-calcium ferrite powders *J. Phys. Chem. Solids* **148** 109681

[53] Mandal B, Panda J, Paul P K, Sarkar R and Tudu B 2020 $MnFe_2O_4$ decorated reduced graphene oxide heterostructures: Nanophotocatalyst for methylene blue dye degradation *Vacuum* **173** 109150

[54] Shahid M 2020 Fabrication of magnesium substituted cadmium ferrite nanoparticles decorated Graphene-Sheets with improved photocatalytic activity under visible light irradiation *Ceram. Int.* **46** 10861–70

Degradation of Organic pollutant by Green Tea Iron Nanoparticles

Gurdev preet Singh[1], Jasvir Singh[1,3], Rishita[1], Khiyati Saini[1], Kanika Chandel[1], Jasmeet Kaur[2], K.J. Singh[1,a]

[1]Department of Physics, Guru Nanak Dev University, Amritsar, Punjab143005, India
[2]Department of Physics, Sri Guru Gobind Singh College, Sector 26, Chandigarh 160019 India
[3]Department of Physics, B.A.M. Khalsa College Garhshankar, Hoshiarpur, Punjab, 144527, India

E-Mail: kanwarjitsingh@yahoo.com

Abstract. Eco Friendly green synthesis method with the use of green tea leaves is used in preparation of Iron nano particles in this research. This method is rarely ever used. Green synthesis methods are one of the recently developed environmentally favorable methods. Green process methods indeed play a crucial role in the advancement of nanotechnology by offering several benefits such as minimizing environmental impact and ensuring the safety of both researchers and end-users. Green manufacturing of catalysts supports sustainable advanced oxidation processes for degradation of organic products. Polyphenol contents which are already in plants prepare the metal nano particles and are revived by non-harmful biodegradable chemical components. High temperature and energy is not required in green synthesis method. The are characterized by X-ray diffraction (XRD) and scanning electron microscopy (SEM) characterized the synthesized nanoparticles. UV-vis spectrophotometer was used for calculating degradation efficiency of Nanoparticles (NPs). Degradation of Methylene Blue (MB) dye was investigated under sunlight with respect to time and catalyst concentration. MB dye was used as a model dye.

Keywords: Iron nanoparticles; Photocatalytic; Malachite Green Dye; Green Synthesis; Photodegradation

1. INTRODUCTION

Nanotechnology has made significant advancements and has had a profound impact on various aspects of human life in recent years [1-2]. The Nanoparticles can be prepared with restrained size, morphology and with enhanced crystalline nature. Therefore, the nanomaterial has tremendous uses in several areas of research [3–5]. Nanoparticles can be synthesized by many physical and chemical routes as given in literature. However, the usage of Toxic organic pollutant in these methods has harmful effect on environment and cost high, still we achieve the controlled size and shape of nanoparticles, these NPs have restriction in certain biological applications due to toxic compounds present [6]. To overcome this drawback researcher focused to develop NPs which are non-toxic, bio-compatible and environment friendly. Moreover, environmental issues today has become reason of concern and forced researchers to find out the new ways for ecofriendly and less toxic methods for the synthesis and application of nanomaterial's, which have lower health risks, easily degradable and can be developed from products which are renewable. Green nanotechnology is good alternate to handle all these issues and green synthesis has become part of this upcoming nanotechnology, which helps in the production of nanomaterial in an eco-friendly manner. Therefore, *green synthesis* method was adopted for clean, non-toxic and environmentally pure materials NPs. For this method biological plant like fungus, roots, yeast, gum, bacteria's and viruses were used across the world to prepare NPs [7-11]. In Green synthesis technique, Natural extracts derived from plants often possess reducing properties, allowing them to facilitate the synthesis of nanoparticles or other materials without the need for conventional reducing agents. These extracts contain bioactive compounds such as polyphenols, flavonoids, terpenoids, and other phytochemicals that can act as reducing agents or stabilizers in the synthesis process. [12] Overall, the presence of flavonoids, polysaccharides, phenolics, and other secondary metabolites with redox capacity in plants contributes to their antioxidant, anti-inflammatory, and protective properties, both for the plants themselves and for potential applications in various industries. [13,14].

In this study the use of green tea extract in the synthesis of Fe-Nanoparticles offers several advantages. It eliminates the need for hazardous reducing agents or stabilizers, thereby reducing the environmental impact of the synthesis process. Additionally, the polyphenols from green tea extract may introduce unique propcrtics or functionalities to the Fe-Nanoparticles, potentially leading to enhanced performance or novel applications [15,16]. These components are biodegradable, non-toxic and are alternative synthesis of chemicals [17-19]. Although in present study photo-catalytic activity of Fe -Nanoparticles was examined for MG dye, almost 71% of dye was degraded within 150 min, which proves its excellent photo-catalytic efficiency.

2 EXPERIMENTAL

2.1. Synthesis

2.1.1. Synthesis of metal nanoparticles

Green tea leaves were purchased from the local market and were used for the preparation of iron (Fe) nanoparticles similar method was adopt by Huang, Weng, Chen, Megharaj and Naidu [20].

2.1.2 Preparation of green leaves extract

Green tea leaves were purchased from market. Extract of green tea leaves was added to the $FeCl_3$ solution for synthesis of Iron nanoparticles. 50 g green tea leaves was taken to prepare extract. 500 ml of De ionized water was taken in 1000 ml of conical flask and 50 g of leave extract was added. Flask containing this mixture was heated at 80^0 C for 20 min as shown in Fig.1. After that mixture was filtered at room temperature. This solution can be used up to one week keeping it at 5^0 C.

2.1.3. Synthesis of Fe-Nanoparticles

Green tea leaves extract was substituted in $FeCl_3$ solution for preparation of Fe-Nanoparticles. Take 16.22 g of $FeCl_3$ granulate in 1000 ml of D.I. water. $FeCl_3$ solution and solution obtained after filtration of tea leaves extract should be mixed in ratio 1:1. Mechanical stirrer at room temperature is used to stirrer the prepared mixture. The arrival of black precipitates reveals the reduction of Fe^{+2} ions. These precipitates were centrifuged to the rpm of 15,000 for 15 min. Finally, collected Fe nanoparticles were dried in oven at 65^0 C for 3hr.

3 PHOTOCATALYTIC ACTIVITIES

In this experiment, 10 mg of Methylene blue (MB) was diluted in 1 L D.I. water for preparation of stock solution. 10 mg Fe nanoparticles were used as catalyst in 250 ml of stock solution. Prior to irradation under sunlight , solution was magnetically stirred in dark place for 30 min to achieve an adsorption-desorption stability. The stock solution containg catalyst was irradiated under sunlight for 120 min. After every 15 min, 5 ml of sample solution was collected and centrifuged to remove the catalyst. The concentration of Methylene blue dye was observed by measuring the abosrption intensity at wavelength of 662 nm.

4 CHARACTERIZATIONS

Structural properties of synthesized Fe-Nanoparticles are determined by X-Ray diffraction (XRD). Surface morphology of the samples were analyzed using Field Emission Scanning Electron Microscopy (FESEM) by ZEISS SUPERA 55, The size is decreased from 20 µm to 200 nm is depicted in Fig.2.(a) and (b).The prepared nanoparticles are in range of 200 nm. The photo catalytic absorption was examined by employing The Shimadzu UV 3600 spectrophotometer. The magnification which was used for the study of particle size for green tea nanoparticles is 50.00 KX. PANalytical X'Pert PRO powder x-ray diffractometer in scan rate of 10° to 80° at room temperature is used to perform powder x-ray diffraction pattern. X-ray diffraction (XRD) is based on the dual wave particle nature of x-rays to reveal the information about the structure of crystalline materials'-ray diffraction structures of Fe-nanoparticles is determined and scanned at 2θ position. The peaks at 2θ values of green tea nanoparticles are 22.31^0, and 30.52^0 are shown in the Fig.3. XRD results are similar to Tejaswi Mareedu et al [21]

Figure 1. Green tea leaves extract

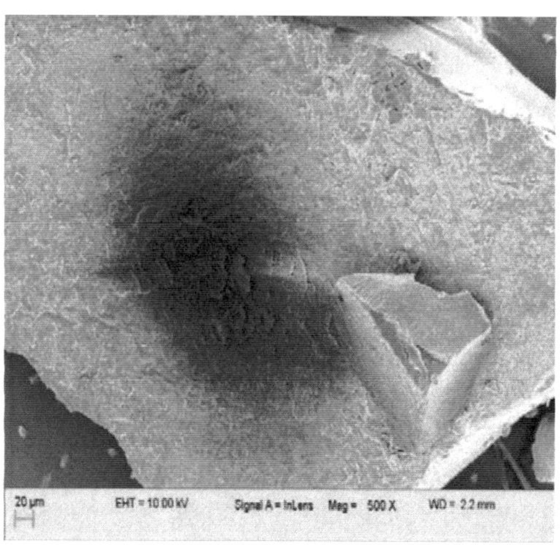

Figure 2. (a) SEM image of Green tea $FeCl_3$ before synthesis

Figure 2. (b) SEM images of Green tea $FeCl_3$ after synthesis

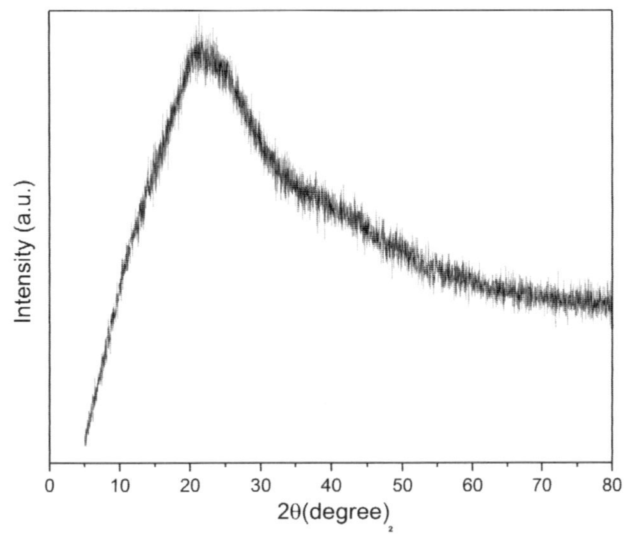

Figure 3. XRD pattern of Green tea

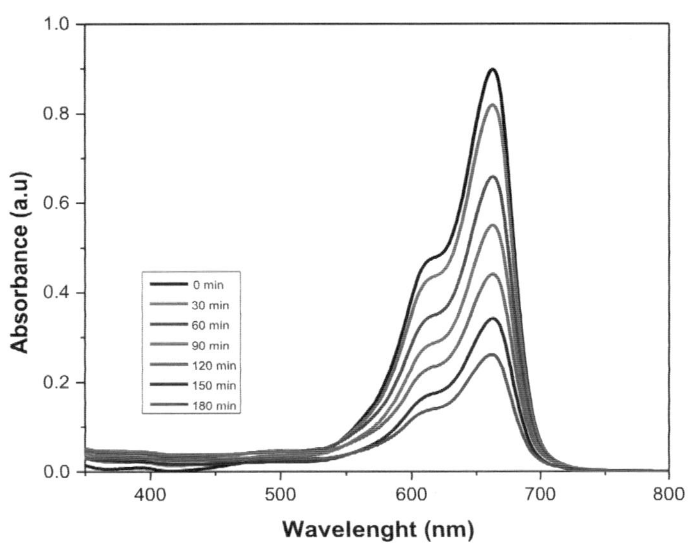

Figure 4. Photo-catalytic absorption spectra of Fe-nanoparticles under sunlight

5 RESULTS AND DISCUSSION

By using green leaves tree extract as reducing agent, we have synthesized the Iron nano particles in smooth and feasible way. Green tea balances an extra quantity of polyphenols and other organic components which are present in it. It has been discovered that almost 33% part contains polyphenols which helps in reducing the salt antecedent to nanoparticles. Reduction technique is carried out in two steps: 1) in primary method, antecedent is added to form a compound by breaking the OH bond so that partial bond can be formed with metal iron. 2) Secondly the partial bond breaks and electron is transmitted to decrease the metal ions to nano particles. This leads them to oxidize to form ortho-quinone.

5.1 Photocatalytic degradation of MB dye

The essential purpose of this research work is to degrade the organic pollutant. In our study Methylene Blue (MB) dye is used as model organic pollutants. The Fe-nanoparticles are used as photo-catalyst to test the degradation of MB dye. The necessary amount of catalyst was amalgamated in stock solution of dye and was irradiated under direct sunlight for different time breaks under continuous stirring. The irradiated samples were centrifuged, filtered and then UV-vis spectrophotometer was used to determine the absorbance spectra. The degradation results of MB dye are shown in Fig. 4. The MB absorbance spectrum was reduced with time; moreover change of color from light blue to light green and then becoming colorless gives information regarding degradation of dye. The degradation efficiency of the catalyst was calculated by

$$Degradation\ (\%) = \frac{A_0 - A_t}{A_0}$$

71% efficiency is shown by Fe-nanoparticles for degradation of MB dye .When the light energy is in comparison or more than the band gap, the electrons attain energy and shifts from Valence band (VB) to the conduction band (CB) which induces holes (h^+) in the VB and electron (e^-) in CB. These holes counter with the molecules of water and OH radicals are produced, which further oxidized the organic pollutant. Electrons from conduction band react with oxygen in the reduction process and induce OH radicals. These radicals decrease the organic pollutant. This oxidation and reduction process degrades the organic pollutant under sunlight.

6. CONCULSION

Green tea leaves extract was used for preparation of Fe-nanoparticles. This is natural and less expensive process to synthesis Iron NPs, further this can enhance the industrial production of metal nanoparticles without using harmful agents. From results we conclude that Fe-NPs exhibits efficient photo-catalytic activity. Applications and future research: Once the Fe-nanoparticles are successfully prepared, they can be investigated for various applications in fields such as catalysis, biomedicine, environmental remediation, and energy storage. The unique properties of nanoparticles at the Nano scale make them attractive for a wide range of research areas. It can be concluded that Fe-NPs synthesis through green tea leaves are sustainable and environmental friendly.

Acknowledgements

Authors would like to thank all colleagues for their guidance and motivation to complete this research work and special thanks to Emerging science department of Guru Nanak Dev University (GNDU)-Amritsar for technical support.

References

[1] D. Saravanakkumar, S. Sivaranjani, K. Kaviyarasu, A. Ayeshamariam, B. Ravikumar, S. Pandiarajan, C. Veeralakshmi, M. Jayachandran, M. Maaza, *J. Semicond.* **39** (3) (2018) 033001.

[2] D. Saravanakkumar, H. Abou Oualid, Y. Brahmi, A. Ayeshamariam, M. Karunanaithy, A. Mohamed Saleem, K. Kaviyarasu, S. Sivaranjani, M. Jayachandran, *OpenNano* **4** (2019) 100025.

[3] R. Shanmuganathan, D. MubarakAli, D. Prabakar, *Environ. Sci. Poll. Res.* **25** (11) (2018) 10362–10370.

[4] N. Geetha, S. Sivaranjani, A. Ayeshamariam, M. Siva Bharathy, S. Nivetha, K. Kaviyarasu, M. Jayachandran, *J. Adv. Microsc. Res.* **13** (1) (2018) 12–19. S.

[5] Sri Rathnakumar, K. Noluthando, A. Jayalatha Kulandaiswamy, J. Bosco Balaguru Rayappan, K. Kaviyarasu, J. Kennedy, M. Maaza, *Sens. Actuators, B* **293** (2019) 100–106.

[6] P.D. Shankar, S. Shobana, S. Karuppusamy, *Enzy. Micro. Technol.* **95** (2016) 28–44.

[7] R.G. Saratale, G. Benelli, G. Kumar, D.S. Kim, G. Saratale, *Environ. Sci. Poll. Res.* **25** (11) (2018) 10392–10406.

[8] K. Poovendran, K.S. Wilson, M.S. Revathy, A. Ayeshamariam, K. Kaviyarasu, *Surf. Interfaces* **19** (2020) 100474.

[9] R.G. Saratale, H.S. Shin, G. Kumar, G. Benelli, G.S. Ghodake, Y. Jiang, D.S. Kim, G. *IOP Conf. Ser.: Mater. Sci. Eng.* **925** (2020) 012055.

[10] D. Saratale, *Environ. Sci. Pollut. Res.* **25** (2018) 10250–10263.

[11] S. Arokiyaraj, V. Kumar, V. Elakya, *Environ. Sci. Poll. Res.* **22** (2015) 9759–9765.

[12] S. Arokiyaraj, S. Vincent, M. Saravanan, *Nanomed. Biotechnol.* **45** (2017) 372– 379

[13] Kuang, Y., q, Chen, Z., Megharaj,M., Naidu, *Journal of colloidal and Interface sceince* . **410** (2014) 67-73

[14] Machado,S.,S.L., Grosso,J.P., Nouws, H.P.A, Albergaria,J.T., Delerue-Matos, *Science of total Envoirment*.445-446 (2013)**1-8**

[15] Madhavi,V., Prasad, T.N.V.K.V., Reddy, A.V.B., Reddy, B.R.,Madhavi,G *Spectrochimica Acta Part A: Molecular and Biomolecular Spectroscopy* **116** (2013) 17-25

[16] Smuleac,C., Varma,R., Sikdar, S., Bhattacharyya,D. *Journal of Membrane Science* **379** (2011)131-137

[17] M.Z. Motamedi, E. MIKHAK, A, Rahnemaie,R. *Chemical Engineering Journal* 166 **166** (2011)490-495

[18] Hoag,G.,Collins,J.,Holcomb,J.,Hoag,J.,Nadagouda,M.,Varma, R., *Journal of materials chemistry* **19** (2009)8671- 8677.

[19] Ting Wang, Jiajiang Lin, Zuliang Chen, Mallavarapu Megharaj, Ravendra Naidu *Journal of cleaner production* **83**(2014) 413-419

[20] Huang, Weng, Chen, Megharaj,and Naidu *Spectrochimica Acta Part A: Molecular and Biomolecular Spectroscopy* **130**(2014)295-301

[21] Tejaswi Mareedu, VenkataRao Poiba, Meena Vangalapati *Materials Today: Proceedings* **42** (2021) 1498–1501.

The DFT study of thermoelectric properties of $CuInS_2$: A first principle approach

Ranjan Kr. Giri[1*], Mitesh B. Solanki[2], Sunil H. Chaki[1,3#], Milind P. Deshpande[1]

[1]P. G. Department of Physics, Sardar Patel University, Vallabh Vidyanagar, Gujarat - 388120, India
[2]Department of Physics, Indian Institute of Information Technology (Gandhinagar Campus), Vadodara – 382028, Gujarat, India
[3]Department of Applied & Interdisciplinary Sciences, CISST, Sardar Patel University, Vallabh Vidyanagar – 388120, Gujarat, India

*ranjankrgiri@gmail.com
#sunilchaki@yahoo.co.in

Abstract. The generalised gradient approximation (GGA) and ultrasoft pseudopotential (US PP) methods to the function of Perdew-Burke-Ernzerhof (PBE) approach are utilised for density functional computation of $CuInS_2$. It enables the prediction of thermoelectric characteristics, including Seebeck coefficient, thermal conductivity, electrical conductivity, magnetic susceptibility, specific heat, power factor and figure of merit by semi-classical Boltzmann approach. At 800 K, the highest magnitude of Seebeck coefficient is estimated as 6.91×10^{-5} $V{\cdot}K^{-1}$. The maximum figure of merit (zT) is predicted as 1.04×10^{-4} at 800 K. Findings from this study suggest that $CuInS_2$ has prospective to be used in the thermoelectric power generating sector.

Keywords: $CuInS_2$, DFT, thermoelectric, Boltzmann computation, first principle approach.

1. Introduction

Recent developments are focusing on to the expanding family of semiconducting materials and their wide variety of potential uses. It covers a wide range of fields, including thermoelectric effect [1,2], thermal study [3–5], superconductivity [6,7], photovoltaic devices [8,9], optical study [10,11], biological applications [12–14] and much more. Semiconducting materials with properties similar to chalcopyrite are of interest due to their potential use and low environmental impact. The high chemical stability and unique optoelectronic properties make the chalcopyrite $CuInS_2$ (CIS) of the I-III-VI_2 group to gain greater attention. The CIS is an excellent candidate for solar-energy conversion due to its significant absorption coefficient (~10^5 cm^{-1}) and direct optical energy bandgap (1.36 eV to 1.54 eV) [15].

Thermoelectric (TE) effect is a source of "green energy" since it uses thermal energy to vigour electrons in converting directly to energy. The TE effect offers a number of benefits for use in industry. The two most important TE effects are the Peltier and Seebeck phenomena. The TE device may chill substances by use of the Peltier effect. Alternatively, the Seebeck effect allows for heat conversion into electricity; this process is termed as TE power production. The zT value is used to characterise TE efficiency of materials. Hence, a high zT value is appreciated to enable effective energy conversion in a TE material.

In this work, looking into the relevance and benefit of TE materials, the authors decided to investigate TE features of CIS. Several aspects of CIS remain unclear, despite the fact that TE properties have been documented in a number of research articles [16,17]. DFT analysis using first-principle approach is used to achieve this aim. TE features like the Seebeck coefficient, thermal conductivity, electrical conductivity, magnetic susceptibility, specific heat, power factor and figure of merit are estimated.

2. First principle computation

The DFT plane waves and pseudopotentials are computed in this investigation using Quantum ESPRESSO programme. Exchange-correlation term is studied using Perdew-Burke-Ernzerhof (PBE) function. Moreover, ultrasoft pseudopotentials (US PPs) and generalised gradient approximation (GGA) are employed. An upper limit of 50 Ry is placed on the energy at which the wave function may expand. Energy threshold value of 500 Ry is used to extrapolate the electronic wave function for charge density. The Monkhorst-Pack $8 \times 8 \times 8$ meshes are used for the Brillouin zone integration. Adjustments are made to the lattice constant of CIS until the resulting energy level reaches 10^8 Ry or higher. The optimum structure is found through the implementation of variable-cell ($v_{c\text{-relax}}$) simulations. For this simulation, four atoms are used because of the molecule's symmetry, which is discussed in further detail. To analyse the TE features of CIS, the BoltzTraP technique is used. The more compact k-mesh of $22 \times 22 \times 22$ is used for predictions of TE properties like the Seebeck coefficient, thermal conductivity, electrical conductivity, magnetic susceptibility, specific heat, and figure of merit. In order to relate Seebeck coefficient (S) to the concentration of carriers, the Mott formula is used [18],

$$S = \frac{\pi^2 k_B}{3e} k_B T \left\{ \frac{1}{n}\frac{dn(E)}{dE} + \frac{1}{\mu}\frac{d\mu(E)}{dE} \right\}_{E=E_F} \quad (1)$$

where E_F denotes Fermi energy, e denotes electronic charge, n denotes carrier concentration, μ denotes chemical potential, T is absolute temperature and k_B is Boltzmann constant.

The relationship between carrier concentration (n) and electrical conductivity (σ) is as follows [19],

$$\sigma = \frac{ne^2\tau}{m} \quad (2)$$

where m denotes electronic mass, and τ denotes relaxation time. The relationship between carrier concentration, chemical potential, energy density (ε) and electronic-specific heat (c) is shown below [20],

$$c(T;\mu) = \int n(\varepsilon)(\varepsilon - \mu) \left[\frac{\partial f_\mu(T;\varepsilon)}{\partial T}\right] d\varepsilon \quad (3)$$

The Pauli magnetic susceptibility is shown to be related to carrier concentration and chemical potential by the equation below [21],

$$\chi(T;\mu) = \mu_0 {\mu_B}^2 \int n(\varepsilon) \left[-\frac{\partial f_\mu(T;\varepsilon)}{\partial T}\right] d\varepsilon \quad (4)$$

where μ_0 denotes the vacuum permeability, and μ_B denotes Bohr magneton.

3. Results and discussion

The TE characteristics are estimated by a semi-classical Boltzmann approach using BoltzTraP software in the constant relaxation time approximation. The computed characteristics are shown for wide temperatures considered from 25 K to 800 K. The Seebeck coefficient (S) quickly approaches zero as S drops when the temperature rises due to an increment in thermal energy. It suggests that CIS has excellent TE features. At 0 K, the minimum magnitude of S is 4.12×10^{-6} V·K^{-1}. The value of S is significantly decreased from -4.12×10^{-6} V·K^{-1} to -6.91×10^{-5} V·K^{-1} at a higher temperature of 800 K. The CIS is predicted to have n-type semiconducting nature.

The S at various temperatures is greater for lower concentrations, according to equation (1). The Figure 1 depicts that as the temperature rises, S drops due to an increment of thermal energy, and hence CIS is predicted to have good TE performance.

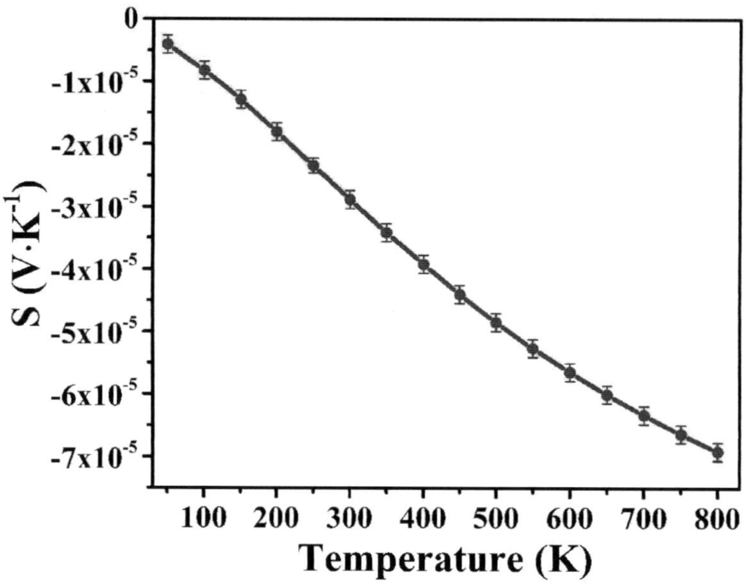

Figure 1. Variation of Seebeck coefficient (S) with T (K).

The Figure 2 depicts electrical conductivity (σ) ($\mho \cdot cm^{-1} \cdot s^{-1}$) related to temperature (K). The nature states that electrical conductivity decreases with temperature.

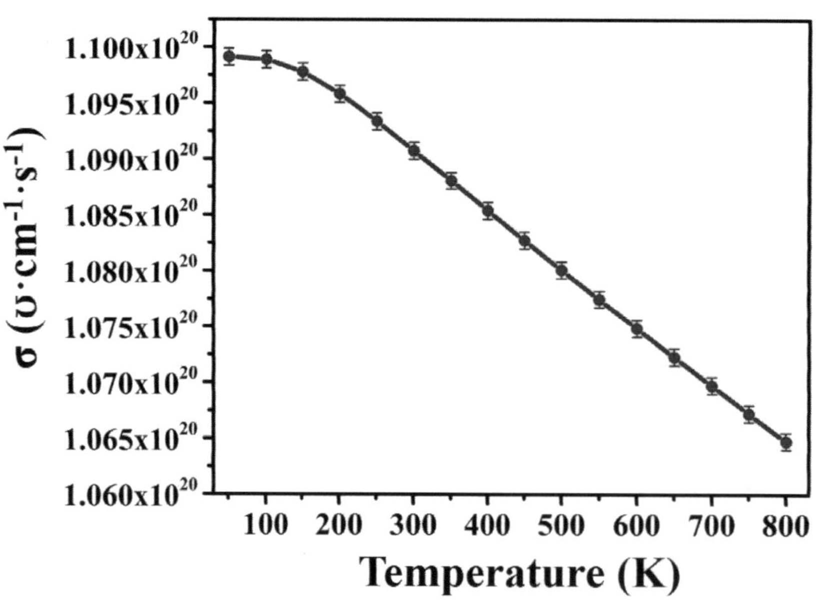

Figure 2. Variation of electrical conductivity (σ) with T (K).

The Figure 3 illustrates electronic thermal conductivity (k) ($W \cdot m^{-1} \cdot K^{-1} \cdot s^{-1}$) related to temperature (K). Thermal conductivity rises with increasing temperature. To achieve a good TE property, materials should exhibit higher values of Seebeck coefficient and electrical conductivity, whilst low thermal conductivity value at an optimal temperature [22]. According to the figure, as temperature rises, thermal conductivity increases linearly.

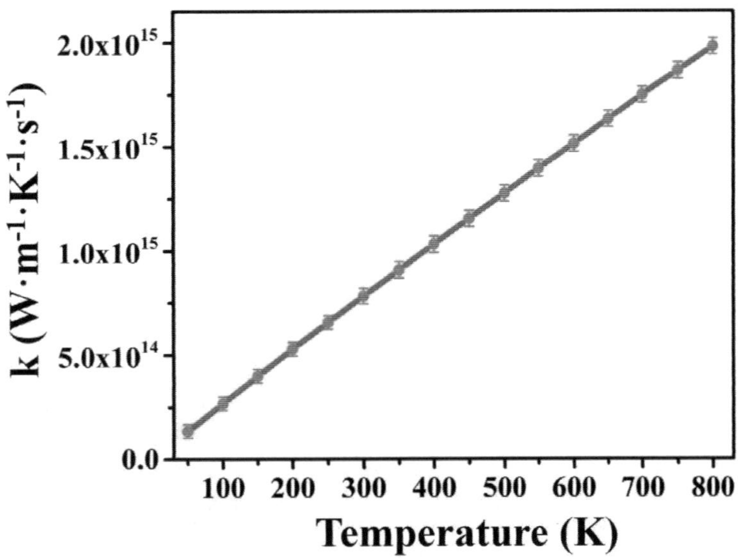

Figure 3. Variation of electronic thermal conductivity (k) with T (K).

The Figure 4 depicts electronic specific heat (c) related to absolute temperature (K). As temperature rises, the specific heat increases in a linear manner according to equation (3). The increasing positive values of the specific heat with increasing temperatures indicate that the carrier concentration is also increasing.

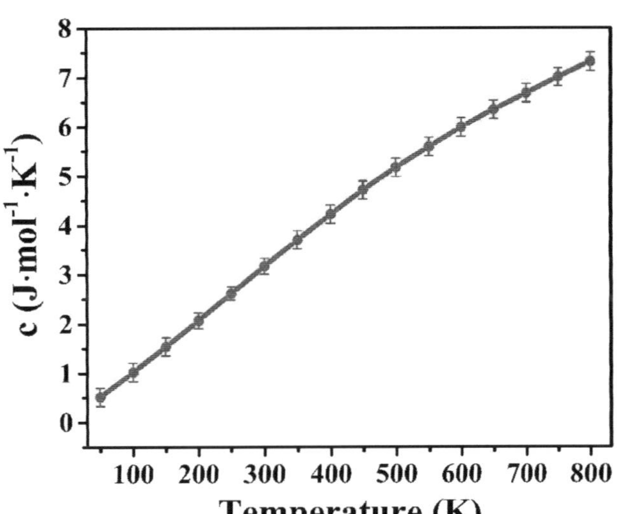

Figure 4. Variation of electronic specific heat (c) with T (K).

The Pauli magnetic susceptibility (χ) varying with temperature is illustrated in the Figure 5. A little variation in the value of χ is observed upto 200 K. Afterwards, the value of χ increases with increasing temperature. The highest recorded value of χ at exhibited Fermi temperature (~550 K) is found to be 1.795 x 10^{-9} $m^3 \cdot mol^{-1}$. A significant drop in χ occurs when the temperature crosses 550 K and reaches to 800 K. This variation in χ values is directly related to carrier concentration, as predicted by equation (4).

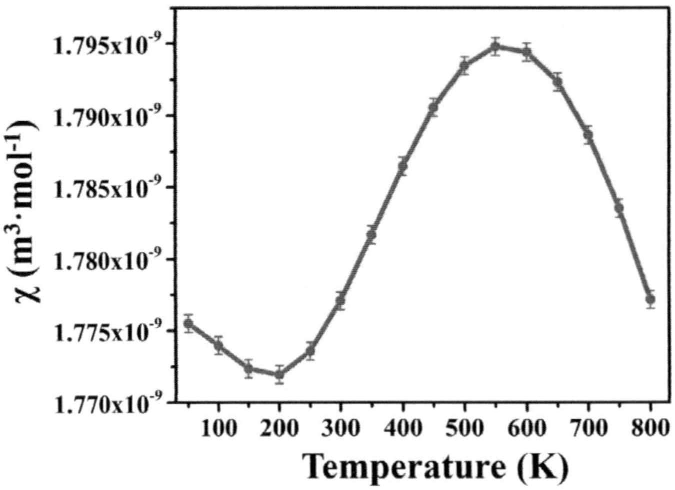

Figure 5. Variation of Pauli magnetic susceptibility (χ) with T (K).

The Figure 6(a) displays power factor values (PF/τ) related to chemical potential with respect to Fermi level. To what extent may TE material functions efficiently is investigated. The Figure 6(a) shows that as temperatures rise, the power factors also rise accordingly. The maximum power factor is estimated as 9×10^{11} $\mu W \cdot cm^{-1} \cdot K^{-2} \cdot s^{-1}$ with a negative chemical potential at 800 K. There has been a little decrement in the factor, which now stands at 2.5×10^{11} $\mu W \cdot cm^{-1} \cdot K^{-2} \cdot s^{-1}$. When comparing the positive and negative regions of the power factor, the former has a higher maxima value of 1×10^{11} $\mu W \cdot cm^{-1} \cdot K^{-2} \cdot s^{-1}$. The electronic thermal conductivity (σ/τ) is shown against chemical potential and carrier concentration in the Figure 6(b) for various temperatures. This graph shows how thermal conductivity increases with temperature. Therefore, 300 K is the optimum temperature for minimising electronic thermal conductivity (σ/τ). Heat conductivity increases with increasing chemical potential. In the $E-E_F$ range of +0.3 Ry to +0.65 Ry, the thermal conductivity is zero at 300 K. The chemical potential (μ) versus S from -0.1 Ry to +0.1 Ry is exhibited in the Figure 6(c). In the vicinity of the Fermi zone, the Seebeck coefficient has two peaks in the range of −300 Ry to +300 Ry; outside of this range, the Seebeck coefficient rapidly approaches zero. Due to an increment in thermal energy, S decreases as temperature rises. It points to the fact that CIS possesses superior TE characteristics. For temperatures between 300 K and 800 K, the maximum value of S is found as 300 $\mu V \cdot K^{-1}$. A reduction in S to −300 $\mu V \cdot K^{-1}$ occurs at higher temperature of 800 K. Thus, CIS exhibits 350 $\mu V \cdot K^{-1}$ for the negative S peak and 750 $\mu V \cdot K^{-1}$ for the positive S peak at 300 K.

The figure of merit (zT) of a material quantifies its TE efficiency. The zT value is related to the absolute temperature as below [23],

$$zT = \frac{S^2 \sigma T}{k} \quad (5)$$

The Figure 7 exhibits the variation of zT values with temperature (K). The zT values are found to increase with temperature. The maximum zT value is predicted as 1.04×10^{-4} at 800 K, while it is 1.79×10^{-5} at ambient temperature which agrees well with experimental findings of zT (~3.99×10^{-4} at 423 K) [24]. It affirms to the fact that CIS is a good TE material to be employed in various TE applications.

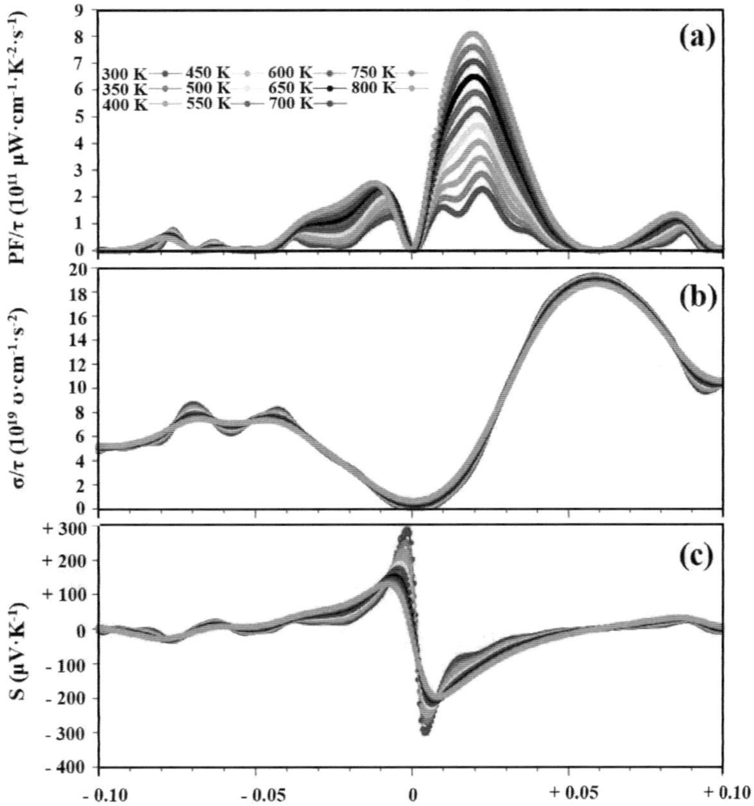

Figure 6. The BoltzTraP plots of (a) power factor, (b) electrical conductivity, and (c) Seebeck coefficient as a function of E-E_F (Ry).

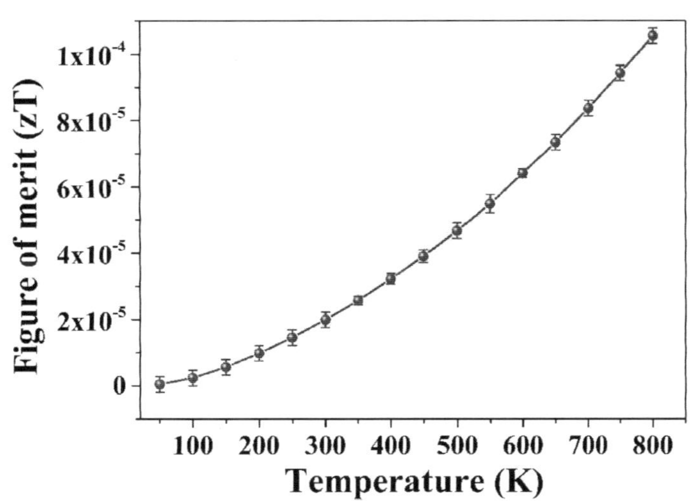

Figure 7. The zT values of CIS as a function of T (K).

4. Conclusion

Boltzmann computation via first principle approach using DFT reveals that the CIS material exhibits higher TE characteristics with the greatest magnitude of S found at 800 K. The material is found to be n-type with electrons as majority charge carriers. Electrical conductivity values decrease as temperature increases, while thermal conductivity values increase with increasing temperature. The highest value of χ at 550 K is estimated as 1.795×10^{-9} $m^3 \cdot mol^{-1}$. The maximum power factor is estimated as 9×10^{11} $\mu W \cdot cm^{-1} \cdot K^{-2} \cdot s^{-1}$ having negative chemical potential at 800 K. The maximum zT value is predicted as 1.04×10^{-4} at 800 K. This research indicates that CIS has a high potential to be applicable in the field of TE power generation. Although further studies are needed to enhance TE features of CIS by using suitable dopants.

References

[1] Lee W Y, Kang M S, Kim G S, Choi J W, Park N W, Sim Y, Kim Y H, Seong M J, Yoon Y G, Saitoh E and Lee S K 2022 Interface-Induced Seebeck Effect in $PtSe_2/PtSe_2$ van der Waals Homostructures *ACS Nano* **16** 3404–16

[2] Kang M S, Lee W Y, Yoon Y G, Choi J W, Kim G S, Kim S H, Park N W and Lee S K 2022 Enhanced Transverse Seebeck Coefficients in 2D/2D $PtSe_2/MoS_2$ Heterostructures Using Wet-Transfer Stacking *ACS Appl. Mater. Interfaces* **14** 51881–8

[3] Kachhia Z S, Chaki S H, Giri R K, Parekh Z R, Kannaujiya R M, Hirpara A B, Deshpande M P and Tailor J P 2023 Thermal decomposition study of cadmium telluride (CdTe) *Mater. Today Proc.*

[4] Khimani A J, Chaki S H, Giri R K, Meena R R, Kannaujiya R M and Deshpande M P 2023 Thermal Exploration of Sonochemically Achieved SnS_2 Nanoparticles: Elemental, Structural, and Morphological Investigations of TG Residual SnS_2 *Chem. Thermodyn. Therm. Anal.* 100104

[5] Kannaujiya R M, Chaki S H, Khimani A J, Hirpara A B, Parekh Z R, Giri R K, Patel S and Deshpande M P 2022 Effect of Sb doping on CVT grown SnTe single crystals electrical and thermal properties *J. Mater. Sci. Mater. Electron.* **33** 20823–36

[6] Lane C and Zhu J X 2022 Identifying topological superconductivity in two-dimensional transition-metal dichalcogenides *Phys. Rev. Mater.* **6** 94001

[7] Xiao F, Lei W, Wang W, Autieri C, Zheng X, Ming X and Luo J 2022 Pressure-induced structural transition, metallization, and topological superconductivity in PdSSe *Phys. Rev. B* **105** 115110

[8] Roy S and Bermel P 2022 Tungsten-Disulfide-Based Ultrathin Solar Cells for Space Applications *IEEE J. Photovoltaics* **12** 1184–91

[9] Pervaiz H, Khan Z S, Shahzad N, Ahmed N and Jamil Q 2022 Synthesis and characterization of $CuInS_2$ nanostructures and their role in solar cell applications *Mater. Chem. Phys.* **290** 126602

[10] Bharucha S, Dave M and Vaidya R 2022 Electronic and optical studies of NbS_2 semiconductor material *Mater. Today Proc.* **55** 118–21

[11] Malek T J, Chaki S H, Giri R K and Deshpande M P 2022 The structural, morphological, and optical study of chemical bath deposition and a spin coating deposited mackinawite FeS thin films *Appl. Phys. A* **128** 830

[12] Giri R K, Chaki S, Khimani A J, Vaidya Y H, Thakor P, Thakkar A B, Pandya S J and Deshpande M P 2021 Biocompatible $CuInS_2$ Nanoparticles as Potential Antimicrobial, Antioxidant, and Cytotoxic Agents *ACS Omega* **6** 26533–44

[13] Hirpara A B, Chaki S H, Kannaujiya R M, Khimani A J, Parekh Z R, Vaidya Y H, Giri R K and Deshpande M P 2022 Biological investigation of sonochemically synthesized CZTS nanoparticles *Appl. Surf. Sci. Adv.* **12** 100338

[14] Kannaujiya R M, Chaki S H, Khimani A J, Giri R K, Hirpara A B, Vaidya Y H and Deshpande M P 2023 Mechanistic Insights into Antibacterial and Anti-biofilm Activities against multidrug-resistant microbes of SnTe nanospheres synthesized by sonochemical method *Chem. Phys. Impact* 100219

[15] Giri R K, Chaki S H, Khimani A J, Patel S R and Deshpande M P 2021 Thermal investigation of nanospheres and nanowhiskers of $CuInS_2$ *Eur. Phys. J. Plus* **136** 320

[16] Puzyrev Y S, Shen X and Pantelides S T 2016 Prediction of Giant Thermoelectric Efficiency in Crystals with Interlaced Nanostructure *Nano Lett.* **16** 121–5

[17] Pachuau Z 2023 First Principle Study of Electronic, Optical and Thermoelectric Properties of $CuInS_2$ and $CuInSe_2$ *Indian J. Pure Appl. Phys.* **61** 108–14

[18] Pakizeh E, Jalilian J and Mohammadi M 2019 Electronic, optical and thermoelectric properties of Fe_2ZrP compound determined via first-principles calculations. *RSC Adv.* **9** 25900–11

[19] Burnett J D, Gourdon O, Ranmohotti K G S, Takas N J, Djieutedjeu H, Poudeu P F P and Aitken J A 2014 Structure–property relationships along the Fe-substituted $CuInS_2$ series: Tuning of thermoelectric and magnetic properties *Mater. Chem. Phys.* **147** 17–27

[20] Bulut P, Beceren B, Yıldırım S, Sevik C and Gürel T 2021 Promising room temperature thermoelectric conversion efficiency of zinc-blende AgI from first principles. *J. Phys. Condens. Matter* **33** 15501

[21] Azam S, Umer M, Saeed U, Khan W, Irfan M, Abbas Z and Kityk I V 2020 DFT modeling of thermoelectric and optical features of novel $Mg_xSn_{1-x}Se$ (x = 6%, 12% & 18%). *J. Mol. Graph. Model.* **94** 107484

[22] Rashid Z, Nissimagoudar A S and Li W 2019 Phonon transport and thermoelectric properties of semiconducting Bi_2Te_2X (X = S, Se, Te) monolayers. *Phys. Chem. Chem. Phys.* **21** 5679–88

[23] Jia K, Yang C L, Wang M S, Ma X G and Yi Y G 2021 First-principles investigation on the thermoelectric performance of half-Heusler compound CuLiX(X = Se, Te). *J. Phys. Condens. Matter* **33** 95501

[24] Giri R K, Chaki S H, Khimani A J and Deshpande M P 2023 Mechanistic insights into transport properties of chemical vapour transport grown $CuInS_2$ single crystal *J. Alloys Compd.* **959** 170487

IOP Conf. Series: Materials Science and Engineering 1291 (2023) 012010 doi:10.1088/1757-899X/1291/1/012010

Antibacterial Activity of Green Synthesized Silver Nanoparticle using Banana Flower Extract

Kajal Francis[1], S Achutha[2], Barakala Pushpa[1] and K Ramya[1a*]

1 Post Graduate and Research Department of Chemistry, St. Joseph's College for Women, Alappuzha-688001, Kerala, India
2 Post Graduate and Research Department of Chemistry, Sanatana Dharma College, Alappuzha-688003, Kerala, India

ramyabraj@gmail.com

Abstract. In recent years, due to unique physical and chemical properties of nanoparticles, its uses to disrupt biofilm of bacteria are revolutionary steps in antimicrobial research. To use in bioprocess the silver nanoparticles should be biocompatible and free from toxic chemicals. In the present study we report a cost effective and environment friendly route for green synthesis of silver nanoparticles using banana flower extract (BFE) as reducing as well as capping agent. This plant has been opted for the present study for its known medicinal properties and it is easily available. The biosynthesized silver nanoparticles are characterized by UV–Vis spectroscopy which showed a broad peak at around 450 nm, indicated the stability of synthesized silver nanoparticles. Antimicrobial potential of silver nanoparticles synthesized were tested against both Gram positive and Gram-negative bacteria. It was found to be effective against both of them and it showed maximum activity against Gram positive bacteria S. aureus. Silver nanoparticles (AgNPs) have been imposed as an excellent antimicrobial agent being able to combat bacteria in vitro and in vivo causing infections. The results show green synthesized silver nanoparticles, using BFE extract, have a potential to inhibit the growth of bacteria.

1. Introduction

Silver nanoparticles refer to particles of silver with a size ranging from 1 to 100 nanometers (nm). These particles exhibit unique properties because of their small size, including high surface area, increased reactivity, and distinctive optical, thermal, and electrical behavior [1]. Silver nanoparticles have variety of applications, including electronics, medical treatments, water purification, and as catalysts in chemical reactions. For example, in electronics, silver nanoparticles are used in conductive inks for printed electronics and in anti-microbial device coating to prevent the growth of bacteria [2,3]. In the medical field, silver nanoparticles have been shown to have antimicrobial properties and are being studied for use in wound healing and in the development of antimicrobial drugs. These silver nano particles can penetrateto cell membrane and they can destroy affected cells in human body [4]. It is crucial to comprehend the possible hazardous effects of silver nanoparticles on human health and the environment due to their distinctive features and ubiquitous use. Therefore, research is being conducted to assess the safety and potential risks associated with the use of silver nanoparticles [5,6].

Now a days green synthesis methods are most popular method because of its non-toxicity, low cost andtheir ease of availability. Many research works are established with this green method nano particle synthesis [7-9]. In this work we are using banana flower extract as a reducing agent. Banana flower ischosen because of its ease of availability. The antibacterial activities of silver nanoparticles could be improved by using banana flower extract as a reducing agent [10]. A prominent fruit grown in India, the banana (Musa spp.) is regarded as a good source of biologically active substances like dopamine, N- acetyl serotonin, nor-adrenaline, derivatives of isochronal-4-one, and polyphenols with high antioxidant characteristics [11]. In addition to their many health advantages over synthetic antioxidant compounds, the isolation and identification of new antioxidants from natural sources, such as phenolic compounds, nitrogen compounds, carotenoids, and other phytochemicals, is attracting particular interest for its potential to prevent diseases [12]. Spathes are boat-shaped bracts that are brightly colored, larger and spirally arranged structures that protect the flowers. They roll back and fall when flowers open-up. The water extract of Banana spathes found to contain wide verity of phytochemicals and showed good antioxidant potential [13,14]. Thus, it has been used for the reduction of silver nitrate for nanoparticle synthesis.

Figure 1. Banana Flower

2. Material and Methods

The chemicals used were of superior grade and glass wares were washed with detergents and rinsed twice with double distilled water before use. For nano synthesis, silver nitrate (Merck) was used as the initial component

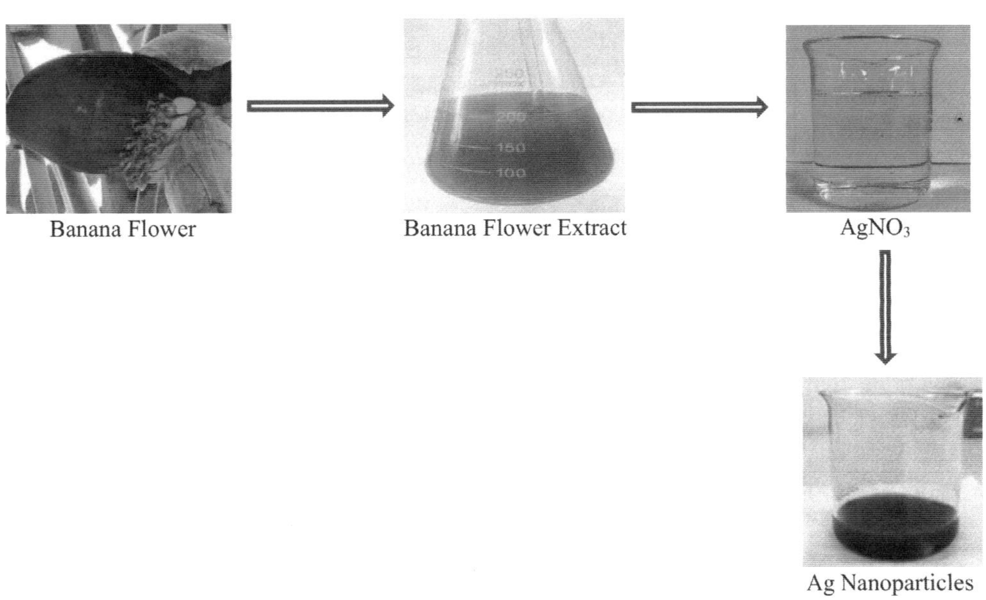

Figure 2. Synthesis of Silver Nanoparticle

2.1 Extract Preparation:

On a hot plate with magnetic stirrer, about 10g of dried banana spathe were extracted with 100 ml distilled water for 1 hour at 60ºC. It was then filtered and centrifuged at 4000 resolutions per minute for 10min. The synthesis and analysis of nanoparticles were done using the supernatant. Phytochemical analysis conducted.

2.2 Synthesis of Silver nanoparticle

A magnetic stirrer set to 45°C was used to continuously mix 5 ml of the extract with 45 ml of silver nitrate solution (0.001M). UV-visible spectral scanning was used to periodically monitor the synthesis of silver nanoparticles. On Systronics UV-Visible spectrophotometer 117, a UV-Visible spectrograph of the colloidal solution of silver nanoparticles was recorded as a function of time [15].

2.3. Evaluation of antimicrobial potential of silver nanoparticles

Method: Kirby-Bauer test: Agar with bacteria growing on it is covered with antibiotic-containing discs, and the antibiotics diffuse into the agar. If an antibiotic stops the bacteria from growing, one can see circular areas around the wafers where they have not developed. The diameter of the inhibitory zone was measured using the well diffusion method using nanoparticles that had been resuspended in DMSO (10 and 15 ul).

Preparation of culture plates for antibacterial assay:
Antibacterial assay was conducted on cultures plates prepared on Mueller Hinton Agar. The composition of the medium is given below. All the chemicals used were from HIMEDIA.

Mueller Hinton Agar (g/L):

Beef infusion	2.0
Starch	1.5
Casein hydrolysate	17.5
Agar	17.0

Final pH 7.4 at 37°C

Directions: All the components were suspended, dissolved, in 1 litre distilled water and sterilized at 121°C for 15 minutes.

3. Results and Discussion

3.1. Phytochemical Analysis

Table 1. Phytochemical Analysis

Sl.No	PHYTOCHEMICALS	TEST PERFORMED	RESULTS
1	Alkaloids	Mayer's test Dragendroff's test	Present
2	Tannins	$FeCl_3$ test	Present
3	Flavanoids	Shibata's reaction	Present
4	Saponins	Froth test	Absent
5	Glycosides	Fehling's test	Present
6	Cardiac Glycosides	Keller Kiliani test	Absent
7	Cyanogenetic	Picrate test	Absent
8	Sugar	Molisch's test	Present
9	Anthraquinones	Borntager's test	Present
10	Steroids	Liebermann-Burchard test	Absent
11	Terpenoids	Phenyl hydrazine test	Present
12	Amino acids	Ninhydrin test	Positive
13	Gums/mucilage	Alcohol-water	Absent

3.2. Nanoparticle synthesis

The formation of silver nanoparticle in the reaction mixture is indicated by the appearance of a yellow colour which gradually changed to reddish brown tinge. Also, it is confirmed by comparing the absorption spectra of produced colloidal solutions with extract blanks. A broad absorption band near the wavelength of 453 nm is obtained, and it was discovered that the intensity increased over time (Figure 1). The phenomenon known as plasmon absorption is what gives silver nanoparticle colloidal solution its characteristic colour. Conduction electrons oscillate on the surface of the nanoparticles as a result of incident light. Surface plasmon resonance allows the scattering and absorption of light at a specific frequency, which gives materials their colour. By centrifuging the colloidal nanoparticle solution at 12000 rpm for 15 minutes, silver nanoparticles were obtained. The precipitate was centrifuged after being rinsed with double-distilled water. The stability of the produced silver nanoparticles was demonstrated by a fragment of the precipitate that was dispersed in water and displayed a broad peak at about 450 nm.

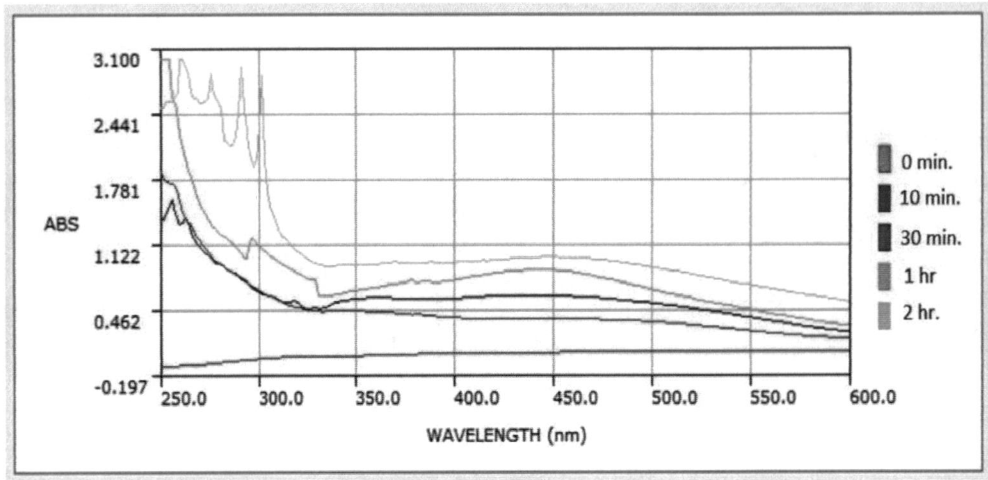

Figure 3. UV-Visible spectrum of the silver nanoparticle synthesized against plant extract blank.

3.3. Antimicrobial Studies

Antimicrobial potential of the synthesized silver nanoparticles was tested against Gram positive and negative bacteria. It was found to be active against both of them and showed maximum efficiency against Gram positive bacteria S. aureus. Silver nanoparticles (AgNPs) have been recommended as a superior antibacterial agent capable of attacking bacteria that cause infections both in vitro and in vivo [16]. Gram-negative and Gram-positive bacteria, including multidrug resistant strains, are all covered by AgNPs' antibacterial property. AgNPs have several, concurrent modes of action, and when combined with antibiotics or other antibacterial agents such as chemical compounds, they have a synergistic effect on pathogenic bacteria like Escherichia coli and Staphylococcus aureus. The characteristics of silver nanoparticles make them suitable for their application in medical and healthcare products where they may treat infections or prevent them efficiently [17].

Table 2. Antimicrobial Studies

Klebsiella pneumonia

Klebsiella pneumoniae is a Gram-negative, non-motile, encapsulated, rod-shaped bacterium. it can lead to a range of illnesses, including pneumonia, bloodstreaminfections, meningitis, and urinary tract infections.

Zone of Inhibition

Banana Flower Extract: 0 mm
AgNP (10μl): 14.2 mm
AgNP (15μl): 16.7 mm
Tetracycline: 22.0 mm

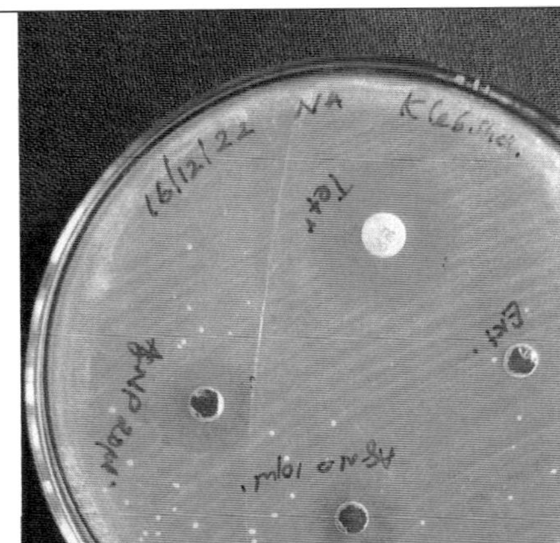

Staphylococcus aureus

Staphylococcus aureus is a Gram-positive spherically shaped bacterium. It is the main contributor to skin and soft tissue infections such cellulitis, furuncles, and abscesses (boils).

Zone of Inhibition
Banana Flower Extract: 0 mm
AgNP (10μl): 22.5 mm
AgNP (15μl): 24.8 mm
Tetracycline:17.2 mm

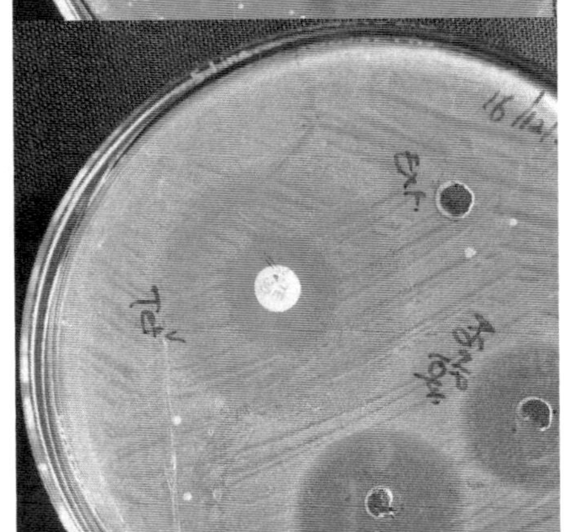

Escherichia coli

E.coli, is a Gram-negative,facultative anaerobic, rod-shaped, coliform bacterium.

Zone of Inhibition
Banana Flower Extract : 0 mm
AgNP (10μl): 13.1 mm
AgNP (15μl): 16.4 mm
Tetracycline: 21.5 mm

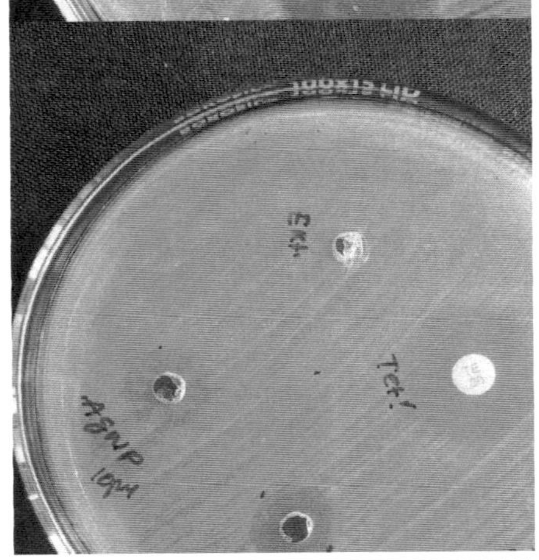

AgNPs are characterised as nanomaterials with all of their dimensions falling between 1 and 100 nm. Compared to silver in its bulk form, they have demonstrated increased surface (area-to-volume ratio) and capacity. This substance displays distinctive electrical, optical, and catalytic capabilities at the nanoscale, which has stimulated research and the creation of goods for targeted drug administration, diagnostics, detection, and imaging. Yet, AgNPs' excellent antibacterial activity is what has drawn researchers' and companies' attention to this nanomaterial. AgNPs have shown antimicrobial activity against a variety of infectious and pathogenic microorganisms, including multidrug-resistant bacteria [18]. Currently, the literature supports principally three mechanisms that have been observed together or separately, by which AgNPs exert their antibacterial action. The first one postulates that AgNPs act at a membrane level as they are able to penetrate the outer membrane, accumulating in the inner membrane where the adhesion of the nanoparticles to the cell generates their destabilization and damage, increasing membrane permeability and inducing leakage of cellular content and subsequently its death [19]. Moreover, there is evidence that AgNPs can interact with sulfur-containing proteins in bacterial cell walls, which may result in structural damage and cell wall rupture. Nanoparticles can break and cross the cell membrane, changing its permeability and structure, similar to how they can damage intracellular machinery, activate the apoptotic pathway, and alter respiration. The second mechanism suggests that nanoparticles can enter cells, where it has been suggested that AgNPs will have an affinity to interact with sulphur or phosphorus groups present in intracellular content such as DNA and proteins, altering their structure and functions. The release of silver ions from the nanoparticles, which because of their size and charge can interact with cellular components and change metabolic pathways, membranes, and even genetic material, is a third process that is hypothesised to happen concurrently with the other two [20].

4. Conclusion

Banana flower extract is used as a reducing agent during the green synthesis process to create silver nanoparticles. Gram positive and gramme negative bacteria were tested against the produced nanoparticles in antimicrobial investigations. Both Gram-negative and Gram-positive bacteria are resistant to AgNPs' antibacterial properties. When exposed to the Gram-positive bacteria S. aureus, both of them displayed their peak activity. AgNPs have several, concurrent modes of action, and when combined with antibiotics or other antibacterial agents such as chemical compounds, they have a synergistic effect on pathogenic bacteria like Escherichia coli and Staphylococcus aureus. Based on the investigation, it is confirmed that both gram-positive and gram-negative bacteria are resistant to the antibacterial action of the produced silver nanoparticles.

Aknowledgement

We thank DST-FIST (Fund for Improvement of S & T infrastructure in Universities and higher Educational Institutions Programme. We thank Management, St. Joseph's College for Women, Alappuzha for the financial support. We thank Dr. Rita Latha D'Couto (Principal), Sr. Barakala Pushpa (HoD, Department of chemistry) and all the faculties in Department of Chemistry for their valuable suggestions and support. Special thanks to Asha Louise for their constant support to complete this research work.

References

[1] Sharma, V. K., Yngard, R. A., & Lin, Y. (2009). Silver nanoparticles: green synthesis and their antimicrobial activities. Advances in colloid and interface science, 145(1-2), 83-96.
[2] Dos Santos, C. A., Seckler, M. M., Ingle, A. P., Gupta, I., Galdiero, S., Galdiero, M.,& Rai, M. (2014). Silver nanoparticles: therapeutical uses, toxicity, and safety issues. Journal of pharmaceutical sciences, 103(7), 1931-1944.
[3] Durán, N., Marcato, P. D., Conti, R. D., Alves, O. L., Costa, F., & Brocchi, M. (2010). Potential use of silver nanoparticles on pathogenic bacteria, their toxicity and possible mechanisms of action. Journal of the Brazilian Chemical Society, 21, 949-959.
[4] Rai, M., Yadav, A., & Gade, A. (2009). Silver nanoparticles as a new generation of antimicrobials. Biotechnology advances, 27(1), 76-83.
[5] Knetsch, M. L., & Koole, L. H. (2011). New strategies in the development of antimicrobial coatings: the example of increasing usage of silver and silver nanoparticles. Polymers, 3(1), 340-366.

[6] Ghramh, H. A., Ibrahim, E. H., Kilnay, M., Ahmad, Z., Alhag, S. K., Khan, K. A., & Asiri, F. M. (2020). Silver nanoparticle production by Ruta graveolens and testing its safety, bioactivity, immune modulation, anticancer, and insecticidal potentials. Bioinorganic Chemistry and Applications, 2020.
[7] Rafique, M., Sadaf, I., Rafique, M. S., & Tahir, M. B. (2017). A review on green synthesis of silver nanoparticles and their applications. Artificial cells, nanomedicine, and biotechnology, 45(7), 1272-1291.
[8] Srikar, S. K., Giri, D. D., Pal, D. B., Mishra, P. K., & Upadhyay, S. N. (2016). Green synthesis of silver nanoparticles: a review. Green and Sustainable Chemistry, 6(1), 34-56.
[9] Ahmad, S., Munir, S., Zeb, N., Ullah, A., Khan, B., Ali, J., & Ali, S. (2019). Green nanotechnology: A review on green synthesis of silver nanoparticles—An ecofriendly approach. International journal of nanomedicine, 5087-5107.
[10] Awwad, A. M., & Salem, N. M. (2012). Green synthesis of silver nanoparticles by Mulberry Leaves Extract. Nanoscience and Nanotechnology, 2(4), 125-128.
[11] Bar, H., Bhui, D. K., Sahoo, G. P., Sarkar, P., De, S. P., & Misra, A. (2009). Green synthesis of silver nanoparticles using latex of Jatropha curcas. Colloids and surfaces A: Physicochemical and engineering aspects, 339(1-3), 134-139.
[12]Abdelghany, T. M., Al-Rajhi, A. M., Al Abboud, M. A., Alawlaqi, M. M., Ganash Magdah, A., Helmy, E. A., & Mabrouk, A. S. (2018). Recent advances in green synthesis of silver nanoparticles and their applications: about future directions. A review. BioNanoScience, 8, 5-16.
[13] Vanlalveni, C., Lallianrawna, S., Biswas, A., Selvaraj, M., Changmai, B., & Rokhum, S. L. (2021). Green synthesis of silver nanoparticles using plant extracts and their antimicrobial activities: A review of recent literature. RSC advances, 11(5), 2804-2837.
[14] Sumathy, V., Lachumy, S. J., Zakaria, Z., & Sasidharan, S. (2011). In vitro bioactivity and phytochemical screening of Musa acuminata flower. Pharmacologyonline, 2, 118-127.
[15]Suganthi S, Kavitha S. Application of banana spathe extracts and compost for improving growth in rice plants. J Appl Biol Biotech, 2022;10(01):112- 119.
[16] Heba I. Mohamed, Jameel M. Al-Khayri (2021). Date palm (Phoenix dactylifera L.) secondary metabolites: Bioactivity and pharmaceutical potential. Phytomedicine. 483-531
[17] Bruna T, Maldonado-Bravo F, Jara P, Caro N. Silver Nanoparticles and Their Antibacterial Applications. Int J Mol Sci. 2021 Jul 4;22(13):7202.
[18] Marambio-Jones C., Hoek E.M.V. A review of the antibacterial effects of silver nanomaterials and potential implications for human health and the environment. J. Nanopart. Res. 2010; 12:1531–1551.
[19] Seong M., Lee D.G. Silver Nanoparticles against Salmonella enterica Serotype Typhimurium: Role of Inner Membrane Dysfunction. Curr. Microbiol. 2017;74:661–670.
[20]Ivask A., Elbadawy A., Kaweeteerawat C., Boren D., Fischer H., Ji Z., Chang C.H., Liu R., Tolaymat T., Telesca D., et al. Toxicity mechanisms in Escherichia coli vary for silver nanoparticles and differ from ionic silver. ACS Nano. 2014;8:374–386.

Modification in Magnetic Properties of $CoFe_2O_4$ Ferrite Nanoparticles Induced by 100MeV O^{+7} Ion Irradiation

Grish Chandra[1], R C Srivastava[2] and V R Reddy[3]

[1]Department of Physics D.S.B. Campus, Kumaun University, Nainital, Uttarakhand, India
[2]Deptt. of Physics, G.B. Pant University of Ag. & Technology, Pantnagar, Uttarakhand, India
[3]UGC-DAE CSR, Khandwa Road, DAVV Campus, Indore 452017, Madhya Pradesh, India

grishphysics@gmail.com

Abstract. In this work $CoFe_2O_4$ ferrite nanoparticles were prepared via sol-gel method using citric acid as chelating agent. The prepared nanoparticles was sintered at 700°C and the crystallite size is found 34 nm. In this investigation the effect of irradiation of 100 MeV O^{+7} beam on these nanoparticles with two fluences 5×10^{12} and 5×10^{13} ions/cm^2 are studied. After irradiation the crystallite size is found to be reduced which is confirmed by transmission electron micrograph. The retentivity and coercivity changes with a measurable value while the saturation magnetization reduces slightly. The value of remanent magnetization increased from 30 emu/g to 34.6 emu/g, it can be attributed to increase of magnetic hardness. The increase in squareness ratio after irradiation leads to increase in anisotropic energy barrier of $CoFe_2O_4$ ferrite nanoparticles. Anisotropy field about doubles after irradiation and corresponding anisotropy constant also doubles with the used fluences.
Key words: irradiation, nanoparticles, magnetization, pelletron, spinel, pristine

1. Introduction

Swift heavy ion irradiation is a tool to produce various types of controlled defects such as point, cluster or columnar defects in material [1]. The ions with energy from fraction of keV to GeV are useful in synthesis, modification and characterization of materials. If the velocity and mass of the ion in the beam is such that $v > v_o Z^{2/3}$ and $M >> m_o$ then ion is called swift heavy ion; where M is the mass of the ion v_o the Bohar velocity (velocity of orbital electron in target atom) and Z is the atomic number of the target atom [2]. The energy loss in the medium is caused mainly by the two processes; one by inelastic collision between ions and electrons (Electronic Stopping) and another by elastic collision with ion and target nuclei (Nuclear Stopping). The electronic energy loss process is dominant at high energies (> 1MeV/nucleon) called threshold energy [3]. This threshold energy varies from material to material beyond which the formation of columnar defect or latent track occurs in the material. The electronic energy loss for swift heavy ions is about two orders of magnitude higher than the nuclear energy loss [4]. The mass and its energy of ions decides the magnitude of the electronic and nuclear energy losses. A swift heavy ion can modify the properties of material such as change in microstructure, increase in electrical resistance; change in magnetic and optical properties etc, this modification is quite different from that of low energy ions. A local phase change could occur, due to the huge amount of energy deposited by swift heavy ions provided that energy is greater than threshold value of electronic stopping power [5]. It has been observed that metallic or insulating materials could be strongly modified by swift heavy ion irradiations [6]. Nanomaterials may be the good candidates to observe the irradiation modifications because the energy density stored in the nanomaterials after the passage of an ion must be very high as compared to the density in the same material in its bulk phase. Consequently the electronic threshold value should be lower than the

IOP Conf. Series: Materials Science and Engineering 1291 (2023) 012011 doi:10.1088/1757-899X/1291/1/012011

corresponding value in the bulk phase. It has been found that the electronic structure of the pristine material is not the main characteristic that explains its behaviour [7,8,9].

Among the transition metal oxides cobalt ferrite is well known hard magnetic material with large magneto-crystalline anisotropy, high coercivity, moderate saturation magnetization and high Curie temperature (~520°C). Chemical stability and high coercivity value makes cobalt ferrite a best material for magnetic recording applications such as audio, video tape and high density digital recording discs. Cobalt ferrite nanoparticles are being used for various biomedical applications such as magnetic resonance imaging magnetic hyperthermia and targeted drug delivery. Cobalt ferrite is an inverse spinel structure with ferric ions equally shared among the tetrahedral A site and octahedral B site while cobalt ions occupy the octahedral site only. The magnetic properties of the cobalt ferrite depends upon types of cations ($Co+^{2}$, Fe^{+3}) and their distributions among the sites. In the present study we reported the swif heavy ions irradiation experiments performed on cobalt ferrite nanoparticles with 100 MeV O^{+7} beams. Investigation on cations distribution between two sites A and B in the crystal lattice and its effect on the magnetic properties of $CoFe_2O_4$ nanoparticles are reported in this article.

2. Experimental

Sol gel technique was used for the preparation of cobalt ferrite nanoparticles. The stoichiometric amounts of $Fe(NO_3)_3.9H_2O$ and $Co(NO_3)_2.6H_2O$ were dissolved in distilled water. In the aqueous salt solution of ferric nitrate and cobalt nitrate citric acid solution was added with the molar ratio 1:3 of cations to citric acid. This solution was heated at 80°C under constant magnetic stirring until the solution got converted into a viscous gel, and was allowed to cool at room temperature. The cooled gel was dried in a furnace at 100°C for overnight to form the precursor material [10,11]. The precursor material was sintered at 700°C for 2 hrs to get the nanoparticles of cobalt ferrite and then grinded to a powder to make particle shape uniform. This fine powder of $CoFe_2O_4$ were irradiated by 100MeV O^{+7} beams with the fluences 5×10^{12} and 5×10^{13} ions/cm^2. The ion accelerator at Inter University Accelerator Centre, New Delhi, India was used for irradiation. After irradiation these samples were characterized for structural and magnetic properties. X-ray diffraction data were collected on Bruker D8 Advanced X-ray diffractometer using Cu K_α radiation (λ=1.54178Å). High-field magnetization measurements at room temperature were carried out with the help of a Quantum Design vibrating sample magnetometer (VSM) (Model 6000, Physical Property Measurement System). Mössbauer spectroscopic measurements were performed using a conventional constant acceleration spectrometer in transmission geometry with ^{57}Co source in Rh matrix at room temperature.

3. Result and discussion

3.1 XRD study

XRD pattern of the cobalt ferrite ($CoFe_2O_4$) nanoparticles irradiated by O^{+7} beam having energy 100MeV is shown in figure 1. Initially this sample was sintered at 700°C and its crystallite size was 34±2 nm. The diffraction peaks are corresponding to the single phase of the ferrite nanoparticles and the phase analysis is done with JCPDS card no. 221086. The diffraction peaks are sharper in the pattern of pristine sample while broader and scattered for irradiated samples. Small deviation in the diffraction angles may be due to some distortion in the unit cell, which indicates that crystallinity is decreasing after irradiation. However, the non appearance of extra peaks reveals that there is no change in cubic spinel structure of ferrite. Measurement of crystallite size was done by Scherer formula [12] using most intense peak (311) i.e.

$$D = \frac{k\lambda}{\beta cos\theta}$$

where λ (1.54Å) is the wavelength of X-ray used, θ is the Bragg's angle, β is the full width at half maximum (FWHM) in radian and 'k' is a constant approximately equal to 0.9. The calculated Crystallite size, x-ray density, lattice parameter and interatomic distances are given table 1. Comparing these results with the results for pristine sample we found that crystallite size is decreasing after irradiation. One cause may be due to breaking of larger particle after striking with high energy ion beam due to which partial amorphization produces in the material [13]. The decrease in crystallite size is almost same for both fluencies. SRIM code (Stopping and Range of Ions in Matter) calculation shows that the electronic stopping $(dE/dx)_e$, Nuclear stopping $(dE/dx)_n$ and projected range of 100Mev O^{+7} ion in $CoFe_2O_4$ are 2.69 keV/nm, 0.0015 keV/nm and 48.74μm respectively. The large projectile range means the ions are implanted much deeper in the substrate and nuclear stopping is negligible. The observe modification is therefore mainly due to electronic stopping induced process. This energy produces different types of defects like cluster/point as well as partial amorphization depending on the energy of ion. Clearly electronic energy loss is about three orders of magnitude larger than nuclear energy loss; therefore the dominant process is the electronic energy loss in the present system. However the threshold energy to create columnar defects is ~13keV/nm [14] indicates that there is no possibility to create columnar defect, but only point defects or defect clusters. The heavy ion irradiation induces strong local distortion and stresses throughout the irradiated region which are radial and isotropic around the spherical defects [8].

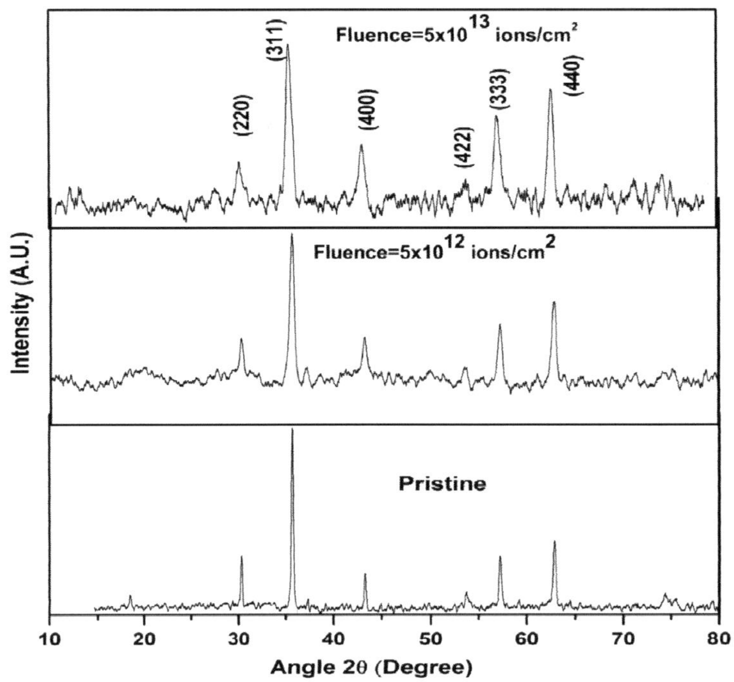

Figure 1. x- ray diffraction pattern

It is observed that the basic crystal structure remains the same after irradiation. However the peak intensities decreased and the widths of the peaks increased slightly. These observations exhibits that the SHI irradiation has generated some defect in the system. The lattice constant and X- ray density are determined by the following relations.

$$a = d_{hkl}(h^2 + k^2 + l^2)^{1/2}$$

where h,k,l are the miller indices of the planes having interplaner spacing 'd_{hkl}'. x-ray density ρ is calculated by using the formula [15]

$$\rho = 8M/(Na^3)\ \text{g/cm}^3$$

where M, N and 'a' are the molecular weight, Avogadro's number and lattice parameter respectively. The factor 8 is due to the fact that there are eight molecules of cobalt ferrite per unit cell. The interatomic distance (hopping length or ionic radius) between magnetic ions at tetrahedral A (L_A) and octahedral B (L_B) sites was calculated by using the following relations [16]

$$L_A = \frac{a\sqrt{3}}{4}, L_B = \frac{a\sqrt{2}}{4}$$

It is clear that after oxygen ion irradiation the hopping length remains almost constant.

Table 1. Structural parameters of samples

Fluences ions/cm^2	Crystallite size (±2 nm)	X-ray density (±0.08 g/cm^3)	a (±0.01 Å)	L_A (Å)	L_B (Å)
Pristine	34	5.29	8.38	3.63	2.96
5×10^{12}	24	5.29	8.38	3.63	2.96
5×10^{13}	23	5.26	8.40	3.64	2.97

The broadening of peaks after irradiation may be due to the effect of increased temperature in the material medium during the passes of oxygen ion beam. The decrease in x-ray density is attributed to increase in the volume of the unit cell. The SHI irradiation of the pristine sample amorphized it, as is clearly indicated by the decrease in line intensity and the increase in background. This result thus supports the applicability of thermal spike model of ion matter interaction.

3.2 TEM Study

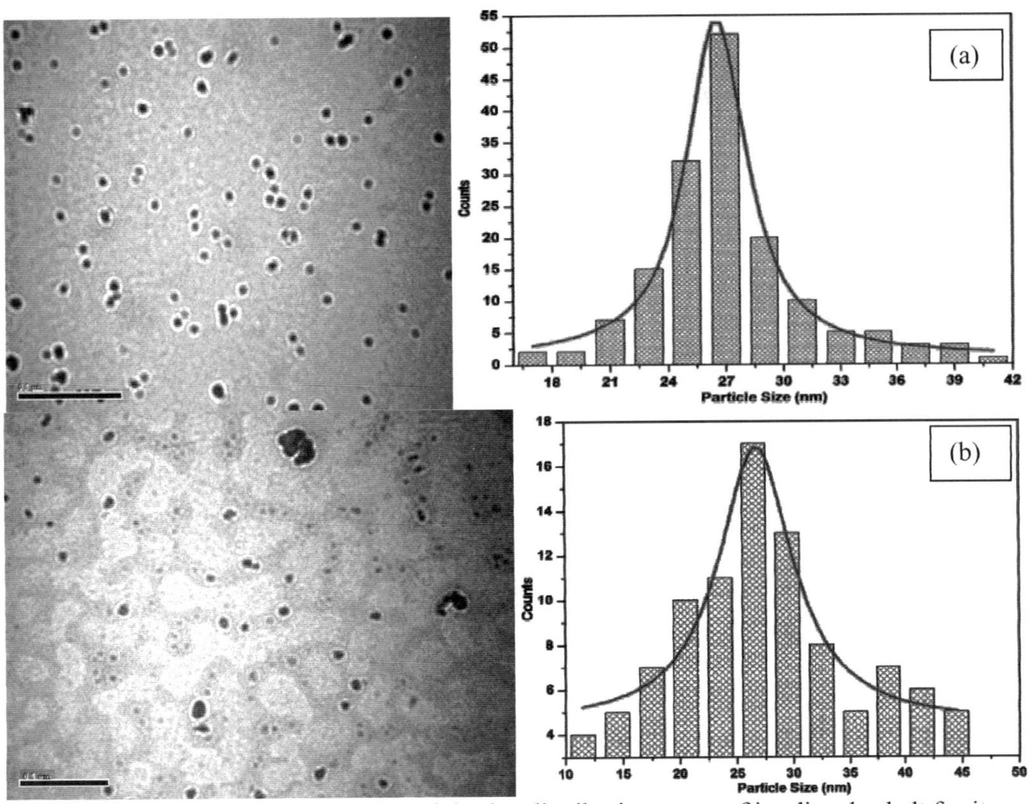

Figure 2. TEM micrograph and Particle size distribution curve of irradiated cobalt ferrite with fluence (a) 5×10^{12} ions/cm^2 and (b) 5×10^{13} ions/cm^2.

Surface morphology and particle size distribution of $CoFe_2O_4$ magnetic nanoparticles was investigated by transmission electron microscopy (TEM). Figure 2 shows the TEM image of cobalt ferrite irradiated with 5×10^{12} ions/cm^2 and 5×10^{13} ions/cm^2 respectively. Their corresponding particle size distribution curves are shown on the right side of TEM images. Measurement of average particle size is done by Lorentzian fit of these histograms and particle sizes were found ~26±1nm.

3.3 VSM Study

Saturation magnetization values are estimated by fitting the high field part of the magnetization curve using the relation [17, 18]

$$M = M_s\left(1 - \frac{a}{H} - \frac{b}{H^2}\right)$$

where H is the field strength, a and b are two parameters called fitting parameters. For high field strength the saturation magnetization varies linearly with 1/H. The plot of M vs 1/H is a straight line whose intercept on M axis gives the value of M_s [19]. The fitted curve is shown in figure 3.

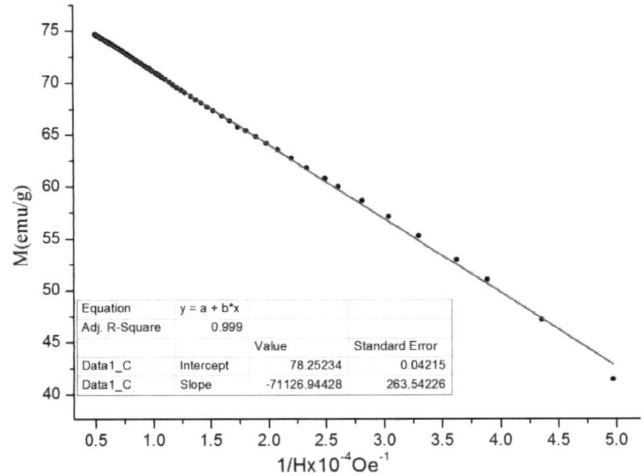

Figure 3. Determination of saturation magnetization using law of approach method.

It is clear from the figure 4 that a measurable change in magnetic parameters saturation magnetization (M_s), remanent magnetization (M_r) and coercivity (H_c) after the irradiation 100 MeV O^{+7} beam are found. The sample shows a large value of coercivity and retentivity at room temperature which is the characteristic of multi domain particles. These observations confirm that both pristine and irradiated nanoparticles are multi domain. VSM parameters for the pristine and irradiated samples are given in table 2. The Saturation magnetization values decreased slightly with fluence and also the remanent magnetization is increased with a substantial amount but theses changes in values of saturation magnetization and remanent magnetization are not unique characteristic of fluence. These are closely related to the microstructure, particle size and residual strain etc. The value of magnetic moment per formula unit (magneton number n_B) was estimated by the following relation [18]

$$n_B = \frac{\text{Saturation Magnetization } \times \text{Molecular weight}}{5585}\ \mu_B$$

In ideal situation when cobalt ferrite sample grows into a pure inverse type spinal structure with all Co^{+2} ions located in the octahedral sub-lattice and all Fe^{+3} ions in the tetrahedral sub-lattice. The magnetic moment of Fe^{+3} cations is fixed to $5\mu_B$ and that of Co^{+2} cations fixed to $3.8\mu_B$ which is corresponds to the saturation magnetization at 0K of bulk cobalt ferrite [20]. A theoretical expression for the net magnetic moment per formula unit can be written as

$$M = |M_B - M_A| = [5 + 3.8] - 5 = 3.8\mu_B$$

where M_A and M_B represents the total magnetic moment of tetrahedral and octahedral sites. Thus saturation magnetization depends upon cation distribution (degree of inversion) which are related to surface spin canting and fine particle size. In present study the observed magnetic moment per formula unit for pristine sample is $3.28\mu_B$ which is less than theoretical value. This signifies the growth of cobalt ferrite crystallites is not pure inverse, but contains some mixed structure. The value of this magnetic moment per formula unit reduces as the sample irradiated by oxygen beam and the value of saturation magnetization also shows reduction in irradiated samples. This reduction in magneton number and saturation magnetization is larger for larger fluence which may be the consequence of disorder produced in the samples due to irradiation.

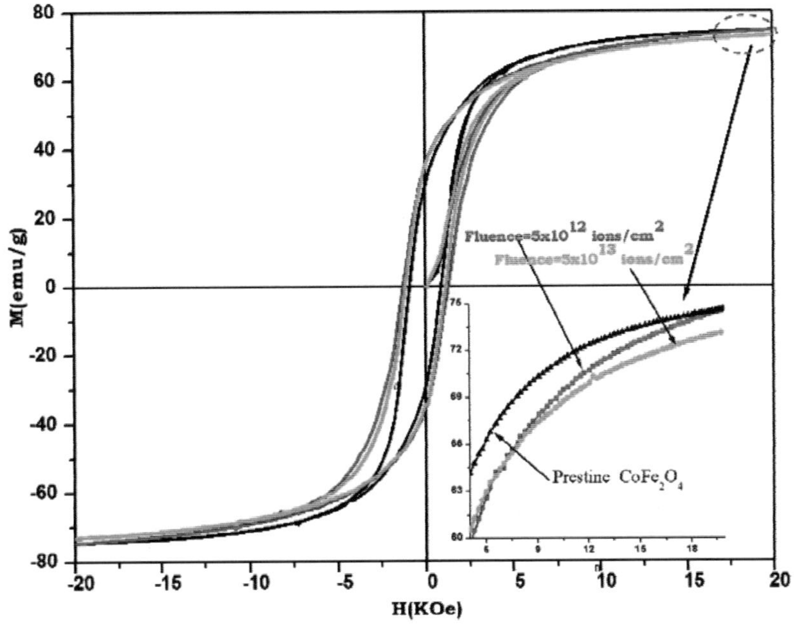

Figure 4. M-H curves for pristine and irradiated cobalt ferrite at different fluencies

Table 2. Magnetic parameters

Fluence Ions/cm^2	M_S(±0.05 emu/g)	M_r(±0.1 emu/g)	H_c (±0.01kOe)	Squareness ratio(M_r/M_s)	H_k (kOe)	K_1 ergs/cc	n_B (μ_B)
Pristine	78.25	30.0	0.93	0.38	5.16	2.02×10^5	3.28
5×10^{12}	77.50	34.8	1.29	0.45	10.17	3.94×10^5	3.26
5×10^{13}	76.90	34.6	1.21	0.45	10.40	3.99×10^5	3.23

As the critical single domain particle size for cobalt ferrite is ~14 nm [21], and in this study the x-ray measurement shows that here particle size is much larger than this single domain limit. In multi domain reason the coercivity of samples is given by a relation $H = a + b/D$ [22], where a and b are the fitting parameters and D is particle size. Thus as the particle size decreases corresponding value of coercivity increased.The reduction in saturation magnetization is smaller than the increase in remanence and coercivity after the irradiation of 100MeV oxygen ions. The squareness ratio for pristine sample, M_r/M_s = 0.38 at room temperature, is smaller than the expected value of randomly oriented isolated particles with uniaxial anisotropy (0.5) and particles with cubic anisotropy (0.8) respectively[23], which implies a tendency towards uniaxial anisotropy. The enhancement in squareness ratio after irradiation leads to increase in anisotropic energy barrier of cobalt ferrite nanoparticles. Anisotropic field is corresponding to the value of field where magnetization and demagnetization curves are bifurcates. The anisotropy constant K_1 of the assembly of particles was calculated using relation $K_1 = M_S H_k/2$ where M_s is saturation magnetization and H_k is anisotropic field [24]. The anisotropy constants for pristine and irradiated samples are given in table 2. Anisotropy energy barrier is defined as K_1V, where V is particle volume. After irradiation of oxygen beam anisotropy constant is increasing and particle volume is decreasing but degree of increment of K1 is higher than decrement in particle volume. Therefore the net value of anisotropy energy barrier K_1V is increasing after irradiation. Actually anisotropy constant is the magnetic energy stored per unit volume of the substance which determines the hardness and softness of magnetic material. The hardness or stability of magnetization depends upon the relaxation time of the assembly of particles which is defined as $\tau = \tau_o exp(K_1V/k_BT)$, where τ is relaxation time, τ_o is relaxation time of non-interacting particles and k_BT is thermal energy. High retentivity and high coercivity corresponds to large relaxation time and at constant temperature relaxation time is strictly dependent on the quantity K_1V which is increasing after irradiation. Therefore the value of relaxation time is increasing after irradiation resulting the enhancement in retentivity and coercivity without affecting much largely the saturation magnetization.

3.4. Mössbauer Study

The Mössbauer spectra were taken using a conventional constant acceleration spectrometer in transmission geometry with a source of about 25 mCi ^{57}Co in an Rh matrix. All three spectra taken at room temperature are shown in figure 5, here the dots represent the data points and the solid line through the data is least square fit. The spectral component A is the contribution from the tetrahedral ferric ions, and the B spectrum is the contribution from the octahedral ferric ions [25]. The spectra are analyzed using least square fitting program NORMOS (SITE/DIST)[26]. Each pattern consists of two normal Zeeman- split sextets; one corresponding to Fe^{+3} at tetrahedral site (**A** site) and another corresponding to Fe^{+3} at octahedral (**B** site) which is the characteristic of ferrimagnetic substances [24]. Evidently, the spectra revels that material is magnetically ordered and there is no change in the ferrimagnetic nature of cobalt ferrite after the irradiation with the used fluencies. From the area ratio r $=A_A/A_B$ of the subspectra A and B the parameter x = (1-r)/(1+r) can be calculated. Assuming that recoil free fraction at room temperature $f_B/f_A=1$, the parameter 'x' gives the Co^{+2} occupation at the A site and hence the deviation from inverse spinal structure may be determined [27]. The fitted Mössbauer parameters are given in the table 3. The value of isomer shift is often associated with particular oxidation state of Mössbauer atom. In this study the isomer shift values are with respect to natural iron foil. Here we found all the isomer shift values less than one (< 1mm/s) which signifies the iron atoms present are in Fe^{+3} state [9]. In powder sample, where the axes of quantization are randomly oriented with respect to gamma ray beam the resulting intensity ratio spectral lines are 3:2:1:1:2:3 for magnetically split sextet and 1:1 for the quadrupole split doublet [22]. In this study the pristine sample exhibits 12 lines which correspond to two sextets. For each sextet the intensities of the spectral lines 1, 2, 3, 4, 5 and 6 are 3, 2, 1, 1, 2 and 3 respectively. On comparing the parameters of pristine sample with irradiated samples we found that the isomer shift and quadrupole splitting values changes slightly after irradiation but there is no explicit pattern with fluence is observed. Cation distribution is calculated using the area ratio of A and B subspectra. In case of pristine $CoFe_2O_4$ sample cation distribution was $(Co_{0.43}Fe_{0.57})[Co_{0.57}Fe_{1.43}]O_4$. After irradiation its cation distribution changes significantly, and becomes $(Co_{0.40}Fe_{0.60})[Co_{0.60}Fe_{1.40}]O_4$ and $(Co_{0.40}Fe_{0.60})[Co_{0.60}Fe_{1.40}]O_4$ for the fluences 5×10^{12} and 5×10^{13} ions/cm^2 respectively. This isotropic distribution of metallic cations between A and B sites induces the ferrimagnetic properties, since the exchange integral J_{AB} between sites is much larger than the exchange integral J_{BB} within the B sites [24,28]. Fraction of concentration of *Fe* at A site is increasing from 0.57 to 0.59 and 0.60 after irradiation which is 2% and 3% migration of ferric ions from octahedral site to tetrahedral site. This reduction of ferric ions at octahedral site may reduce the magnetic moment per formula unit M_B-M_A which is being confirmed by VSM measurements, where we found the reduced values of saturation magnetization after irradiation.

The values of magneton numbers calculated via Mössbauer measurements are not similar as the values obtained from VSM studies and this discrepancy may be due to the spin canting at B site. Spin canting is defined as a lack of full alignment of spins; the canting angle Ø can be calculated using the assumption $M = M_B cos\emptyset - M_A$. The expected values of canting angles are found 33.15° and 33.37° after the irradiation of the fluences 5×10^{12} and 5×10^{13} ions/cm^2 respectively, while the value of canting angle for pristine sample was 33.63°[11]. Hyperfine fields at A and B site for pristine sample are ~51.18T and ~48.80T, after irradiation these values changes to ~51.34T, ~48.46T and ~51.56T, ~48.64T respectively. A non collinear structure should be visible in Mössbauer spectra by way of the intensities of lines 2 and 5. A hyperfine field parallel to the direction of gamma ray corresponds to zero intensity for spectral lines 2 and 5 [22]. In this study the lines 2 and 5 exhibit sufficient intensity in all irradiated as well as pristine samples which may be in agreement with a canted sub lattice and gamma ray is not parallel to the hyperfine field.

Table 3. Mössbauer parameters

Cation Distribution	Site	IS (± 0.02 mm/s)	QS (mm/s)	LW (±0.02 mm/s)	B_{hf} (± 0.10T)	% area	M_B-M_A (μ_B)
$(Co_{0.43}Fe_{0.57})[Co_{0.57}Fe_{1.43}]O_4$ Pristine	B	-0.27	0	0.53	48.80	71.85	4.85
	A	-0.37	0	0.39	51.18	28.15	
$(Co_{0.41}Fe_{0.59})[Co_{0.59}Fe_{1.41}]O_4$ Fluence = 5×10^{12} ions/cm^2	B	-0.33	0.05	0.61	48.46	70.53	4.78
	A	-0.33	-0.02	0.48	51.34	29.47	
$(Co_{0.40}Fe_{0.60})[Co_{0.60}Fe_{1.40}]O_4$ Fluence = 5×10^{13} ions/cm^2	B	-0.30	0.06	0.58	48.44	70.14	4.76
	A	-0.37	0	0.46	51.56	29.86	

Note: IS-Isomer Shift; QS-Quadrupole Splitting; LW - Line Width; B_{hf}-Hyperfine Field

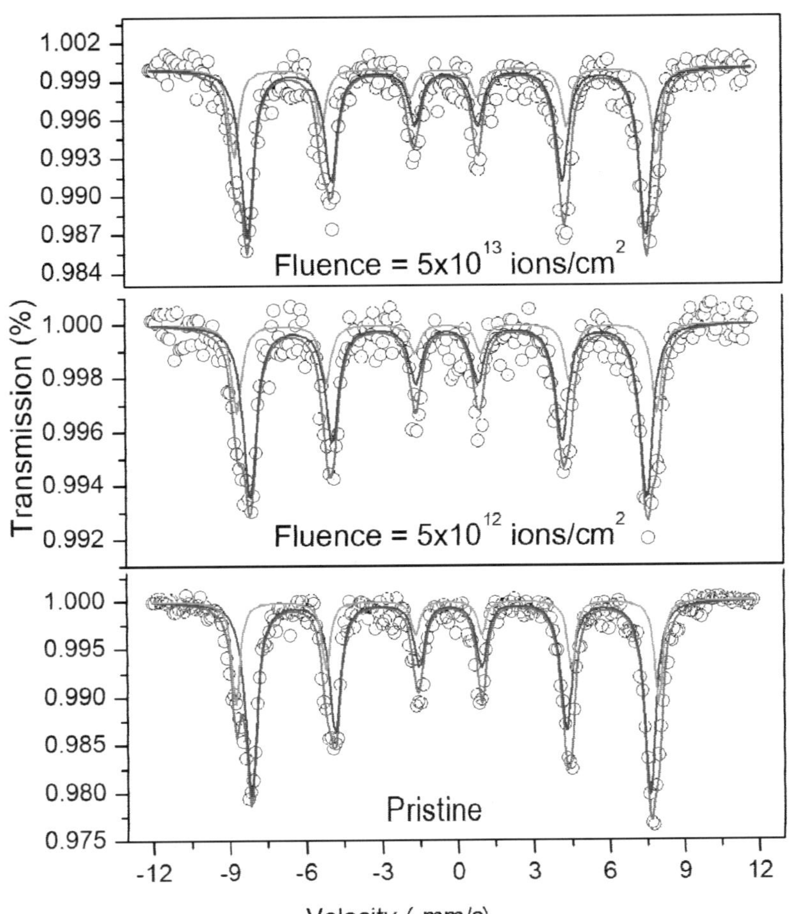

Figure 5. Mössbauer spectra of pristine and irradiated cobalt ferrite nanoparticles.

Under ion irradiation ferric ions undergo displacement from octahedral to tetrahedral site simultaneously same amount of Co^{+2} ions moves from tetrahedral to octahedral site. Taking into account the low value of magnetisation measured in the irradiated samples, and it is obvious that these displacements are stable and involve large concentration of iron atoms at tetrahedral site. This outcome of Mössbauer measurement verifies the results of our above VSM study.

4. Conclusion

In this study we discussed the heavy ion irradiation modifications on the nanoparticles of cobalt ferrite prepared using sol gel method. These nanoparticles were sintered at 700°C for two hours and cooled slowly to room temperature. The pristine sample was irradiated by 100 MeV O^{+7} beam with two different fluencies 5×10^{12} and 5×10^{13} ions/cm^2. X-ray diffraction pattern shows that all three samples are purely crystalline but irradiated sample has scattered background and broader peaks, which is because of reduction in particle size after irradiation. Particle size directly measured by TEM is closed to the size calculated using Scherer's formula. After irradiation the saturation magnetization and magneton numbers reduced very slightly while the coercivity and remanence increased significantly. Mössbauer spectra shows that after irradiation the migration of Fe^{+3} ions undergoes from octahedral to tetrahedral sites which reduces the magnetization of octahedral site consequently the difference of magnetization of both the sites becomes smaller. The non-zero intensity of spectral lines 2 and 5 exhibits the possibility that hyperfine field at B site is not parallel to the gamma ray.

5. References

[1] Thompson M. W., *Defects and Radiation Damage in Metals*, Cambridge University Press, Cambridge 1969.
[2] D. Kanjilal, *Current Science*, Vol 80 No.12, 25 June 2001.
[3] D.K. Avasthi, *Defence Science Journal,* Vol 59, No.4, July 2009, pp 401-412.
[4] Emile Knystautas 2005CRC Press, *Engineering thin films and Nanostructures with Ion Beams*.
[5] A. V. Krasheninnikov et al, *Journal of Applied Physics* 107, 071301 (2010).
[6] E. Balanzat and S. Bouffard, *Solid State Phenomena*, pp30-31, (1993).
[7] E. Balanzat, Radiat. *Effect Defects in Solids*, 126(1993)97.
[8] Meillon, S. F. Studer, M. Hervieu, H. Pascard. *Nuclear Instruments and Methods in Physics Research B*, (1996) 363-367.
[9] F. Studer, Ch. Houpert, D. Groult, J. Yun. Fan., A. Meftah and M. Toulemonde, *Nuclear Instruments and Methods in Physics Research* B 82, (1993) 91-102.
[10] Gangothri S. Kumar, T. Raguram, K. S. Rajni, *Journal of Superconductivity and Novel magneitism*, (2018) https://doi.org/10.1007/s10948-018-4867-5.
[11] Grish Chandra, R.C. Srivastava, V. R. Reddy and H. M. Agrawal; *J. Magn. Magn. Mat.* 427(2017) 225-229.
[12] H. Kaur, A. Singh, V. Kumar, D.S. Ahlawat, *J. Magn. Magn. Mat.* 474(2019) 505-511.
[13] Rao, B. P.; Rao, K. H.; Rao, P.S.V.S.; Kumar, A. M.; Murthy, Y. L. N.; Ashokan, K.; Kumar V.V.S.; Kumar R.; Gajbhiye, N. S. and Caltum, O.F. 2006. *Nucl. Instr. Meth. Phys. Res. B* 244: 27- 30.
[14] Ravi Kumar, S. K. Sharma, Anjana Dogra, V.V. Shiva Kumar, S.N. Dolia, A. Gupta, M. Knoble and M. Singh, *Hyperfine Interactions* (2005)160:143-156.
[15] M.V. Chaudhary, S.E. Shirsath, A.B. Kadam, R.H. Kadam, S.B. Shelke, B.R. Mane, J. Alloys Compd. 552(2013)145-151.
[16] Swati Tapdiya, Ashwini K. Shrivastava, Sarika Singh, *Advanced Material Proceeding,* 2017, 2(9), 547-551.
[17] Richard J. Harrison, Andrew Putnis, *American Mineralogist* Vol. 80, pages 213-221, 1995.
[18] Suman, N. D. Sharma, *Indian Journal of Pure and Applied Physics*, Vol.15 June 2007, pp. 549-554.
[19] R. N. Panda, J.C. Shih, T. S. Chin, *J. Magn. Magn. Mat.* 257(2003)79-86.
[20] L. Ben Tahar, M Artus, S. Ammar, L.M. Smiri, F. Herbst, M. J. Vaulay, V. Richard, M. Greneche, F.Villain, F. Fievet, *J. Magn. Magn. Mater.* 320(2008) 3242.

[21] Pawan Kumar, S.K. Sharma, M. Knoble and M. Singh, *Journal of Alloys and compounds* 508(2010) 115- 118.
[22] Dominic P.E. Dickson, Frank J. Berry, *Mössbauer Spectroscopy, Cambridge University Press*, 1986.
[23] Kamlesh V Chandekar, K. Mohan Kant, *Advanced Material Letters,* 2017,8(4), 435-443.
[24] Cullity, B. D. 1972. *Introduction to Magnetic Materials* (Addision-Wesley, Newyork).
[25] H. Topsøe, J.A.Dumesic and Boudart, *Journal De Physique*, Colloque C6, Supplement au n° 12, Tome 35, December 1974, page C6-411.
[26] Brand R.A., Lauer J. and Herlach D. M., *J. Phys.,* F14 (1984), 55.
[27] V. Rusanav, V. Gushterov, S. Nikolov, A.X. Trautwein. *Hyperfine Interactions* 191(2009) 67-74.
[28] Kittle C. *Introduction to Solid State Physics*. John Wiley & Sons, 1986.

Acknowledgement
The authors are thankful to Nuclear Science Centre New Delhi, India, for providing experimental facilities in the Beam Line to perform the irradiation experiment. Authors are also thankful to Dr. Alok Banerjee Scientist UGC DAE CSR Indore India for providing VSM measurement facility.

Molybdenum powder coating on steel substrate to study three body wear resistance

Darshan Umashankar, Thakshak D S, Deekshith J Nayaka, Shivakumar R Hiremath, Suhas R D[1],

Department of Mechanical engineering, PES University, Bangalore, Karnataka, India

E-mail: darshdb15@gmail.com

Abstract. Steel rollers are most commonly used in conveyor system across different industries like miningand construction. In the conveyor rollers used in mining industry, the steel specimen of the roller is in continuous contact with another metal of the same or different material rolling and sliding on that surface. In this motion of the roller, the third body i.e., the sand, grits and dust of various sizes may be entrapped in the interface of the roller material. The third body induced causes the wear due to its abrading action. To reduce this abrading action, a layer of molybdenum coating is developed on the steel substrate to improve its wear resistance by using the plasma spray technique, a thermal spray method. The experimental procedures were performed according to standard ASTM G65 method. The results were compared and tabulatedin full factorial method. The SEM and EDAX results were obtained. The average specific wear rate [Ks] of the coated specimen is reduced by 30% when compared with the uncoatedspecimen.

Keywords: ASTM G65, SS304, three body wear, Molybdenum coating, Plasma spray process, SEM, EDAX.

1. Introduction.

Wear is the gradual removal or deformation of material at solid surfaces. Causes of wear can be mechanical (e.g., erosion) or chemical (e.g., corrosion). The study of wear and related processes is referred to as tribology. Abrasion wear is one of the most common failures observed in many industrial applications. The worn-out material from the softer material surface is called the wear debris. Based on the movement of the wear debris formed during the abrasive wear, the abrasive wear can be divided into two groups namely:1) two body abrasive wear and three body abrasive wear.

Our primary focus is on three body abrasion wear. Certain applications like conveyor roller operation, the steel specimen is in continuous contact with another material which results in wear. Molybdenum is widely used material in the thermal spray industries for developing coating due to its excellent wear and scuff resistance properties. Mo has a higher hardness and can be used as a coating material to reduce the friction considerably and helps in reducing wear. This project focuses on developing Molybdenum coating on the SS304 steel specimen and study its behavior towards three body abrasion wear. The ASTM G65, the rubber wheel abrasion test is the most popular method for estimating the three- body abrasion wear of components.

2. Literature Review and Objectives.

C.S. Ramesh et.al:[1] Authors work include mild steel specimen coated with Mo and Mo-Si carbide composite using HVOF technique. The authors used SEM and EDAX studies for the microstructure. The authors got a result of improvement in microhardness and reduction in porosity. The Mo coated specimen showed 8% increase in hardness whereas the composite coated specimen showed 23% of Improvement in hardness. They came to a conclusion that the average thickness of the Mo coating was of range of 15μm to 25μm and Mo-Sic was of range 15μm and 20μm.

Manjunath Patel G C et.al:[2] In this research work Mo-Ni-Cr was used as coating material on super duplex stainless steel for minimum loss in wear. The factors used for microhardness were current, powder feed rate and standoff distance of the plasmin spray coating. Taguchi method was used to betteroptimization of microhardness. The results that the author got were that current was the major factor that showed effect on microhardness, then comes the factors such a powder feed rate and standoff distance. The optimized plasma spray condition showed microhardness of 764.33 which was 2.78 times higher than that of super duplex stainless steel. The results of wear loss of uncoated specimen and optimized coated specimen was found out to be 18mg and 2.8mg respectively.
Byoungchul Hwang et.al:[3] Authors have discussed the five types of spray powder were pure molybdenum powder and the others were, blends of brass bronze and aluminium alloy powder with Mopowder were deposited on a low carbon steel substrate by atmospheric plasma spray (APE). The wear test concluded that blend coating showed better wear resistance than No coating. The reason for that being blend coating contained hard phases. The Mo blend coating exhibited a lower hardness a significantly lower wear rate in both the of coating and the counterpart material then pure MO coating. The wear rate of coating and counterpart material showed an increase with increase in load but the friction coefficient decreased. The Mo blend coating with bronze and Al-Si powder coating showed better hardness and resistance than the other blend coatings and its counterparts wear rate were lower.
Cristian Puscasu et.al:[4] In this research thermal sprayed Mo coating were deposited on Steel support by the use of both electric and atmospheric plasma spray method and was investigated. Microstructural analysis showed that in both the methods plates formed by rapid solidification. In this experiment methodology the assessment of surface roughness, hardness and bond strength properties of Mo coating on a steel substrate were done. The conclusion from the experiment were all properties such as surface roughest, hardness and bond strength were influenced by the splat size. lower the size of splat, lower the surface roughness and higher the hardness and bond strength of the coating.
H Adarsha et.al:[5] In this research Stainless Steel304 and low carbon Steel A36 were coated with Molybdenum using HVOF spray technique. The coated and uncoated sample went through an abrasive test using pin-on disc apertures and SEM and EDAX were used for microstructure analysis. The results authors obtained was that uncoated specimen has higher wear rate whereas Molybdenum Coated specimen had 50% less wear rate, they also came to conclusion that because of the coating hardness of the material also got increased with decrease in wear rate. Lang Huang et.al:[6] here Author discussed about the wear behavior of the developed coating on the steel substrate, rubber wheel abrasive taste (ASTM G65) was performed and the results were compared with the standard NM 500. The wear mechanism was of surface fatigue and formation of pits compared to the deep. R Riastuti et.al: [7] here the author has compared the mechanical properties and corrosion resistance of molybdenum coating and aluminium coating, base material used here is stainless steel 316L.This research applied two thermal spray method, namely high velocity oxygen fuel (HVOF) for molybdenum coating and electric arc spray for aluminium coating. Several test methods such as PMI using X ray beam principle to identify the coating composition. The metallographical test using SEM method to identify the corrosion resistance, the wear test using the standard ASTM G99 Ogashi machine and hardness test using Vickerhardness method to characterize the mechanical properties of coating, salt spray method test to identify the corrosion resistance. Based on this research it is concluded that the Mo coating and Al coating on stainless steel 316L acts as an insulation to protect the substrate, but comparing both the coatings Mo coating has higher hardness value than Al layer due to the present of Mo oxide on the surface of Mo coating. Stainless steel coated by Mo has higher wear resistance than 316L than Al coating because Mocoating has smaller value of abrasive than Al coating.
Sivakumaran Ilaiyavel et.al : [8] The Authors have discussed about the industrial coating to reduce wear, friction behavior of D2 steel. D2 was use as the base material and the coating used was Manganese Phosphate with Molybdenum disulphide (MoS_2). They performed the standard pin on disk wear test asperASTMG99 standards and the results were compared with the uncoated material. Wear loss of Manganese Phosphate coated withMoS2 is lesser than only Manganese Phosphate coated pins because of very less Coefficient of friction that was exhibited. From the overall tribology study, the wear behavior of the coated substrate was improved and the coefficient of friction was reduced.

Alexandru Paraschiv et.al: [9] The author as investigated the thermal coated steel substrate using electric arc, atmospheric plasma spray and high velocity oxy-fuel spray. Mo coatings deposited by three thermalspray techniques such SEA, APS and HVOF methods are investigated by tensile bond strength and Vickers indentation tests in order to evaluate and correlate the two of most important mechanical properties of the thermal sprayed coatings such as, cohesion strength and fracture toughness. They concluded that the tensile bonding tests indicate that the Mo coatings deposited by HVOF (HV) method had the highest value of bonding strength (43.3 MPa), followed by APS method (26.7 MPa) and by EA method (19.2 MPa). L. Bourithis et.al: [10] The author investigated the wear performance of four different coatings applied on a low carbon steel substrate by the plasma transferred arc technique was investigated under low stress abrasion conditions, using the dry sand/rubber wheel apparatus. The coatings belonged to two major categories: two of them were tool steels with very hard particles (respectively, TiC and M2C/M2C carbides), the other two are boride coatings(belonging to the Fe–B and Fe–Cr–B systems, respectively) It was said that the wear rate of all coatings increases linearly withthe applied load. However, above a certain load the wear rate of the coating belonging with the Fe–B system increases, mainly due to its brittleness associated to the presence of FeB type bodies. The wear rate of the other three coatings diminishes above a critical load.

The proposed project is on attempt to evaluate the performance of the coated steel substrate in the wear environment and its resistance towards different grainsizes. Main objectives of this project are:

1) To evaluate the performance of the Molybdenum coated SS304.
2) To compare performance of the uncoated and coated system subjected towear environment.
3) To provide optimum range of Molybdenum coating on the substrate.

3. Methodology

3.1. Material procurement.

The following material are selected for the experiment and analysis based on the literature review. The below list follows the important aspects of the material involved. 1.Base material -SS304 SS304 is an austenitic T300 series steel, it is a chromium-nickel austenitic alloy with the nickel content around (2-10%), chromium content of minimum 18% and max of 0.08% carbon. It has variety of commercial application due to its properties like high tensile strength, corrosion resistance, high temperature resistance. The material required for the experimentation were procured from Sri Durga sales, Bengaluru. The dimension of the procured material was 75*25 scale in mm with 8mm thick. A steel strip with 25mm*8mm were cut into required dimension of each equal to length 75mm using the Bandsaw-Double column machine. The cut specimen was further subjected to hand grinding to get flat surface across the edges to ensure proper fit into the specimen holder.

Figure 1. dimension specification of specimen as mentioned in the above paragraph.

3.2. Experimentation

The standard test method used for evaluating the 3-body abrasion is the dry sand-rubber wheel abrasion test according to the standard ASTM G65.

Figure 2. Actual experimental setup of rubber abrasion tester as per standard ASTMG65.

The abrasive grains are introduced between the test specimen and rotating wheel made of chlorobutyle rubber, the standard test specimen (75*25*8) is placed and mounted on the clamp such that the rubber wheel is pressed against the it at a specified force by a mean of the lever arm. The characteristics of the setup is shown in chart below ,

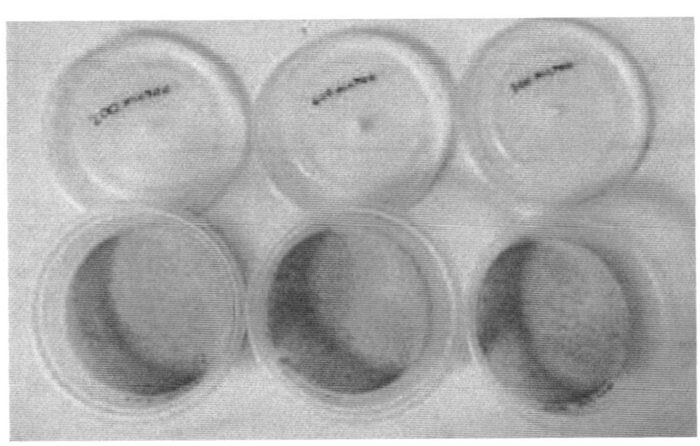

Figure 3. Images of the various sand grains used during experimentation, 200 grain, 400 grain and 600grain.

IOP Conf. Series: Materials Science and Engineering 1291 (2023) 012012 doi:10.1088/1757-899X/1291/1/012012

he characteristics of the setup is shown in chart below:

Table 1. The above table shows the various characteristics of the apparatus used for experiment. Thedata present in the table is used in calculations of various properties.

Test parameter	Details
Specimen size	75*25*8[mm]
Wheel	Chlorobutyl rubber A-60, diameter 228.6*32.7mm
Sand	AFS50/70, Quartz sand
Sand Flow rate	350 gms/min
Pre-load on rubber wheel	2.62 kg
Loading lever ratio	1:2.42
Hopper capacity	15 kg

The loading lever ratio and specimen size values used in calculations are derived from the chart.

3.3 Process Parameters

The flow rate of grain is controlled to abrade the test surface. The rubber wheel rotates in the direction of the sand flow, the experimentation has been done on the based on full factorial method. The parameter which was taken into account during the experimentation are as follows

1)speed/sliding distance

2)grain size

3) Load

The parameter taken were as follow Speed:

- 50rpm
- 100rpm
- 150 rpm

Time(t) is taken as constant. Each specimen was subjected to abrading action for a duration of 5 minutes. (t=5 minutes). The sliding distance for 3 different speeds thus becomes:

For 1 revolution of the rubber wheel, the sliding distance on the specimen will equal to:

Sliding distance= pi (^) * D, where Dis the diameter of the rubber wheel(D=228.6mm),

= 3.1428 * 228.6= 718.44mm

- for 50 rpm, total revolutions = 250 rev.
  Sliding distance=250 * 718.44 = 179.61m.
- for 100 rpm, total revolutions = 500 rev
  Sliding distance = 500* 718.44 = 259.22m.
- for 150 rpm, total revolutions = 750 rev
  Sliding distance = 750 * 718 = 538.83m

The different grain sizes chosen for the experiment:
1) 212 um ------ fine sand
2) 300 um ------- medium sand
3) 600 um -------- coarse sand.

Since in the open environment, grit size ranges from very fine to heavy coarse grains, the above three parameters were considered for experimentation which gives the behavior in various ranges. The load parameters:
1) self-weight of the load arm= 2.62 kg
2) self-weight of the load arm along with 1 kg additional load
3)self-weight of the load arm along with 2 kg additional load.

The loading lever ratio of the abrasion tester is 1:2.41 with initial weight of 2.62 kg. Load on the rubber wheel = [(dead weight * loading lever ratio) + initial weight]. Therefore, the three parameters result in load of 1)25.702N 2)49.344N 3)72.986N .
Will be effective in studying the load behavior of the specimen as in its application purpose where in the roller system, a series of rollers will be present which results in a distributed loading network and thus the load on roller will not be constant but will be varying with respect to the load and the material carried by the roller. The parameter taken above results in developing a full factorial experimental design with 27 tests being performed on both the coated and uncoated specimen.

3.4 Experimental Procedures.

The test was performed based on the setup in the previous section. The tabular column of the experiment design is as below:

No Trials	Load(kg)	Speed(rpm)	Initial weight(g)	Final weight(g)	Mass Loss(g)	Volume loss(mm^3)

Figure 4. the tabular column designed for taking the readings during experiment

The specimen is weighted before and after the test, the specimen were cleaned, dried and weighted using electro weighting scale (model with accuracy of 0.001g weight) The difference in the weight before and after test given the mass loss, it is necessary to convert mean loss to volume loss in cubic millimeters as abrasion is reported as volume loss herespecified procedure.
The calculation procedure is defined below: Mass loss(g) = initial weight(g) – final weight(g)

Volume loss(mm^3) = [mass loss(g) * 1000 / density(g/cm^3)]
Specific wear rate, Ks = [(Volume loss) / (Load * abrading distance)]
The following calculation procedure are performed on each of the individual data obtained after the test procedure of both the uncoated and the coated specimen and their characteristics are compared.

3.5 SEM and EDAX of coated substrate:

The coated substrate characteristics such a coating composition, constituents of elements, the coating thickness, bonding between the coating and substrate. All these characteristics can be defined by the SEM & EDAX analysis. The coated specimen is cut into required dimension to fit into the of equipment which performs the EDAX. In test was performed on the TESCAN VEGA3 machine at BMS college of engineering, Bangalore. The sample size of 20mm*10mm were cut from the available specimen size by wire EDM process. Wire EDM process is defined as a wire electrode which is electrically charged, so that when it comes alone to a conductive material, electricity jumps across it, which results in a spherodising the material [SS304]. The cut specimen is then finally polished using Emery paper of different grades. First it was rubbed against rough finish emery paper of 500 grit.

Then polished against same rough of 800 grits. The final finish must be highly smooth and polished, so it was finally rubbed against 1500 grit paper.

4. Results and Discussions.

4.1. SEM and EDAX results

Figure 5. SEM image of coated sample at 500x.

Figure 6. SEM image of coated sample at 1000x

The inference that can be obtained from the SEM image is that there are no cracks formed on the surface at 500X (magnifications) and 1000X (magnifications). Hence it clearly indicates that there is no stress generated and no cracks are formed after coating.

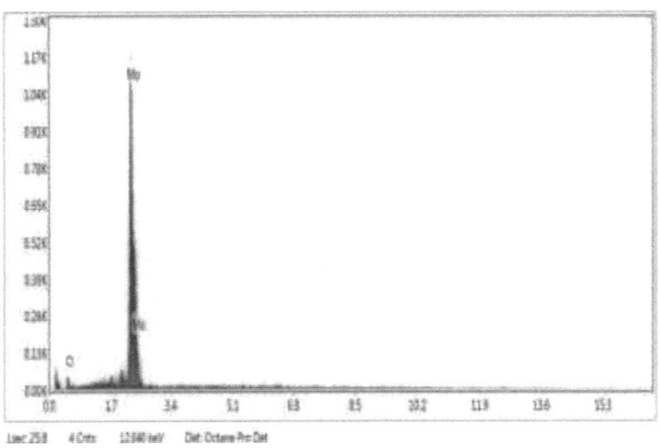

Figure 7. EDAX spectrum of the coated sample

The EDAX spectrum obtained confirms the presence of Molybdenum as the coating material. Oxygen presence is due the high temperature during the plasma process where molybdenum may have oxidized. The atomic and weight percentage of the elements obtained through EDAX analysis is as shown in table

IOP Conf. Series: Materials Science and Engineering 1291 (2023) 012012 doi:10.1088/1757-899X/1291/1/012012

Table 2. Atomic and Weight Percentage of Elements in Coated Sample

Element	Weight %	Atomic %
Mo L	89.28	58.13
O K	10.72	41.87

4.2Coating Thickness Measurement:

To measure the length of the coating thickness, the polished specimen was kept in a cross-sectional manner on the TESCAN VEGA 3 machine. The thickness is measured using the scale on the microstructural image. Five different coating thickness values along the length of coated specimen was measured. The average coating thickness of Mo was found to be 110 microns as shown in figure.

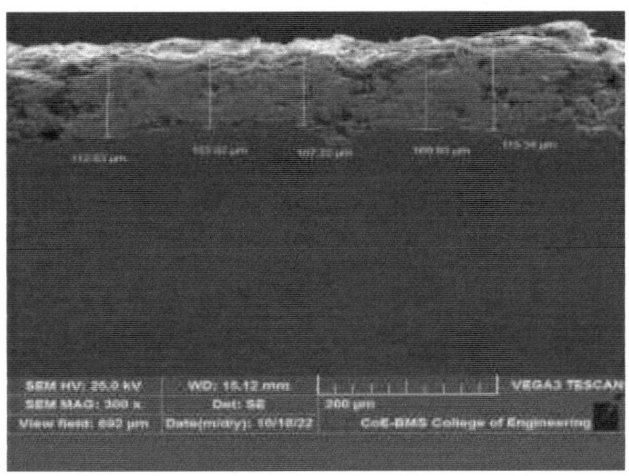

Figure 8. Coating thickness measured using SEM

4.3. Graphical Representation of results:

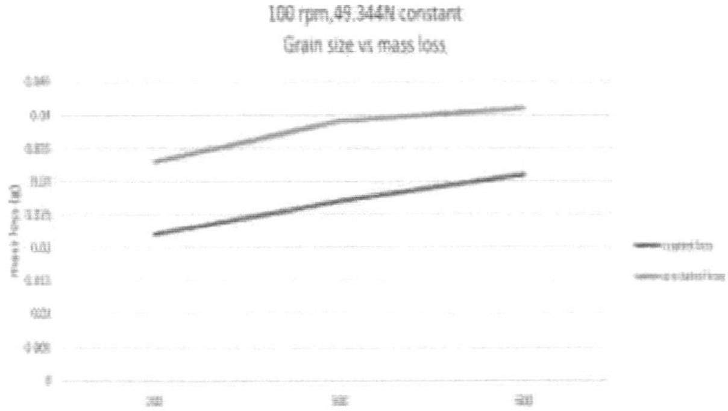

Figure 9. Graphical comparison of coated and uncoated specimen wear with respect to grain size.

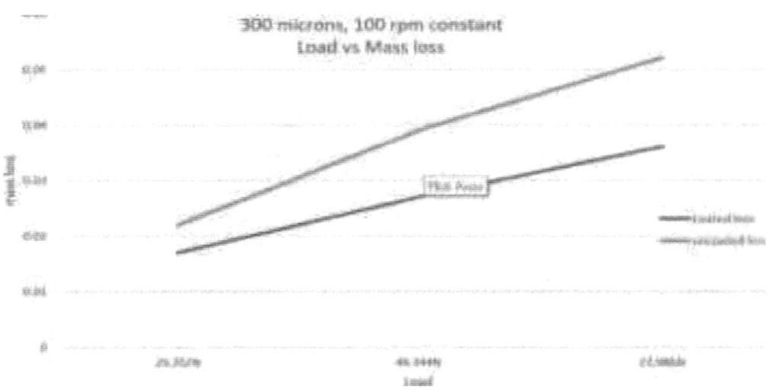

Figure 10. Graphical comparison of coated and uncoated wear with respect to varying load.

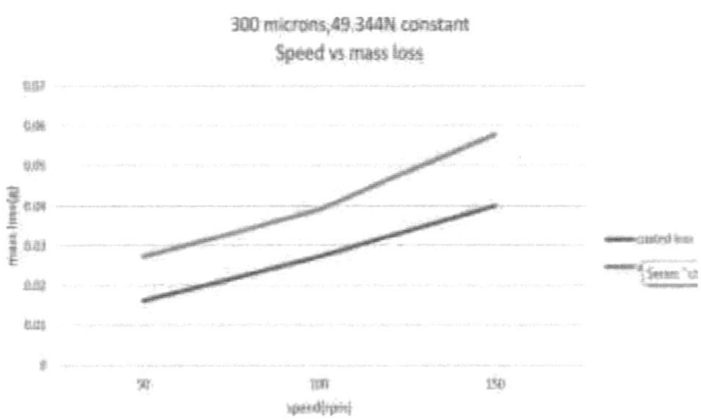

Figure11 : Graphical comparison of coated and uncoated wear with respect to varying speed.

The above graphs are the comparison of mass loss at mid- range characteristics i.e., 50 rpm speed, 300 microns sand grains and 49.344N load. The variation of mass loss with respect to grain size and load showed similar line curves. The curve with respect to speed is seen increasing linearly with increased levels whereas vs Load, it had a linear increase till the middle range load and then the curve is seen converging slightly towards the coated curve. Though the difference is significant, it can be said that the coated specimen's resistance towards speed improved much better than compared to the grain size and load at higher levels. The specific wear rate of the coated specimen when compared to the uncoated specimen at the same levels of operation had decreased which signifies lesser volume loss per unit load and per unit abrading distance. The specific wear rate [Ks] are obtained from the abovetest parameters and levels and the overall improvement in the coated specimen was found to be 30%.

5. Conclusion

The coated specimen improved its wear resistance compared to the uncoated specimen and the variation was linear with increasing value of parameters i.e., the load on the specimen, rotating speed of the rubber wheel and the grain size. The graphical comparison of results implied that the trend variation was linear and the coated substrate performance towards wear improved compared to uncoated substrate. The specific wear rate calculation obtained after calculating the volume loss showed that the coated specimen improved its wear rate by around 30% compared to the uncoated specimen.

REFERENCES

[1].CS Ramesh, Adarsha H, Nikitha Nair, 2018, “Investigations on the effect of molybdenum (Mo) and molybdenum silicon carbide (Mo- 10%SiC) composite coating on mild steel substrate using HVOF technique.

[2]. Manjunath Patel GC, Pradeep N B, Girisha L,2020, “Experimental analysis and optimization of plasma spray parameters on micro hardness and wear loss of Mo-Ni-Cr coated super duplex stainless steel.”

[3].Byoungchul Hwang, Sunghak Lee, Jeehon Ahn, 2003,”Correlation of microstructural and wear resistance of molybdenum blend coatings fabricated by atmospheric plasma spraying”.

[4]. Gheorge Matache, Cristian Puscasu, 2015, “Investigations of some intrinsic property of thermal spray molybdenum coating for railway axle applications”.

[5]. Adarsha H, C S Ramesh,2017, “Investigation on the abrasive wear behaviour of molybdenum coating on SS304 and A36 using HVOF technique”.

[6]. Long Huang, Xiangtao Deng and Yi Jia, “effects of using (Ti,Mo)C particles to reduce3 body abrasive wear of a Low alloy steel.” State key laboratory of rolling and automobiles.

[7]. R Riastuti et al 2018 IOP Conf. Ser.: Mater. Sci.Eng. 420 012058. “The study of mechanical properties and corrosion resistance of Molybdenum coating and Aluminium coating using Thermal Spray method on stainless steel316L”.

[8]. Sivakumaran Ilaiyavel, Alankaram Venkatesan, Experimental Investigation of Wear Characteristics on Manganese Phosphate Coated AISI D2 Steel

[9]. Carlos Alberto Guevara Chávez, Jorge Leobardo Acevedo Dávila, Pedro Hernandez Gutierrez, Jose Jorge Ruiz Mondragon & Patricia del Carmen Zambrano-Robledo “Tribological and microstructural behavior in coatings applied using GTAW and HVOF (thermal spraying process) and used to recover AISI/SAE D2 grade tool steels”, Intl. Journal of Welding, Taylor & Francis publications 2020.

[10]. L Bourithis & G. Papadimitriou “Three body abrasion of low carbon steel modified surfaces”, Elsevier Publications, 2005.

Structural, Morphological, and Dielectric Properties of Gd^{3+} modified $LaPO_4$ Ceramics

Kanchan Khanna[a], Akshay Kumar[a], Sunil K. Dwivedi[a] and Sunil Kumar[b*]
[a]Department of Physics, SAS, OM Sterling Global University, Hisar-125001
[b]Department of Physics, Guru Nanak College, Batala, Punjab-143505

*Corresponding Author: sunil_nano@yahoo.com

Abstract. Gd^{3+} modified $LaPO_4$ ($Gd_xLa_{1-x}PO_4$, where $x = 0.02$, 0.04 & 0.06) ceramics have been prepared using Auto Combustion. X-Ray diffractograms analysis shows that samples exhibits pure monoclinic structural phase upto $x = 0.04$ without presence of any other secondary phases whereas in x = 0.6, some secondary phases appears. The FESEM micrographs show grain growth, densification along with compactness of grains whereas EDS micrographs confirm the presence of elements according to stoichiometric proportion. Elemental mapping gives information about uniform distribution of elements. The dielectric relaxation behavior of prepared ceramics confirms from Cole-Cole model whereas dielectric loss gives information about conductivity of samples which has also be confirmed from conductivity profile.

Keywords: Phosphates, Light Emitting Diode, Cole-Cole Model, Elemental Mapping

1. Introduction

Rare earth modified phosphate materials commonly known as orthophosphates materials attracts focus of researchers for their wide usage in industrial applications such as light emitting diodes, laser technology, plasma display panels, compact fluorescent lamps etc. Apart from electronic applications, the orthophosphate materials also used for other industrial application such as in medical industry as a biological implant materials and host materials for nuclear waste management. In solid state illumination, light emitting diode is one of a milestone for its high resolution. Orthophosphate materials are important for above said industrial applications due to 4f orbitals electrons. Among such category of industrial applications, lanthanum phosphate is one of most promising materials due to its high physical and thermal stability and also for water insolubility. Rare earth modified phosphate with monoclinic , monazite or xenotime phase become are very much important as requirement of emitting diodes as a refractory application, high temperature photonic conductors, fuel cell etc. Lanthanum-Phosphate exhibits monoclinic crystal phase in two polymorphic phase such as monoclinic monazite and quardratic xenotime phase [1-8].

In this paper, we report the effect of partial substitution of Gd^{3+} on structural, optical as well as dielectric properties $LaPO_4$ (Gd^{3+} at La^{3+} site in $LaPO_4$) ceramics prepared using Combustion method with citric as a fuel agent. The prepared modified $LaPO_4$ with rare earth ion ceramics sintered at 1250°C for 2 hours.

2. Synthesis and Characterization

2.1 Synthesis Procedure:

Rare earth ion modified $LaPO_4$ ($Gd_xLa_{1-x}PO_4$) where $x = 0.02$, 0.04 & 0.06 ceramics have been synthesis using Auto Combustion method. The required metal nitrates such as $GdNo_3$, $LaNo_3$, and phosphates have been first weighed in above mentioned stoichiometric proportion and made their clear solution with distilled water. Solution of Citric acid with water has also been prepared for using as fuel agent. The individually prepared solutions of above mentioned metal nitrates mixed one by one with continuous stirring and heating. The fuel agent mixed in last and heated solution along with continuous stirring till the liquid get converted into powder. The powder was then calcined at 1000 °C for 1 hour for crystal phase formation. The PVA (Poly Vinyl Alcohol) mixed calcined powder was pressed in circular disc for sintering and characterization. The PVA acts as binder. The circular disc were sintered at 1200 °C for 2 hours and termed as 'a', 'b' for analysis of various properties that have been characterized.

2.2 Characterization Procedure:

The powder of sintered disc was used for structural characterization using X-ray diffractometer from Shimadzu Maxima equipped with anode of Cu Kα (λ = 1.54 Å) whereas morphological as well as elemental analysis was carried out over pallets using Scanning Electron Microscope from Carl Zeiss Supra 55 equipped with X-ray detector from Oxford Instruments. For dielectric measurements such as dielectric permittivity, dielectric loss and ac conductivity measurements have been performed over sintered pallets using Impedance Analyzer (E4990A) from KEYSIGHT Technologies.

3 Results and Discussion

3.1 Structural Analysis

X-Ray diffraction data of $Gd_xLa_{1-x}PO_4$, where $x = 0.02$, 0.04 & 0.06 has been shown in figure 1 (a) & (b). Sharp, high intense crystalline peaks clearly validate the crystalline nature of prepared samples. Primarily, all the three graphs of prepared samples has merged to a single graph for confirmation of single phase formation. It has been clearly seen that diffraction peaks in prepared samples for $x = 0.02$, 0.04 overlapped each other reveals the single phase formation whereas in sample $x = 0.06$, beyond $2\theta = 70$, peaks does not overlapped meaning presence of some secondary phases.

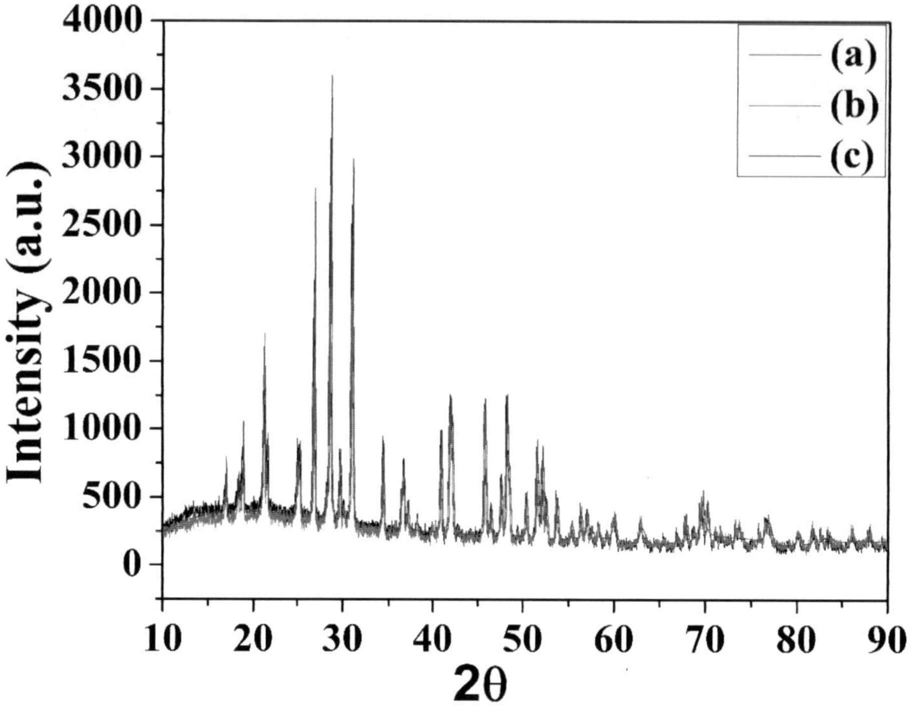

Figure 1 (b): X-Ray Diffraction Pattern of $Gd_xLa_{1-x}PO_4$, (a) = 0.02, (b) = 0.04 & (c) = 0.06

Further, for studying the crystal phase of prepared samples, diffraction data has been matched with reported Crystallographic data published in JCPDS card. The diffraction peaks clearly shows matching with diffraction peaks reported in JCPDS card no. 84-0600 for $x = 0.02$ & 0.04. Since it has been clearly visualized from figure 1(a) that peaks for $x = 0.02$ & 0.04 clearly overlapped reveals presence of similar crystal structure whereas in $x = 0.06$, some secondary phases appears. So all peaks matched with above mentioned JCPDS card for $x = 0.02$ & 0.04. This JCPDS card belongs to Monoclinic phase having space group $P2_1/n$ (14). All the peaks upto $x = 0.04$ indexed according to above mentioned JCPDS card whereas in $x = 0.06$, some secondary phases also appear which reveals that the soluble – solubility limit of Gd^{3+} substitution in $LaPO_4$ is only upto $x = 0.04$. The graphs with indexed peaks have been shown in figure 1(b).

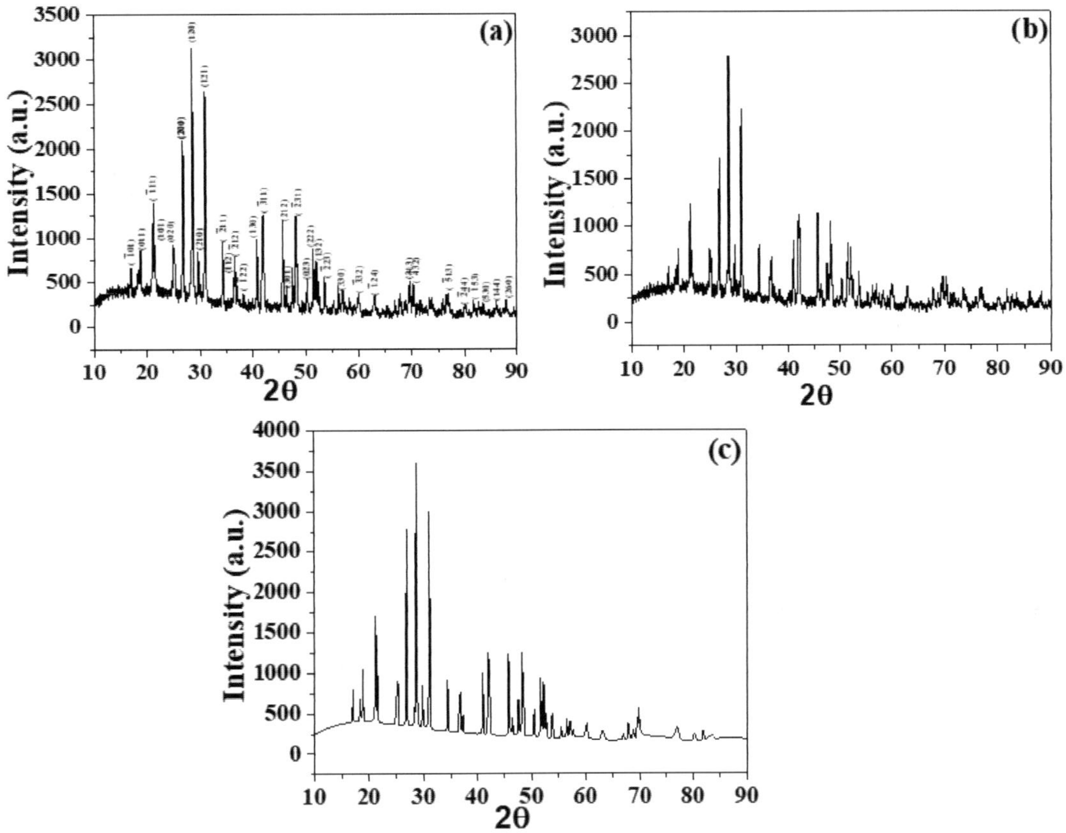

Figure 1 (b): X-Ray Diffraction Pattern of $Gd_xLa_{1-x}PO_4$, (a) = 0.02, (b) = 0.04 & (c) = 0.06

3.2 Morphological l Analysis

FESEM Micrographs of $Gd_xLa_{1-x}PO_4$, where x = 0.02, 0.04 have been shown in figure 2. The irregular, cross linked, vary in size shows increase in grain size, grain growth, and decrease in porosity along with enhanced densification. The all micrographs were taken at same electric voltage of 15kV and magnification of 30KX using InLens detector. It has been clearly visualised that grain size increases with increase in content of substitution. To confirm the exact grain size, grain size has been calculated using length of scale bar given in SEM micrographs. For this approximately 8-10 grain have been selected randomly and calculate the size. From calculation, it appears that grain size increases from 3.21 μm to 4.56 μm as substitution of Gd^{3+} has been increased from $x = 0.02$ to $x = 0.04$.

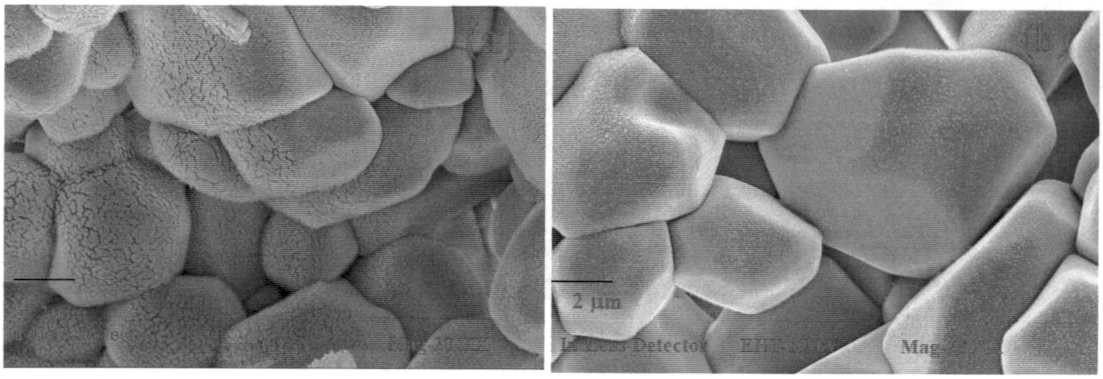

Figure 2: SEM Micrographs of $Gd_xLa_{1-x}PO_4$, (a) = 0.02 & (b) = 0.04

To confirm the element as per stoichiometric proportion, Energy Dispersive X-ray spectroscopy has also been performed. The peaks accordingly binding energy (eV) are direct evidence of presence of elements as per stoichiometric proportion in prepared samples. The table, electro micrograph and graphs shown in figure 3. The table shows presence of elements in calculated wt% by software attached with EDS is direct relevance of presence of elements as per stoichiometric proportion. The show uniform distribution of elements according to stoichiometric formula has been confirmed from elemental mapping shown in figure 3. In elemental mapping, different-different colour assigned to different ion. The mixed micrograph shows uniform distribution.

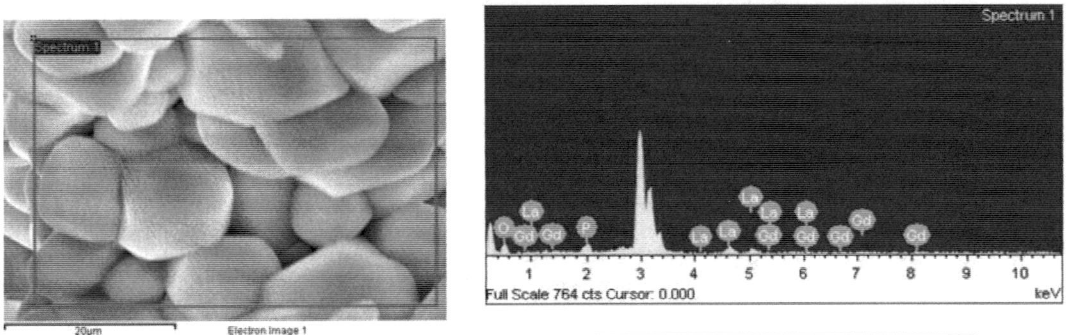

Element	Weight (%)
O K	35.84
P K	15.43
La L	45.23
Gd L	3.50
Total	100

Figure 3: EDS and Elemental Map Micrographs of $Gd_xLa_{1-x}PO_4$, (a) = 0.02 & (b) = 0.04

3.3 Dielectric Properties:

Frequency dependent dielectric data (Real Part of dielectric permittivity) i.e. ε' vs. Frequency at room temperature of $Gd_xLa_{1-x}PO_4$, $x = 0.02$, 0.04 in frequency range100Hz-100KHz have been shown in figure 4. Real part of dielectric permittivity (ε') starts decreases with increasing frequency reveals that sample show behaviour similar to general dielectrics. In the lower range of frequency, sample exhibits maximum value of dielectric constant value and decreases continuously upto certain value of frequency and beyond that, dielectric constant become almost constant. The maximum value of dielectric constant results due to contribution of all polarization such as electronic, ionic, dipolar, space charge polarizations but in higher frequency range, some of these polarization do not responds results in lower value of dielectric constant. The phenomenon of dipole relaxation can be used to explain these observations [9]. The relaxation phenomena has been explained on basis of various relaxation model such as Cole-Cole relaxation model, a modified version of the Debye relaxation model [10]. According to this model, the ε′ and ε″ vary with frequency as:

$$\varepsilon'(\omega) = \varepsilon_\infty + (\varepsilon_s - \varepsilon_\infty)\frac{1+(\omega\tau_0)^{1-\alpha}\sin\frac{1}{2}\alpha\pi}{1+2(\omega\tau_0)^{1-\alpha}\sin\frac{1}{2}\alpha\pi+(\omega\tau_0)^{2(1-\alpha)}}$$

$$\varepsilon''(\omega) = (\varepsilon_s - \varepsilon_\infty)\frac{(\omega\tau_0)^{1-\alpha}\cos\frac{1}{2}\alpha\pi}{1+2(\omega\tau_0)^{1-\alpha}\sin\frac{1}{2}\alpha\pi+(\omega\tau_0)^{2(1-\alpha)}}$$

where ε_∞ = dielectric constant in higher frequency regime, ε_s = dielectric constant in lower frequency regime, ω is angular frequency and τ = characteristics relaxation time of the medium. The value of exponent parameter (α) usually varies from 0 and 1, and gives information about shape of spectral curves. The value of exponent parameter (α =0) means Cole-Cole model changes to the Debye model. The fitted dielectric data using Cole-Cole dispersion formulas has been shown in figure 3.4(a & b). The graphs clearly shows that in lower range of frequency, dielectric constant remains almost constant that shows that substitution Gd^{3+} at La^{3+} site in $LaPO_4$ decreases the insulating behaviour of $LaPO_4$ results in increase in conductivity as well as dielectric loss. The decreases in insulating behaviour has been confirmed from lower value of dielectric constant in comparison with pure $LaPO_4$ is an evidence of this behaviour.

IOP Conf. Series: Materials Science and Engineering 1291 (2023) 012013 doi:10.1088/1757-899X/1291/1/012013

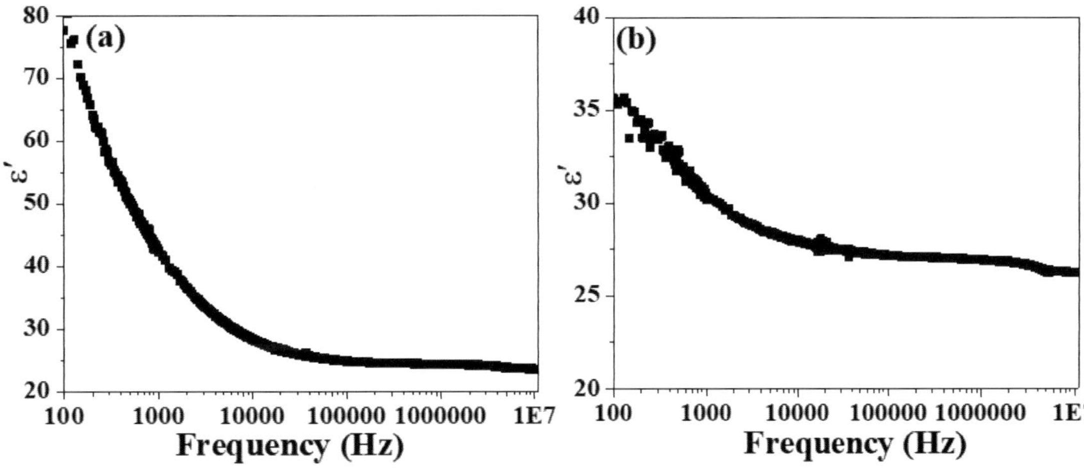

Figure 4: ε' vs. Frequency (Hz) of $Gd_xLa_{1-x}PO_4$, (a) = 0.02 & (b) = 0.04

3.4 Dielectric Loss (Tanδ)

The figure 5 shows the variation of dielectric loss (Tanδ) vs. frequency of $Gd_xLa_{1-x}PO_4$, $x = 0.02$, 0.04. The graphs clearly shows that substitution of Gd^{3+} increase in dielectric loss in pure $LaPO_4$. This increase in dielectric loss is due to increase in oxygen vacancies which increases the conduction behaviour results in increase in dielectric loss.

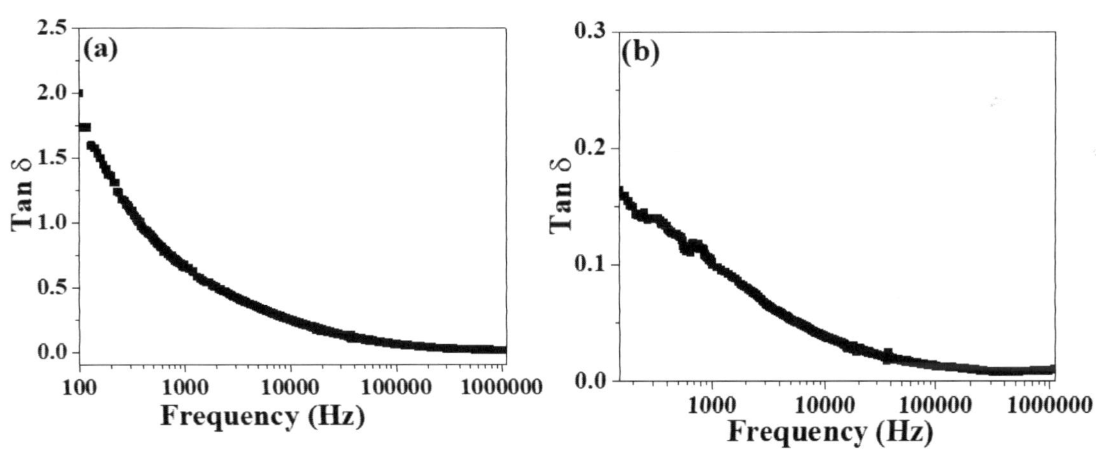

Figure 5: Tanδ vs. Frequency (Hz) of $Gd_xLa_{1-x}PO_4$, (a) = 0.02 & (b) = 0.04

3.5 Conductivity Properties:

Room temperature σ_{ac} vs. Frequency profile of of $Gd_xLa_{1-x}PO_4$, $x = 0.02$, 0.04 have been shown in figure 6. The conductivity from dielectric data has been calculated using relation: $\sigma_{ac} = 2\pi f\varepsilon'\varepsilon_o \tan\delta$

[12], where the parameters have their usual meaning. Graphs clearly reveals that the conductivity profile shows two regimes plateau and dispersion region. The plateau region gives value of dc conductivity where dispersion region corresponds to the ac conductivity. The plateau region is lower frequency region where as dispersion region lies in higher frequency regime. The conductivity has been found to be independent of frequency in plateau region whereas in dispersion region, conductivity increases with increasing frequency. The ac conductivity profile has been in materials has been analysed by Jonscher's power law [13-14]; $\sigma_{ac}=\sigma_{dc} + A\ \omega^n$, where "A" is the dispersion parameter representing the strength of polarizibilty and "n" is the dimensionless frequency exponent representing the interaction between mobile ions with the lattice around them [15-17]. It has been clearly visualized from graphs that conductivity increases with Gd^{3+} substitution in $LaPO_4$ in compare to pure $LaPO_4$ samples.

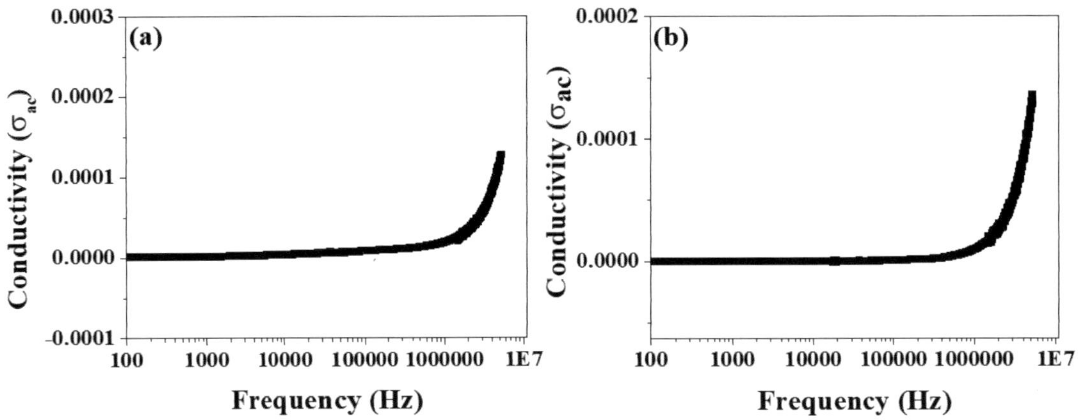

Figure 6: σ_{ac} vs. Frequency (Hz) of $Gd_xLa_{1-x}PO_4$, (a) = 0.02 & (b) = 0.04

4. Conclusion:

Gd^{3+} modified $LaPO_4$ ($Gd_xLa_{1-x}PO_4$, where x = 0.02, 0.04 & 0.06) ceramics has been successfully prepared using Auto Combustion. X-Ray diffractograms confirms that solid solubility limit of Gd^{3+} substitution in $LaPO_4$ at La^{3+} site is upto x = 0.04. SEM micrographs shows the grain growth with increase in grain size whereas EDS and elemental micrograps confirm presence of elements according to stoichiometric proportion uniform distribution of elements. The dielectric constant first decreases with increasing Gd^{3+} substitution. The dielectric loss and conductivity profile confirm the decrease in insulating behavior with increase in conduction behavior.

References

1. I. S. Cho, G.K. Choi, J.S. An, J. R. Kim and K.S. Hong, Sintering, microstructure and microwave properties of rare earth orthophosphate, RePO4 (Re = La, Ce, Nd, Sm, Tb, Dy, Y, Yb), Material Research Bulletin 44 (2009) 173-178
2. V. Kumar, S. Singh, R.K. Kotnala and S. Chawla, GdPO4:Eu3+ nanoparticles with intense orange red emission suitable for solar spectrum conversion and their multifunctionality, Journal of Luminesence 146 (2014) 486-491
3. R. Priya, R. Mariappan, A. Karthikeyan, E. Palani, E. Krishnamoorthy and G. Gowrisankar, Review on rare earth metals doped $LaPO_4$ for optoelectronics, Solid State Communications 399 (2021) 114457
4. I.S. Cho, G. K. Choi, J.S. An, J.R. Kim and K.S. Hong. Sintering, microstructure and microwave dielectric properties of rare earth orthophosphate, $RePO_4$, (Re = La. Ce, Nd, Sm, Tb, Dy, Y, Yb), Material Research Bulletin 44 (2009) 173-178
5. M. Ferhi, K.H. Naifer and M. Ferid, Combustion synthesis and luminescence properties of LaPO4: Eu (5%), Journal of Rare Earths, Vol. 27 No. 2 (2009) 182-186
6. A.K. Bedyal, V. Kumar and H.C. Swart, Influence of an adjoining cation on the luminescence performance of Dy^{3+} doped A3Gd(PO4)2; (A = Na, K) Phosphors, Journal of Alloys and Compounds 845 (2020) 156352
7. B.R. Verma and R.N. Baghel, A Review on Luminescence properties of Eu Dopes Phosphate Phosphors, International Journal of Material Science, Vol. 12 No. 3 (2017) 483-489
8. S. Khan, Y. R. Parauha, D. K. Halwar and S. J. Dhoble, Rare Erath (RE) doped color tunable phosphors for white light emitting diodes, International Conference on Research Frontiers in Sciences (ICRFS2021): Journal of Physics: Conference Series, 1913 (2021) 012017.
9. Kumar M, Yadav KL. J. Phys.: Condens. Matter. 2002;19:242202.
10. Cole KS, Robert H. Dispersion and Absorption in Dielectrics:- Alternating Current Characteristics. Journal of Chemical Physics. 1941;9:341–351.
11. Havriliak S, Negami S. A complex plane representation of dielectric and mechanical relaxation processes in some polymers. Polymer. 1967;8:161-210. DOI: 10.1016/0032-3861 (1967) 90021-3.

12. Prakash V, Choudhary SN, Sinha TP. Dielectric relaxation in complex perovskite oxide BaCo1/2W1/2O3. Physica B. 2008;403:103–108.
13. Jonscher AK. The Universal dielectric response. Nature. 1977;267:673-679.
14. Pradhan DK, Behera B, Das PR. Studies of dielectric and electrical properties of a new type of complex tungsten bronze electroceramics. J. Mater. Sci. Mater. Electron. 2012;23:779.
15. Tan M, Koseoglu V, Alan F, Senturk E. Overlapping large Polaron tunneling conductivity and giant dielectric constant in $Ni_{0.5}Zn_{0.5}Fe_{1.5}Cr_{0.5}O_4$ nanoparticles (NPs). J Alloys Compd. 2011;509:9399-9405.
16. Vaish R, Varma KBR. Dielectric properties of Li_2O-$3B_2O_3$ glasses. J. Appl. Phys. 2009;103:064106.
17. Megdiche M, Pellegrino CP, Gargouri M. Conduction mechanism study by overlapping large Polaron tunneling model in $SrNiP_2O_7$ ceramic compound. J Alloys Compd. 2014;584:209-215.

Dye-sensitized solar cell employing chitosan-based biopolymer electrolyte

Simantini Majumdar[1,2], Archita Mondal[2], Ankita Mahajan[3,4], Swapan Kumar Bhattacharya[3] and Ruma Ray[2*]

[1]Chemistry and Physics of Materials Unit, Jawaharlal Nehru Centre for Advanced Scientific Research (JNCASR), Bengaluru-560064, India
[2]Department of Physics, Jadavpur University, Kolkata-700032, India
[3]Department of Chemistry, Jadavpur University, Kolkata-700032, India
[4]Max-Planck-Institut für Eisenforschung GmbH, 40237 Düsseldorf, Germany

*Corresponding author.
Email address: juphyruma@gmail.com

Abstract. Conductivity and transport properties of a cost-effective and environment friendly chitosan based solid biopolymer electrolytes which form mechanically robust thick film, have been reported here. A maximum ionic conductivity of ~ 10^{-4} S/cm has been achieved by optimizing the concentrations of the salt ($LiClO_4$) and the plasticizer (EC) in the biopolymer electrolyte. Ion transport properties of the biopolymer electrolytes are studied from Raman spectroscopy. A dye-sensitized solar cell (DSSC), with a sandwich structure, is fabricated with chemically synthesized ZnO (~ 60 nm) as the nanoporous semiconductor material coated with Rose Bengal dye as a photosensitizer, the chitosan biopolymer as electrolyte and platinum as counter electrode. Linear Sweep Voltammetry analysis of the DSSCs illustrates the photovoltaic performance of these cells. Without any external addition of redox couple in the biopolymer electrolytic system, a maximum short-circuit current density of J_{sc} = 0.556 mA/cm^2 and open-circuit voltage V_{oc} = 0.605 V with power conversion efficiency 0.051 % is achieved by the DSSC.

Keywords: Chitosan, Biopolymer electrolyte, Dye-sensitized solar cell, Photovoltaic performance.

1. Introduction

The rapid worldwide consumption of fossil resources (such as coal, oil, gas) and the undesirable consequences of environmental pollution are emerging drastically which lead to the development of green, sustainable and highly efficient electrochemical energy storage and energy conversion devices such as batteries [1-2], supercapacitors [3-4] and dye-sensitized solar cells (DSSC) [5-6]. Polymer electrolytes have attracted huge attention among the researchers worldwide since the past few decades for their good electrochemical properties along with the improved safety which ensures their potential applications in the energy industry.

Dye-sensitized solar cells (DSSCs) are photovoltaic (PV) cells that convert any visible light into electrical energy. DSSCs have attracted great interest for their low fabrication cost and simple structure, compared to the silicon-based p-n junction photovoltaic devices [7-8]. A standard nanostructured DSSC (first developed by O'Regan and Gratzel) comprises of a wide band gap nanocrystalline semiconductor film (TiO_2) coated with ruthenium dye complex on a transparent conducting glass electrode (ITO or FTO), a redox couple electrolyte (I^-/I_3^-) and a counter electrode. The basic operation of the DSSC involves the injection of photo-excited electrons from dye molecules into the conduction band of the semiconductor electrode that subsequently diffuse toward the transparent conducting glass substrate, leaving oxidized dye molecules behind. The oxidized dye molecules are regenerated by oxidizing I^- to I_3^- ions in the electrolyte. The electrons finally reach the counter electrode through external circuit where I_3^- reduced to I^- and make the circuit complete. Recently, ZnO which has a similar band gap (~ 3.3eV) like TiO_2, appears to be an alternative semiconductor material for the fabrication of high efficiency DSSCs [9]. Comparing with TiO_2, ZnO has much higher electron diffusivity and mobility [10-11] which provides a direct conduction pathway for rapid collection of electrons at the substrate of the photoanode.

The electrolytes play a very important role in getting highly efficient photo-electrochemical DSSCs [12]. Various biopolymers such as starch, chitosan, cellulose etc. have attracted considerable interest owing to their biodegradability, non-toxicity, cost-effective and environmentally friendly properties [13-15]. Hence the use of these materials can contribute in minimizing the environmental waste, and thereby lowering the cost of the energy conversion devices. Among all the natural biopolymer hosts, chitosan which is composed of two main groups i.e., *N*-acetyl glucosamine and glucosamine units linked by β-1,4-glycosidic bonds, has attracted considerable attention of researchers in the development of SPE since the past few decades. The presence of polar functional groups in chitosan such as hydroxyl and amine can give the material a dipole moment that affects the dielectric polarization. Hence, intermolecular interactions are created between these polar groups of chitosan and alkali metal salt ions, leading to the solvation of salt. Also, chitosan has some other outstanding features such as its abundant availability, low cost of extraction, non-toxicity, biodegradability, biocompatibility and excellent mechanical strength. Hence, chitosan can be considered as a fascinating polymer host for developing polymer electrolyte. But chitosan has very low conductivity [16] which can be tuned by doping with suitable ionic salts and plasticizers. The most popular ionic salt used for incorporating charge carriers in the polymer matrix is lithium perchlorate ($LiClO_4$) [16-18]. Ethylene carbonate (EC) has been chosen as plasticizer since it has high dielectric constant which may assist in the dissociation of salt by weakening the Coulombic force between its cations and anions and thus can tailor the ionic conductivity and provide mechanical stability to the film [19].

In this article, the conductivity and transport properties of the chitosan based solid biopolymer electrolytes (SPE) containing different concentrations of salt ($LiClO_4$) and plasticizer (EC) are reported at room temperature. The ion transport parameters in the biopolymer electrolytes are studied by Raman spectroscopy. A dye-sensitized solar cell (DSSC) is assembled with Zinc Oxide (ZnO) nanoparticles as the semiconductor material, Rose Bengal (RB) dye as the photosensitizer and chitosan-based biopolymer electrolyte. The constructed photovoltaic cells are characterized using linear sweep voltammetry (LSV) analyses and the power conversion efficiencies of the cells are estimated for the chitosan biopolymer electrolytes containing different concentrations of $LiClO_4$ and EC. So far as found in literature, this is the first report of direct application of the chitosan biopolymer electrolyte without any external introduction of redox couple (I^-/I_3^-), in photovoltaic cells.

2. Experimental

2.1. Synthesis of Zinc Oxide (ZnO) nanoparticles

ZnO nanoparticles are synthesized in de-ionized water ambiance and $(CH_3COO)_2Zn.\ 2H_2O$ is taken as Zn^{2+} source. ZnO nanoparticles are prepared by following the protocol as reported by Mondal *et al* [20]. The prepared nanoparticles were characterized by X-ray diffraction (XRD) and UV-VIS Spectroscopy. XRD was done by X-ray diffractometer (model: Bruker D8 Advance) using the CuKα radiation of wavelength 1.5406 Å. The optical transmission spectrum of the ZnO nanoparticles has been studied in the wavelength region 200 – 800 nm by using UV-VIS a diffuse reflectance spectrophotometer (Shimadzu, UV 2401PC).

2.2. Synthesis of chitosan-based biopolymer electrolytes

The solid biopolymer electrolytes are synthesized by solution casting technique. Chitosan (0.5 g) is dissolved in 50 ml 1% aqueous acetic acid solution. Then different weight percentages of $LiClO_4$ (40, 60, 80 wt.%) as ionic salt and ethylene carbonate (EC) (30, 45, 50 wt.%) as plasticizer are added in the chitosan solution and stirred continuously until complete dissolution. All the resulting homogeneous solutions are cast in separate polypropylene petridishes and allowed to dry at room temperature for the thin polymer films to form. The solid biopolymer electrolyte films (SPE) are cut into suitable sizes and characterized using computer interfaced impedance analyzer (Agilent 4294A-Precision Impedance Analyzer) in the frequency range between 40 Hz and 2 MHz at room temperature, by sandwiching them between two gold electrodes, each of diameter 1 cm, at room temperature. Raman spectroscopy measurements were performed in backscattering geometry using LabRAM HR (JobinYvon) spectrometer equipped with a Peltier-cooled charge-coupled-device (CCD) detector.

2.3. Development of the dye-sensitized solar cell (DSSC)

The photoanode of the dye-sensitized solar cell (DSSC) was fabricated by depositing ZnO nanoparticles on indium tin oxide (ITO) glass and thereafter, coating it with 10 mM Rose Bengal (RB) dye solution using drop casting technique and dried for 24 hours at room temperature. The chitosan-based biopolymer electrolyte was then sandwiched between the ITO/ZnO/RB (photoanode) and platinum (Pt) (counter electrode). Linear Sweep Voltammetry (LSV) responses of the photovoltaic cells are performed with AUTOLAB PG STAT 12 Potentiostat/Galvanostat (Eco Chemic, Netherlands). The photovoltaic performance of the DSSC is measured from current-voltage (I–V) characteristics under light illumination (intensity 100 mW/cm^2) using the Keithley 2635B source meter interfaced with a computer.

3. Results and discussion

3.1. Structural characterization: X-ray diffraction (XRD)

ZnO nanoparticles have been characterized using XRD. The diffraction peaks of ZnO nanoparticles as illustrated in Fig. 1(a), exhibit a single-phase wurtzite structure (space group *p63mc*, JCPDF #36-1451). The crystallite size (D) and microstrain (ε) of ZnO nanoparticles have been calculated from Williamson and Hall's modified Scherrer's formula given by [21],

$$\beta\cos\theta = \frac{0.89\lambda}{D} + 2\varepsilon\sin\theta \quad (1)$$

The inset of the Fig. 1a shows the $\beta\cos\theta$ vs $\sin\theta$ plot revealing the average size of the ZnO nanoparticles as 58 nm.

3.2. UV-Vis Spectroscopy

UV-VIS absorption experiments are performed to investigate the optical band gap of ZnO. The fundamental absorption, which corresponds to the transition from the valence band to the conduction band, can be used to determine the optical band gap of a material. The absorption coefficient for direct band gap systems, can be related with the incident photon energy by the following equation,

$$(\alpha h\nu)^2 = A(h\nu - E_g) \quad (2)$$

where, E_g is the band gap of the material. The band gap of ZnO has been evaluated using this relationship, in the standard manner from $(\alpha h\nu)^2$ versus $h\nu$ plot as illustrated in Fig. 1b. The calculated value of optical band gap of ZnO nanoparticles is 3.32 eV.

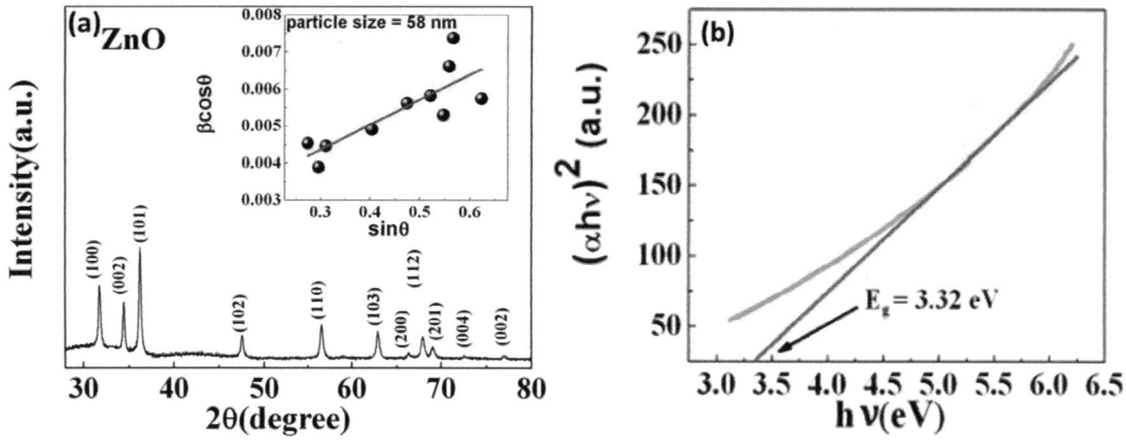

Fig. 1. (a) XRD spectrum of ZnO and the inset figure shows the Williamson–Hall plot, (b) Plot of $(\alpha h\nu)^2$ versus $h\nu$ to determine the band gap of the ZnO sample.

3.3. Ionic conductivity studies: Impedance Spectroscopy

The Cole-Cole plots of the chitosan based solid biopolymer electrolytes (SPE) with different concentrations of salt ($LiClO_4$) (henceforth denoted as CL_x, where x = 40, 60, 80 wt. percentages of the polymer) and plasticizer (EC) (henceforth denoted as $CL_{80}E_y$, where y = 30, 45, 50 wt. percentages of the polymer) are shown in Figs. 2 (a) and (b) respectively. The DC ionic conductivity (σ_{DC}) of the SPE is calculated using the relation, $\sigma_{DC} = L/R_bA$ where, L is the SPE thickness, A is the effective contact area of the electrode and the SPE surface and R_b is the bulk resistance. R_b is determined from the high frequency intercept on the real Z axis of the Cole-Cole plot. The variation of DC conductivities for different $LiClO_4$ and EC concentrations are shown in the insets of Figs. 2 (a) and (b) respectively. It is found that the ionic conductivity increases on increasing the $LiClO_4$ concentrations in the SPE system and CL_{80} shows the highest conductivity value of ~ 10^{-5} S/cm at room temperature (300 K). This indicates that the mobile charge carrier ions (Li^+) play a significant role for the enhancement of conductivity. To increase the mobility of Li^+ further, different concentrations of EC as plasticizer, have been added in the matrix of the biopolymer electrolyte CL_{80}. It is observed that the ionic conductivity is increased by one order of magnitude (from 10^{-5} to 10^{-4} S/cm) on addition of EC in the $CPL_{80}E_y$ (y = 45). This enhancement in the ionic conductivity may be attributed to the increase in salt dissociation on the incorporation of EC in the SPE. Further increase in EC content (y > 45) decreases the conductivity of the SPE by inhibiting the fast ionic transport through salt recrystallization.

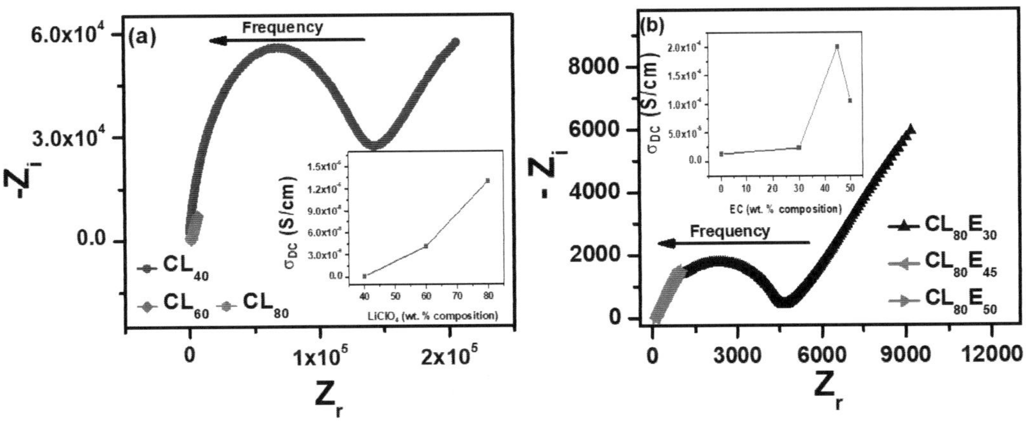

Fig. 2. Cole-Cole plots of chitosan based solid biopolymer electrolytes (SPE) with different (a) salt ($LiClO_4$) and (b) plasticizer (EC) concentrations. Insets of (a) and (b) shows the variation of DC conductivities for different $LiClO_4$ and EC concentrations, respectively.

3.4. Transport properties: Raman Spectroscopy

The Raman spectra for CL_x ($0 \le x \le 80$) and $CL_{80}E_y$ ($30 \le y \le 50$) SPEs are shown in Fig. 3 in the range of 250 cm^{-1} to 1500 cm^{-1}. Raman analysis is carried out to study the interactions of the chitosan biopolymer with $LiClO_4$ salt and EC plasticizer for the estimation of transport parameters such as number density of free charge carriers responsible for the ionic conduction in the CL_x ($40 \le x \le 80$) and $CL_{80}E_y$ ($30 \le y \le 50$) SPE systems. Raman deconvolution technique has been adopted to determine the ion transport parameters in the SPE systems.

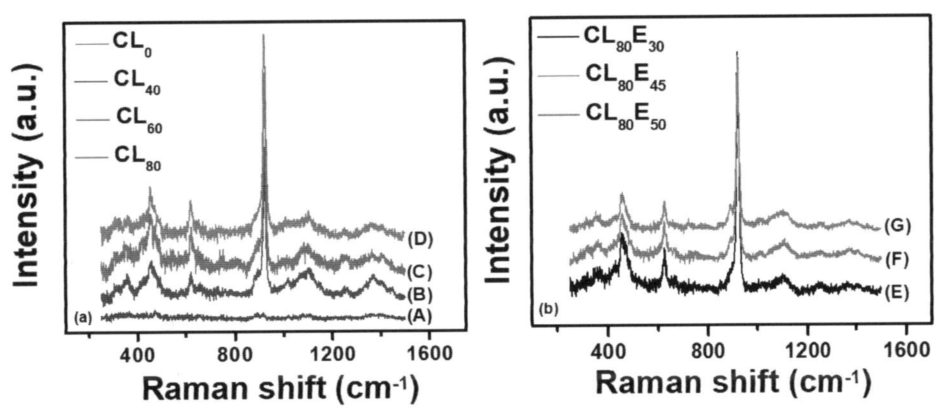

Fig. 3. The Raman spectra over the range of 250 cm^{-1} to 1500 cm^{-1} for (a) CL_x ($0 \le x \le 80$) [(A) CL_0, (B) CL_{40}, (C) CL_{60} and (D) CL_{80}] and (b) $CL_{80}E_y$ ($30 \le y \le 50$) [(E) $CL_{80}E_{30}$, (F) $CL_{80}E_{45}$ and (G) $CL_{80}E_{50}$].

The number density of free charge carrier (n) in the SPE system is directly determined by deconvoluting the Raman spectrum in the range of 900 cm^{-1} to 950 cm^{-1} [22]. Consequently, the ionic mobility (μ) and the diffusion coefficient (D) of the ions are calculated using the Eqs.5 and 6, respectively. The peak close to 919 cm^{-1} is assigned due to the free ions and peak close to 927 cm^{-1} is assigned to contact ion pairs in the SPE system [22]. The deconvoluted Raman spectra in the range 900 cm^{-1} to 950 cm^{-1} for for CL_x ($40 \leq x \leq 80$) and $CL_{80}E_y$ ($30 \leq y \leq 50$) SPEs are shown in Fig. 4.

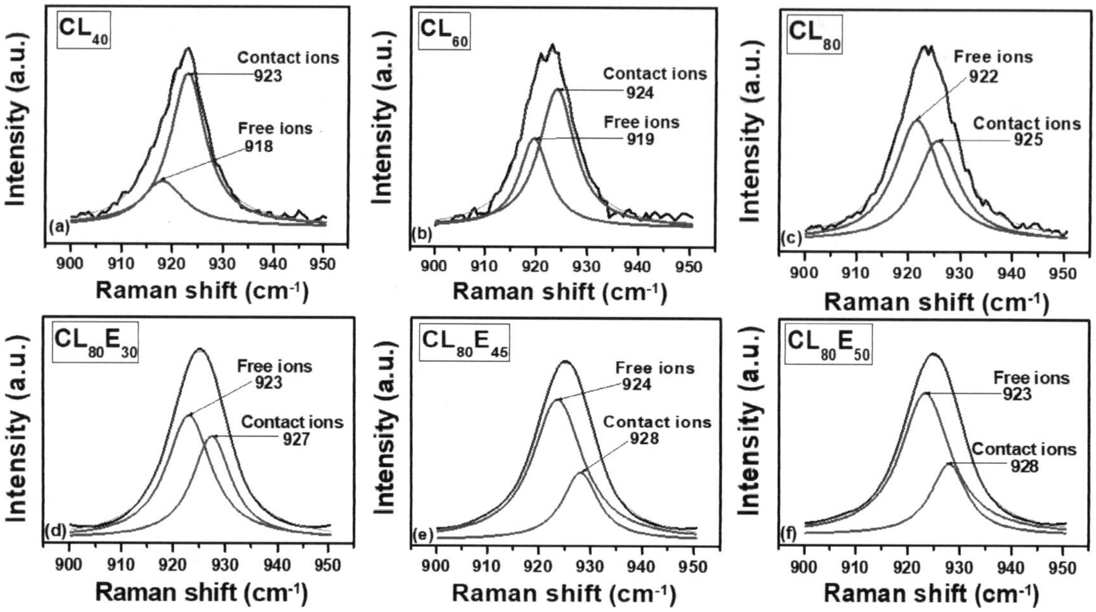

Fig. 4. The deconvoluted Raman spectra over the range of 900 cm^{-1} to 950 cm^{-1} for CL_x ($0 \leq x \leq 80$) and $CL_{80}E_y$ ($30 \leq y \leq 50$) SPEs.

The percentage of free ions (P_f) is estimated as,

$$P_f = \frac{A_f}{A_f + A_c} \times 100\% \quad (3)$$

where, A_f is the area under the peak corresponding to the free ions and A_c is the area under the peak corresponding to the contact ion pairs [18]. The percentage of free ions (P_f), as shown in Fig. 5, follows the conductivity trend (Fig. 2) of the SPEs. Thus, the main contribution in conductivity enhancement of these SPEs are due to the free ions, rather than the contact ions.

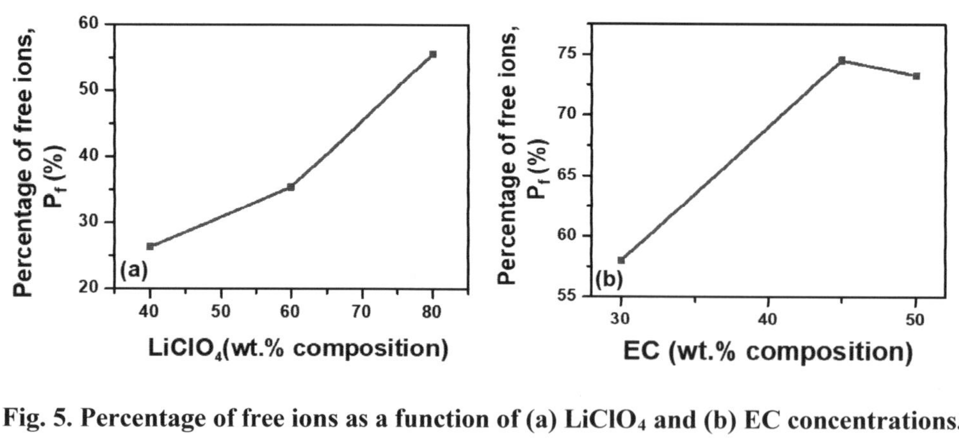

Fig. 5. Percentage of free ions as a function of (a) $LiClO_4$ and (b) EC concentrations.

The values of n, μ and D are calculated using the Eqs. 4, 5 and 6 respectively, as given below [18],

$$n = \frac{M \times N_A}{V_{total}} \times P_f \tag{4}$$

$$\mu = \frac{\sigma_{DC}}{ne} \tag{5}$$

$$D = \frac{\mu K_B T}{e} \tag{6}$$

where, M is the number of moles of $LiClO_4$ salt present in each SPE, V_{total} is the total volume of the SPE and σ_{DC} is the DC conductivity obtained from impedance spectroscopy. The calculated values of the transport parameters n, μ and D for the SPE systems are shown in Fig. 6. It is noteworthy to mention that the highest conducting SPE, $CL_{80}E_{45}$ exhibits the maximum mobility (3.125×10^{-6} $cm^2V^{-1}s^{-1}$) as well as the maximum number of charge carriers (43×10^{19} /cm^3).

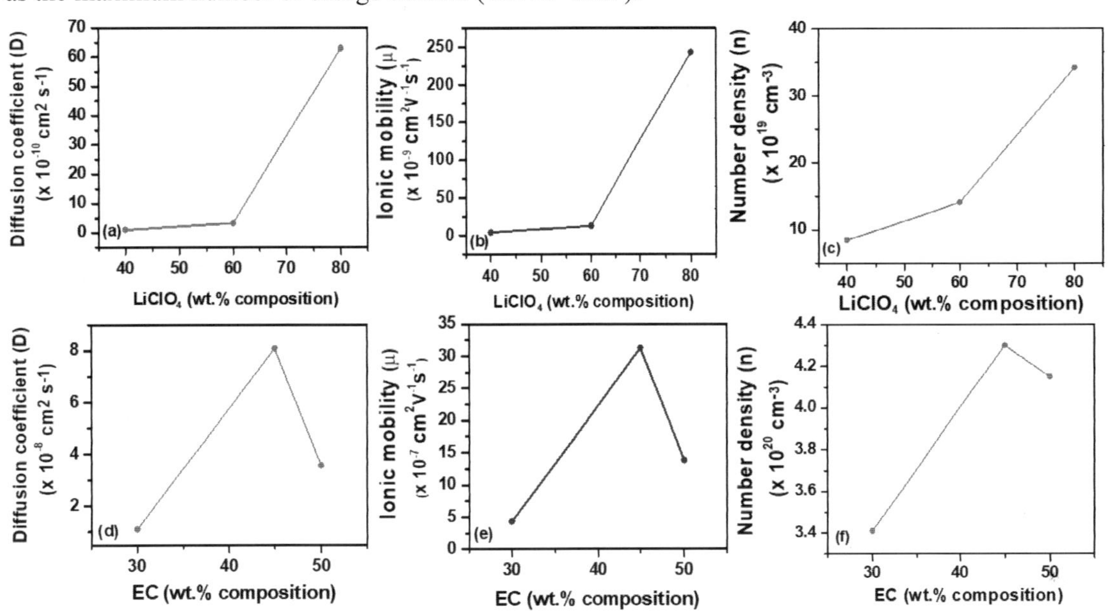

Fig. 6. Ion transport parameters as a function of (a-c) $LiClO_4$ and (d-f) EC concentrations.

3.5. Electrochemical Characterization: Photovoltaic performance of DSSC

Linear sweep voltammetry (LSV) responses for the photo-electrochemical dye-sensitized solar cells (DSSC) cells are performed at room temperature under the illumination of 200 W yellow light having intensity 11300 lux. All the LSV curves have been performed with the DSSC assembled with CL_x ($40 \le x \le 80$) and $CL_{80}E_y$ ($30 \le y \le 50$) biopolymer electrolytes at a scan rate of 5 mV/s with ITO/ZnO/RB photo-anode as the working electrode, platinum as the counter electrode and saturated aqueous calomel electrode taken as reference electrode. The schematic set-up and the corresponding band diagram demonstrating the working principle of the constructed DSSC are shown in Figs. 7a and 7b, respectively.

LSV responses of the DSSCs assembled with CL_x ($40 \le x \le 80$) and $CL_{80}E_y$ ($30 \le y \le 50$) biopolymer electrolytes are shown in Fig. 8. An increase in current density is noticed under illumination, compared to that in the dark for both the sets of polymer electrolytes. This indicates the photovoltaic performance of the cells. The current has been produced due to the diffusion of monomer units of chitosan in the polymer electrolyte on the application of voltage.

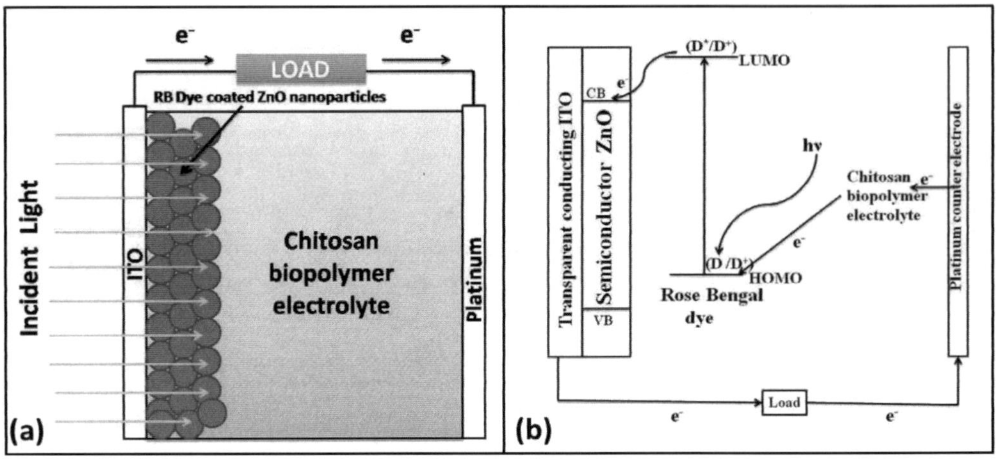

Fig. 7. Schematic set-up of the constructed dye-sensitized solar cell (DSSC), (b) Band diagram demonstrating the working principle of the photovoltaic cell.

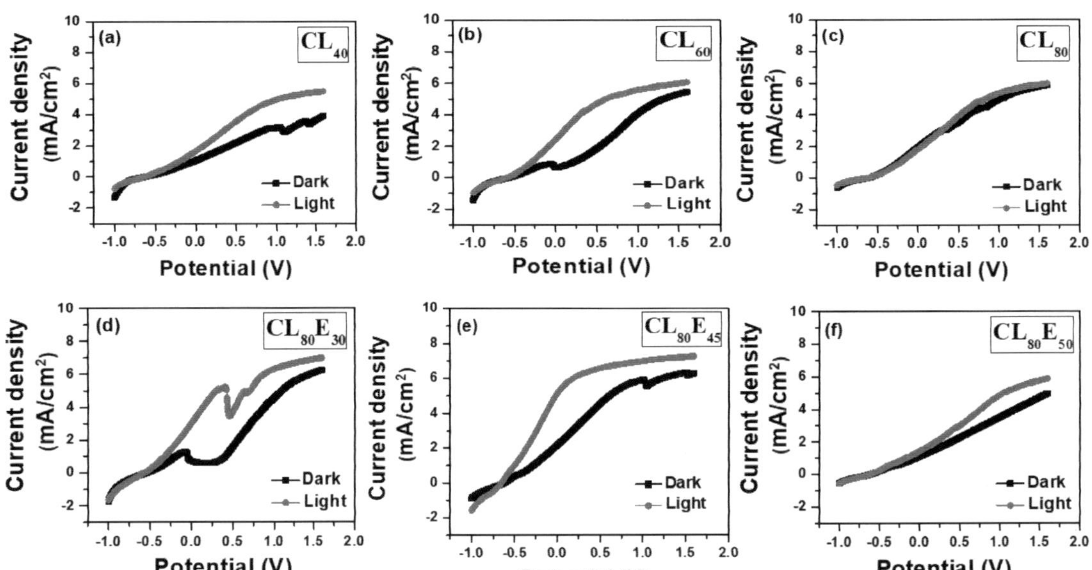

Fig. 8. LSV responses of the DSSCs assembled with (a-c) CL_x ($40 \leq x \leq 80$) and (d-f) $CL_{80}E_y$ ($30 \leq y \leq 50$) biopolymer electrolytes.

The apparent diffusion coefficients (**D′**) of the chitosan monomer units which are assumed to be responsible for the redox reactions in the photovoltaic cells, have been calculated using the following equation [23-24],

$$\mathbf{D}' = \frac{J_{lim}\delta}{nFC_0} \tag{7}$$

where, **D′**, the diffusion coefficient, has unit in cm²/s, J_{lim} is the limiting current density in A/cm², δ is the ZnO/RB film thickness in cm, n is the number of electrons required for the redox reaction to happen,

F is the Faraday constant which is equal to 9.65×10^4 C/mol and C_0 is the initial molar concentration of chitosan monomer unit in moles/mL. The limiting current densities (J_{lim}) and the apparent diffusion coefficients ($\mathbf{D'}$) of the chitosan monomer units for CL_x ($40 \le x \le 80$) and $CL_{80}E_y$ ($30 \le y \le 50$) biopolymer electrolytes based DSSCs are presented in Table 1. It is revealed that the diffusion coefficient ($\mathbf{D'}$) of monomer unit of chitosan increases in presence of light for each working electrode (photoanode). This is due to removal of the products like -CHO unit and H^+ ion from the electrode surface, which facilitates the diffusion of another $-CH_2OH$ unit of another monomer to the surface. In presence of light, the oxidation and the associated product removal occurs quickly, with an effect of increased diffusion. Moreover, the local temperature of the electrode surface may increase in presence of light and thus the diffusion coefficient which is proportional to the absolute temperature, increases.

Table 1. Values of J_{lim} and D′ for the photovoltaic DSSC cells with different salt and plasticizer compositions in chitosan biopolymer electrolytes

Biopolymer electrolytes	Conditions	J_{lim} (mA/cm^2)	$\mathbf{D'}$($\times$ 10^{-6} cm^2/s)
CL_{40}	DARK LIGHT	3.90 5.50	1.95 2.75
CL_{60}	DARK LIGHT	5.40 6.06	2.70 3.03
CL_{80}	DARK LIGHT	5.80 5.97	2.90 2.99
$CL_{80}E_{30}$	DARK LIGHT	6.20 6.94	3.10 3.47
$CL_{80}E_{45}$	DARK LIGHT	6.29 7.25	3.15 3.63
$CL_{80}E_{50}$	DARK LIGHT	4.98 5.88	2.49 2.94

In the DSSCs, the working electrode (photoanode) absorbs incident photon energy from the illuminated light and thereafter the electrons in the RB dye molecules become excited from ground state to the excited state (D^*). The excited photoelectrons are then injected into the conduction band (CB) of ZnO which results in oxidation of the RB dye photosensitizer (D^+) (Fig. 7). These injected electrons are collected by the conducting ITO glass and finally reach the counter electrode (CE) through external circuit. Then the chitosan polymer electrolyte donates electrons to the dye restoring it to its ground state (D), oxidizing the $R\text{-}CH_2OH$ (primary alcohol) functional group of chitosan to R-CHO (aldehyde) group. The possible redox reactions for CL_x ($40 \le x \le 80$) and $CL_{80}E_y$ ($30 \le y \le 50$) biopolymer electrolytes based DSSCs are given by,

$R\text{-}CH_2OH \longrightarrow R\text{-}CHO + 2H^+ + 2e^-$

$2H^+ + 2e^- \longrightarrow H_2$

$R\text{-}CHO + H_2 \longrightarrow R\text{-}CH_2OH$

At the counter electrode, this H^+ ion by accepting electrons from the external circuit is getting reduced to H_2 which may reduce R-CHO to regenerate $R\text{-}CH_2OH$, restoring the chitosan polymer electrolyte to its initial state.

The photocurrent density (J) versus the potential of the cell (V) under light illumination (intensity 100mW/cm^2) is measured for CL_x ($40 \le x \le 80$) and $CL_{80}E_y$ ($30 \le y \le 50$) biopolymer electrolytes based DSSCs to estimate the power conversion efficiency (η) of the photovoltaic cell. Their J-V characteristics are shown in Fig. 9.

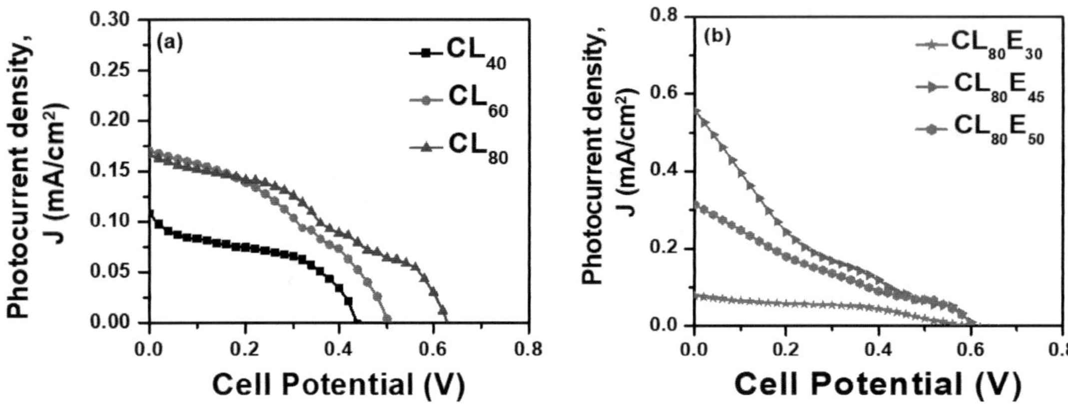

Fig. 9. J-V characteristics of the DSSCs assembled with (a) CL_x ($40 \leq x \leq 80$) and (b) $CL_{80}E_y$ ($30 \leq y \leq 50$) biopolymer electrolytes.

The fill factor (ff) and efficiency (η) are calculated by,

$$ff = \frac{P_{max}}{V_{oc}J_{sc}} \quad (8)$$

$$\eta = \frac{V_{oc}J_{sc}ff}{P_{in}} \quad (9)$$

where, P_{max} is the maximum power output and P_{in} is the input power. The fill factor (ff) is a measure of the quality of the dye-sensitized solar cell and its power conversion efficiency (η) represents the percentage of solar (light) power that the cell can transform to electrical power. The DSSC performance parameters for CL_x ($40 \leq x \leq 80$) and $CL_{80}E_y$ ($30 \leq y \leq 50$) biopolymer electrolytes based DSSCs are listed in Table 2. It is found that the power conversion efficiency (η) of the photovoltaic cells follows the conductivity trend of the biopolymer electrolytes used to construct the cells. The DSSC fabricated with CL_{80} biopolymer electrolyte exhibits an efficiency of 0.037%, which is increased to 0.051 % for $CL_{80}E_{45}$ biopolymer electrolyte based DSSC. This observation is analogous to that reported by Buraidah *et al* [25] with chitosan-polyethylene oxide polymer blend electrolytes by adding iodine crystals as a redox couple. The highest conducting biopolymer electrolyte, $CL_{80}E_{45}$ based DSSC demonstrates a large short-circuit current density (J_{sc}) of 0.556 mA/cm^2 even without any introduction of external redox couple. This experimental result may suggest that the redox couple is intrinsic within the chitosan biopolymer complex which results an increased current density under illumination.

Table 2. DSSC parameters for the photovoltaic DSSC cells with CL_x ($40 \leq x \leq 80$) and (b) $CL_{80}E_y$ ($30 \leq y \leq 50$) biopolymer electrolytes

Biopolymer electrolytes	J_{sc} (mA/cm^2)	V_{oc} (V)	*ff*	η (%)
CL_{40}	0.107	0.438	0.422	0.019
CL_{60}	0.169	0.505	0.369	0.031
CL_{80}	0.166	0.630	0.362	0.037
$CL_{80}E_{30}$	0.078	0.580	0.379	0.017
$CL_{80}E_{45}$	0.556	0.605	0.1515	0.051
$CL_{80}E_{50}$	0.314	0.603	0.209	0.039

4. Conclusions

Solid biopolymer electrolytes based on chitosan with $LiClO_4$ as a supplier of charge carriers and ethylene carbonate (EC) as a plasticizer are prepared using solution-casting technique. Introduction of salt and the plasticizer helps to tune the ionic conductivity of chitosan from ~ 10^{-9} S/cm to ~ 10^{-4} S/cm at room temperature. The transport properties of the SPEs are estimated by Raman deconvolution technique. All the ion transport parameters follow the trend of ionic conductivity of the biopolymer electrolytes. The highest conducting SPE ($CL_{80}E_y$ with $y = 45$) shows the maximum charge carrier mobility (3.125×10^{-6} $cm^2V^{-1}s^{-1}$) and highest number of mobile charge carriers (43×10^{19} /cm^3). Dye-sensitized solar cells (DSSC) are developed with CL_x ($40 \leq x \leq 80$) and $CL_{80}E_y$ ($30 \leq y \leq 50$) biopolymer electrolytes respectively with ITO/ZnO/RB as the working electrode and platinum as the counter electrode. Without any external addition of redox couple in the electrolyte, the maximum power conversion efficiency of 0.051% is obtained in $CL_{80}E_y$ (with $y = 45$) biopolymer electrolyte based DSSC, which also shows highest short-circuit current density $J_{sc} = 0.556$ mA/cm^2 and an open-circuit voltage $V_{oc} = 0.605$ V. It is found that the power conversion efficiencies of the DSSCs follow the conductivity trend of the biopolymer electrolytes used to fabricate the cells. LSV responses of the cells confirm the photovoltaic performance of the DSSCs. The current is produced due to the diffusion of monomer units of chitosan in the polymer electrolyte on the application of voltage.

Acknowledgements

Simantini Majumdar is thankful to DST for the research fellowship. Authors are thankful to Dr. Partha Pratim Ray (Jadavpur University) and Dr. Achintya Singha (Bose Institute) for rendering the facilities of Kiethley source meter and Raman spectroscopy.

References

[1] C. Polo Fonseca, S. Neves, 2006 *J. Power Sources* **159** 712-716.
[2] N.A. Johari, T.I.T. Kudin, A.M.M. Ali, T. Winie, M.Z.A. Yahya, 2009 *Mater. Res. Innov.* **13** 232-234.
[3] Y.N. Sudhakar, M. Selvakumar, D.K. Bhat, 2013 *Ionics* **19** 277-285.
[4] S. Majumdar, P. Sen, R. Ray, 2022 *Materials Research Bulletin* **151** 111814
[5] R. Singh, B. Bhattacharya, H.W. Rhee, P.K. Singh, 2014 *Int. J. Electrochem. Sci.* **9** 2620-2630.
[6] G.J. Liang, Z.C. Zhong, J. Xu, Z.C. Zhang, M.H. Chen, Z.F. Li, 2012 *Acta Phys.Chim. Sin.* **28** 2852-2860.
[7] T. Stübinger, W. Brütting, 2001 *J. Appl. Phys.* **90** 3632.
[8] C.J. Brabec, N.S. Sariciftci, J.C. Hummelen, 2001 *Adv. Funct. Mater.* **11** 15.
[9] C.P. Lee, C.Y. Chou, C.Y. Chen, M.H. Yeh, L.Y.Lin, R. Vittal, C.G.Wu, K.C. Ho, 2014 *Journal of Power Sources* **246** 1-9.
[10] M. Law, L.E. Greene, J.C. Johnson, R. Saykally, P.D. Yang, 2005 *Nat. Mater.* **4** 455.
[11] B. Pradhan, S.K. Batabyal, A.J. Pal, 2007 *Solar Energy Materials & Solar Cells* **91** 769-773
[12] M.H. Khanmirzaei, S. Ramesh, K. Ramesh, 2015 *Materials and Design* **85** 833-837.
[13] T. Tiwari, N. Srivastava, P.C. Srivastava, 2011 *Ionics* **17** 335-360.
[14] S. Majumdar, P. Sen, R. Ray, 2022 Ionics **28** 1403-1418
[15] M. H. Buraidah, L. P. Teo, S. R. Majid, A. K. Arof, 2009 *Physica B* **404** 1373-1379.
[16] S. Majumdar, R. Ray, 2021 *Journal of Polymer Research* **28** 157.
[17] S. Majumdar, P. Sen, R. Ray, 2022 *Journal of Solid State Electrochemistry*, **26** 527-547.
[18] S. Majumdar, R. Ray, P. Sen, 2021 *Electrochimica Acta* **385** 138295.
[19] P. Pal, A. Ghosh, 2016 *Journal of Applied Physics* **120** 045108.
[20] A. Mondal, N. Giri, S. Sarkar, S. Majumdar, R. Ray, 2019 *Materials Science in Semiconductor Processing* **91** 333–340.

[21] G. K. Williamson, W. H. Hall, 1953 *Acta Metall.* **1** 22–31.
[22] B. Klassen, R. Aroca, M. Nazri, G. A. Nazri, 1998 *J. Phys. Chem. B* **102** 4795-4801.
[23] M.H. Buraidah, S. Shah, L.P. Teo, F.I. Chowdhury, M.A. Careem, I. Albinsson, B.-E. Mellander, A.K. Arof, 2017 *Electrochimica Acta* **245** 846-853.
[24] A. G. Kontos, M. Fardis, M. I. Prodromidis, T. Stergiopoulos, E. Chatzivasiloglou, G. Papavassiliou, P. Falaras, 2006 *Phys. Chem. Chem. Phys.* **8** 767-776.
[25] M.H. Buraidah, L.P. Teo, C.M. Au Yong, S. Shah, A.K. Arof, 2016 *Optical Materials* **57** 202-211.

IOP Conf. Series: Materials Science and Engineering 1291 (2023) 012015 doi:10.1088/1757-899X/1291/1/012015

Physical and electrical properties of Sm_2O_3 doped Boro-Zinc-Vanadate glasses

Ashwini Devidas[1], T Sankarappa[1,*], Amarkumar Malge[1], Mohansingh Heerasingh[1], Raghavedra B[1], Jamadar Pallavi[1] and Arvind Dyama[1]

[1]Physics Department, Gulbarga University, Kalaburagi

*Author for correspondence, e-mail: sankarappa@rediffmail.com

Abstract. Sm_2O_3 doped boro-zinc-vanadate glass systems were synthesised by following the melt quenching method. XRD patterns indicated largely non-crystalline nature with few nano-crystallites. Room temperature density was measured. Molar volume and various polaron parameters were estimated. Density and molar volume are found to vary non-linearly with samarium concentration. Conductivity has been measured by two probe technique for temperature range 303K - 573K. High temperature conductivity obeyed the small polaron hopping (SPH) theory. Activation energy for conduction in the temperature regime of small polaron theory is found to vary from 0.249 eV to 0.368 eV non-linearly with Sm_2O_3 concentration. The conductivity data at low temperature deviated has been looked into using Mott's VRH model and the density of states at Fermi level were determined. Shimakawa's multiphonon tunnelling model has also been applied to the low temperature conductivity and found linearity between logarithmic conductivity, ln(σ) and logarithmic temperature ln(T) as predicted by the model. The temperature exponent values obtained from Shimakawa's model fit are found to be in good agreement with literature. Therefore, it is concluded that at low temperature, carrier multiphonon tunnelling is the charge transport mechanism in the present glasses.

Key words. Zinc-vanadate glasses, density, dc conductivity, polaron hopping.

1. Introduction

Many researchers have discussed importance of transition metal oxides (TMO) as modifiers in the glass formation. One of the most conventional TMOs is vanadium pentoxide (V_2O_5), which is a conditioned glass former and has good electrical properties [1-2]. Almost all V_2O_5 doped glasses demonstrated semiconducting characteristics due to hopping of electrons from a low valence state (V^{4+}) to the high valence state (V^{5+}) [3-4]. These glasses are known as n-type semiconducting glasses and are useful in opto-electronic, memory switching, solid state devices and lasers [1, 3]. Boric oxide (B_2O_3), a glass former has been incorporated into various glasses as a flux material to attain low melting temperatures, thermal stability, etc. [5]. Addition of ZnO enhances glass forming, chemical resistance and reduces crystallization rate. It also acts as a modifier and as well as network former [6, 7]. Samarium oxide (Sm_2O_3) is a rare earth oxide used for enhancing physical, electrical, dielectric and optical properties of glasses [8, 9]. A limited literature exists on Sm_2O_3 and TMI doped borate glasses. Available results indicated significant changes in electrical properties with increasing rare-earth ion concentrations caused by fluctuations in the mobility of rare-earth ions and their blocking effect to polaronic and ionic motions [10-14]. There are reports suggesting that addition of Sm_2O_3 improves glass performance in solid-state devices [15].

IOP Conf. Series: Materials Science and Engineering 1291 (2023) 012015 doi:10.1088/1757-899X/1291/1/012015

The technological applications of zinc vanadate glasses as cathode materials in batteries, electrical and optical memory switching devices makes them fascinating. Few researchers recently studied physical, structural and electrical properties of ternary zinc borovanadate glasses and reported that their electrical properties got enhanced [16-18].

Understanding the influence of rare earth and transition metal ions on glasses enables to develop suitable glass compositions for desired applications with required electrical properties and chemical durability [10, 14]. The zinc-boro vanadate glasses containing samarium oxide (Sm_2O_3) have not been previously received much attention. The present study was aimed at investigating the role of Sm_2O_3 on density, structure and conductivity of zinc-boro vanadate glasses. These glasses provide a unique opportunity to learn about the conduction process in these scientifically important set of glasses. Activation energy for electrical conductivity, phonon frequency and Debye's temperature were also determined. Density of states at Fermi level in the light of Mott's VRH model and temperature exponent in Shimakawa's model at low temperature have been determined.

2. Experimental

Samarium doped zinc borovanadate glasses were produced by melt quenching route. Chemicals of AR grade (V_2O_5, ZnO, H_3BO_3 and Sm_2O_3) in fixed concentrations were mixed in an agate mortar and taken in silica crucibles and melted at 1323K in a furnace and quenched between SS plates. The samples so obtained were annealed at 523K for 24 hours to relieve the thermal strains caused by the quenching process. The composition of the samples is fixed to be, $(ZnO)_{0.3} - (V_2O_5)_{0.3} - (B_2O_3)_{0.4-x} - (Sm_2O_3)_x$ where x = 0.002, 0.005, 0.007, 0.01, 0.02 mole % labeled as ZVBS1, ZVBS2, ZVBS3, ZVBS4 and ZVBS5 respectively.

The XRD patterns were recorded in a Smartlab X-ray diffractometer having CuK_α radiation of wavelength 1.541 Å, in the 2θ range 5° – 90°. Density, ρ, was measured as per Archimedes principle using, a VIBRA makes HT analytical balance of 0.1mg precession and Xylene for immersing the samples. Samples were cut to 3mm x 3mm x 4mm size and their two large surfaces were silver painted. The electrical resistance measurements were carried out for the temperature range 303-573K by adopting the two point method. Before measuring the current and voltage, the ohmic contacts have been confirmed. The method described in [13] has been followed to determine conductivity.

3. Results and discussion

3.1. *X-ray diffraction studies*

The recorded XRD patterns of Sm_2O_3-doped boro-zinc-vanadate glass nanocomposites are presented in figure1. The patterns exhibited a single small peak. The diffraction peak for all the samples is appearing exactly at 28°. This crystalline peak is indexed by comparing with JCPDS card. As per JCPDS card no. 00-009-0142, the space lattice of VO_2 is monoclinic with a = 5.743 Å, b = 4.517 Å, and c = 5.375 Å. The Miller indices (*h k l*) of this diffraction peak corresponds (0 1 1) [19, 20]. The average grain size (D) due to the observed peak has been determined using the Scherrer formula [10],

$$D = \frac{0.89\ \lambda}{\beta\ cos\theta} \tag{1}$$

Where, β is the full width at half maxima (FWHM) of the peak, λ the wavelength of X-rays (1.54Å) and θ the Bragg diffraction angle. The average crystallite sizes are found to be 1.995 nm, 0.036 nm, 1.901 nm, 1.895 nm and 2.671 nm for samples of x = 0.002, 0.005, 0.007, 0.01 and 0.02 respectively. Since the grain sizes determined are of few nanometers, the current samples are called glass nanocomposites. The ZVBS2 sample is observed to be less crystalline compared to other samples as, the peak corresponding to it is of less intensity.

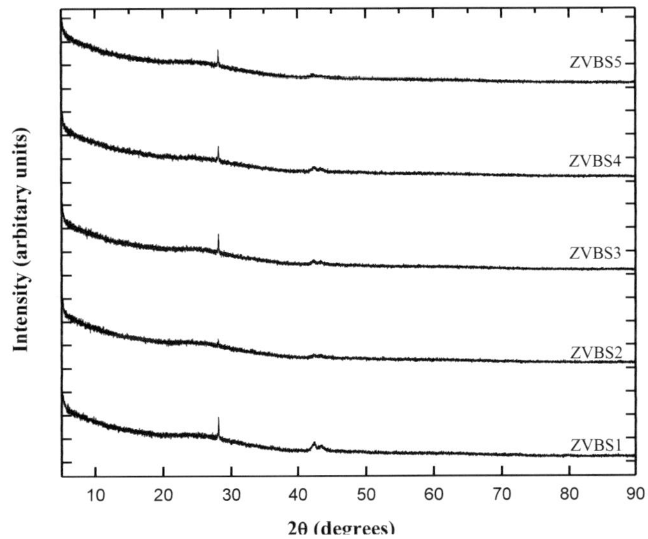

Figure1. The XRD patterns of ZVBS glass nanocomposites.

3.2. Density and Molar volume

Density, ρ is a significant physical parameter that can be utilised to understand network packing. It also explains the degree of structural compactness. Density of the present glass nanocomposites was measured by following Archimedes method and using the following relation [5],

$$\rho = \rho_b \times \frac{W_a}{W_a - W_b} \quad (2)$$

Where, ρ_b is the density of the buoyant (Xylene = 0.865gm/cm^3), W_a and W_b are weight of the samples in air and buoyant respectively. The experiments were repeated four times for reproducibility and the error on the measured density has been estimated to be ±0.005 g/cm^3. It is noticed that density and molar volume behave in opposite fashion to each other with Sm_2O_3 content and they vary in a zig-zag way. In fact, in the present glasses high dense Sm_2O_3 (8.35gm/cm^3) is substituted for low dense B_2O_3 (2.46gm/cm^3). Therefore, the samples are expected to show increasing trend of density with Sm_2O_3 content. Variations in both density and V_M with Sm_2O_3 are depicted in figure 2 and recorded in Table 1. Density showed increasing trend with Sm_2O_3 content except for ZVBS2 sample for which density is bit high. It may be due to its less crystalline in nature.

Using the measured density and composition, the molar volume, V_M has been estimated as per equation [6],

$$V_M = \sum \frac{x_i M_i}{\rho} = \frac{(0.3)M_{V_2O_5} + (0.3)M_{ZnO} + (0.4-x)M_{B_2O_3} + (x)M_{Sm_2O_3}}{\rho} \quad (3)$$

Where, x_i is the mole fraction, M_i the molecular weight of the ith component and ρ the density of glass sample. Increase of ρ and decrease of V_M with increase of Sm_2O_3 content except for ZVBS2 sample indicates that glass network is getting tight packed with Sm_2O_3 concentration. Such changes in ρ and V_M have been noticed in the previous works of samarium doped zinc-tellurite glasses [21]. The concentration of TMI (N), internuclear distance (R) and polaron radius (r_p) are determined using the following mentioned expressions [15],

Concentration of TMI, $N = (mol\%\ of\ RE)\left[\frac{\rho}{M}\right] N_A$ (4)

Internuclear distance of TMI, $R = \left[\frac{1}{N}\right]^{\frac{1}{3}}$ (5)

Polaron radius $r_p = \frac{1}{2}\left(\frac{\pi}{6N}\right)^{\frac{1}{3}}$ (6)

The obtained values of N, R and r_p are tabulated in Table 1. Both R and r_p are decreased with increase of Sm^{3+} ion concentration (N) and this result agree with reports on Sm_2O_3 doped lithium-borate glasses [14] and lithium borosilicate glasses [15]. Concentration of TMI (N) in the present case is increased as Sm_2O_3 content is increased. Sm_2O_3 content is increased at the expense of B_2O_3 and this should lead to decrease in the concentration of TMI (N). Instead, N is increasing with Sm_2O_3. This may be due to dynamical changes in network with Sm_2O_3 content. This agrees with reports on Sm_2O_3–ZnO–TeO_2 [21] and Co_3O_4–V_2O_5–B_2O_3 [22] glasses.

Table 1: Density, ρ, molar volume, V_m, concentration of TMI, N, internuclear distance, R and polaron radius, r_p.

Glass	Density, ρ (gm/cm^3) ±0.005	Molar volume, V_m (cm^3/mol) ±0.004	N x10^{20} (ions/cm^3)	R (10^{-7} m)	r_p (10^{-8} m)
ZVBS1	3.0944	41.6526	0.2892	3.2578	13.1289
ZVBS2	3.2143	40.2276	0.7486	2.3727	9.5619
ZVBS3	3.1400	41.4061	1.0182	2.1415	8.6303
ZVBS4	3.1604	41.3528	1.4564	1.9006	7.6594
ZVBS5	3.2087	41.4319	2.9074	1.5094	6.0828

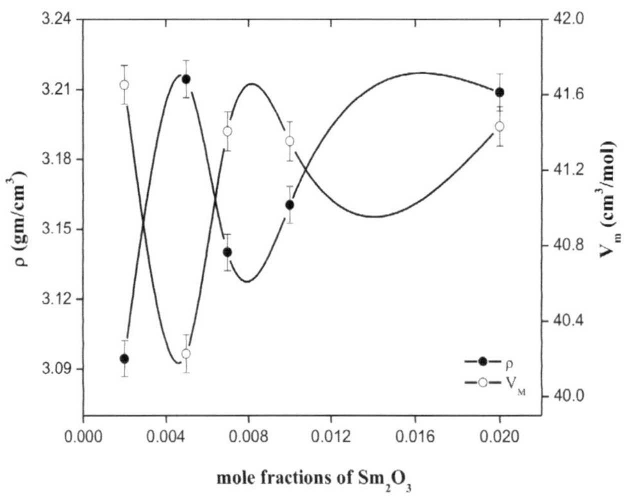

Figure2. Compositional dependence of density, D and molar volume, Vm.

3.3. DC electrical conductivity

As depicted in figure 3, the electrical conductivity, σ, was found to vary between 0.5x10^{-4} and 3.2x10^{-4} $(\Omega m)^{-1}$ for the measured temperature range. Due to positive change in σ with temperature, these samples may be treated as semiconducting glasses. It can be observed that conductivity was very minimal for temperature upto 400K and started increasing then onwards. The conductivity of the present glasses is more than reported values for Sm_2O_3–B_2O_3–V_2O_5 –Li_2O glasses [8] and Sm_2O_3–

B_2O_3–Al_2O_3 –Na_2O–$CuCl_2$ glasses [9] by a few orders of magnitude. However, it is less than that reported for V_2O_5–SeO_2–ZnO glasses [2] and V_2O_5–ZnO and glasses [4]. The SPH model of Mott has been used to know changes occurring in σ with temperature.

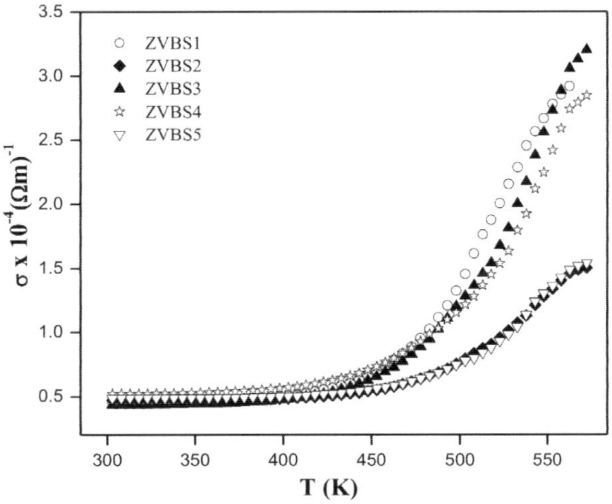

Figure 3. Variation of dc conductivity as a function of temperatures.

In Mott's SPH model [23, 24] the conduction is predicted to be caused by phonon-assisted hopping of (small) polarons between localised states. It predicts conductivity due to nearest-neighbour hopping mostly in the non-adiabatic regime for temperature above T_D to be,

$$\sigma = (\sigma_0/T) \exp(-W/K_B) \quad (7)$$

Where, W being the activation energy required for conduction and σ_0 the pre-exponential quantity. The graphs of ln(σT) versus (1/T) are plotted in figure 4. It is observed that ln(σT) decreases linearly with $(T)^{-1}$ up to T_D and then deviates from this behavior at lower temperature. The temperature at which ln(σT) departs from linear behavior with $(T)^{-1}$ is related to Debye's temperature, θ_D as $\theta_D = 2T_D$. So, the linear lines were fit to data above T_D. The W value was estimated using the slope and is found in the range 0.2489 – 0.3675eV. The W values so obtained are found comparable with literature value for Fe_2O_3–Sm_2O_3–ZnO–P_2O_5 glasses [10].

As per reference [2], $h\nu_0 = k_B\theta_D$ where, ν_0 is the optical phonon frequency, h the planck's constant and k_B Boltzmann constant. The calculated θ_D and ν_0 values are listed in Table 2. With increasing Sm_2O_3 concentration, the Debye temperature (θ_D) and optical phonon frequency (ν_0) gradually increases. The increased Sm_2O_3 content results in opening up of the glass network structure, which in term increases thermal vibrations due to a rise in optical phonon frequency and hence θ_D. Variation of θ_D with Sm_2O_3 content is shown in figure 5.

Figure 6 shows observed changes in σ at 563K and W with mole fractions of Sm_2O_3. A close observation of conductivity and W variations with Sm_2O_3 content revealed that they are decreasing with increase of Sm_2O_3 up to 0.005 mole fractions and increased for higher amounts of Sm_2O_3 and then decrease. Both σ and W show the same trends with increment in Sm_2O_3 content. This indicates that the structure of the present nano-composites is changing such that an obstacle to polaronic hopping develops in increasing order with increment in Sm_2O_3. Obtained results are useful to gauge possible mechanism of conductivity. In general, the increase in conductivity due to the presence of Sm_2O_3 can be attributed to the successive jumping of Sm^{+3} ion from one non-bridging oxygen (NBO) to another [23].

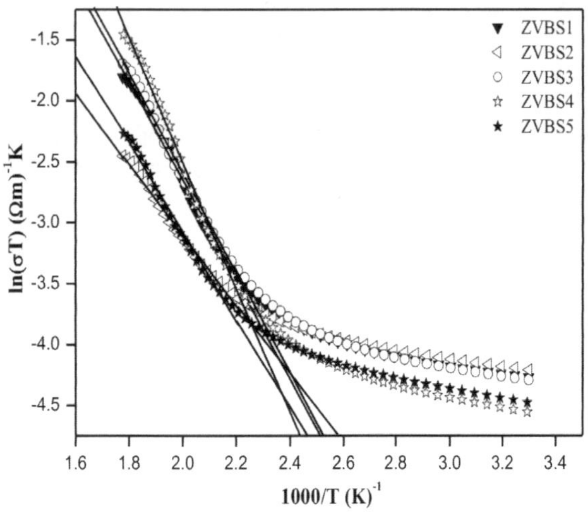

Figure 4. ln(σT) variation with reciprocal temperature, (T^{-1}) for ZVBS glass nanocomposites.

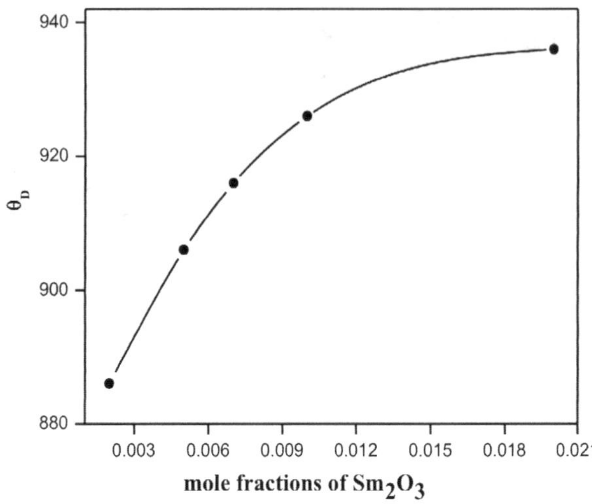

Figure 5. Effect of Sm_2O_3 concentration on Debye Temperature for ZVBS glass nanocmposites.

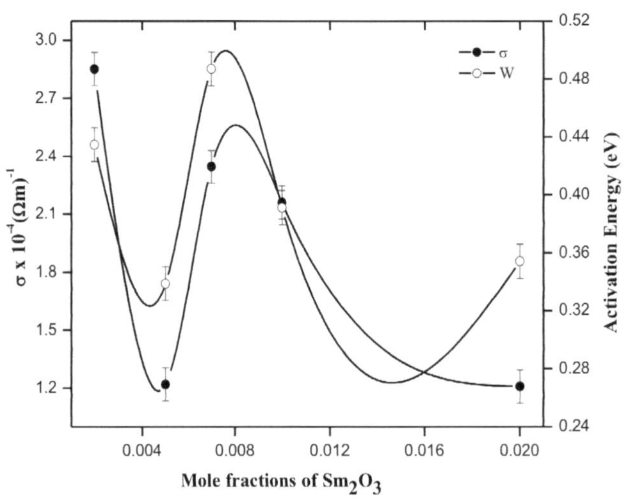

Figure 6. Variation of σ at 563K and activation energy (W) with mole fractions of Sm_2O_3.

Table 2: Conductivity σ at 563K and Debye's temperature, θ_D and activation energy, W from SPH fits.

Glass	T_D (K)	θ_D (K)	ν_0 x 10^{13} (S^{-1}) optical phonon frequency	σ at 563K x10^{-4} (Ωm)$^{-1}$	W ±0.01 (eV)
ZVBS1	443	886	1.846	2.9203	0.3469
ZVBS2	453	906	1.887	1.4596	0.2489
ZVBS3	458	916	1.908	3.0578	0.3675
ZVBS4	463	926	1.929	2.7406	0.3267
ZVBS5	468	936	1.950	1.4880	0.2719

Mott's VRH model was also used for analyzing conductivity for below T_D. This model predicts positive variation of conductivity with n temperature as follows [24],

$$\sigma = A \exp(-B/T^{-1/4}) \quad (8)$$

Where, $A = [e^2/2(8\pi)^{1/2}]\upsilon_0[N(E_F)/\alpha k_B T]^{1/2}$ And $B = 4[2\alpha^3/9\pi k_B N(E_F)]^{1/4}$ and $N(E_F)$ the density of states at Fermi level.

The plots of ln(σ) vs $T^{-1/4}$ below T_D are plotted as per Mott's Variable Range Hopping model in figure 7. The computed values of $N(E_F)$ are of the order of 10^{23} - 10^{24} eV^{-1} cm^{-3} are listed in (Table 3). These $N(E_F)$ values are in close agreement with the values mentioned in [23, 24]. These $N(E_F)$ are higher than those obtained for La_2O–V_2O_5–P_2O_5 glasses [25] and La_2O and CeO_2 doped vanadotellurite glasses [26] and, are smaller than reported values for V_2O_5–SeO_2–ZnO [2] and Fe_2O_3–Sm_2O_3–ZnO–P_2O_5 [10] glasses.

Shimakawa and Miyake [24, 27] suggested a different model for explaining conductivity at low temperatures in glass nanocomposites based on hopping of localized carriers due to multi phonon tunneling with weak electron-lattice coupling. It involves absorption and/or emission of phonons to overcome the site separation energy.
According to this model, conductivity, σ is given as [27],

$$\sigma = \sigma_0' T^P \quad (9)$$

Where, the constant σ_0 depends on composition of the glass and exponent p. Fig. 8 shows Shimakawa's plot of ln(σ) versus ln(T) and linear fits gave p values in the range from 4.034 to 9.307 for ZVBS glasses. A linear relationship between ln(σ) and ln(T) is clearly is in figure 8. A comparable values of p are reported in [27, 28] for different V_2O_5 based glasses. The temperature exponent, p values obtained are nearer to reported values for two TMI doped tellurite glasses [24]. It is possible to draw the conclusion that the carrier transport below T_D may involve large polarons (electrons which are weakly coupled) tunnelling between the nanocrystallites through multiplephonons.

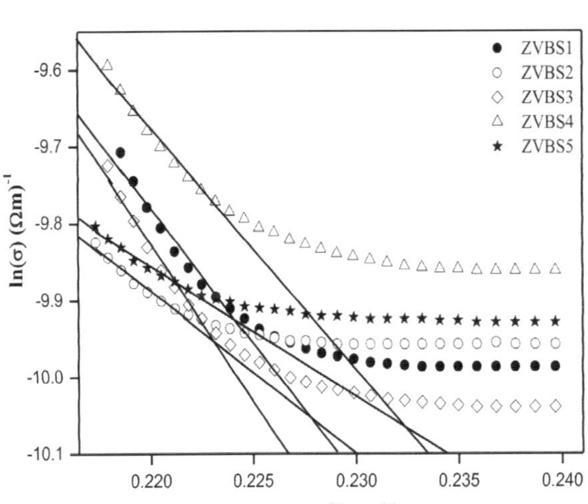

Figure 7. Mott's VRH plots of ln(σ) versus $T^{-1/4}$ for ZVBS glass nanocomposites.

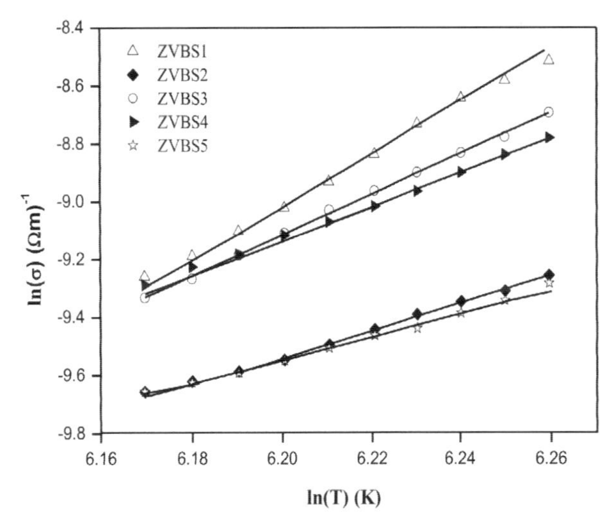

Figure 8. Plots of ln(σ) versus ln(T) for T<T_D. Shimakawa's multiphonon tunnelling model fits are shown as solid lines.

Table 3: The fit parameters due to Mott's VRH and Shimakawa's models.

Glass	Mott's VRH model		Shimakawa's model	
	$N(E_F)$ eV^{-1} cm^{-3}	Linear Regression coefficient	p	Linear Regression coefficient
ZVBS1	4.5828 x 10^{23}	0.984	9.307	0.997
ZVBS2	5.2685 x 10^{24}	0.980	4.817	0.998
ZVBS3	1.2271 x 10^{23}	0.989	7.057	0.998
ZVBS4	2.4233 x 10^{23}	0.982	5.960	0.997
ZVBS5	3.4125 x 10^{24}	0.984	4.034	0.995

Compared to VRH model fit, Shimakawa's model gave better degree of fits with higher regression coefficient and quality. Also, since there are nanocrystallity present in the samples, Shimakawa's model of multiphonon tunnelling may be considered to be the conduction mechanism at low temperature.

4. Conclusions

Samarium oxide (Sm_2O_3) doped boro-zinc-vanadate glasses were prepared and investigated for structure, density and electrical conductivity.

- The XRD patterns exihibited presence of nanocrystalline phases.

- The density and molar volume of the glasses are found to be inversely related to each other. Density increased from 3.0944 to 3.2143gm/cm^3 and molar volume decreased from 41.6526 to 40.2276cm^3/mol with Sm_2O_3 content except for ZVBS2 sample for which density is slightly high and molar volume is bit low. It may be due to its less degree of crystallinity.

- The conductivity of the present glass nano composites is found increased with temperature. It is few orders of magnitude high compared to Sm_2O_3–B_2O_3–V_2O_5–Li_2O glasses.

- The small polaron hopping is found to explain conductivity at high temperature ($T>T_D$). The conductivity and activation energy of the present systems are observed to be varying non-linearly with Sm_2O_3 content. This behaviour is attributed to variations in polaron hopping distance and polaron radius with Sm_2O_3. The activation energy, W for conduction is found in the range 0.2489 – 0.3675eV.

- Mott's VRH model has been applied for analyzing conductivity data at low temperature and from which the density of states at Fermi level were determined to be in the range 1.227x 10^{23}–5.268 x 10^{24} eV^{-1} cm^{-3}.

- Shimakawa's multiphonon tunnelling model has also been employed to analyze conductivity at low temperature. At these temperatures, conductivity was found to be proportional to T^p. The exponent p was found varied from 4.034 to 9.307 for the present glasses, which indicates that the charge transport is governed by a multi-phonon process of weak electron-lattice coupling.

- For the first time Sm_2O_3 doped zinc-boro vanadate glasses have been explored thoroughly for physical, structural and conductivity and conduction mechanisms have been understood.

Reference

[1] Anindya Sundar Das, Madhab Roy, Debasish Roy and Sanjib Bhattacharya, *Indian Journal of Pure and Applied Physics*, **57** (2019) 803–811
[2] Anindya Sundar Das, Madhab Roy, Debasish Roy and Sanjib Bhattacharya, *International Journal for Research in Applied Science and Engineering Technology,* **5** (2017) 1940–1951
[3] R. Punia, R. S. Kundu, S. Murugavel and N. Kishore, *J. Appl. Phys.* **112**, (2012) 113716
[4] Aloka Ghosh, S. Bhattacharya, D. P. Bhattacharya and A. Ghosh, *J. Appl. Phys.* **103** (2008) 083703
[5] Gokhan kilic, *Adiyaman University Journal of Science,* **10** (1) (2020) 307–325
[6] N. Sivasankara, *Materials Todays: Proceedings*, **49** (3) (2022) 771–776
[7] Gokhan kilic, Erkan Ilik, Mohmoud, R. El-Mallawany, F. I. El-Agawamy and Y. S. Rammah, *Journal of Ceramics International*, **46** (11) (2020) 19318–19327
[8] M. P. F. Graça, H. Fawzy, Y. Badr, M. M. Elokr, C. Nico, R. Soares, L. C. Costa and M. A, Valenta, *Phys. Status Solidi C*, **8** (2011) 3107–3110
[9] M.S. Sadeq and H.Y. Morshidy, *Journal of Rare Earths*, **38** (2020) 770–775
[10] Yumnam Bonney Singh , Dipankar Biswas , Subham Kumar Shah , Subham Shaw, Rittwick Mondal, Anindya Sundar Das, Soumyajyoti Kabi and Loitongbam Surajkumar Singh *Journal of Materialie,* **15 (**2020) 100963
[11] Amarkumar Malge, T. Sankarappa, T. Sujatha, P. Abdul Azeem, G. B. Devidas and Santhoshkumar Kori, *Materials Today: Proceedings,* **26** (2020) 1960–1963
[12] A.S. El-Deeb, M.M. Abd elkader, A. M. Ismail and M. Y. Hassaan, *Journal of Optik - International Journal for Light and Electron Optics*, **221** (2020) 165358
[13] Ashwini Devidas, T. Sankarappa, Amarkumar Malge, Mohansingh Heerasingh, B. Raghavendra, Jamadar pallavi and Arvind Dyama, DC conductivity in La_2O_3 doped zinc vanadate glasses, *Materials Todays: Proceedings*, 2023.
[14] Sunil Thomas, Sk. Nayab Rasool, M. Rathaiah, V. Venkatramu, Cyriac Joseph and N. V. Unnikrishnan, *Journal of Non-Crystalline Solids,* **376** (2013) 106–116
[15] V. Y. Ganvir and R. S. Gedam, *Journal of Integrated Ferroelectrics* **185** (1) (2017) 102–108
[16] Ashwini Devidas, T. Sankarappa , Amarkumar Malge, Mohansingh Heerasingh and B. Raghavendra, *Journal of the Australian Ceramic Society* **59** (2023) 391–402
[17] S. Sindhu, S. Sanghi, A. Agarwal, N. Kishore and V. P. Seth, *Journal of Alloys and Compounds*, **428** (2007) 206–213
[18] B. Sujatha, R. Viswanatha, H. Nagabushana and C.N. Reddy, *Journal of Materials Research and Technology*, **6** (1) (2017) 7–12
[19] Changhong Chen, Ypng Zhao, Xuan Pan, V. Kuryatkov, A. Bernussi and M. Holtz, *Journal of Applied Physics,* **110** (2011) 023707
[20] Georg Andersson, *Journal of Acta Chemica Scandinavica*, **10** (1956) 623-628
[21] B. Eraiah, *Journsl of Bull. Mater. Sci.,* **29** (4) (2006) 375–378.
[22] G. Rajashekara, J. Sangamesh, B. Arunkumar, N. Nagaraja and M. Prashant Kumar, *Journal of Non-crystalline solid*, **481** (2017) 289–294
[23] D Souri, Z E Tahan and S A Salehizadeh, *Indian J Phys,* **90** (4) (2016) 407–415
[24] H. Sakata and K. Sega, B. K. Chaudhuri, *Physical Review B*, **60** (1999) 3230–3236

[25] G.B. Devidas, T. Sankarappa, T. Sujatha, M. prashant Kumar, P. J. Sadashivaiah and P. S. Naik, *J. Solid State Sciences* **13** (2011)1385–1390
[26] M. Prashant Kumar and T. Sankarappa, *J. Non-Crystalline Solids,* **355** (2009) 295-300
[27] E. M. Ahmed, N. A. El-Ghamaz, and A. M. Edres, *Journal of Materials Science: Materials in Electronics* **29** (2018) 15191–15202
[28] F. Abdel-Wahab, A. Merazga and A. A. Montaser, *J Low Temp Phys* **186** (2017) 372–384

IOP Conf. Series: Materials Science and Engineering 1291 (2023) 012016 doi:10.1088/1757-899X/1291/1/012016

DFT study of Hydrogen storage capacity on Hf doped graphene

Ravindra Karde and Baliram Lone*

Nanomaterials Research Laboratory, Department of Physics, Vinayakrao Patil Mahavidyalaya, Vaijapur-423701 Dist. Aurangabad, Maharashtra, India

* *E-mail: drbglone@gmail.com*

Abstract. We explored the molecular hydrogen storage capacity of complex system with Hafnium doped graphene (Gr-Hf) using the Density Functional Theory (DFT) method. We efficiently adsorbed molecular six hydrogens on the Hf doped graphene surface. The quantum chemically calculated adsorption energy is found negatively in the range of -344.433 Ry to -333.836 Ry this implies as increasing the adsorbed hydrogen molecule on hafnium doped graphene (Gr-Hf) sheet the adsorption energy decreases continuously. The binding energy of after adsorbing second hydrogen molecule too much larger than the next adsorbed H2 molecules i.e., the binding energy per hydrogen molecule highly decreases when we increase adsorbed atom (2.197 Ry in 2H2 to 1.048 Ry in 3H2) then small decreases for next adsorbed H2 molecules. The extracted binding energy found in the range 2.197 Ry to 2.120 Ry, fermi energy found minimum for 1H2 shows the minimum electron occupancy at the different energy levels. The fermi energy increases accordingly, the electron occupancy also increases and evaluates higher electron occupancy with fermi energy 2.850 eV for 6H2 and density of states (DOS) confirm the weak interaction between σ bonding electrons of H2 molecule with Hf doped graphene complex system. The proposed system opens a new insight for hydrogen storage-based devices.

1. Introduction

There is an abrupt decrease in the reserves of fossil fuel which causes environmental pollution. It's our prime responsibility to find a suitable alternative of sustainable renewable energy resources. We know that hydrogen is abundantly available in the molecular form on the Earth [1-2]. However, hydrogen has specific characteristics due to which it makes it complicated to store in large quantities and lack a specified amount of space. Since, last few years scientists across the globe have investigated that binding ability between hydrogen molecules (H2) and pristine nanomaterials is weak [2-6]. Graphene is a two-dimensional (2D) honeycomb crystal structure with vast applicability in optical, magnetic, electrical and electronic properties [7-8]. Acik M et al investigated the optical, magnetic, electrical, and electronic properties of graphene. Sun Y reported Al doped with Hf molecule can generate adsorption energy hundred times greater than intrinsic graphene [9-10]. Ao Z M found average adsorption energy -0.193 eV/H2 [11], increases while increasing adsorption of H2 molecule as calculated in the range -0.31, -0.28, -0.21 eV/H2 [12], the adsorption energy found in the negative range with Sb monolayer decorated on different atoms [13]. Lone B. reported Boron doped Mg decorated graphene on hydrogen molecule. The binding energy evaluated in the range -0.566 to -0.689 eV/H2 [14-15]. Whereas, the increase in Titanium doped graphene [16] between -0.534 to -0.626 eV/H2. In our previous findings the s and d orbitals of H2 molecule and Cs atom at -0.2eV overlaps of main peaks represent strong hybridization and binding of s and d orbitals of H2 and Cs atom [17]. Increasing the binding energy correspondingly decreases the binding distance [18]. However, adsorption energy of molecular hydrogen on Mo doped graphene complex system explored in the range -0.534 eV/H2 to -0.624 eV/H2 [19]. The bio-molecular system studied with adsorption of thymine with graphene evaluated the adsorption energy -6.01 kJ/mol [20]. The cytosine adsorbing on SWNT shows the binding energy -0.38 eV [24] with good agreement

IOP Conf. Series: Materials Science and Engineering 1291 (2023) 012016 doi:10.1088/1757-899X/1291/1/012016

to charge transfer [25]. Furthermore, in the present study, we calculated the structural and electronic properties of Hf doped graphene with up to 6H2 molecules by adsorption phenomenon. These results indicate that the hydrogen storage capacity is extensively improved by doping of Hf atom.

2. Computational details

We considered the model [26] by performing Density Functional Theory (DFT) [27] calculations using Quantum ESPRESSO (QE) package [28-29] to study the adsorption of H_2 molecules on Hf doped graphene sheet. For valence electron interaction pseudopotential used. The modelled and optimized structure visualized with Xcrysden visualization code [30]. The general gradient approximation (GGA) used to system and the approximation Perdew, Burke and Ernzerhof (PBE) applied [31]. For self-consistent calculation for the wave function the kinetic energy cut off (ecutwfe) 20 Ry and charge density cut off (ecutrho) 200 Ry are used (1 Ry = 10^{-4} Ry) to perform calculation. The k-points used $4 \times 4 \times 4$ Monk-horst Pack. The described kinetic energy cut off around ten times of charge density cut off. The mesh k-points $12 \times 12 \times 12$ were used to fermi energy in non-self-consistent calculation.[32]. The separately degauss and ecutwfc optimized in self-consistent and non-self-consistent calculations. The degauss value taken is 0.01. Number of bands 50 were used to carry non-self-consistent calculation.

The modelled graphene sheet 3×3 supercell containing 18 carbon atoms. The adsorption energy (E_{ad}) of Hf doped graphene (Gr-Hf) and pristine (Gr) calculated using following expression

$$E_{ad\text{-}Hf} = E_{Hf/Gr} - E_{Hf} - E_{Gr} \quad (1)$$

Where $E_{Hf/Gr}$ is calculated total energy of the complex Hf doped graphene sheet, E_{hf} energy of isolated *Hf atom,* E_{Gr} energy of pristine graphene sheet

The average binding energy of Hf atoms in pristine graphene is calculated using expression

$$E_b = [E_{nHfGr} - E_{Gr} - nE_{Gr}]/n \quad (2)$$

Where the E_{nHfGr}, E_{Gr}, nE_{Gr} *are* the total energy of Hf doped graphene with n atoms, pristine graphene (PGr) and pristine graphene with n atoms respectively.

3. Results and Discussion

3.1. *Structural Analysis*

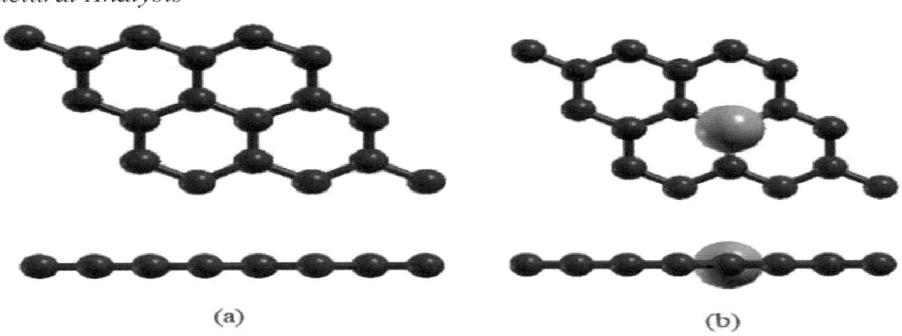

Figure (1). The optimized geometries of (a) Pristine Graphene (PGr) and (b) Hf-Doped graphene (top view and side view)

The optimized molecular geometries of pristine graphene and complex system Hf doped Graphene (GrHf) sheet and Hafnium (side view and top view) shown in figure 1 (a) and figure 1 (b) respectively. The pristine graphene (PGr) containing 18 carbon atoms having electronic configuration $1s^2\,2s^2\,2p^2$, the doping atom (Hafnium) which have outermost electronic configuration is $5d^2\,6s^2$. When a carbon atom was replaced by single Hf atom in graphene sheet the change in the bond length is observed. After optimization of the considered complex system the bond length of Hf-C, C-C is found 2.6927 Å and 1.4249 Å respectively.

3.2. *Adsorption of Hf doped on graphene surface*

We performed a complex system of hafnium doped graphene (Hf-Gr) sheet and decorated six hydrogen molecules (H_2) successively. Figure 2. shows optimized geometries of Hafnium doped graphene with adsorbed hydrogen molecules (a-f) top view and figure 3 shows side view of same.

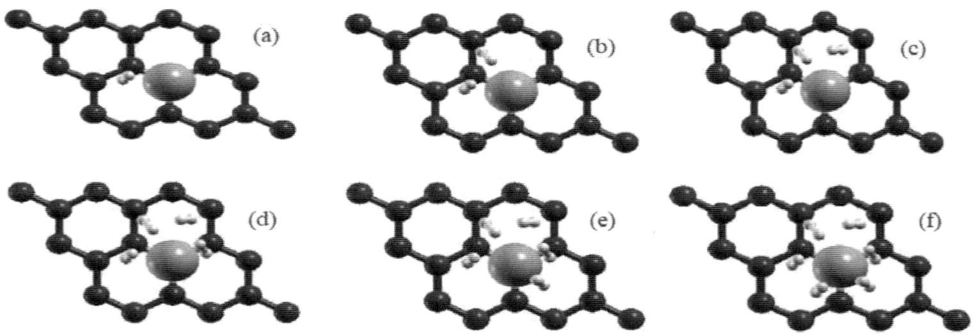

Figure (2). The DFT optimized geometries (top view) of one (a), two (b), three (c), four (d), five (e), six (f) H_2 molecules adsorbed on Hf- Doped Graphene

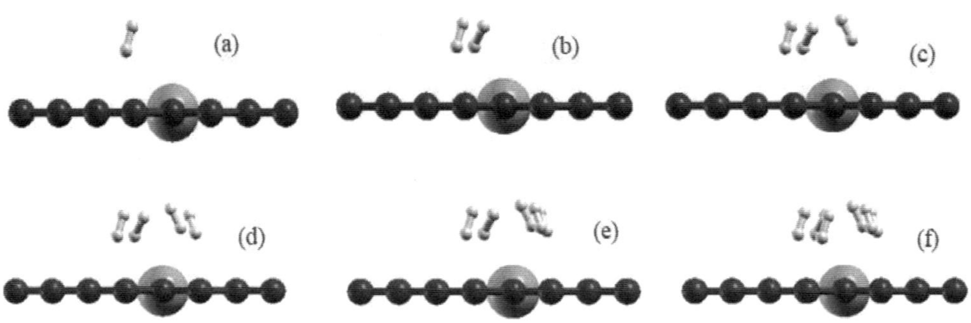

Figure (3). The DFT optimized geometries (side view) of one (a), two (b), three (c), four (d), five (e), six (f) H_2 molecules adsorbed on Hf- Doped Graphene

3.2.1. *Mechanism of possible adsorption sites of H_2 molecule on GrHf*

Table 1 shows the estimated adsorption energy (E_{ads}), binding energy (B.E.), binding energy per H_2 molecule (B.E./H_2) and fermi energy (E_f) of graphene hafnium doped adsorbed hydrogen molecules ($1H_2 - 6H_2$). As increasing the adsorbed hydrogen molecule on hafnium doped graphene (Gr-Hf) sheet the adsorption energy decreases continuously from -333.836 to -344.433. As comparing another theoretical result [33]. Experimentally, the gravimetric capacity of hydrogen storage is found to be 4% [34]. The average energy is calculated when Hf is doping -408.8 eV with ZrCo alloy. While after doping Hf and Ti atoms the cohesive energy found -7.518 eV/atom and -7.531 eV/atom for Zr_7Co_8Ti and Zr_7Co_8Hf respectively [35]. The binding energy of after adsorbing second hydrogen molecule too much large than the next adsorbed H_2 molecules i.e., the binding energy per hydrogen molecule highly decreases when we increase adsorbed atom (2.197 Ry in $2H_2$ to 1.048 Ry in $3H_2$) then small decreases for next adsorbed H_2 molecules. The evaluated fermi energy increases with same difference as while adsorbed H_2 molecules increases, fermi energy found minimum for $1H_2$ shows the minimum electron occupancy at the different energy levels. The fermi energy increases accordingly, the electron occupancy also increases and evaluated higher electron occupancy with fermi energy 2.850 eV for $6H_2$.

Table 1. Physical parameters of complex system GrHf sheet adsorbed adsorption energy (E_{ads}), binding energy and binding energy per H_2 molecule, of H_2 molecule adsorbed on Hf doped graphene, Fermi energy (in eV)

System	E_{ads} in Ry	Binding Energy in Ry	Binding energy /H_2 molecule in Ry	Fermi energy in eV
Gr-Hf-1H_2	-333.836			2.368
Gr-Hf-2H_2	-336.032	2.197	2.197	2.457
Gr-Hf-3H_2	-338.128	2.096	1.048	2.649
Gr-Hf-4H_2	-340.288	2.159	0.720	2.812
Gr-Hf-5H_2	-342.313	2.025	0.506	2.832
Gr-Hf-6H_2	-344.433	2.120	0.424	2.850

3.2.2. Bond length in complex system Gr-Hf-H_2

Table 2. optimized bond lengths of Hafnium Doped Graphene (Hf/Gr) sheet for one to six adsorbed H_2 molecules per Hf atom

Serial Number	Number of H_2 molecule	Bond length of Hf/Gr (in Å)
1	1H_2	2.397
2	2H_2	2.673
3	3H_2	2.791
4	4H_2	2.257
5	5H_2	2.387
6	6H_2	2.156

The estimated bond length listed in table 2, it shows the bond length of adsorbed H_2 molecules (1H_2 to 6H_2) of Hafnium Doped Graphene (Hf/Gr) sheet in Å, bond length between hafnium (Hf) and one adsorbed hydrogen molecule (1H_2) is 2.397 Å it increases gradually for 2H_2 is 2.673 Å and for 3H_2 is 2.791 Å. When adsorbed fourth H_2 molecule the bond length shows 2.257 Å, and finally decreases for 6H_2 is 2.156 Å.

3.3. *Density of States of complex system*

The density of states (DOS) for (3 × 3) hafnium doped graphene (Gr-Hf) indicated in figure 4. The studied system revels that the change in surface area due to increasing the adsorbed hydrogen molecule (H_2). In the DOS plots vertical dotted line represents fermi-level set at zero which shows the variation in DOS. We analysed the density of states (DOS) considering hafnium doped graphene with decorating hydrogen molecules (1H_2-6H_2). The gaps between DOS plots depicts the electrical conductivity of Hf-doped Graphene.

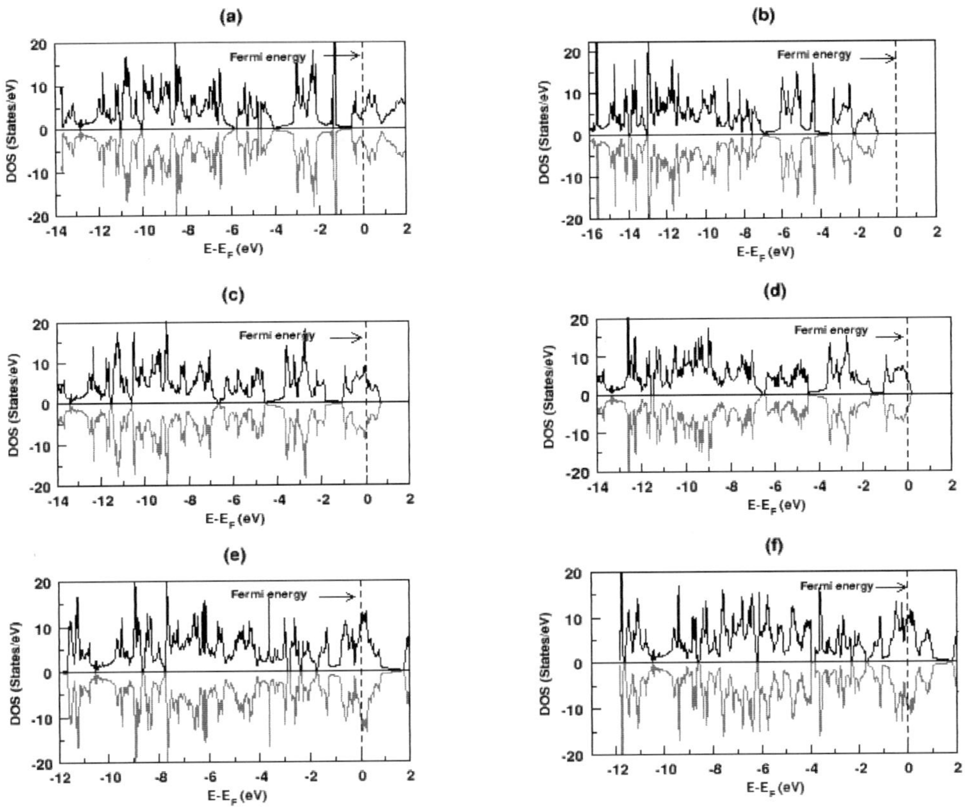

Figure (4). Density of states plots for complex system, H_2 molecules ($1H_2$-$6H_2$) on Hf doped graphene GrHf. The Fermi level set to zero

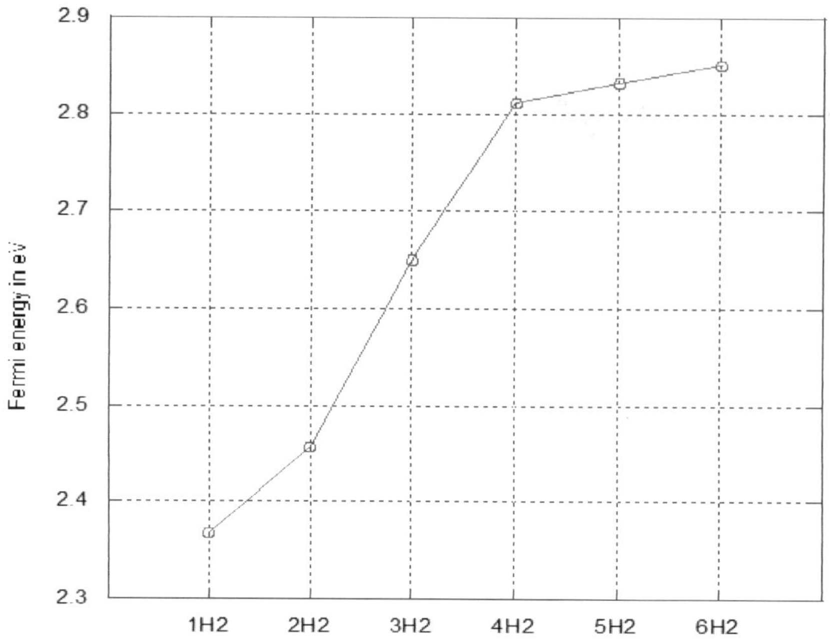

Figure (5). Plot of Fermi energy versus adsorbed H_2 molecule on Hf doped graphene Gr-Hf. The Fermi level set to zero

4. Conclusion

We theoretically performed a hafnium doped graphene complex system using the density functional theory method. The adsorbed hydrogen molecule on graphene doped with the Hf atom gradually increases the hydrogen molecule storage capacity effectively. We adsorbed hydrogen molecules (1H2-6H2) on the surface of the considered complex model. Investigated results show hydrogen molecules exhibit weak interaction with pristine graphene. Furthermore, the hydrogen molecule chemically adsorbs on the Hf-doped graphene with adsorption energy in the range of -344.433 Ry to -333.836 Ry. The evaluated binding energy found in the range 2.197 Ry to 2.120 Ry. The Fermi energy is observed in the range 2.357 eV to 2.827 eV. The density of states (DOS) confirms the weak interaction between σ bonding electrons of H2 molecule with the Hf doped graphene complex system. The proposed system opens a new insight for hydrogen storage-based devices.

5. Acknowledgement

We acknowledge the Department of Science and Technology and University Grants Commission, New Delhi, India, for providing supercomputing facilities through grants No.SR/FT/LS/020/2009(OYS 2009), Grant No.33-16-2007(SR) respectively. We Acknowledge departmental personnel from, Nanomaterials Research Laboratory. The simulation work was conducted in High Performance Computing at Nanomaterials Research Laboratory, Department of Physics, Vinayakrao Patil Mahavidyalaya, Vaijapur, Maharashtra, India.

Note: The authors declare no competing financial interest.

References

[1] Jain I P 2009 *Int. J. of hyd. ene* **34** 7368-7378
[2] Schlapbach L and Züttel A 2002 *Nature* **414** 353-358
[3] Züttel A 2004 *Springer-Verlag* **91** 157-172
[4] Momirlan M and Veziroglu T N 2005 *Int. J. of hyd. ene.* **30** 795-802
[5] Edwards P P, Kuznetsov V L and David W I F 2007 *Phil. Trans. R. Soc.* **365** 1043-1056
[6] Preuster P, Alekseev A and Wasserscheid P 2017 *Annu. Rev. Chem. Biomol. Eng.* **8** 445-71
[7] Geim A K and Novoselov K S 2007 *Nat. Mater.* **6** 183-191
[8] Novoselov K S, Geim A K, Morozov S V, Jiang D, Zhang Y, Dubonos S V, Grigorieva I V and Firsov A A 2004 *Science* **306** 666-669
[9] Acik M and Chabal Y J 2011 *Japanese J. of Appli. Phys.* **50** 070101
[10] Sun Y, Chen L, Zhang F, Li D, Pan H and Ye J 2010 *Solid Sta. Commu.* **150** 1906-1910
[11] Ao Z M and Peeters F M 2010 *Physical Rev. B* **81** 205406
[12] Chu S, Hu L, Hu X, Yang M and Deng J 2011 *Int. J. of hyd. ene.* **36** 12324-12328
[13] Xi P, Tan L, Yang S, Xu H, Lan Z, Wang Z and Gu H, 2021 *Vacuum* **183** 109868
[14] Lone B 2020 *Int.J. Res. In Appli. Sci. engi. Tech.* **8** 181-190
[15] Susilo D N A, Ganta M, Sunnardianto G K and Handayani M 2019 *IOP Conf. Series: Materials Science and Engineering* **541** 012004
[16] Liu Y, Ren L, He Y and Cheng H 2010 *J. Phys.: Condens. Matter* **22** 445301
[17] Karde R and Lone B 2022 *Int. J. Res. And Ana. Rev* **9** 544-554
[18] Qasem J, Pardeshi P, Ingle A, Karde R, Ali S and Lone B 2021 *J. Phys. Conf. Ser* **2070** 012012
[19] Karde R and Lone B 2022 *Int. J. Sci. Res. In Sci. and Tech.* **9** 238-245
[20] Qasem J and Lone B G 2021 *AIP Conference Proceedings* **2369** 020034
[21] Lone B 2009 *Journal of Computational and Theoretical Nanoscale* **6** 2146-2151
[22] Lone B Scheiner S and Kar T 2014 *Carbon* **80** 405-418
[23] Lone B 2009 *Advanced Sci. engi. And med* **11(3)** 182-186
[24] Lone B 2016 *J Nanomed Nanotechnol* **7** 354
[25] Ali S and Lone B 2021 *AIP Conference Proceedings* **2335** 080001
[26] Choudhary A, Malakkal L, Siripurapu R K, Szpunar B and Szpunar J 2016 *Int. J. of hyd. ene.* **41** 17652-15656
[27] Kohn W and Sham L J 1965 *Physical Rev.* **140**
[28] Giannozzi P, Baroni S, Bonini N, Calandra M, Car R, Cavazzoni C, Ceresoli D, Chiarotti G L, Cococcioni M, Dabo I, Corso A D, Gironcoli S, Fabris S, Fratesi G, Gebauer R, Gerstmann U, Gougoussis C, Kokalj A, Lazzeri M, Martin-Samos L, Marzari N, Mauri F , Mazzarello R, Paolini S, Pasquarello A, Paulatto L, Sbraccia C, Scandolo S, Sclauzero G, Seitsonen A P, Smogunov A, Umari P and Wentzcovitch R M 2009 *J. Phys. Condence. Matter* **21** 395502
[29] Scandolo S, Giannozzi P, Cavazzoni C, Gironcoli S, Pasquarello A and Baroni S 2005 *Z. Kristallogr.* **220** 574-579
[30] Kokalj A 1999 *Jou. of Mole. Graphi. And Modell.* **17** 176-179
[31] Perdew J P Burke K and Ernzerhof M 1996 *Phys, Rev, Lett.* **77** 3865-3868
[32] Monkhorst H J and Pack J D 1976 *Phys. Rev. B* **13** 5188-5192
[33] Pantha N, Thapa S and Adhikari N P 2020 *J. of Institute of Sci. and Tech.* **25(1)** 15-23
[34] Spyrou K, Gournis D and Rudolf P 2013 *ECS J. Sol. State Sci. and Tech.* **2**(10) M3160-M3169
[35] Wang L, Ding J, Huang X, Song K, Liu B and Zeng X 2018 *Int. J. hyd. ene.* **43** 13328-13338

IOP Conf. Series: Materials Science and Engineering 1291 (2023) 012017 doi:10.1088/1757-899X/1291/1/012017

Effect of water retting on the physical and mechanical properties of extracted pine needle fibers

Rajesh Kumar[1], Pawan Kumar Rakesh[1] and Dungali Sreehari[1]

[1]Department of Mechanical Engineering & Design Innovation Center,
National Institute of Technology Uttarakhand Srinagar, Pauri (Garhwal),
Uttarakhand, India
Email: rajeshkumaruit1@gmail.com

Abstract: Due to the negative impact of synthetic fibers on the environment, natural fibers are being increasingly used in composite preparation. The hilly area of Uttarakhand state in India has an abundance of pine needle fibers (PNFs). This study focuses on extracting the PNFs using water retting, varying the retting duration (2 weeks, 4 weeks and 6 weeks) and measuring its physical and mechanical properties. After 4-weeks of retting, the extracted PNFs exhibited the lowest moisture content, resulting in improved hydrophobicity and rendering them suitable for reinforcement in composites. The tensile strength and Young's modulus of the 4-weeks retting PNFs were 52.89 ± 12.77 MPa and 1.48 ± 0.39 GPa, respectively. Mechanical characteristics of extracted fibers have been assessed using Weibull's statistical method in terms of tensile strength and Young's modulus. Based on the findings, PNFs are predicted to serve as a reinforcement in environmentally friendly composites.
Keywords: Pine needle fiber, Natural Fiber, Water Retting, Extraction, Mechanical Properties

1. Introduction

Glass, carbon, and aramid fibers are popular reinforcing materials in composites. These synthetic materials, however, need a lot of energy to process and may be dangerous to workers and users. Additionally, due to environmental regulations and concerns, their non-biodegradable nature restricts their usage. As a result, plant based natural fiber are increasingly used as substitutes for synthetic materials [1]. Nowadays, there is a growing focus on using natural fibers in composite fabrication, as they offer numerous benefits, such as being biodegradable, renewable, lightweight, non-abrasive, innocuous, and affordable [2]. At present, the industry's demand for cellulosic fibers exceeds the current production capacity. To address this issue, researchers have introduced various natural fibers like cymbopogon flexuosus fiber, cocania grandis, acacia arabica, grand naine penduncle, mulberry barks fiber, kusha grass and althaea officinalis L. fiber to serve as reinforcement materials for composite [3]. The local people in the Himalayan region utilize pine needles in various ways, such as for decorative items, animal bedding, and energy feedstock [4]. In mountainous regions across the globe, there are a lot of pine needle trees. Environmental pollution occurs throughout the summer when the forest catches fire and decimates a large number of pine trees [5]. Pine needles from pine trees can be utilised in composite fabrication, preventing wasteful burning and reducing the consumption of synthetic fiber. The qualities of a biocomposite are governed by the properties of its fibers, matrix, and interfacial bonds [6]. Thus, it is vital to understand the fiber's qualities before constructing a green composite, which majorly depends on its extraction process. Plant-based natural fibers can be extracted using a variety of methods, including mechanical, chemical, water retting, and steam explosions, each process have their own advantage and disadvantage [7]. Water retting involves the separation of individual fibers from the stem or leaf by eliminating amorphous material attached to the fibers. This process requires the removal of pectin and other cementing compounds that bind the fibers, resulting in the production of single fibers. The success of the process depends on the condition and duration, which determine the yield and quality of the fibers [8]. The retting process involves using enzymatic activity to remove non-cellulosic materials from the fiber bundle. In the case of water retting, anaerobic bacteria fermentation is employed to produce enzymes that break down the components which bind the fiber together [6]. This method is simplest and eco-friendly in all, hence selected for the current study.

This study aimed to examine the effect of various water retting periods on the physical and mechanical properties of the extracted PNFs. Pine needles were collected from Pauri Garhwal, Uttarakhand, and their physical properties, such as average diameter, density and moisture content, were evaluated. Furthermore, single fiber tensile strength and Young's modulus were determined for different retting periods and subsequently compared. Minitab 16 software was utilized to examine the Weibull distribution of mechanical properties in the extracted samples.

2. Materials and method

2.1 PNFs extraction

Natural fibers like kenaf, hemp, and jute are obtained through conventional water retting. If the process is under-retted or over-retted, it can negatively impact the quality of the fiber [7]. So, the optimal duration for water retting is a crucial factor in achieving high-quality fiber. The pine needle was taken from the Pauri Garhwal area in Uttarakhand, India (Figure 1(a)). After collecting the Pine needles from the pine tree, the leaves are separated from stem (Figure 1(b)). The needles were then dipped in water for microbial degradation (Figure 1(c)). Microorganisms are main cause in water retting which separates the plant-based fiber's layers, producing high-quality fibers without causing any physical damage [6]. Three different times were chosen for water retting after lots of trials. The duration for the retting was selected as 2 weeks (fiber retted for two weeks in water), 4 weeks (fiber retted for four weeks in water) and 6 weeks (fiber retted for 6 weeks in water). After completion of water retting, the fiber was separated with the help of a wire brush [9].

Figure 1 (a) Pine Trees (b) Pine needles and (c) Pine needles in water

2.2 Diameter measurement

A total of 25 PNFs specimens were chosen randomly to calculate the mean diameter of 2, 4, and 6 weeks water-retted PNFs. An optical microscope (Radical-RSMr-3) with a 40X magnification was used to measure the fiber diameters. The average diameter of the 2, 4, and 6 weeks water-retted PNFs is provided in Table 1.

2.3 Density measurement

The density of PNFs was measured using pycnometer. For this arrangement, a liquid with a known density is utilised as a reference liquid (toluene = 866 kg/m^3) [10]. Equation 1 calculates extracted PNFs density.

$$\rho_{PNFs} = \frac{(m_2 - m_1)}{(m_3 - m_1) - (m_4 - m_2)} \times \rho_{toluene} \quad (1)$$

Where m_1 = pycnometer weight, m_2 = weight of pycnometer filled with PNFs, m_3 = weight of toluene with the pycnometer, and m_4= weight of pycnometer filled with toluene and PNFs.

2.4 Moisture Content

The oven-dry weight method was employed to determine the moisture content of PNFs, where initial and oven-dry weights of the fiber were utilized in the testing process [1]. Three repetitions were carried out to ensure data accuracy, and the resulting mean values were reported. Equation (2) calculates PNFs moisture content.

$$\%\text{ of moisture in PNFs} = \frac{(W_1 - W_2)}{W_1} \times 100 \quad (2)$$

The weight of the fiber before drying is denoted by W_1, whereas weight of the fiber after drying in a hot air oven is denoted by W_2.

2.5 Tensile Test of PNFs

The mechanical properties of single PNFs were evaluated using the ASTM D 3822 standard test method [11]. To maintain a tight and straight hold on the test clamp, the fiber was pasted onto a paper frame. A 25 mm gauge length was utilized with a paper frame as depicted in Figure 2 (a) to conduct the test.

Figure 2 (a) Paper frame Specimen and (b) Specimen with tensile testing machine

The paper frame was then clamped between the top and bottom jaws of the Universal Testing machine as shown in Figure 2(b). The Instron 5982 was used to measure the pine needle's tensile strength and Young's modulus, with a cross-head speed of 15 mm/min and a load cell of 5 kN. Just before testing, the paper was carefully cut down the middle.

3. Results and discussion

3.1 Diameter, density and moisture analysis

The diameter, density and moisture content of 2, 4 and 6 weeks water-retted PNFs were analysed as shown in Table 1. Figure 3 (a), (b) and (c) shows the diameter measurement of PNFs. The PNFs average diameter of 2 weeks of retted fiber is higher than 4 weeks and 6 weeks of retted fiber (Table 1). The result indicated that the average diameter of the fibers was reduced with respect to increasing the duration of water retting. The fiber density of 6 weeks retted fiber shows the highest value of 1040 kg/m^3 followed by 2 weeks (923 kg/m^3) and 4 weeks (1011 kg/m^3) retted fiber which indicates that fiber from 6 weeks is highly dense, Senwitz reported similar result et al. [12] for bast fiber. However, all retted PNFs confirmed a lower density than synthetic fibers like E-glass (2500 kg/m^3) and S-glass (2500 kg/m^3), making them a potential substitute for lightweight applications [3].

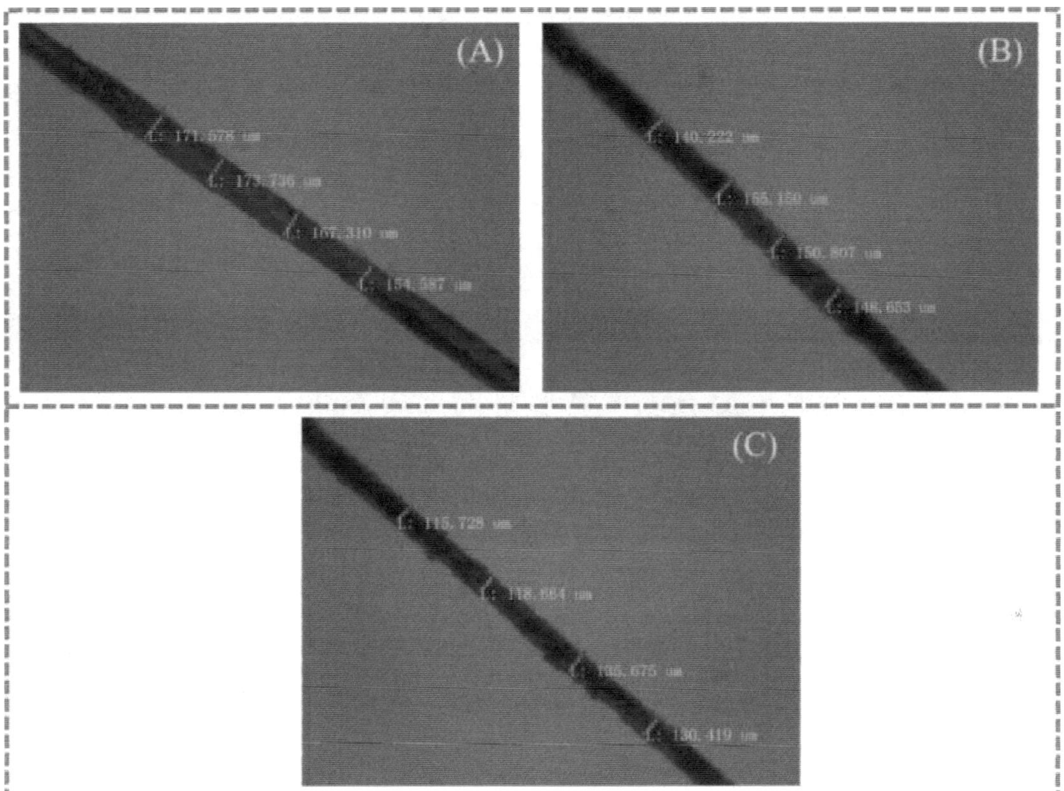

Figure 3 Microscope image of (a) 2 week (b) 4 weeks and (c) 6 weeks retted PNFs

Table 1 Comparison of physical and mechanical properties of PNFs with other natural fibers

Fiber	Diameter (μm)	Density (kg/m^3)	Moisture content (wt.%)	Tensile strength (MPa)	Young's Modulus (GPa)	References
PNFs (2 Weeks)	**161 ± 11.10**	**923**	**8.80**	**38.74 ± 9.37**	**1.29 ± 0.33**	**This work**
PNFs (4 Weeks)	**148 ± 9.65**	**1011**	**8.45**	**52.89 ± 12.77**	**1.48 ± 0.39**	**This work**
PNFs (6 Weeks)	**129 ± 10.27**	**1040**	**17.04**	**46.76 ± 11.74**	**1.34 ± 0.36**	**This work**
Areca palm leaf	285-330	1090	9.35	334.66	7.64 ± 1.13	[13]

Banyan root fiber	0.09 - 0.14	1234	10.21	19.37 ± 7.72	1.8 ± .40	[14]
Coccinia grandis L.	27.33 ± 0.3	1243	5.6	273 ± 27.74	10.17 ± 1.261	[15]
Cymbopogon flexuosus	118	1270	9.17	431.1 ± 23.9	53.88	[3]
Ficus religiosa root	25.62 ± 0.9	1246	9.33	421.25 ± 18	5.11 ± 1.4	[10]
Furcraea foetida	12.8	778	5.43	623.52 ± 45	6.52 ± 1.9	[16]
Juncus effuses L	1.139	280	-	113 ± 36	4.38 ± 1.37	[17]
Parthenium fiber	381.67	1251	8.6	24 ± 2	-	[9]
Pineapple leave fiber	20–80	1440	11.8	13–1627	34.5 – 82.5	[18]
Prosopis Juliflora	20	580	9.48	558 ± 13	-	[19]
Tridax procumbens	233.1 ± 9.9	1160	11.2	25.75	0.94 ± .09	[20]

Natural fibers reinforced composites have a significant disadvantage of having a greater tendency to absorb moisture, resulting in dimensional instability of the composites and decreased mechanical properties [13]. Moisture absorption in 4 weeks of extracted PNFs is less than in 2 weeks and 6 weeks of extracted PNFs. After 2 weeks of water retting, the PNFs contained 8.80% moisture, while PNFs retted for 4 weeks had a moisture content of 8.45%. However, PNFs that were retted for 6 weeks had a higher moisture content of 17.04%.

3.2 Analysis of mechanical properties

At each period of water retting, 25 sample fibers were randomly selected for mechanical properties testing. The tensile strength and Young's modulus of water-retted PNFs (2, 4, and 6 weeks) are shown in Figures 4 (a) and (b). The maximum tensile strength and Young's modulus were observed for the 4 Weeks retted PNFs, with a value of 52.89 ± 12.77 MPa and 1.48 ± 0.39 GPa, respectively and the lowest tensile strength and Young's modulus were found in the 2 weeks retted PNFs, with a value of 38.74 ± 9.37 MPa and 1.29 ± 0.33 GPa, respectively. 4 Weeks retted PNFs have a tensile strength of 52.89 ± 12.77 MPa, which is higher than other natural fibers such as Banyan tree root fiber (19.37 ± 7.72 MPa), Parthenium hysterophorus (24 ± 2 MPa), Tridax Procumbens (25.75 MPa), and lower than Juncus effuses L (113 ± 36 MPa), Coccinia grandis L. (273 ± 27.74 MPa).

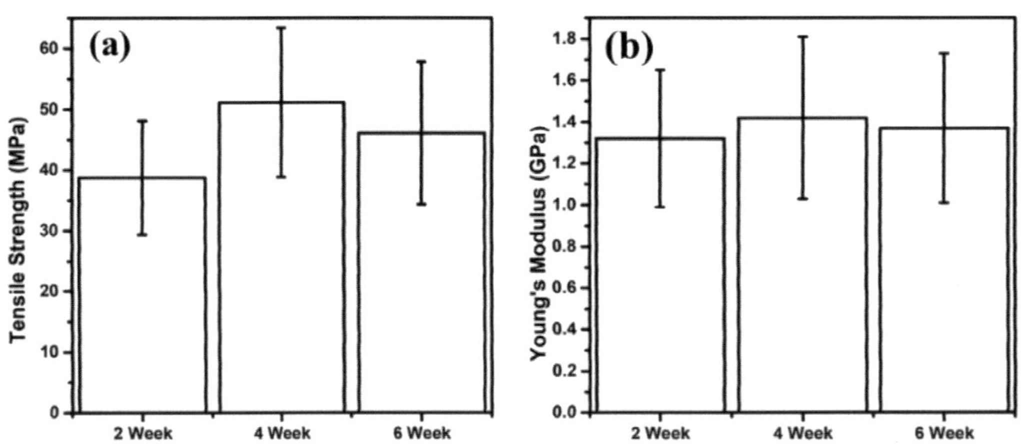

FIGURE 4(a) Tensile strength and (b) Young's modulus of 2, 4 and 6 weeks water-retted PNFs

Tensile testing on a single lignocellulosic fiber to analyse natural fiber mechanical characteristics are difficult owing to the high degree of variation. This variety is caused by the fiber's source, plant's age, extraction method, and flaws. To address this issue, a two-parameter Weibull distribution was used to analyze natural fiber mechanical properties [21]. By utilizing the Weibull distribution to plot the data, it becomes possible to estimate statistical parameters such as the shape and scale factors. Additionally, this method can help in validating the accuracy of the collected data [22]. The probability distribution plot in Figure 5 represents the data obtained from 25 trials of fiber tensile strength and Young's modulus. In Figure 5, the values lie within the line and are a perfectly fit to the distribution, indicating that they closely correspond to the experimental outcomes.

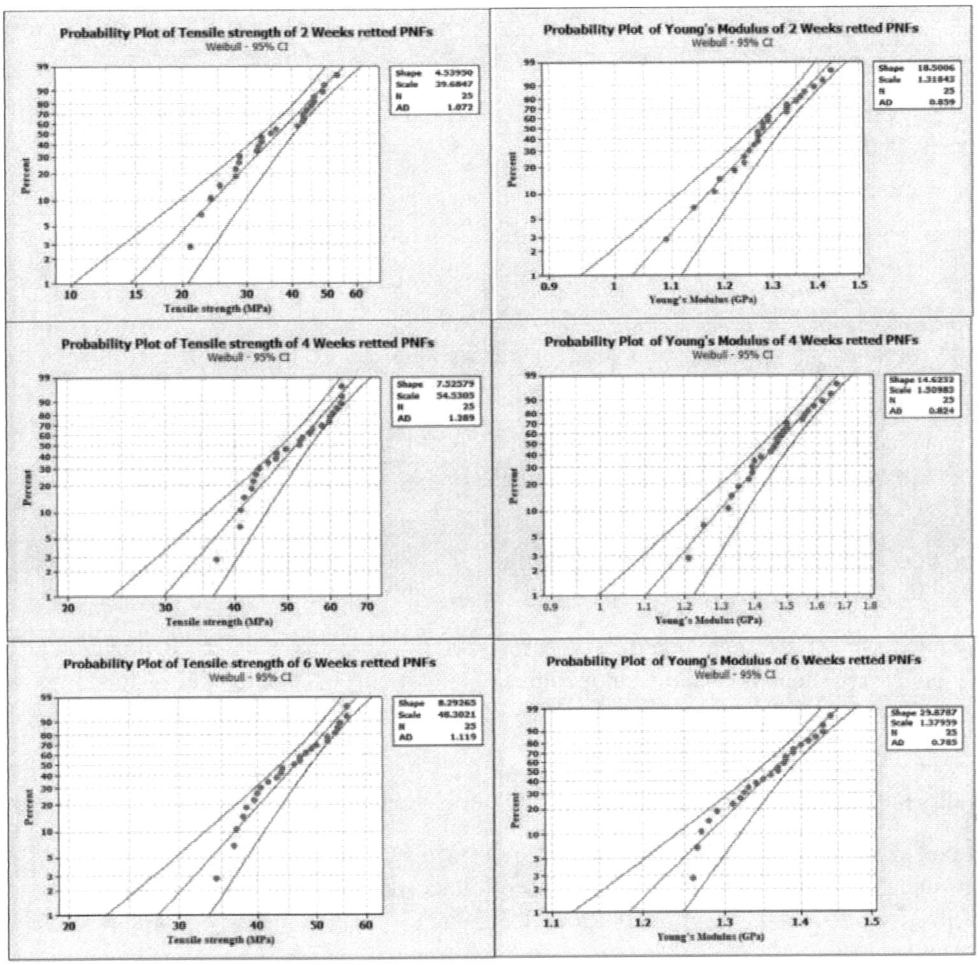

Figure 5 Weibull distribution plot for (a) Tensile Strength (c) Young's modulus of 2, 4 and 6 Weeks retted PNFs

4 Conclusions

This study examined how the duration of water retting PNFs' physical and mechanical properties. The water-retted extracted PNFs after 2, 4, and 6 weeks have a lower density than synthetic fiber. Hence they can be used as reinforced material in lightweight composites. Compared to 2 weeks and 6 weeks of extracted PNFs, 4 weeks extracted PNFs absorb less moisture. The less moisture content could potentially improve the adherence of the fibers to the matrix in composites when used as reinforcement. The significant differences in tensile strength and Young's modulus, were seen in PNFs retted for 2, 4, and 6 weeks. The extracted fibers, after the retting process, offer good mechanical properties. Water retting process for 4 weeks gives the best tensile strength and Young's modulus compared to 2 weeks and 6 weeks water-retted fiber. The results above suggest that pine needle fibers retted for 4 weeks are the best suitable reinforcement for the composite fabrication.

5. References

1. Nagaraja Ganesh B, Ganeshan P, Ramshankar P, Raja K 2019 Assessment of natural cellulosic fibers derived from Senna auriculata for making light weight industrial biocomposites. *Ind Crops Prod* **139** 111546.

2. Faruk O, Bledzki AK, Fink H, Sain M 2012 Progress in Polymer Science Biocomposites reinforced with natural fibers : 2000 – 2010. *Prog Polym Sci* **37** 1552–1596.

3. Raja S, Rajesh R, Indran S, et al 2021 Characterization of industrial discarded novel Cymbopogon flexuosus stem fiber: A potential replacement for synthetic fiber. *J Ind Text* **51** 1207S-1234S.

4. Kala LD, Subbarao PMV 2018 Estimation of pine needle availability in the Central Himalayan state of Uttarakhand, India for use as energy feedstock. *Renew Energy* **128,** 9–19.

5. Chandran M, Sinha AR, Rawat RBS 2011 Replacing controlled burning practice by Alternate methods of reducing fuel load in the Himalayan Long leaf Pine (Pinus roxburghii Sarg.) forests. *5th Int Wildl.*

6. Lee CH, Khalina A, Lee SH, Liu M 2020 A Comprehensive Review on Bast Fibre Retting Process for Optimal Performance in Fibre-Reinforced Polymer Composites. *Adv Mater Sci Eng* **2020** 1-27.

7. Tahir PM, Basher AB, Saifulazry S, Ahmed Z 2011 Retting process of some bast plant fibres and its effect on fibre quality: a review. *BioResources* **6** 5260–5281.

8. Rozyanty AR, Zhafer SF, Shayfull Z, et al 2021 Effect of water and mechanical retting process on mechanical and physical properties of kenaf bast fiber reinforced unsaturated polyester composites. *Compos Struct* **257** 113384.

9. Vijay R, Singaravelu DL, Vinod A, et al 2021 Characterization of Alkali-Treated and Untreated Natural Fibers from the Stem of Parthenium Hysterophorus. J Nat Fibers **15** 80-90

10. Marcel AA, Ravindran D, Bharathi SRS, et al 2020 International Journal of Biological Macromolecules Characterization of a new cellulosic natural fiber extracted from the root of Ficus religiosa tree. *Int J Biol Macromol* **142** 212–221.

11. Kumar R, Rakesh PK, Sreehari D 2023 Investigation on physicochemical, mechanical and thermal properties of extracted novel pinus roxburghii fiber. *J Nat Fibers* **20** 2157924.

12. Senwitz C, Kempe A, Neinhuis C, et al 2016 Almost forgotten resources - Biomechanical properties of traditionally used bast fibers from northern Angola. *BioResources* **11** 7595–7607.

13. Shanmugasundaram N, Rajendran I, Ramkumar T 2018 Characterization of untreated and alkali treated new cellulosic fiber from an Areca palm leaf stalk as potential reinforcement in polymer composites. *Carbohydr Polym* **195** 566–575.

14. Ganapathy T, Sathiskumar R, Senthamaraikannan P, et al 2019 Characterization of raw and alkali treated new natural cellulosic fibres extracted from the aerial roots of banyan tree. *Int J Biol Macromol* **138** 573–581.

15. Senthamaraikannan P, Kathiresan M 2018 Characterization of raw and alkali treated new natural cellulosic fiber from Coccinia grandis.L. *Carbohydr Polym* **186** 332–343.

16. Manimaran P, Senthamaraikannan P, Sanjay MR, et al 2018 Study on characterization of Furcraea foetida new natural fiber as composite reinforcement for lightweight applications. *Carbohydr Polym* **181** 650–658.

17. Maache M, Bezazi A, Amroune S, et al 2017 Characterization of a novel natural cellulosic fiber from Juncus effusus L. *Carbohydr Polym* **171** 163–172.

18. Asim M, Abdan K, Jawaid M, et al 2015 A review on pineapple leaves fibre and its composites. *Int J Polym Sci* **2015** 950567.

19. Saravanakumar SS, Kumaravel A, Nagarajan T, et al 2013 Characterization of a novel natural cellulosic fiber from Prosopis juliflora bark. *Carbohydr Polym* **92** 1928–1933.

20. Vijay R, Singaravelu DL, Vinod A, et al 2018 Characterization of raw and alkali treated new natural cellulosic fibers from Tridax procumbens. *Int J Biol Macromol* **125** 99-108.

21. Pokhriyal M, Rakesh P K Rangappa, S M, & Siengchin S 2023, Effect of alkali treatment on novel natural fiber extracted from Himalayacalamus falconeri culms for polymer composite applications. Biomass Convers. Biorefin. 1-17.

22. Fiore V, Scalici T, Valenza A 2014 Characterization of a new natural fiber from Arundo donax L. as potential reinforcement of polymer composites. *Carbohydr Polym* **106** 77–83.

Synthesis and characterization of graphene via electrochemical exfoliation technique and study its electrochemical properties

S R Kumar[1] and Kumar Anurag[1]

[1] National Institute of Advanced Manufacturing Technology(Formerly NIFFT), Department of Applied Sciences and Humanities, Thin Film Laboratory, Ranchi-834003, India

srk.nifft@gmail.com

Abstract. In this study, we present a one-pot approach for the large-scale synthesis of multilayer to few-layer graphene nanosheets in an aqueous medium that is cost-effective, environmentally friendly, high-yielding, and simple. This electrochemical exfoliation method produces low defect and high yield graphene nanosheet products and is more efficient than chemical exfoliation methods. Two highly oriented pyrolytic graphite plates (HOPG) are utilized as the anode and cathode in this method, which also incorporates two electrode geometry configurations. XRD, HRTEM, AFM, XPS, and Raman spectroscopy are used to describe graphene nano-sheets. Raman characterization supports graphene's inherent qualities. By using cyclic voltammetry, the electrochemical performance of the produced nano-sheets is evaluated.

1. Introduction

In the last four to five years, graphene, the substance that underlies all graphitic forms has emerged as one of the most fascinating areas of study [1-6]. This two-dimensional substance is a brand-new type of nanocarbon made up of layers of six-member rings of carbon atoms. In contrast to carbon nanotubes and fullerenes, it differs noticeably and possesses unusual features that have intrigued scientists. Generally, the fractional quantum hall effect at ambient temperature [7-9], the ambipolar electric field effect together with the ballistic conduction of charge carriers [10], and high elasticity [11] are essential characteristics of graphene. Despite expectations that graphene would be perfectly flat, heat fluctuations generate ripples[1]. Although a single sheet of graphene is the ideal structure, samples with two or more layers are also being studied with interest. Graphene can be classified into three main types: single-layer, few-layer (four layers or less), and multi-layer (ten layers or more). Several methods have been used to create graphene with various numbers of layers [4,5,12]. In addition, there is no established method for producing graphene with a specific number of layers. A rapidly developing area of study, graphene is applicable to a wide range of technological applications, including transparent conducting films, catalytic supports, lithium-ion batteries, solar cells, electrochemical capacitors, and more[13]. With the isolation of graphene nanosheets using the scotch tape method, researchers have been working to create ways for producing graphene nanosheet products on a large scale[13]. The chemical oxidation of

graphene powders (Hummer's process), chemical vapour deposition, liquid-phase exfoliation of graphite, sublimation of Si from SiC, and mechanical cleavage from graphite paper are examples of well-studied graphene nanosheet production methods, according to research [13]. The creation of an effective technique for producing high-quality graphene nanosheets by exfoliating graphite powders, however, remains a significant hurdle. For instance, the Hummer's approach and its modified version are often used to chemically exfoliate graphite powder, but the presence of oxide functionalities on the GO sheets considerably lowers their electrical conductivity due to an inevitable disruption of long range conjugation[14,15]. About CVD, its high energy consumption because it needs a high temperature, high material consumption since it needs a metal catalyst, and a difficult transfer process to the required substrate are its limitations. The yield and controllability of the mechanical cleavage process are too low to be commercially feasible, despite the ability to manufacture high-quality graphene[13]. Under the aforementioned parameters, the electrochemical exfoliation approach has been used to begin the process of producing high-yield and high-quality graphene nanosheets. Due to its quick synthesis, ease of use, and environmentally benign approach, the electrochemical exfoliation process has gradually gained popularity[13]. Different types of electrolytes, such as LiCl in propylene carbonate (PC)[14], $LiClO_4$ in PC[16], LiCl in dimethyl sulfoxide[17], $(NH_4)_2SO_4$ aqueous solution[15], and H_2SO_4 aqueous solution[13], can be used to electrochemically exfoliate graphite into graphene with one or more stacks of graphene-like sheets. It has been demonstrated that the electrochemical exfoliation approach using electrolytes in the presence of SO_4^{2-} ions has enhanced exfoliation efficiency (using 20 min). This is because SO_4^{2-} ions, which are produced during the electrolysis of water, tend to intercalate faulty sites at the margins or grain boundaries of graphite. The release of SO_2 gas and anion depolarization are made possible by SO_4^{2-} ions, increasing the distance between graphite layers.
In this study, we present a two-step, effective electrochemical approach for producing high-quality graphene, as well as the basic reaction kinetics of this method's electrochemical exfoliation in the presence of sulphate ions.

2. Experimental

During the electrochemical exfoliation process, two HOPG with a size of 3×0.5×0.2 cm^3 were utilized as an electrode and a source of graphene nanosheets. In order to make the electrolyte solution, 1.1g of $FeSO_4$ were dissolved in 40ml of distilled water. The two electrodes were dipped into the electrolyte solution at a distance of 2cm after stirring the mixture for 10 minutes. The scientific HM5041 dual output power supply's negative and positive sources were linked to the graphite rods. A 15V power supply was used to begin the exfoliation process on positive rods, which lasted for 6 hours. The anode graphite rod corroded once the potential was applied, and black precipitate gradually developed at the reaction cell's bottom. The experimental setup used in this work is displayed in figure 1(a-e). To get rid of any ferrous ions that might have remained in the samples, they were filtered and repeatedly washed with distilled water and ethanol. The samples were then dried for 10 hours at 60° C.

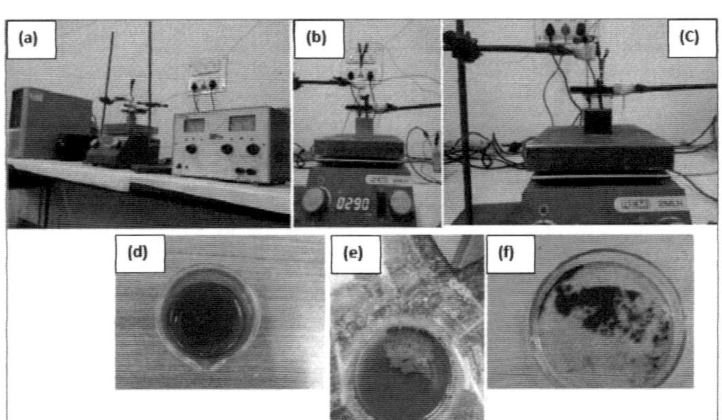

Figure 1. (a) Experimental Setup (b) Colour of the solution at 0min (c) colour of the solution after 30 min (d) Solution after 6hr (e) After sonication EG appears at the top surface

3. Results and Discussion

Typical XRD patterns of electrochemically exfoliated natural graphite powder and graphene nanosheets are shown in figure 2. Both graphene and graphite exhibit a reflection peak at 2θ = 26.4° with (002) however, as a result of the breakdown of interplanar carbon inside the graphitic structure, the intensity of exfoliated graphene is significantly lower than that of natural graphite. In addition to (002) plane (004) is also seen at 2θ=54.4°, in both cases with a low intensity value. However, the absence of graphene oxide characteristic peak around 16°confirms that the exfoliated samples majorly contained reduced graphene layers. According to the Debye-Scherer equation[18], the average crystallite width (D) is the perpendicular dimension within which the graphite ordering is maintained.

$$\boldsymbol{D} = \frac{K\lambda}{\beta cos\theta} = \frac{0.9\lambda}{\beta cos\theta} \quad (1)$$

where, λ is the x-ray wavelength, β is the full width at half maximum (FWHM) expressed in radians, θ is the half diffraction angle of peak corresponding to inter-layer spacing (2θ~26°) for graphene and graphite), and k is a parameter related to crystal structure. In a similar manner, in plane periodicity peak (2θ~54°)[18] can be used to express in plane crystallite size (L) as follows:

$$\boldsymbol{L} = \frac{1.84\lambda}{\beta cos\theta} \quad (2)$$

The average crystallite size has been determined to be substantially smaller in EG (D = 27 nm; L = 61 nm) than in graphite (D = 59 nm; L = 120 nm). As a result of the expulsion of graphene layers from domains during reduction, the graphitic domain may contract, and extra grain boundaries or lateral deflects may arise.

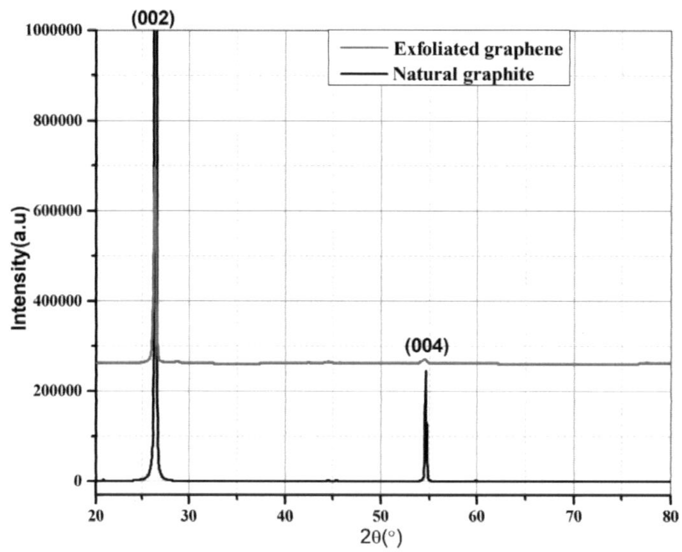

Figure 2. XRD spectra of exfoliate graphene and natural graphite

Absorption or reflectance spectroscopy in the UV-Visible range is referred to as UV-Visible spectroscopy. The absorption bands of a molecule's functions are mapped onto the UV-visible spectrum. Figure 3 displays the UV-Visible absorption spectrum of EG. The electrochemically exfoliated samples showed the absorbance peak at the wavelength (max) of 267 nm, which is associated with the π* transition of the aromatic C-C bonds in graphene [19]. The exfoliated products were proven to be devoid of graphite oxide by the absence of the absorbance peak about 230 nm, which is associated with the π* transition of the aromatic C-C bond of graphene oxide [19].

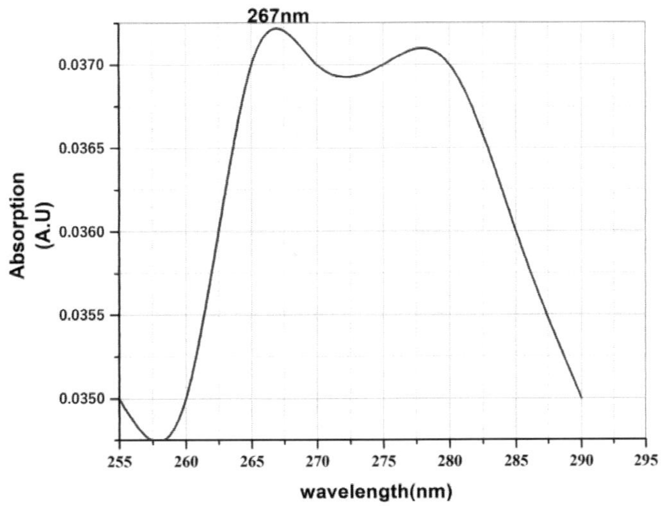

Figure 3. UV-Visible spectra of exfoliated graphene

Figure 4 shows a HRTEM image of exfoliated graphene. The HRTEM findings offer more evidence that the graphene nanosheets are made up of isolated graphene layers or several layers. The interaction of H_2O with expanded graphite, which was electrochemically peeled by the production of O_2 and development of SO_2 gas, is the main source of the synthesis of graphene layers [13]. As hydroxyl (OH) and oxygen ($O^{\cdot}$) radicals are produced when water is oxidized, these active radicals can cause expanded graphite to corrode at edges, grain boundaries, or defect sites, which can lead to the dissolution of carbon crystals [20]. The following steps are included in the reaction kinetics for electrochemical exfoliation: I ionic intercalation (ii) anionic insertion and polarisation (iii) electrolysis of water and SO_2 evolution (iv) bubble expansion which is formulated as follows:

$$H_2O \rightarrow H^+ + OH^- + e^- \rightarrow 2H^+ + O^- + e^- \quad (3)$$

$$SO_4^{2-} + 4H^+ + 4e^- \rightarrow SO_2\uparrow + 3H_2O \quad (4)$$

$$H_2O \rightarrow 2H^+ + 2e^- + 1/2\ O_2\uparrow \quad (5)$$

The graphene sheet's circle-marked area was where the SAED pattern was captured, and the resulting SAED pattern is displayed in figure 4(b). The SAED pattern exhibits hexagonal diffraction spots that come from sp^2 bonded carbon frameworks, indicating that the majority of the produced material is composed of few-layered graphene with few flaws. These flaws primarily result from the electrochemical process' introduction of oxygen functions. Strong 6-fold symmetric diffraction in the designated area is evident from the intensity of the spots with Bravais-Miller(hkil) indices-(1-210) plane and later from the (0101) plane, demonstrating the strong crystallinity of graphene [21]. Moreover, it was discovered that the inner and outer spots' relative intensities were 1, confirming multi-layered graphene [21]. The AFM 3D image of exfoliated graphene is shown in figure 4(c). Spin coating of graphene dissolved in NMP on a glass substrate was used to prepare the AFM sample. The fluffy morphology caused by heat fluctuation is shown in the AFM image.

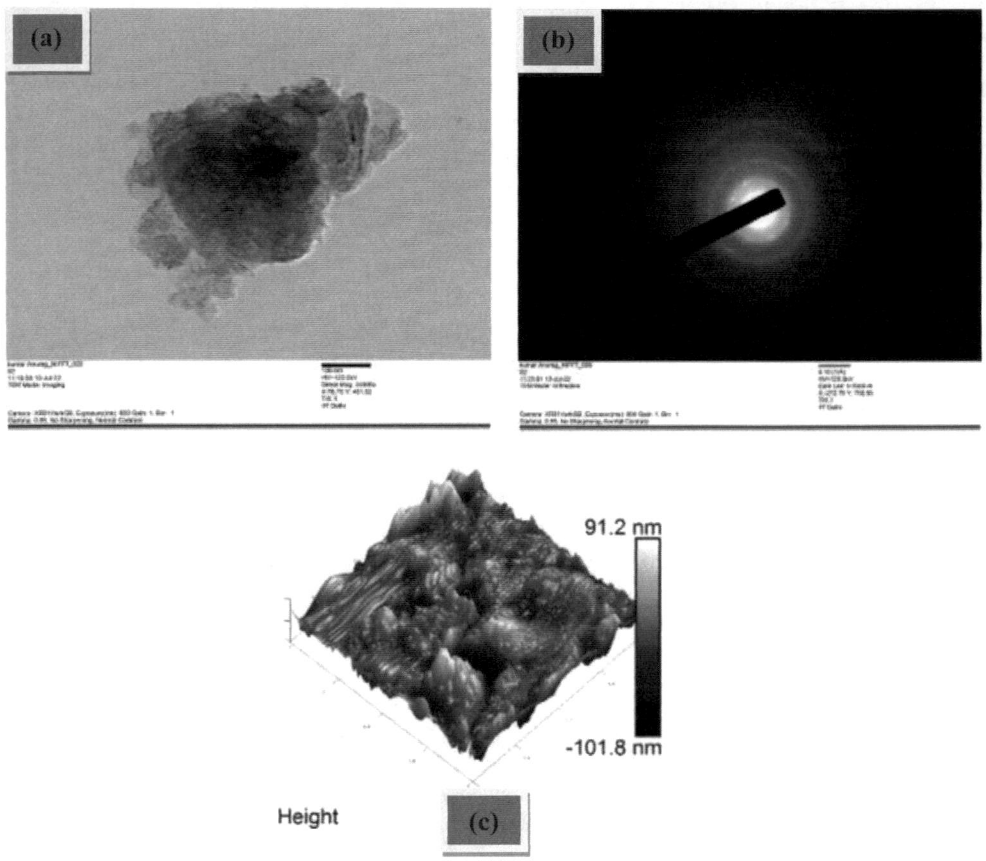

Figure 4. (a) HRTEM image of EG at 20KX magnification (b) SAED pattern of EG and (c) AFM 3D image of EG

To assess the quality of EG as it is made, X-ray photo electron spectroscopy (XPS) is used. In figure 5 the C1s spectrum of EG is displayed. The sample's C1s spectra reveal three different types of carbon bonds, with peaks at 287.4 eV, 285 eV, and 285.9 eV for C=C, C-O, and C-O-C, respectively. The sample has a significant concentration of sp^2 carbon, as evidenced by the strong peak at 284.7 eV[22]. The sample's C/O atomic ratio, which is approximately 10.13, demonstrates the high calibre of the graphene sheets in their as-prepared state. In comparison to those reported by Badrayyana Subramanya et. al [22], the reported C/O value is much greater.

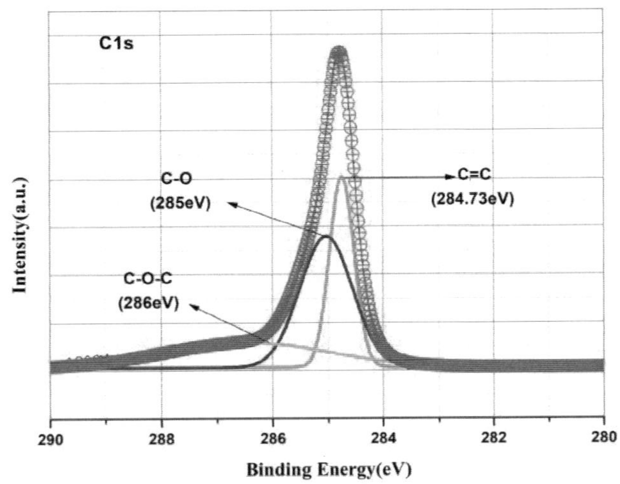

Figure 5. Deconvoluted C1s spectra of EG

The EG's Raman spectra are shown in figure 6. Three prominent peaks were visible in the Raman spectra, which correspond to the D-band at 1343 cm^{-1}, the G-band at 1590 cm^{-1}, and the 2D band at 2706 cm^{-1}. The existence of sp3 defects is indicated by the D-band, the G-band by the in-plane phonon mode vibration of sp^2 carbon atoms, and the 2D band by the two-phonon lattice vibration. The ratio of the integrated intensity of the 2D band (I_{2D}) to the G-band (I_G) reveals the quantity of graphene layers. According to the intensity ratios I_{2D}/I_G of >2, 1-2, and <1, respectively, graphene is one layer, a few layers, and multiple layers. Graphene nanosheets with many layers can be verified by an I_{2D}/I_G intensity ratio of 0.437.

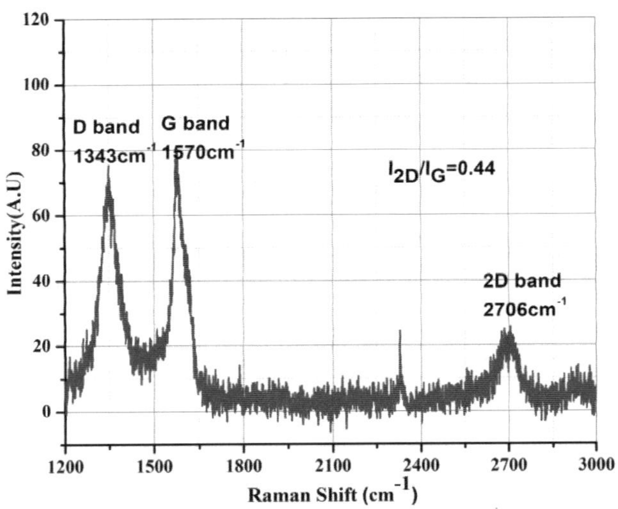

Figure 6. Raman spectra of EG

We also show how to employ EG as conductive ink, which is a crucial component of printable electronics of the future. The preparation of conductive graphene ink involved scattering EG flakes in NMP. Using a paintbrush and produced graphene ink, A-4 size paper was quickly converted into an electrically conductive sheet. Paper's high degree of porosity gives the EG ink a strong capillary force that enhances solvent absorption and results in conformal coatings of EG ink on paper. We investigated the potential of EG coated paper for flexible supercapacitors to show the multifunctionality of the EG paper. To achieve this, an overnight solidification of a PVA/H_2SO_4 gel was drop cast over the top surface of EG coated paper. After that, a supercapacitor was constructed utilizing two EG paper electrodes without the use of an additional current connection. By using cyclic voltammetry and an electrochemical window of 0.4V, the device's electrochemical properties were examined. The CV curves of EG electrodes at a scan rate of 10-90 mVs^{-1} are shown in figure 7. Calculating a specific gel electrolyte's specific capacitance requires knowledge of the CV curve's shape. At various scan rates, the EG paper-based supercapacitor that had been binder- and additive-free manufactured displayed normal double capacitive performance. With an increase in scan rate, it was discovered that the specific capacitance decreased. The range of the specific capacitance values is from 126 to 52.5μFcm^{-2}. A faster redox reaction occurs at a higher scan rate, which reduces electrolyte ion diffusion into the active material and, as a result, lowers capacitance.

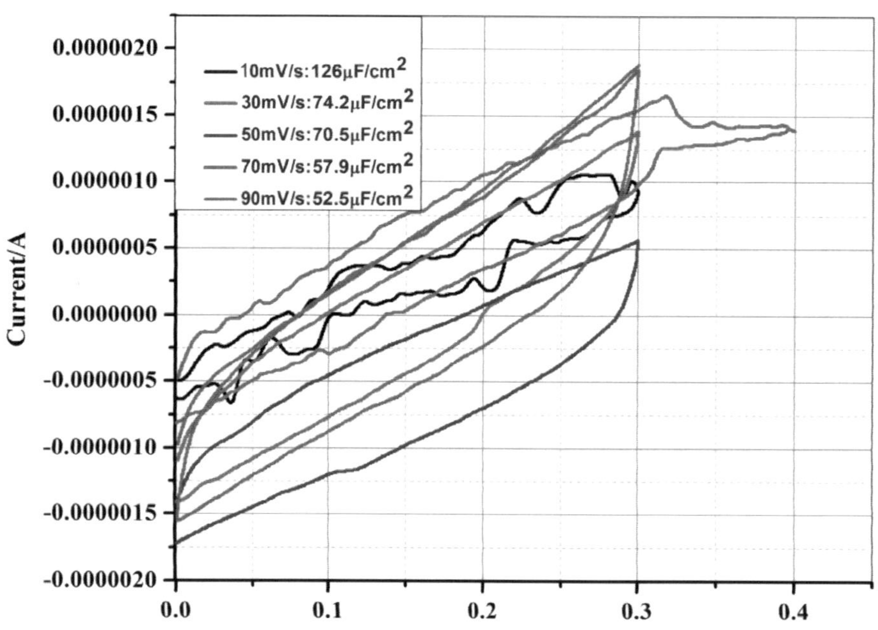

Figure 7. Cyclic Voltammetry curves of EG paper-based supercapacitor at different scan rates

4. Conclusion

Graphene was synthesized via a simple electrochemical exfoliation method from graphite anode. The best DC voltage for creating high-quality graphene nanosheets was discovered to be 15V. The creation of graphene nanosheets was verified by the HRTEM, UV visible, Raman, and XPS. Flexible paper supercapacitors were created using as-prepared graphene and a PVA/H_2SO_4 gel electrolyte. With an increase in scan rate, it was discovered that the specific capacitance decreased.

Acknowledgements:
We are thankful to CIF,BIT Mesra, CIF, IIT-Delhi and Material Research Centre, MNIT-Jaipur for the characterization of graphene samples.

References

[1] Geim A K and Novoselov K S 2007 *Nat. Mater.* **6** 183
[2] Li D and Kaner R B 2008 *Science* **320** 1170
[3] Katsnelson M I 2007 Mater. Today **10** 20
[4] Rao C N R., Biswas K, Subrahmanyam K S and Govindaraj A 2009 *J. Mater. Chem.* **19** 2457
[5] Rao C N R, Sood A K, Subrahmanyam K S and Govindaraj A 2009 *Chem. Int.* **48** 775
[6] Rao C N R, Sood A K, Voggu R and Subrahmanyam K S 2010 *J. Phys. Chem. Lett.* **1** 572
[7] Novoselov K S, Geim A K, Morozov S V, Jiang D, Katsnelson M I, Grigorieva I V, Dubonos S V and Firsov A A 2005 *Nature* **438** 197
[8] Zhang Y, Tan J W, Stormer H L and Kim P 2005 *Nature* **438** 201
[9] Novoselov K S, Jiang Z, Zhang Y, Morozov S V, Stormer H L, Zeitler U, Maan J C, Boebinger G S, Kim P and Geim A K 2007 *Science* **315** 1379
[10] Novoselov K S, Geim A K, Morozov S V, Jiang D, Zhang Y, Dubonos S V, Grigorieva I V and Firsov A A 2004 *Science* **306** 666
[11] Lee C, Wei X, Kysar JW and Hone J 2008 *Science* **321** 385
[12] Park S and Ruoff RS 2009 *Nature Nanotech.* 4 217
[13] Hsieh CT and Hsueh JH 2016 *RSC Adv.***6** 64826
[14] Wang J, Magnga KK, Bao Q and Loh KP 2011 *J.Am. Chem. Soc.* **133** 8888
[15] Parvez K, Wu ZS, Li R, Liu X, Graf R, Feng X and Mullen K 2014 *J. Am. Chem. Soc.* **136** 6083
[16] Yuan GD, Zhang WJ, Yang Y, Tang YB, Li YQ, et.al. 2009 *Chem. Phys. Lett.* **467** 361
[17] Dato A, Radmilovic V, Lee Z, Phillips J and Frenklach M 2008 *Nano Lett.* **8** 2012.
[18] Sharma Rahul, Chadha N and Saini Praveen 2017 *IJPAP* **55** 625
[19] Guardia L, Fernandez-Merino MJ, Paredes JI, Fernandez PS, Villar-Rodil S, et.al. 2011 *carbon* **49** 1653
[20] Ravula S, Baker S, Kamath G and Baker G 2015 *Nanoscale* **7** 4338
[21] Ali B, Qayoum A, Saleem S, Mir FQ 2022 *Res. Eng. Struct. Mater.* **8** 447
[22] Subramanya B, Bhat DK 2015 *New J. Chem.* **39** 420

IOP Conf. Series: Materials Science and Engineering 1291 (2023) 012019 doi:10.1088/1757-899X/1291/1/012019

Comparative analysis of MgB_2 superconducting wire for various cooling treatments through a sintering process

S Herbirowo[1,2], A W Pramono[2], N Sofyan[1], A H Yuwono[1], A Imaduddin[2]

[1]Department of Metallurgical and Materials Engineering, Universitas Indonesia, Indonesia
[2]Research Center for Advanced Materials, National Research and Innovation Agency, Indonesia
Satrio.herbirowo91@ui.ac.id
ahyuwono@eng.ui.ac.id

Abstract. This paper presents a comparative analysis of MgB_2 superconducting wire for various cooling treatments. The study investigates the effect of different cooling rates on the phase transformation, hardness, microstructure, and properties of the MgB_2 wires. The samples were prepared by the powder-in-tube method using a mixture of magnesium and amorphous-boron powders through the sintering and hot rolling process The wires were cooled down at three different conditions, followed by characterization using X-ray diffraction, scanning electron microscopy, and hardness measurements. The results show that the cooling rate after sintering significantly impacted the microstructure and hardness strength of the MgB_2 wires. A slower cooling rate led to a denser microstructure and higher hardness properties, while a faster cooling rate resulted in a more porous microstructure and lower hardness. However, it can be recommended further deformation for smaller wire sizes. These findings could provide insights into optimizing the sintering process to produce high-performance MgB_2.

1. Introduction

Superconducting materials have revolutionized various fields of science and technology, including but not limited to medical imaging, transportation, and energy storage. Among the various superconducting materials, magnesium diboride (MgB_2) has emerged as a promising candidate for high-temperature superconducting (HTS) applications due to its relatively high critical temperature (Tc) of approximately 39 K and simple crystal structure[1]. The potential of MgB_2-based superconductors has led to extensive research on their synthesis and properties, especially in the form of wires that can be used for practical applications[2].

One of the critical factors affecting the properties of MgB_2 wires is the cooling treatment used during the sintering process. Sintering is a process of heating a powdered material below its melting point to form a solid mass by bonding the particles together. In the case of MgB_2 wires, sintering is an essential step in the fabrication process, where the powdered MgB_2 is compacted into a wire shape and heated to high temperatures to form a solid mass[3].

Various cooling treatments, including furnace cooling, quenching, and slow cooling, have been employed to optimize the properties of MgB_2 wires. Furnace cooling involves allowing the material to cool down naturally inside the furnace after sintering. Quenching, on the other hand, involves rapid cooling of the material using a coolant such as liquid nitrogen or helium. Slow cooling involves allowing the material to cool down slowly in the air or under controlled conditions.

Several studies have investigated the effect of cooling treatment on the properties of MgB_2 wires. For instance, Shcherbakova et al. reported that to obtain the best superconducting properties for pure and MgB_2 wires with SiC addition, cooling rates must be considered and will affect to mechanical properties[4][5] [6]. The objective of this research is to study the effect of the cooling method after the hot rolling process to form a single-phase superconducting MgB_2 with minimal oxides to characterize its microstructure and hardness properties for the wire drawing process.

2. Experimental

The MgB_2 wires were fabricated using the sinter and hot rolling method. The starting materials used were magnesium (Mg) and boron (B) powders with a purity of 99.9%. The Mg and B powders were weighed in the stoichiometric ratio of Mg:B=1:2, mixed thoroughly, and then put into the SUS316 tube using a hydraulic press. The tubes were then sintered in a vacuum furnace at 790°C for 2 hours to obtain a dense MgB_2 precursor. The lower reaction temperature is crucial because it reduces the likelihood of powder-barrier reactions[7].

The sintered precursor was then hot-rolled into a wire using a rolling mill at a temperature of 500°C and a reduction ratio of 50%[8]. The wire was then cooled using three different methods: normalizing, full anneal, and water quenching. For the normalizing method, the wire was allowed to cool in the air to room temperature. For the full anneal method, the wire was slowly cooled in the furnace to room temperature over a period of 24 hours. For the water quench method, the wire was rapidly cooled by immersion in water immediately after hot rolling.

The microstructure of the fabricated MgB_2 wire was analyzed using scanning electron microscopy (SEM). The wire samples were sputter-coated with gold to avoid charging effects during SEM analysis. The grain size, texture, and distribution of MgB_2 grains were investigated. The phase transformation of the MgB_2 wire was measured using an X-Ray diffraction with a Cu source. The mechanical properties of the wire, such as hardness strength were also determined using a Rockwell testing machine.

3. Results

The XRD analysis revealed that the MgB_2 wire has a hexagonal crystal structure with lattice parameters a = 3.084 Å and c = 3.523 Å. The SEM analysis showed that the MgB_2 wire has a uniform microstructure with an average grain size of approximately 1 μm.

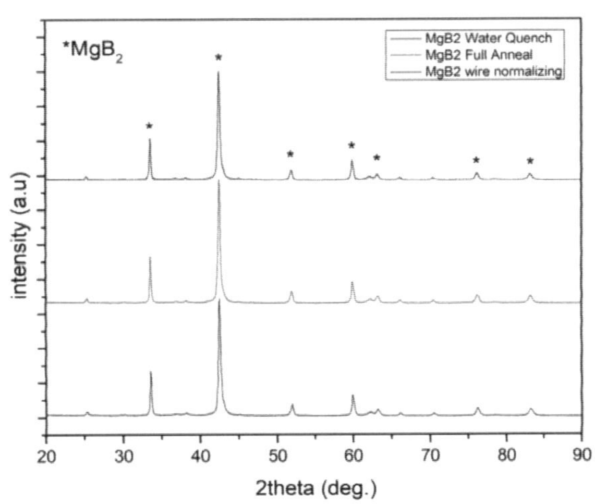

Figure 1. XRD diffraction analysis

The XRD analysis results are consistent with the previous studies on MgB_2 wires, which also showed a hexagonal crystal structure. The SEM analysis showed that the MgB_2 wire has a uniform microstructure with a small grain size, which is important for high critical current density[9]. The Vickers hardness test result showed that the MgB_2 wire has excellent mechanical properties, which is crucial for its application in various devices[10].

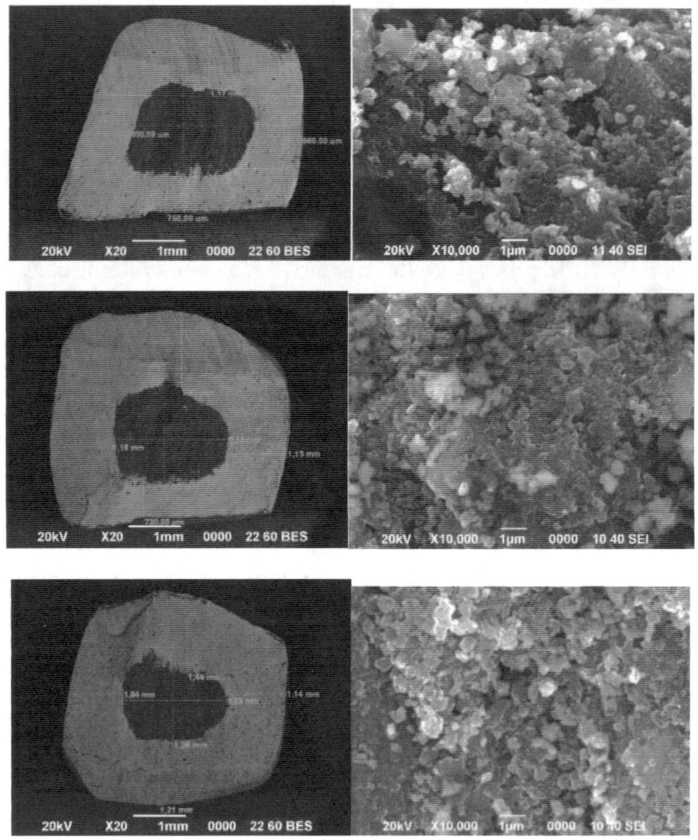

Figure 2. SEM image using backscattered and secondary electron (a) normalizing (b) full anneal (c) water quench

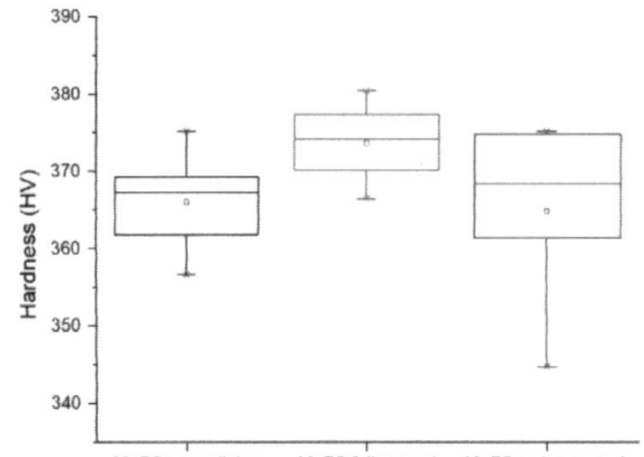

Figure 3. Hardness Vickers result

The results indicate that the cooling treatment significantly influenced the hardness of the MgB_2 samples. The highest average hardness was observed in the full anneal treatment, followed by normalizing, and lastly, the water quench treatment. This process allows for the formation of larger grains and reduces internal stresses, which results in a higher hardness value. In the case of the MgB_2 samples, the full anneal treatment resulted in an average hardness of 373.5 HV, demonstrating the effectiveness of this cooling treatment in enhancing the material's hardness.

In comparison, the normalizing treatment involves heating the material and then allowing it to be cooled in air. This process results in a more uniform microstructure and moderate hardness values. The MgB_2 samples subjected to normalizing displayed an average hardness of 366.0 HV, slightly lower than the full anneal treatment but still relatively high. The water quench treatment, on the other hand, involves rapidly cooling the material by immersing it in water. Although this method is effective in producing a fine grain structure and increasing hardness in some materials[11]–[13], the MgB_2 samples exhibited the lowest average hardness value of 360.0 HV. The rapid cooling may have introduced internal stresses and generated a less uniform microstructure, leading to reduced hardness in comparison to the other treatments.

In summary, the full anneal treatment resulted in the highest hardness values among the MgB_2 samples, followed by normalizing and water quenching. These findings suggest that the choice of cooling treatment can significantly impact the hardness of MgB_2 materials, and understanding the effects of each method can aid in optimizing the desired material properties for specific applications.

4. Conclusion

The characterization of MgB_2 wire using XRD, SEM, and hardness tests revealed that the wire has a single phase of MgB_2, uniform microstructure, and excellent mechanical properties. These properties make the MgB_2 wire a promising candidate for various applications, such as superconducting magnets, power transmission cables, and thin film/wire superconductors.

Acknowledgement

We would like to express our deepest gratitude to Universitas Indonesia PUTI Pascasarjana for their support and assistance throughout this project. This work was made possible through the funding provided under contract number NKB-301/UN2.RST/HKP.05.00/2022 and ELSA BRIN has facilitated material characterization.

References

[1] Y. Shirai *et al.*, "Cooling stability test of MgB2 wire immersed in liquid hydrogen under external magnetic field," *J. Phys. Conf. Ser.*, vol. 507, no. PART 2, 2014.

[2] S. Herbirowo and A. W. Pramono, "Fabrikasi Kawat Resistansi Nol Berbahan MgB2/Stainless Steel dengan Variasi Reduksi Ukuran melalui Pengerolan Dingin terhadap Karakteristik Mekanik dan Struktur Mikro," *Energi & Kelistrikan*, vol. 13, no. 2, pp. 242–249, 2021.

[3] S. Herbirowo, N. Sofyan, A. H. Yuwono, and A. Imaduddin, "Manufacturing and Development of MgB2 Superconducting Wire Through Hot Rolling Process," *ARPN J. Eng. Appl. Sci.*, vol. 17, no. 7, pp. 727–732, 2022.

[4] V. Puspasari, S. Herbirowo, A. M. Habieb, D. P. Utama, R. Roberto, and B. Adjiantoro, "Effect of sub-zero treatments on hardness and corrosion properties of low-alloy nickel steel," *AIMS Mater. Sci.*, vol. 10, no. 1, pp. 55–69, 2023.

[5] A. Sendya, H. Marzuki, S. Herbirowo, B. Adjiantoro, and S. A. Chandra, "Study of Cr/Mo Alloys Lateritic Steel with the Soaking Time Cryogenic Treatment on Hardness, Toughness, Abrasive Resistant, and Microstructure," *Met. Indones.*, vol. 42, no. 1, pp. 21–27, 2020.

[6] O. V. Shcherbakova, A. V. Pan, S. Soltanian, S. X. Dou, and D. Wexler, "Influence of the cooling rate on the main factors affecting current-carrying ability in pure and SiC-doped MgB2 superconductors," *Supercond. Sci. Technol.*, vol. 20, no. 1, pp. 5–10, 2007.

[7] L. Ye, M. Majoros, A. M. Campbell, T. Coombs, and D. Astill, "Experimental studies of the quench behaviour of MgB2 superconducting wires for fault current limiter applications," *Supercond. Sci. Technol.*, vol. 20, no. 7, pp. 621–628, 2007.

[8] S. Herbirowo, N. Sofyan, A. Herman Yuwono, and A. Imaduddin, "Properties of low-cost MgB2 superconducting wires fabricated by high reduction cold rolling," *Mater. Today Proc.*, no. xxxx, pp. 1–7, 2023.

[9] M. Fu, Z. Pan, Z. Jiao, H. Kumakura, and K. Togano, "Quench characteristics and normal zone propagation of an MgB2 superconducting coil," *Supercond. Sci. Technol.*, vol. 17, no. 1, pp. 160–163, 2004.

[10] Hendrik, M. N. Farhanudin, N. Darsono, and S. Herbirowo, "Synthesis and Characterization of MgB2 SUperconductors with Carbon Nanotubes (CNTs) and Tin (Sn) Addition," *EUREKA, Phys. Eng.*, vol. 2022, no. 3, pp. 91–100, 2022.

[11] F. Karaboğa, A. T. Ulgen, H. Yetiş, M. Akdoğan, M. Pakdil, and I. Belenli, "Mechanical properties and uniformity of Fe-MgB2 wires upon various wire drawing steps," *Mater. Sci. Eng. A*, vol. 721, no. December 2017, pp. 89–95, 2018.

[12] E. P. Utomo, S. Herbirowo, V. Puspasari, and Y. N. Thaha, "Characteristics and corrosion behavior of Ti–30Nb–5Sn alloys in histidine solution with various NaCl concentrations," *Int. J. Corros. Scale Inhib.*, vol. 10, no. 2, pp. 592–601, 2021.

[13] S. Herbirowo, A. H. Yuwono, N. Sofyan, D. Tinambunan, H. Nugraha, and A. Imaduddin, "Fabrication and characterization of MgB2/SS 316L superconducting wire with amorphous boron prepared by sintering and cold rolling," *Mater. Sci. Energy Technol.*, vol. 6, pp. 409–416, 2023.

Design and synthesis of silver bio- nanocomposite for photocatalytic applications.

Ramya M[a], Jyothika V G[b], Rajita Ramanarayanan*[a]

[a] Department of Physics, Government Victoria College, Palakkad -678001
[b]International school of Photonics, CUSAT
* Corresponding Author
E mail address: rajithar@gvc.ac.in

ABSTRACT:
One of the major concerns faced by humanity since Industrial Revolution is the contamination of water by the effluents released to the water bodies as a part of modernization and socio-economic development. Water purification is the most researched topic and numerous researches carried out in the past and are still going on in the present which pinpoints the seriousness of the grave issues raised by contamination of water bodies. In this work we have tried to develop a bio-nanocomposite utilizing colloidal silver nanoparticles synthesized via green protocol and arrowroot powder. Nano Silver well known for its photocatalytic properties has been incorporated into biodegradable and ecofriendly arrowroot powder for enhanced photocatalytic action and carried out degradation studies on the common contaminant dye Methylene Blue (MB). The synthesized bio nanocomposite exhibits promising photocatalytic action to be used for water purification.

Keywords: nanocomposite, water purification

1. INTRODUCTION:

One of the biggest challenges faced by human society in the recent decades is energy crisis and pollution of land, water and air. Water pollution is one of the serious concerns faced by human society in the past few decades. The main cause of water pollution is the chemicals released from the industrial sector and from textile dyeing in garment production. Among the different approaches to find solution to this grave problem, photocatalysis is a promising method to provide clean energy while simultaneously solving environmental issues using renewable solar energy.

Metal nanoparticles have contributed greatly in this regard because of their absorption in visible region of sunlight which corresponds to about 45% of the total sunlight. Noble metals like gold, silver in their nano regime exhibit strong absorption in the visible region which has been exploited to manipulate light to suit various Photocatalytic applications [1-8].

Green synthesis is a bottom-up strategy aiming to replace expensive, dangerous chemical reducing agent with natural materials from plants like tree leaves, plants, fruits, and bark. It is a non-toxic procedure that is friendly to the environment with less chance of contamination.

The goal of green nanoparticle synthesis is to reduce hazardous waste production and use environmentally friendly procedures. In the past decade environment friendly green synthesis of silver nanoparticles has gained momentum and is being widely reported for its simple andcost effective protocol.

The noble metal nanoparticles like gold and silver owing to their surface plasmon resonance can broaden the absorption region and enhance the local electromagnetic field to facilitate visible light driven photocatalysts with high efficiency and low cost [9-10].

In this investigation we have synthesized silver nanoparticles using a green protocol using guava leaf extract at room temperature without heating and stirring. We have made an attempt to enhance the well-established photocatalytic activity of silver nanoparticles by synthesizing a novel nanocomposite using arrowroot starch.

Nanocomposites are a combination of two or more phases containing different compositions or structures where atleast one of the phases is in the nanometer range of 10-100 nm. They offer enhanced properties than conventional composites in terms of mechanical strength, thermal stability, electrical properties etc. [11]. Due to its renewability, biodegradability and low cost, starch, a bio polymer has attracted interest in waste water treatment [12]. Arrowroot starch consists of a mixture of two polymers of anhydroglucose units, amylase and amylopectin [13].

In this work we have synthesised bio-nanocomposite using arrowroot starch andsilver colloidal nanoparticles using simple lab technique. The synthesized silver arrowroot nanocomposite has been tested for the degradation of Methylene Blue dye under sunlight.

2. MATERIALS AND METHODS

2.1 Synthesis of colloidal silver nanoparticles

Silver nitrate [$AgNO_3$] was purchased from Merck. Guava leaves extract was prepared using the method employed in our previous report by boiling fresh guava leaves in distilled water which serves as both reducing and capping agent for synthesis of colloidal silver nanoparticles [14].2 ml of the guava extract is added to 20 ml of freshly prepared $AgNO_3$ solution which turned reddish brown in colour confirming the formation of silver colloidal nanoparticles (Fig.1b).

2.2 Synthesis of Ag-arrowroot bio-nanocomposite

5 ml of Ag colloidal solution was added to1 gram of arrowroot powder and heated and stirred at 60° C till the solvent evaporated. The powder was air dried and grinded to obtain the bio-composite Figure 1(d) to be used as photocatalyst in our dye photodegradation studies.

The scheme of the synthesis of Ag arrowroot composite is illustrated in Fig 1.In this mechanism in the presence of plant reducing agent silver ions undergo reaction to form stabilized silver nanoparticles which when heated with arrowroot powder become Ag arrowroot composite.

IOP Conf. Series: Materials Science and Engineering 1291 (2023) 012020 doi:10.1088/1757-899X/1291/1/012020

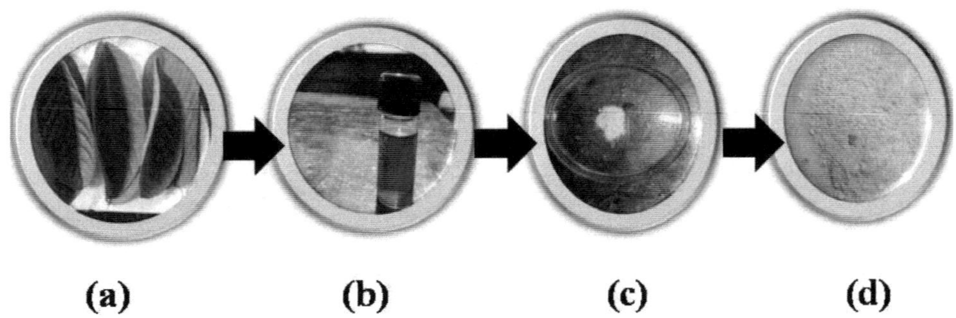

Figure 1(a) Fresh Guava leaves, (b) Ag nanoparticlecolloidal solution, (c) Arrowrootpowder (d) Ag-arrowroot nanocomposite

2.3 Photocatalysis

20 mg of Ag arrowroot nanocomposite was mixed with 20 ml of 20 ppm methylene blue (MB) dye solution, stirred for fifteen minutes to ensure uniform mixing and placed in sunlight (870-880 Watt/m^2) for 180 minutes. At intervals of 60 minutes the supernatant solution was taken and UV - Visible absorption studies were done for observing dye degradation.

2.4 Measurements

The UV-Vis absorption measurements were recorded using spectrophotometer in the visible range (Ocean Optics Maya 2000). The sample morphology of was observed through Scanning Electron Microscope (Jeol). Xray diffraction (XRD) measurements of pure arrowroot powder and Ag-arrowroot composite were performed using X-ray diffractometer (Bruker with monochromatic Cu-K radiation (1.54A°)). The size of the synthesized nanoparticles was determined by transmission electron microscopy (TEM) Jeol/ JEM 2100 using LaB6 source operated at 200 kV.

3. **RESULTS AND DISCUSSION**

3.1 TEM measurements

The TEM analysis of the synthesized silver nanoparticles is shown in the figure 2. It is seen that the particles are spherical in shape and in the range of around 14-25 nm.

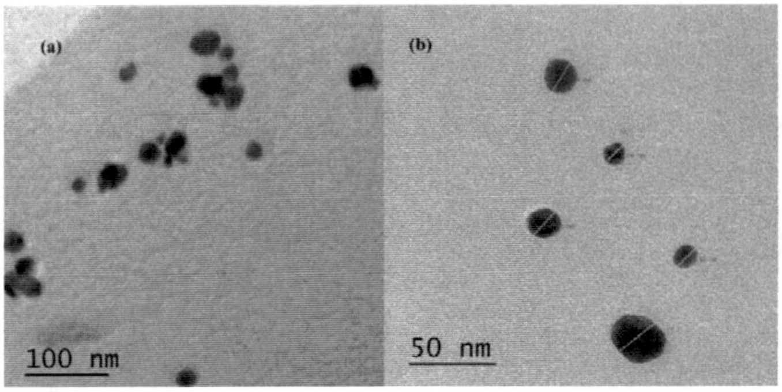

Figure 2 TEM images of different magnifications

3.2 XRD Measurements

The phase and crystal nature of the synthesized samples were investigated by the XRD pattern as shown in Figure3.

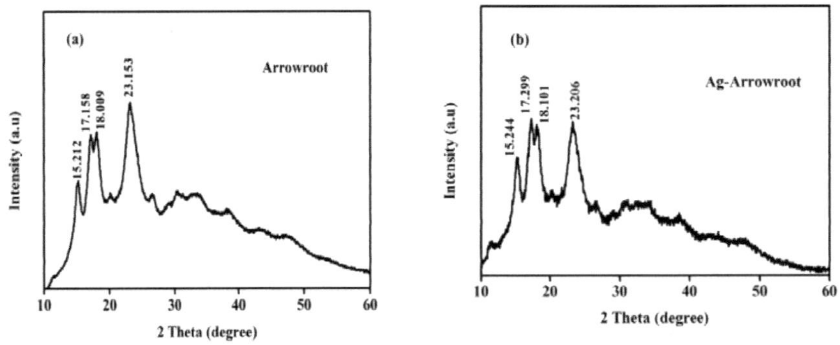

Figure 3 XRD analysis (a) Pure arrowroot (b) Ag-arrowroot nanocomposite

The addition of synthesized Ag nanoparticles is observed to cause a shift in peaks in the XRD pattern of the arrowroot powder. The peaks observed at 2 θ values of 15.212°, 17.158°, 18.009° and 23.153° for pure arrowroot powder have shifted to 15.244°, 17.299°, 18.101°, and 23.206° in the Ag arrowroot composite confirming its structural change. The dominant peaks at approximately $2\theta = 15, 17, 18$ and 23 can be related to A type crystalline arrangement in arrowroot powder [15]. The crystallite size is calculated using by applying the Scherrer formula $D=0.89\ \lambda / \beta \cos(\theta)$ where λ is wavelength of the X-ray source (0.154 nm), β is Full Width at Half Maximum (FWHM), θ is the Bragg angle. It is found that the crystallite size of arrowroot powder is 0.31 nm and that of Ag arrowroot composite is decreased to 0.26 nm.

3.3 SEM Measurements

Morphology of the synthesized Ag arrowroot composite is studied using scanning electron microscope (SEM) as shown in Figure 4. Pure arrowroot granules (Fig 4a) have a compact spherical or elongated shape with smooth surfaces, few irregularities and no porosities exhibiting crystalline nature whereas an irregular structure is exhibited by Ag arrowroot nanocomposite (Fig 4b). It shows that the formation of nanocomposite by heating and stirring and subsequent grinding has degraded the structure of starch granule resulting in the reduction in size and increase in surface area. Present results are in agreement with the earlier results reported in the study of starch [16]

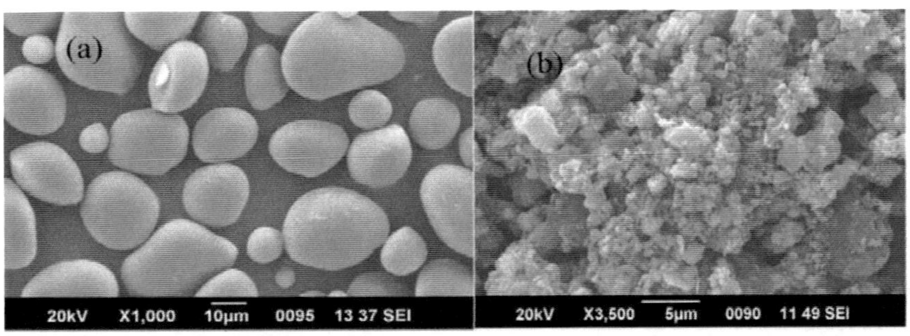

Figure 4 SEM image of (a) Pure arrowroot powder (b) Ag-arrowroot nanocomposite

3.4 Photocatalytic studies

The Photocatalytic activity of Ag and Arrowroot nanocomposite in the degradation of Methylene blue dye under natural sunlight is taken. The result is compared with the degradation profile of pure arrowroot powder (inset) by drawing the plot of C/C_o versus time as shown in Figure 5.

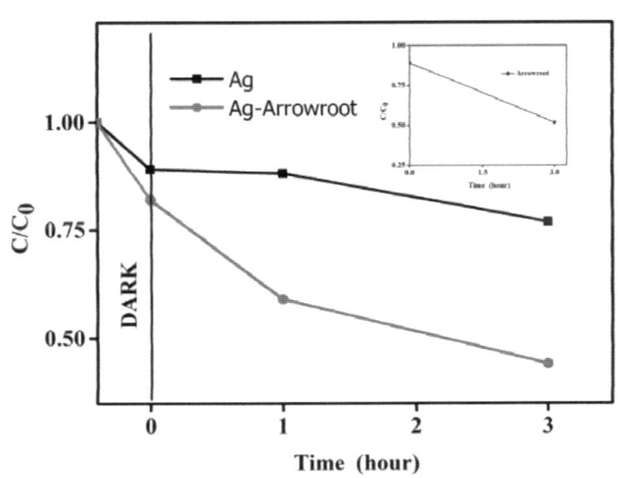

Figure 5 Photocatalytic studies of synthesized samples.

The degradation efficiency (%) is calculated using the formula $E\% = (1-C/C_0)100$, where C and C_0 represent the final and initial concentrations of dye, respectively, before and after photo-irradiation. It is seen that the silver nanoparticles show 23% degradation after three hours and arrowroot starch shows about 48% in three hours. It is observed that that the degradation efficiency is maximum for the nanocomposite showing about 56% degradation efficiency.

Mechanical treatment by grinding refines the structure of starch, increases the amount of the amorphous phase of the starch, thereby enhancing the sample's solubility and reactivity [16] as shown by XRD, SEM and photocatalytic studies in this work.

The presence of silver nanoparticles in the nanocomposite has also contributed to the improvement of photocatalytic efficiency due to its light harvesting capability in visible region. Both these factors have contributed to the enhanced photocatalytic ability of the Ag-arrowroot nanocomposite.

4. CONCLUSIONS

Colloidal silver nanoparticles are synthesized via sustainable green route using guava leaf extract. Further biodegradable and ecofriendly arrowroot starch has been used for the synthesis of bio nanocomposite of silver. Photocatalytic studies clearly demonstrate the superior activity of the synthesized nanocomposite over colloidal silver nanoparticles in the degradation of the Methylene blue dye. Further research in this area has the potential to engineer diverse multifunctional bio nanocomposites which can replace the non-biodegradable materials for various applications including water purification.

ACKNOWLEDGEMENTS

The authors acknowledge Sophisticated Test and Instrumentation Centre (STIC), Cochin University of Science and Technology for XRD, SEM and TEM analysis.

REFERENCES

[1] Jamkhande, P.G., Ghule, N.W., Bamer, A.H. and Kalaskar, M.G., 2019. Metal nanoparticles synthesis: An overview on methods of preparation, advantages and disadvantages, and applications. *Journal of drug delivery science and technology*, **53**, p.101174.

[2] Chen, S., Zhang, Q., Hou, Y., Zhang, J. and Liang, X.J., 2013. Nanomaterials in medicine and pharmaceuticals: nanoscale materials developed with less toxicity and more efficacy. *European Journal of Nanomedicine*, **5(2)**, pp.61-79.

[3] Kaabipour, S. and Hemmati, S., 2021. A review on the green and sustainable synthesis of silver nanoparticles and one-dimensional silver nanostructures. *Beilstein Journal of Nanotechnology*, **12(1),** pp.102-136.
[4] Iravani, S., Korbekandi, H., Mirmohammadi, S.V. and Zolfaghari, B., 2014. Synthesis of silver nanoparticles: chemical, physical and biological methods. *Research in pharmaceutical sciences*, **9(6)**, p.385.
[5] Jeyaraj, M., Gurunathan, S., Qasim, M., Kang, M.H. and Kim, J.H., 2019. A comprehensive review on the synthesis, characterization, and biomedical application of platinum nanoparticles. *Nanomaterials*, **9(12)**, p.1719.
[6] Zhang, X.F., Liu, Z.G., Shen, W. and Gurunathan, S., 2016. Silver nanoparticles: synthesis, characterization, properties, applications, and therapeutic approaches. *International journal of molecular sciences*, **17(9)**, p.1534.
[7] Kumar, J.A., Krithiga, T., Manigandan, S., Sathish, S., Renita, A.A., Prakash, P., Prasad, B.N., Kumar, T.P., Rajasimman, M., Hosseini-Bandegharaei, A. and Prabu, D., 2021. A focus to green synthesis of metal/metal-based oxide nanoparticles: Various mechanisms and applications towards ecological approach. *Journal of Cleaner Production*, **324**, p.129198.
[8] Liu, L., Zhang, X., Yang, L., Ren, L., Wang, D. and Ye, J., 2017. Metal nanoparticles induced photocatalysis. *National Science Review*, **4(5)**, pp.761-780.
[9] Kamat, P.V., 2002. Photophysical, photochemical and photocatalytic aspects of metal nanoparticles. *The Journal of Physical Chemistry B*, **106(32)**, pp.7729-7744.
[10] Cozzoli, P.D., Fanizza, E., Comparelli, R., Curri, M.L., Agostiano, A. and Laub, D., 2004. Role of metal nanoparticles in TiO2/Ag nanocomposite-based microheterogeneous photocatalysis. *The Journal of Physical Chemistry B*, **108(28)**, pp.9623-9630.
[11] Thomas, S., Bandyopadhyay, S., Thomas, S. and Stevens, R., 2006. Polymer Nanocomposites: -preparation, properties and applications. *Gummi FascrnKunststoffe*, ***9.***
[12] Teh, C.Y., Wu, T.Y. and Juan, J.C., 2014. Potential use of rice starch in coagulation–flocculation process of agro-industrial wastewater: treatment performance and flocs characterization. *Ecological engineering*, **71**, pp.509-519.
[13] Tarique, J., Sapuan, S.M., Khalina, A., Sherwani, S.F.K., Yusuf, J. and Ilyas, R.A., 2021. Recent developments in sustainable arrowroot (Maranta arundinacea Linn) starch biopolymers, fibres, biopolymer composites and their potential industrial applications: A review. *Journal of Materials Research and Technology*, **13,** pp.1191-1219.
[14] Ramanarayanan, R., Meethal, B.N., Pullanjiyot, N., Niveditha, C. and Swaminathan, S., 2019. Metal photocatalysis by red and black nano-gold colloids: a rational design through green pathway. *Materials Research Express*, **6(8)**, p.085033.
[15] Villas-Boas, F. and Franco, C.M., 2016. Effect of bacterial β-amylase and fungal α-amylase on the digestibility and structural characteristics of potato and arrowroot starches. *Food hydrocolloids*, **52**, pp.795-803.
[16] Dome, K., Podgorbunskikh, E., Bychkov, A. and Lomovsky, O., 2020. Changes in the crystallinity degree of starch having different types of crystal structure after mechanical pretreatment. *Polymers*, **12(3)**, p.641.

IOP Conf. Series: Materials Science and Engineering 1291 (2023) 012021 doi:10.1088/1757-899X/1291/1/012021

Preparation and characterisation of nickel oxide and nickel oxide codoped with magnesium and zinc

Erin Ann Sunny[1], Nahan Nazar[1], M Bhagyalakshmi[1], S Sooryanarayanan[1], A Chithra Mohan[1], Jomol Mariyam Thomas[2], Varsha Nair[2], G Sivasubramanian[3], K M Sreekanth[3*] and K M Sreedhar[1*]

[1]Department of Chemistry, Amrita Vishwa Vidyapeetham, Amritapuri, Kerela, India
[2]Department of Chemistry, Devaswom Board Pampa College, Parumala, Mannar, Kerala, India
[3]Department of Sciences, Amrita School of Physical Sciences, Amrita Vishwa Vidyapeetham, Coimbatore, India
*Corresponding authors: sreedharkm@am.amrita.edu
km_sreekanth@cb.amrita

Abstract - The present work focuses on the formulation and characterisation of undoped NiO and NiO codoped with Mg and Zn. The samples were characterized by x-ray diffractometer and ultraviolet-visible spectrophotometer. For the detection of elements present in the samples, energy dispersive x-ray analysis was employed. Two sets of codoped samples were prepared varying the amount of Mg and Zn. A set of codoped samples was prepared keeping the quantity of Mg at a fixed value and allowed the quantity of Zn to fluctuate. The second set of codoped samples was prepared keeping the extent of Zn at a fixed value and allowed the quantity of Mg to fluctuate. The bandwidth of undoped and codoped materials was computed raising Tauc plot. Crystallite size was computed following Scherrer formula.
Keywords: codoping, coprecipitation, bandwidth, nickel oxide, codoped nickel oxide

1. Introduction

The significance of undoped, doped and codoped metal oxides in various disciplines like medical field, photocatalysis, electronics and spintronics is illustrated by the copious number of research publications in various national and international reputed journals [1-7]. Zinc oxide codoped with silicon and manganese was investigated for its photocatalytic activity in the reduction of hazardous chromium [8]. Zinc oxide and zinc oxide doped with manganese prepared under different conditions were studied for the application in photocatalysis [9]. Foxtail millet husk was used in the preparation of silica-silver composite with antimicrobial and free radicle scavenging activities [10]. Zinc oxide codoped with nickel and magnesium was reported along with the influence of the quantity of of the co-dopants on the bandwidth of zinc oxide [11]. There is a report on another photocatalyst based on silica incorporated magnesium-zinc oxide [12]. A report is available on electrochemical sensing based on mixed metal oxide [13]. A microcidal, non-cytotoxic and free radical scavenging composite was reported based on opaline silica [14]. Zinc oxide doped with magnesium was studied for its structural, optical, and photocatalytic properties [15]. A detailed report is available on the investigation of synthesis, structural and optical properties of zinc oxide doped with manganese [16]. Magnesium oxide codoped with nickel and cobalt was investigated for its electrochemical biosensor applications

[17]. A review available in public domain is more than enough to illustrate the significance of nickel oxide in different areas [18]. Effect of annealing on the properties of nickel oxide thin film was investigated and reported [19]. A study on the magnetic properties of nickel oxide nanoparticles was reported in addition to the study on thermal conductivity [20, 21]. Present work aims at the formulation and characterisation of undoped NiO and NiO codoped with Mg and Zn. Two sets of codoped samples were prepared varying the amount of Mg and Zn. While preparing first set of NiO codoped with Mg and Zn, masses of $NiSO_4.6H_2O$ and $ZnSO_4.7H_2O$ were kept constant, while mass of the other dopant, $MgSO_4.7H_2O$ was varied. The second set of NiO codoped with Mg and Zn was prepared keeping masses of $NiSO_4.6H_2O$ and $MgSO_4.7H_2O$ constant, while mass of the dopant, $ZnSO_4.7H_2O$ was varied. Tauc plot was employed for computing bandwidth.

2. Experimental

2.1 Chemicals used in the work

Chemicals used include $NiSO_4.6H_2O$, NaOH, $ZnSO_4.7H_2O$ and $MgSO_4.7H_2O$. For preparing solutions and washing of products, double deionised water was used.

2.2 Formulation of Nickel oxide

Weighed out exactly 4 g of $NiSO_4.6H_2O$ into 250 ml beaker and dissolved in 100 ml of deionised water. It was kept for agitating over magnetic stirrer at normal temperature. While stirring, added dropwise, solution of 2.0679 g NaOH in 50 ml of deionised H_2O. The amount of NaOH was determined based on the balanced chemical equation representing the reaction. After addition, continued stirring for 10 minutes. The precipitate of $Ni(OH)_2$ formed was then sorted out of G4, splashed with deionised H_2O till washings freed from Ni^{+2}, SO_4^{-2} and NaOH. Chemical test was conducted to check presence or absence of these kinds in the washings. After washing with deionised water, the precipitate was ultimately bathed with acetone then dried for 10 minutes at 60 ^{0}C in an oven. Dried $Ni(OH)_2$ was transferred into silica crucible, then kept in a muffle furnace at normal temperature. Raised temperature of furnace to 450 ^{0}C from normal temperature then stayed at 450 ^{0}C for 2 hours. Following 2 hours, switched off furnace, then permitted to cool to room temperature. Once cooling to normal temperature, silica crucible containing NiO was taken out. The material, NiO was powdered and used for further study.

2.3 Formulation of NiO codoped with Mg and Zn

Two sets of codoped samples were prepared varying the amount of Mg and Zn. While preparing first set of NiO codoped with Mg and Zn, masses of $NiSO_4.6H_2O$ and $ZnSO_4.7H_2O$ were kept constant, while mass of other dopant, $MgSO_4.7H_2O$ was varied. Weighed out $NiSO_4.6H_2O$, $ZnSO_4.7H_2O$ and $MgSO_4.7H_2O$, as given in table 1, into a 250 ml beaker and dissolved in 100 ml of deionised water. It was kept for agitating over magnetic stirrer at normal temperature. While stirring, added dropwise, a solution of the required amount of NaOH, calculated based on the balanced chemical equation representing the reaction, in 50 ml of deionised water. After addition, continued stirring for 10 minutes. The precipitate formed was then sorted out of G4. Washed precipitate using deionised water till washing was free from Mg^{2+}, SO_4^{-2}, Zn^{2+}, Ni^{+2} and NaOH. Chemical test was conducted to check presence or absence of these kinds in the washings. After washing with deionised water, precipitate formed was ultimately bathed with acetone then dried for 10 minutes at 60 ^{0}C in an oven. After drying at 60 ^{0}C, the precipitate was transferred into silica crucible, then kept in muffle furnace at normal temperature. Raised temperature of furnace to 450 ^{0}C from normal temperature, then remained at 450 ^{0}C for 2 hours. Following 2 hours, switched off furnace, then permitted to cool to normal temperature. Following cooling to normal temperature, silica crucible containing NiO codoped with Mg and Zn was taken out. The doped sample was powered and used for further study. All samples as given in table 1 were prepared taking masses as mentioned in this table. The second set of NiO codoped with Mg and Zn was prepared keeping masses of $NiSO_4.6H_2O$ and $MgSO_4.7H_2O$ constant, while mass of the dopant, $ZnSO_4.7H_2O$ was varied. Following the same procedure adopted for the preparation of the first set, the second set was also prepared considering the masses as given in table 2.

3. Characterisation of the samples - Instrumental analysis

3.1 Energy Dispersive X-ray analyses (EDX)

Elements in materials were detected with EDX mentioned as EDS or EDAX.

3.2 X-Ray Diffraction (XRD) analyses

To study details of crystals of prepared materials, powder X-ray diffraction analysis was performed.

3.3 Ultraviolet-visible spectroscopy analyses

For determining band gap of materials, solid-state ultraviolet-visible spectral analyses were made.

Table 1: Set 1-Preparation of NiO and NiO codoped with Mg and Zn

Sl. No.	Sample code	Mass % of the dopant $ZnSO_4.7H_2O$ with respect to the mass of $NiSO_4.6H_2O$	Mass % of the dopant $MgSO_4.7H_2O$ with respect to the mass of $NiSO_4.6H_2O$	Mass of $NiSO_4.6H_2O$	Mass of $ZnSO_4.7H_2O$	Mass of $MgSO_4.7H_2O$
1	G	0	0	4	0	0
2	H	20	2	4	0.80	0.08
3	I	20	6	4	0.80	0.24
4	J	20	10	4	0.80	0.40
5	K	20	15	4	0.80	0.60
6	L	20	20	4	0.80	0.80

Table 2: Set 2-Preparation of NiO and NiO codoped with Mg and Zn

Sl. No.	Sample code	Mass % of the dopant $MgSO_4.7H_2O$ with respect to the mass of $NiSO_4.6H_2O$	Mass % of the dopant $ZnSO_4.7H_2O$ with respect to the mass of $NiSO_4.6H_2O$	Mass of $NiSO_4.6H_2O$	Mass of $MgSO_4.7H_2O$	Mass of $ZnSO_4.7H_2O$
1	G2	0	0	4	0	0
2	H2	20	2	4	0.80	0.08
3	I2	20	6	4	0.80	0.24
4	J2	20	10	4	0.80	0.40
5	K2	20	15	4	0.80	0.60
6	L2	20	20	4	0.80	0.80

4. Results and discussion

4.1 Elemental analyses - EDAX result

Results of elemental analysis of materials H, I, J, K, L H2, I2, J2, K2, and L2 are presented in figure 1 to figure 10. The samples are all codoped nickel oxides, codoped with Mg and Zn. The expected elements from all these samples include Ni, Zn, Mg and O. The results confirm occurrence of the four elements in the samples. The peak, in all samples, at about 0.5 keV corresponds to oxygen. There are peaks in all samples at about 1 keV, 8.5 keV and 9.5 keV, corresponding to zinc. The peak, in all samples, at about 7.5 keV corresponds to nickel. The peak, in all samples, at about 1.5 keV corresponds to magnesium.

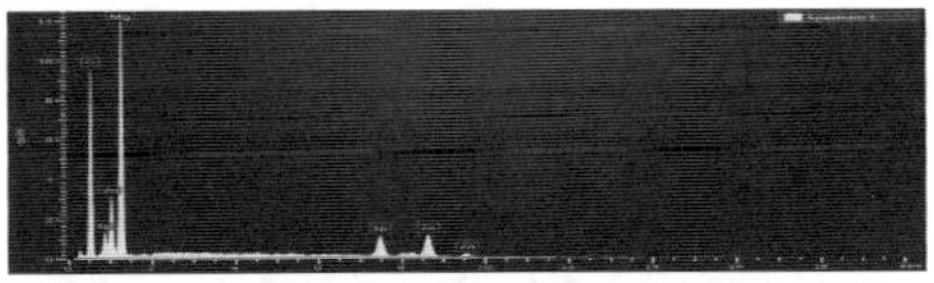

Figure 1. EDAX result of the sample H

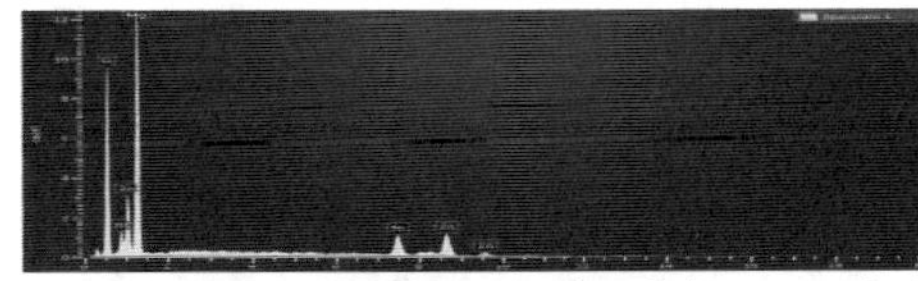

Figure 2. EDAX result of the sample I

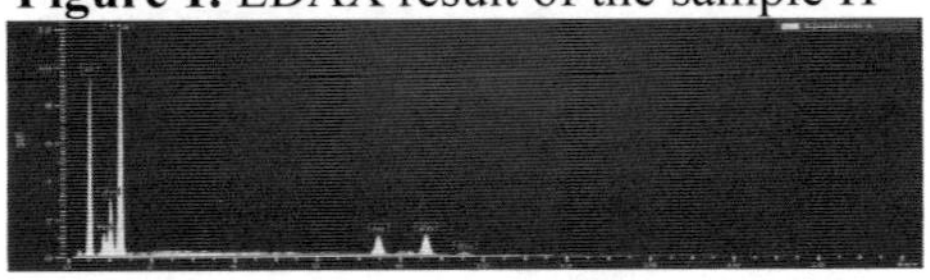

Figure 3. EDAX result of the sample J

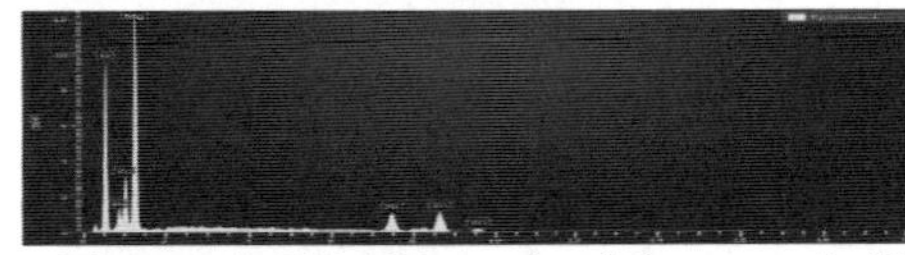

Figure 4. EDAX result of the sample K

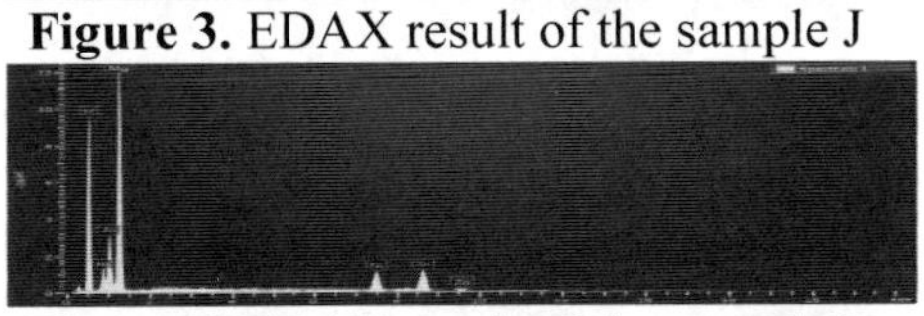

Figure 5. EDAX result of the sample L

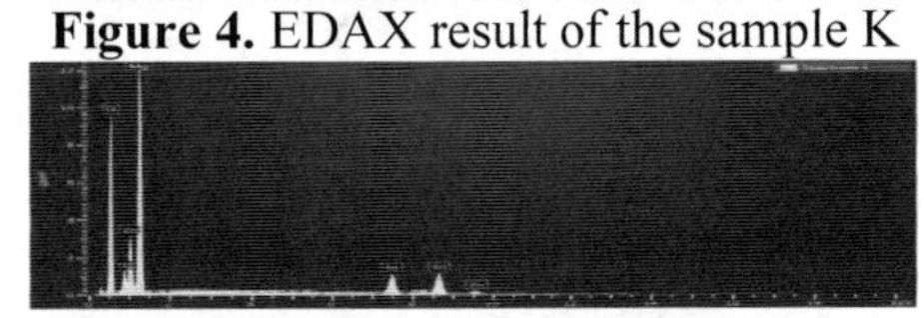

Figure 6. EDAX result of the sample H2

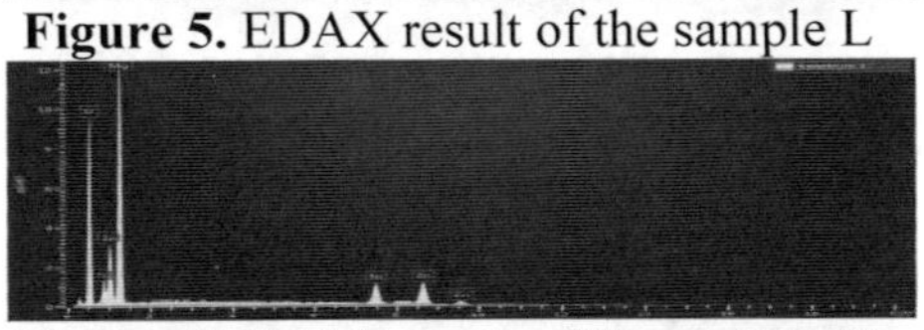

Figure 7. EDAX result of the sample I2

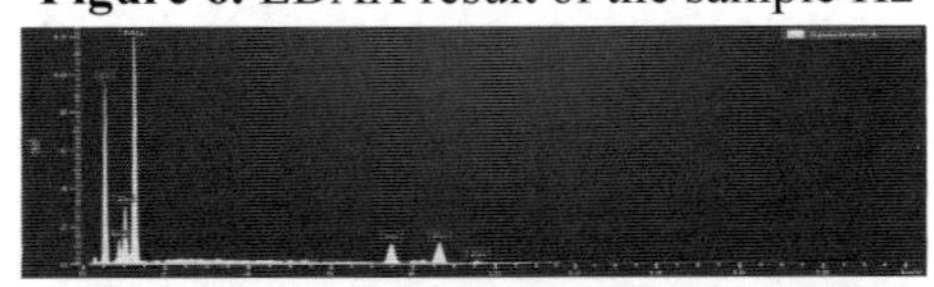

Figure 8. EDAX result of the sample J2

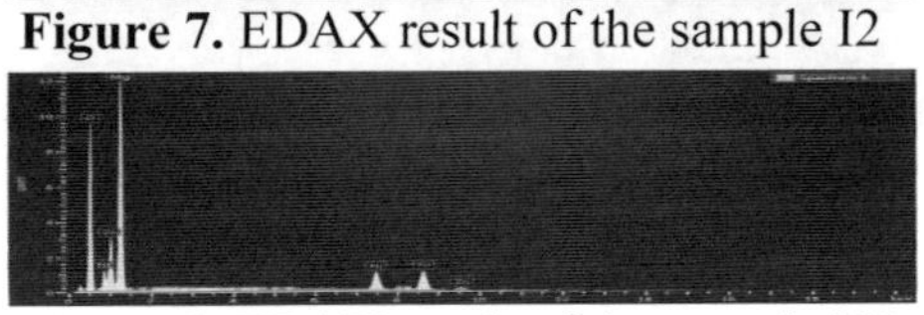

Figure 9. EDAX result of the sample K2

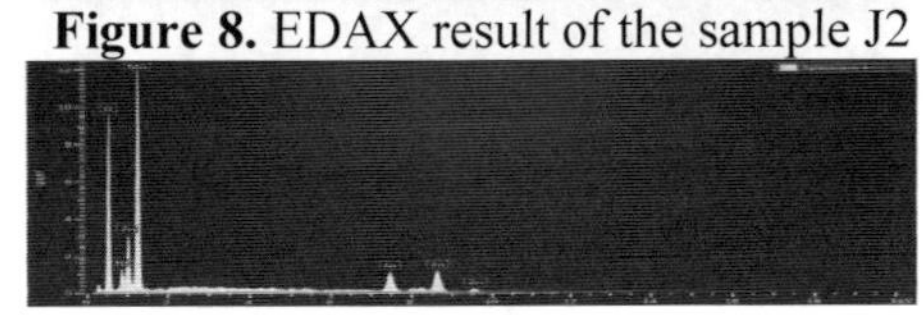

Figure 10. EDAX result of the sample L2

4.2 Results of the powder XRD analysis

Findings of powder XRD analyses of the prepared materials, G, H, I, J, K, L, G2, H2, I2, J2, K2 and L2 are presented in the figures, figure 11 to figure 23. Sample G and G2 are undoped nickel oxide, and all other samples are Mg and Zn codoped nickel oxides. Using raw data, created powder XRD patterns of all materials in 2θ range of 10^0 - 80^0. All samples, both codoped and undoped, produced XRD lines at 2θ of 37.303, 43.287, 62.837, 75.406, and 79.370, all values are in degree. These lines could be indexed as (003), (202), (220), (223), and (042), respectively. They correspond to rhombohedral crystal system of NiO, with respect to Joint Committee on Powder Diffraction Standards (JCPDS) with JCPDS File Number 01-089-3080. Therefore, both undoped and codoped samples possess rhombohedral crystal structure, and the codoping did not lead to the distortion of rhombohedral crystal lattice of NiO. Nickel oxide belongs to cubic crystal system. However, it undergoes a distortion from cubic to rhombohedral [22-24]. With variation of codoping percentage, we observed a shift in peak as given by figure 24 and figure 25 . As codoping percentage increases, it could be observed that the peak is shifting towards lower 2θ values. This shift in the peak, with respect to the undoped sample, is a good support for the conclusion that the codopants, Mg and Zn, entered the crystal lattice of nickel oxide, that is, codoping occurred. Since the shift in peak is exceedingly small, as one can observe from figure 24 and figure 25, it can be concluded that codoping could not distort the rhombohedral structure of nickel oxide, and so all the samples, codoped and undoped, possess rhombohedral crystal structure.

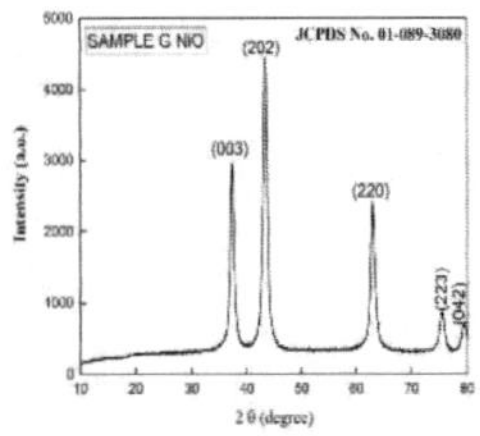

Figure 11. XRD of G

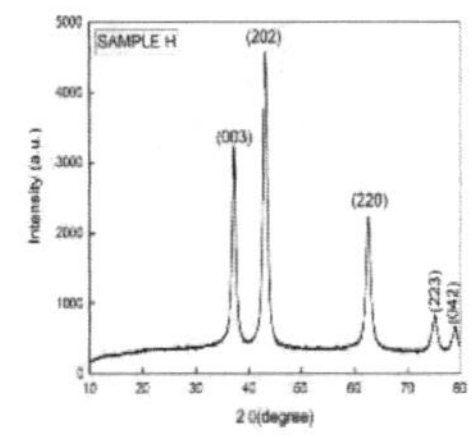

Figure 12. XRD of H

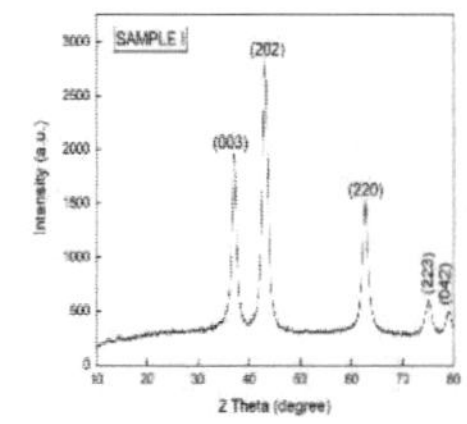

Figure 13. XRD of I

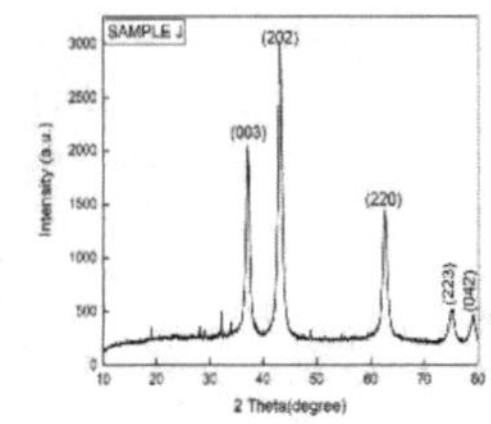

Figure 14. XRD of J

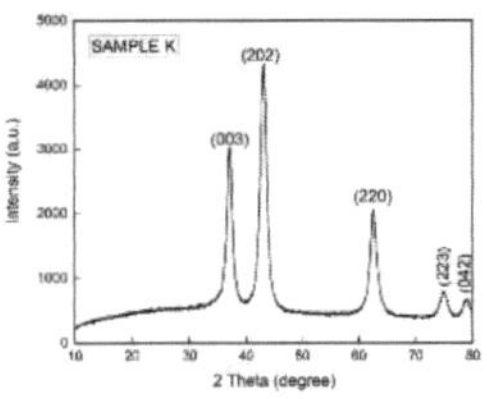

Figure 15. XRD of K

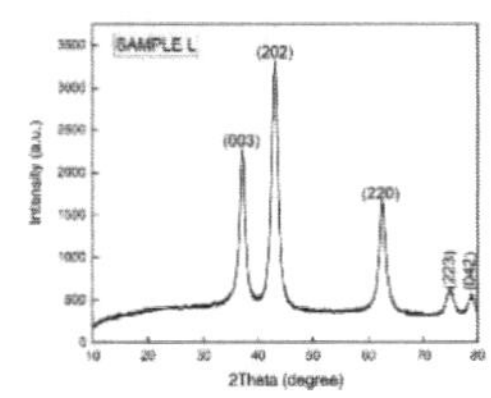

Figure 16. XRD of L

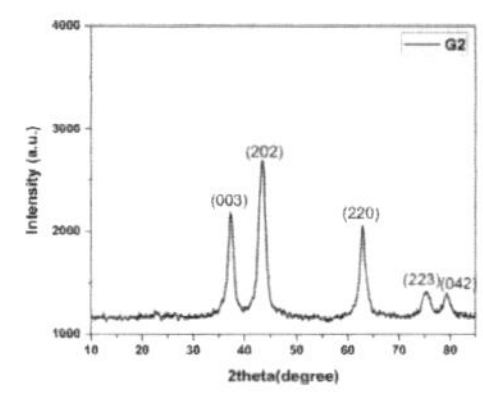

Figure 17. XRD of G2

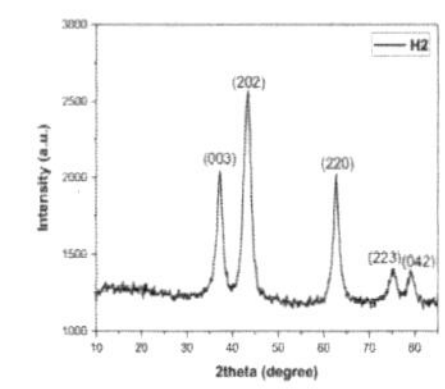

Figure 18. XRD of H2

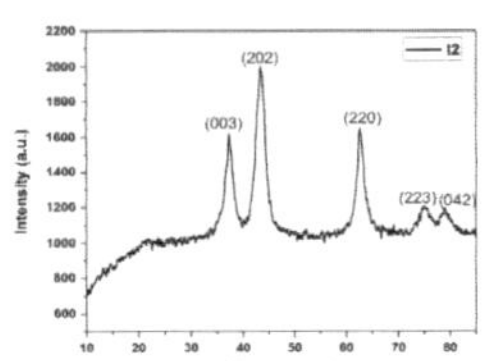

Figure 19. XRD of I2

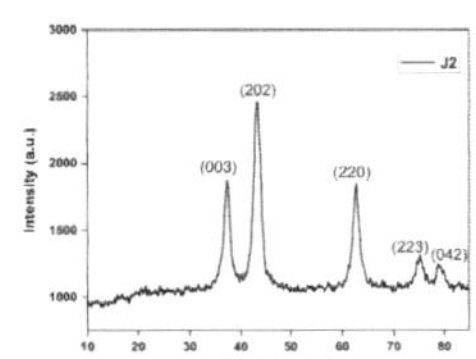

Figure 20. XRD of J2

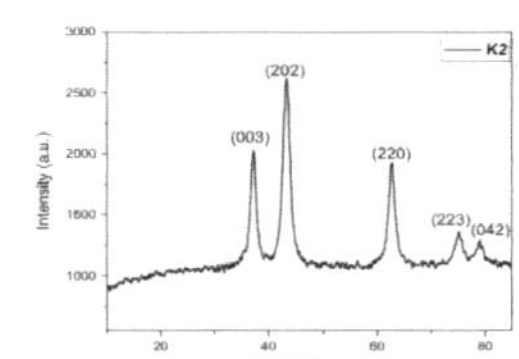

Figure 21. XRD of K2

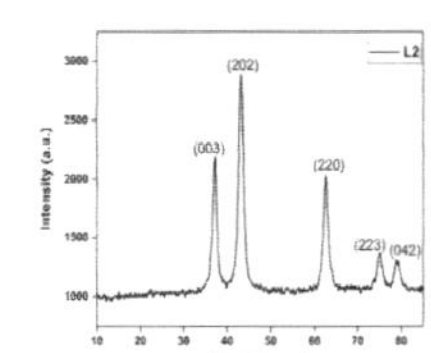

Figure 22. XRD of L2

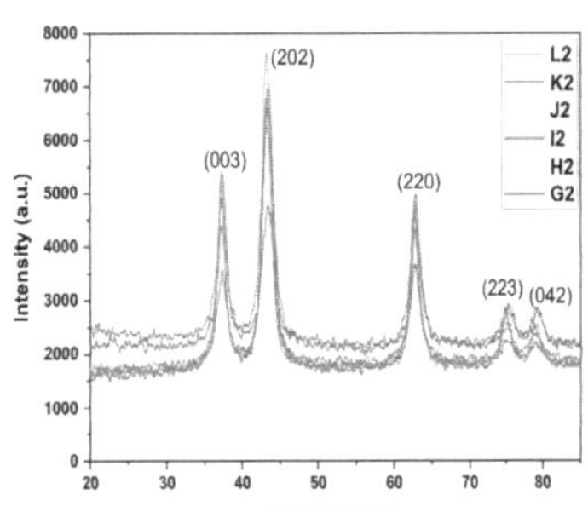

Figure 23. XRD of G2, H2, I2, J2, K2, L2

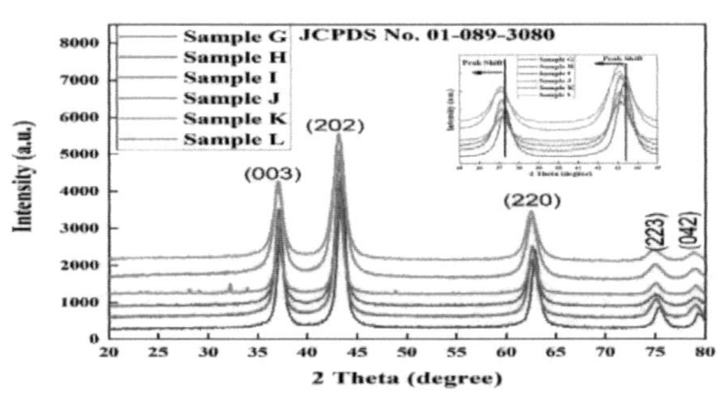

Figure 24. XRD of G, H, I, J, K, L with peak shift

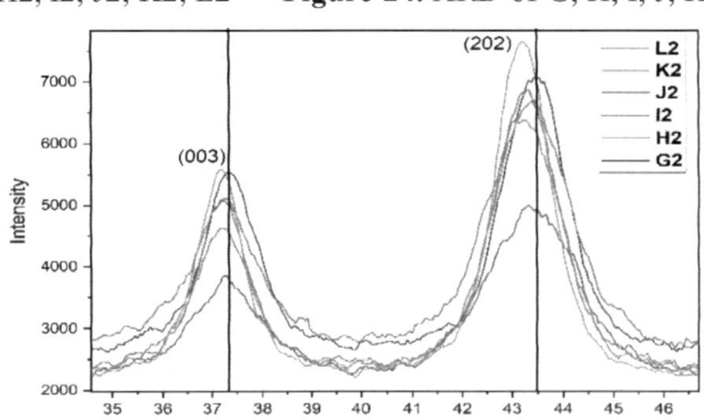

Figure 25. Peak shift in H2, I2, J2, K2, L2

IOP Conf. Series: Materials Science and Engineering 1291 (2023) 012021 doi:10.1088/1757-899X/1291/1/012021

4.3 Band gap calculation from uv-visible data

All prepared samples were subjected to uv-visible spectral analysis, and data collected were used to calculate band width of the undoped and codoped oxides. Uv-visible spectroscopy is a tool for the characterisation of semiconductor objects. Optical absorption features of semiconductor can be explained based on excitation of the valence electrons to conduction band, producing an electron-hole pair. Theoretical band gap energy can be calculated by means of Tauc relation as $\alpha h \nu = D (h \nu - E)^n$. In current job, n takes ½ because NiO involves direct permitted transitions [25, 26]. So, Tauc relation for NiO is $(\alpha h \nu)^2 = D^2 (h \nu - E)$. Because of Tauc relation for NiO, it is feasible to have plot of $(\alpha h\nu)^2$ on ordinate vs. hν on abscissa. When straight-line portion of plot is extrapolated to abscissa, it will meet some point on abscissa, at this point $(\alpha h\nu)^2$ is zero. Value of hν, usually in eV, at this point on abscissa will be equal to band gap energy of the semiconductor. Such a plot is known as Tauc plot. Here in this work, the determination of the band gap or band gap energy of cooked samples was brought out creating Tauc plot, using the data from uv-visible analysis of samples. The Tauc plots are presented for set 1 samples G, H, I, J, K, and L, figure 26 to figure 31. Band-gap energies of set 1 samples are summarised in table 3. Band gap energy of NiO, G, is 3.05 eV. Value available in literature varies from 3.6 eV to 4 eV. Upon codoping, the value decreases. As codoping percentage increases, that is, from sample G to sample L, it gradually decreases as shown in the table 3.

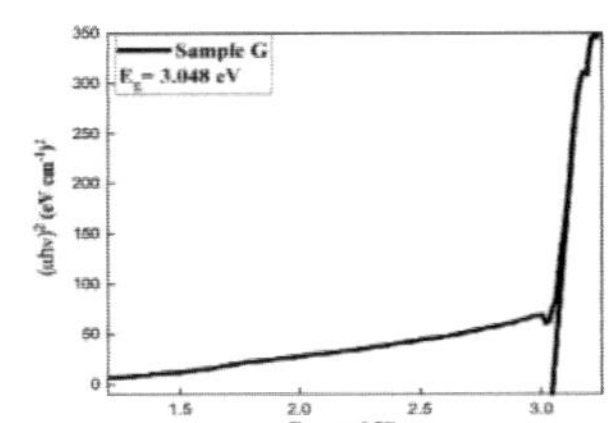

Figure 26. Tauc plot for G

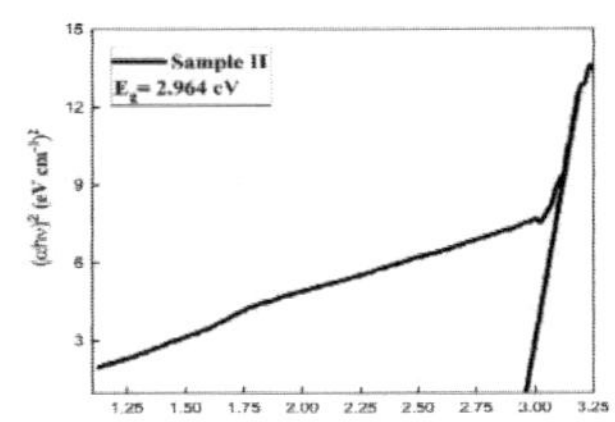

Figure 27. Tauc plot for H

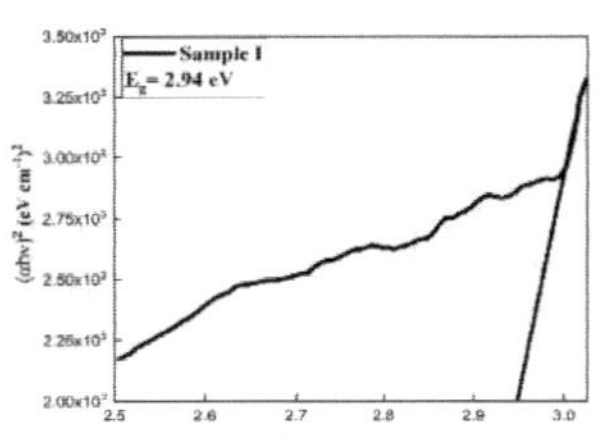

Figure 28. Tauc plot for I

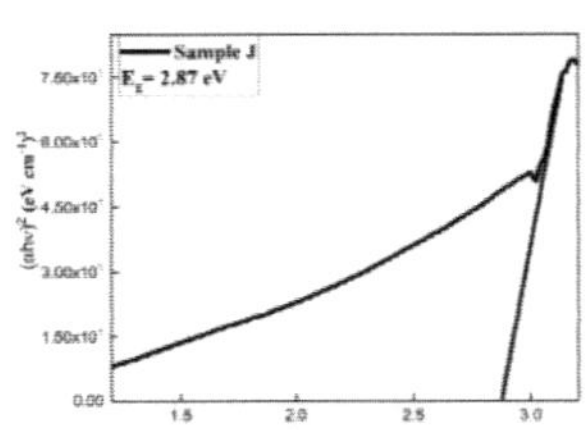

Figure 29. Tauc plot for J

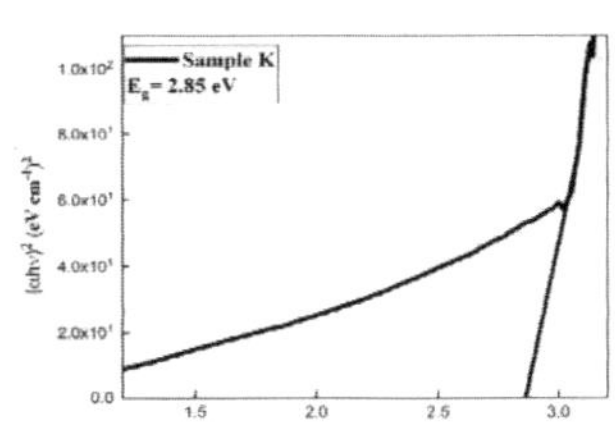

Figure 30. Tauc plot for K

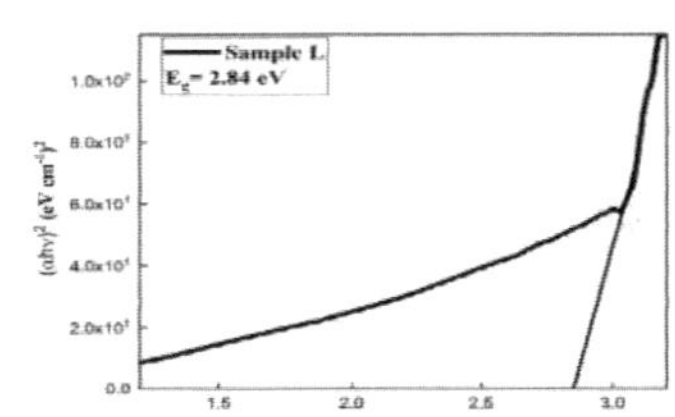

Figure 31. Tauc plot for L

Table 3: Set 1-Band gap energies in the undoped and codoped NiO

Sl. No.	Sample code	Mass % of the dopant $ZnSO_4.7H_2O$ with respect to the mass of $NiSO_4.6H_2O$	Mass % of the dopant $MgSO_4.7H_2O$ with respect to the mass of $NiSO_4.6H_2O$	Band gap Energy (eV)
1	G	0	0	3.05
2	H	20	2	2.96
3	I	20	6	2.94
4	J	20	10	2.87
5	K	20	15	2.85
6	L	20	20	2.84

The Tauc plots are presented for set 2 samples G2, H2, I2, J2, K2, and L2, figure 32 to figure 37. Band-gap energies of set 2 samples are summarised in table 4. Band gap energy of NiO, G2, is 3.06 eV. Value available in literature varies from 3.6 eV to 4 eV. Upon codoping, the value increases. As codoping percentage increases, that is, from sample G2 to sample K2, it gradually increases as shown in the table 4. But sample L2 shows a decrease.

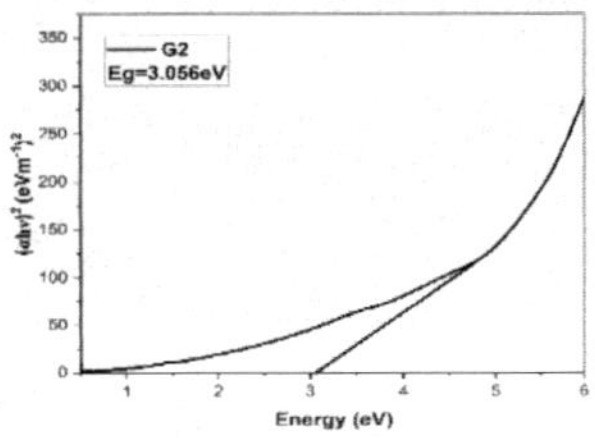

Figure 32. Tauc plot for G2

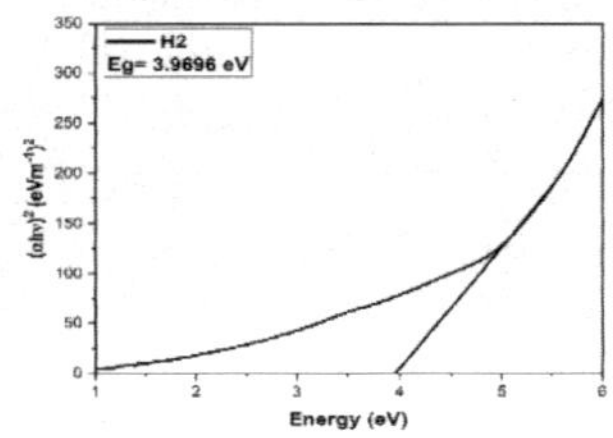

Figure 33. Tauc plot for H2

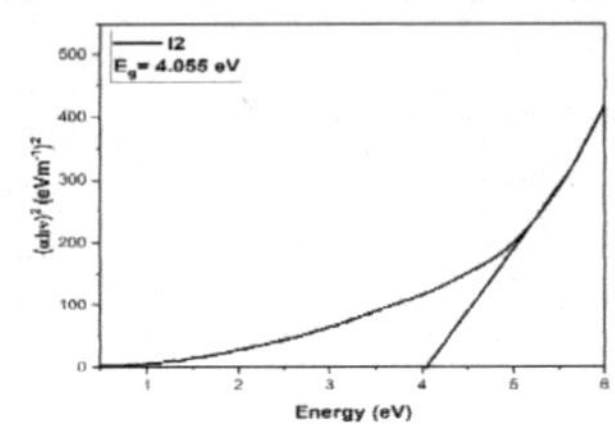

Figure 34. Tauc plot for I2

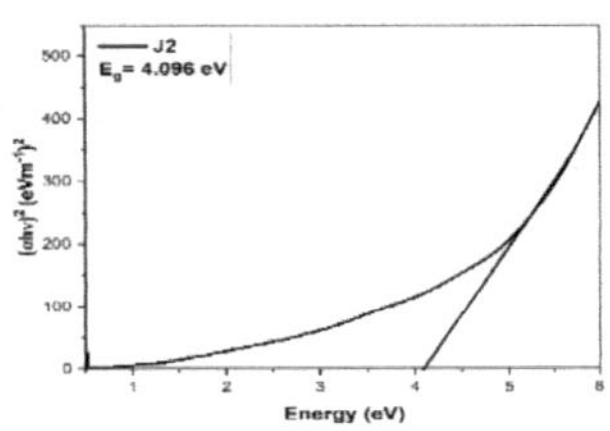

Figure 35. Tauc plot for J2

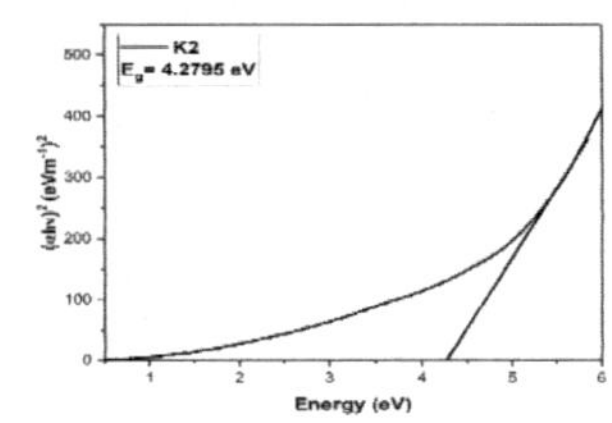

Figure 36. Tauc plot for K2

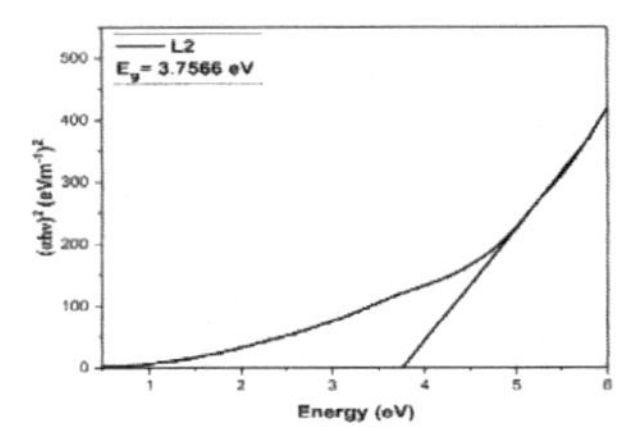

Figure 37. Tauc plot for L2

Table 4: Set 2-Band gap energies in the undoped and codoped NiO

Sl. No.	Sample code	Mass % of the dopant $MgSO_4.7H_2O$ with respect to the mass of $NiSO_4.6H_2O$	Mass % of the dopant $ZnSO_4.7H_2O$ with respect to the mass of $NiSO_4.6H_2O$	Band gap Energy (eV)
1	G2	0	0	3.06
2	H2	20	2	3.97
3	I2	20	6	4.06
4	J2	20	10	4.10
5	K2	20	15	4.28
6	L2	20	20	3.76

4.4 Full Width Half Maxima (FWHM) and crystallite size

Crystallite size computed following Scherrer formula, and full width half maxima (FWHM) obtained are summarised in table 5 and table 6.

Table 5: Set 1-Full width half maxima and crystallite size

Sl. No.	Sample code	Peak Position (2θ)	FWHM	crystallite size (nm)	Average crystallite size (nm)
1	G	37.30612	0.93081	9.007978155	
		43.36328	1.0179	8.398793064	
		62.9158	1.06271	8.763530255	9.525843982
		75.43867	1.11975	8.969083762	
		79.41242	1.02813	10.04370981	
2	H	37.11362	0.9905	8.460349221	
		43.15081	1.0919	7.823837928	

		62.62122	1.16249	7.998776744	9.487837884
		75.07745	1.12516	8.904293935	
		79.03149	0.97932	10.51532655	
3	I	37.11437	1.28465	6.523173095	
		43.15838	1.42381	6.000148769	
		62.60623	1.47689	6.295498141	7.586459222
		75.04327	1.3975	7.167413291	
		78.98476	1.19014	8.64974535	
4	J	37.1043	0.92276	9.081177271	
		43.13823	0.97617	8.751014524	
		62.61087	1.07723	8.631385308	10.85319118
		75.05606	0.96715	10.35756474	
		79.03156	0.81566	12.62520509	
5	K	37.06929	1.44463	5.800023813	
		43.12035	1.5297	5.584069479	
		62.56159	1.49538	6.216183839	8.333824893
		74.97902	1.19602	8.371223516	
		78.96207	0.9471	10.86762597	
6	L	37.02923	1.50051	5.583372968	
		43.07559	1.59333	5.360241503	
		62.50935	1.52415	6.097158466	8.202473228
		74.90032	1.24893	8.012363847	
		78.87415	0.95053	10.82157349	

Table 6: Set 2-Full width half maxima and crystallite size

Sl. No.	Sample code	Peak Position (2θ)	FWHM	crystallite size (nm)	Average crystallite size (nm)
1	G2	37.35577	1.50756	5.562593509	
		43.42976	1.64926	5.184813334	
		62.94711	1.55769	5.979783284	5.565566362
		75.40973	1.9253	5.215379701	
		79.38701	1.75427	5.885261982	
2	H2	37.2249	1.86321	4.499070204	
		43.31789	2.03869	4.192783608	
		62.74336	1.63606	5.687160423	5.8211544
		75.22079	1.72303	5.820213384	
		79.17053	1.44307	7.143238596	
3	I2	37.28683	2.24965	3.726908554	
		43.38677	2.38958	3.577962839	
		62.78908	2.06568	4.50544154	3.659829667
		75.1045	2.95417	3.392009678	
		79.19397	2.86965	3.592750781	
4	J2	37.19949	1.7875	4.689279349	
		43.25781	1.87114	4.567273342	
		62.69666	1.82593	5.094513482	4.702351583
		75.09256	2.36385	4.238750625	
		79.18012	2.18621	4.715423817	
5	K2	37.20739	1.49378	5.611456394	
		43.26088	1.71495	4.983293174	

		62.69416	1.70887	5.443422484	4.625153945
		75.05125	2.15493	4.648408896	
		79.00114	2.82936	3.638851496	
6	L2	37.14903	1.31876	6.355095457	
		43.1994	1.44613	5.90837757	
		62.62414	1.52126	6.112460793	6.037486058
		75.02772	1.83961	5.444315093	
		79.03954	1.80047	5.71987666	

5. Conclusion

Two sets of codoped samples were prepared varying the amount of Mg and Zn. A set of codoped samples was prepared keeping the extent of Zn at a fixed value and allowed the quantity of Mg to fluctuate, set 1. Another set of codoped samples was prepared keeping the quantity of Mg at a fixed value and allowed the quantity of Zn to fluctuate, set 2. The presence of the elements Ni, Zn, Mg and O in the samples was confirmed by EDAX. All samples were exposed to powder XRD analysis. XRD lines could be indexed as (003), (202), (220), (223), and (042). They correspond to rhombohedral crystal system of NiO. Both undoped and codoped samples possess rhombohedral crystal structure, and the codoping did not lead to the distortion of rhombohedral crystal lattice of NiO. With variation of codoping percentage, we observed a shift in peak. As codoping percentage increases, it could be observed that the peak is shifting towards lower 2θ values. This shift in the peak, with respect to the undoped sample, is a good support for the conclusion that the codopants, Mg and Zn, entered the crystal lattice of nickel oxide, that is, codoping occurred. Since the shift in peak is exceedingly small, it can be concluded that codoping could not distort the rhombohedral structure of nickel oxide, and so all the samples possess rhombohedral crystal structure. Band-gap energies of set 1 samples shows a decrease as codoping percentage increases, that is, from sample G to sample L. Band-gap energies of set 2 samples shows an increase as codoping percentage increases, that is, from sample G2 to sample K2. FWHM and crystallite size were also obtained.

References

[1] Xuan J et al 2020 Low-temperature operating ZnO-based NO_2 sensors: a review *RSC Advances* **10** 39786-807 https://doi.org/10.1039/D0RA07328H

[2] Ong C B, Ng L Y and Mohammad A W 2018 A review of ZnO nanoparticles as solar photocatalysts: synthesis, mechanisms and applications *Renewable and Sustainable Energy Reviews* 536-51 https://doi.org/10.1016/j.rser.2017.08.020

[3] Ben Elkamel I, Hamdaoui N, Mezni A, Ajjel R and Beji L 2019 Synthesis and characterization of Cu doped ZnO nanoparticles for stable and fast response UV photodetector at low noise current *Journal of Materials Science: Materials in Electronics* **30** 9444-54 https://link.springer.com/article/10.1007/s10854-019-01276-2

[4] Jiang S, Lin K and Cai M 2020 ZnO nanomaterials: current advancements in antibacterial mechanisms and applications *Frontiers in Chemistry* **8** 580 https://www.frontiersin.org/articles/10.3389/fchem.2020.00580/full

[5] Maiti R P, Basu S and Chakravorty D 2009 Synthesis of nanocrystalline $YFeO_3$ and its magnetic properties *Journal of Magnetism and Magnetic Materials* **321**(19) 3274-77 https://doi.org/10.1016/j.jmmm.2009.05.061

[6] Ahmed W and Iqbal J 2022 Mn doped ZrO_2 nanoparticles: an optically tuned photocatalyst with superior structural, magnetic and dielectric characteristics *Journal of Physics and Chemistry of Solids* **160** https://doi.org/10.1016/j.jpcs.2021.110285

[7] Azzaza S et al 2014 Structural, optical and magnetic characterizations of Mn-doped MgO nanoparticles *Materials Chemistry and Physics* **143**(3) 1500-07 https://doi.org/10.1016/j.matchemphys.2013.12.006

[8] Amal N, Athira Chandran M, Sreekanth K M, Sivasubramanian G, Chithra Mohan A, Gowri S L, Rishika Narendran and Sreedhar K M 2022 ZnO codoped with Si and Mn-preparation, characterization and photocatalytic reduction of hazardous chromium (VI) by the codoped ZnO *Materials Today: Proceedings* https://doi.org/1 0.1016/j.matpr.2022.11.333
[9] Malavika Menon, Daphne Mary John, Sreekanth K M, Sivasubramanian G, Sajna P J, Santhy S Dharan, Silpa Sathyan, Devika L, Aiswarya S, Karthik Raja and Sreedhar K M 2022 ZnO and Mn doped ZnO generated under three different synthesis conditions and photocatalytic reduction of hazardous hexavalent chromium by these photocatalysts *Materials Today: Proceedings* **66**(4) 2374-80 https://doi.org/10.1016/j.matpr.2022.06.333
[10] Darshitha P P, Athul Ravi, Lasya P, Maneesha Menon, Sivasubramanian G, Sreekanth K M and Sreedhar K M 2022 Foxtail millet husk as an innovative biomass in the preparation of silica-silver composite with antimicrobial and free radicle scavenging activities *Materials Today: Proceedings*, **66**(4) 1830-36 https://doi.org/10.1016/j.matpr.2022.05.285
[11] Sreedhar K M, Sangeeth Sivan, Karthik Raja, Kirti Suresh, Sreelekshmi R, Appu Palat and Sreekanth K M 2021 ZnO codoped with Ni and Mg: preparation by coprecipitation, characterisation and effect of amount of the co-dopants on the bandwidth of ZnO *Rasayan Journal of Chemistry* **14**(2) http://rasayanjournal.co.in/admin/php/upload/3198_pdf.pdf
[12] Athira Chandran M, Amal N, Sreedhar K M, Sivasubramanian G and Sreekanth K M 2020 Synthesis and characterizations of a versatile silica incorporated magnesium-zinc oxide photocatalyst *Rasayan Journal of Chemistry* **13**(4) 2616-25 https://www.scopus.com/record/display.uri?eid=2-s2.0-85099462734&origin=resultslist&sort=plf-f
[13] Rajasree G Krishnan, Beena Saraswathyamma, Sreedhar K M, Sofia David, Alisha P Prakash and Greeshma S 2020 Mixed metal oxide based electrochemical sensing of inhibitory neurotransmitter serotonin in the presence of dopamine *2020 IOP Conference Series: Materials Science and Engineering 872 012105* **872** https://iopscience.iop.org/article/10.1088/1757-899X/872/1/012105
[14] Lekshmi Mohan, Archana P, Muktha M Varma, Murlidhar Kocherla, Sreedhar K M, Sreekanth K M and Sivasubramanian G 2021 Micro circular economy conceptualized though the sustainable synthesis of a valuable opaline silica based microcidal, non-cytotoxic and free radical scavenging, composite from the dung of vechur cattle–a rare breed of bos taurus indicus *Materials Today: Proceedings* **46**(8) 2960-68 https://doi.org/10.1016/j.matpr.2020.12.422
[15] Sachin Varma U, Gautham P, Ravi Kumar D V, Sreekanth K M, Sivasubramanian G and Sreedhar K M 2018 Co-precipitation as a tool for effective doping of magnesium in zinc oxide: studies on structural, optical and photocatalytic properties *Rasayan Journal of Chemistry* **11**(4) 1491-1500 http://rasayanjournal.co.in/admin/php/upload/508_pdf.pdf
[16] Gautham P, Sachin Varma U, Sreekanth K M, Ravi Kumar D V, Sreedhar K M and Sivasubramanian G 2018 Synthesis, structural and optical properties of Mn doped ZnO with its photocatalytic activity, *Asian Journal of Chemistry* **30**(12) 2631-37 https://asianjournalofchemistry.co.in/User/ViewFreeArticle.aspx?ArticleID=30_12_7
[17] Anusiya A et al 2019 Transition-metal element (Ni, Co)-doped MgO microflowers for electrochemical biosensor applications *Jom* **71**(1) 279-84 10.1007/s11837-018-3211-7
[18] Sani Garba Danjumma, Yakubu Abubakar and Sahabi Suleiman 2019 Nickel oxide devices and applications: a review *International Journal of Engineering and Technical Research* **8**(4) 461-67 https://www.ijert.org/nickel-oxide-nio-devices-and-applications-a-review
[19] Vikas Patil, Shailesh Pawar, Manik Chougules, Prasad Goduse, Ratnakar Sakhare, Shashwati Sen and Pradeep Joshi 2011 Effect of annealing on structural, morphological, electrical and optical studies of nickel oxide thin films *Journal of Surface Engineered Materials and Advanced Technology* **1**(2) 10.4236/jsemat.2011.12006

[20] Marin Tadic, Dobrica Nikolic, Matjaz Panjan and Graeme R Blake 2015 Magnetic properties of NiO (nickel oxide) nanoparticles: blocking temperature and Neel temperature *Journal of Alloys and Compounds* **647** 1061-68 https://doi.org/10.1016/j.jallcom.2015.06.027

[21] Pranati Sahoo, Dinesh Misra, Girija Shankar Chaubey, James Salvador, Nathan J Takas and Pierre F P Poudeu Thermal conductivity of nickel oxide nanoparticles synthesized by combustion method *MRS Online Proceedings Library* **1256**(621) https://link.springer.com/article/10.1557/PROC-1256-N06-21

[22] Mironova-Ulmane N, Kuzmin A, Grabis J, Sildos I, Voronin V I, Berger I F and Kazantsev V A 2011 Structural and magnetic properties of nickel oxide nanopowders *Solid State Phenomena* **168-169** 341-44 http://www.dragon.lv/oxidephys/papers/ak147.pdf

[23] Rooksby H P 1948 A note on the structure of nickel oxide at subnormal and elevated temperatures *Acta Crystallographica* **1** 226 https://doi.org/10.1107/S0365110X48000612

[24] Glen A Slack 1960 Crystallography and domain walls in antiferromagnetic NiO crystals *Journal of Applied Physics* **31** 1571 https://doi.org/10.1063/1.1735895

[25] Nasser Mohammed Hosny 2011 Synthesis, characterization and optical band gap of NiO nanoparticles derived from anthranilic acid precursors via a thermal decomposition route *Polyhedron* **30(**3) 21 470-76 https://doi.org/10.1016/j.poly.2010.11.020

[26] Muyang Shi, Tian Qiu, Biao Tang, Guanguang Zhang, Rihui Yao, Wei Xu, Junlong Chen, Xiao Fu, Honglong Ning and Junbiao Peng 2021 Temperature-controlled crystal size of wide band gap nickel oxide and its application in electrochromism *Micromachines (Basel)* **12**(1) 80 https://www.mdpi.com/2072-666X/12/1/80

Cooling channel selection for big rectangular plastic parts in injection molding

M S M Adzhar[1,2,a)], **M N Harun**[1] **and A P M Saad**[1]

[1] Faculty of Mechanical Engineering, Universiti Teknologi Malaysia, Johor, Malaysia
[2] Canon Opto (M) Sdn Bhd, Selangor, Malaysia

[a)] Corresponding author: syafiq_adzhar@yahoo.com

Abstract. There are various of types cooling channel designs that have been explored by many researchers to find the best cooling channel that resulted in the optimum results in the injection molding process. The previous research reported that the correct selection of cooling channels in mold design affects the quality of the product and the total cycle time, thus it affected the quality and profit in the manufacturing [1]. The problem in mold design is selecting the correct cooling channel design to ensure the heat transfer distribution is in uniform condition. This paper analysed the effectiveness of four types of different cooling channel designs, the diameter size of the cooling channel, and the variable of coolant temperature in the big rectangular plastic product with the same injection parameters and material type. The objective is to determine the effect of using different cooling channel designs on the cooling time and warpage effect with different coolant temperatures. The commercial mold flow software was used to simulate the injection behaviour to determine the quality of the warpage and total cooling time in the injection molding process. The result found that using a cooling channel type with arrays of the baffle, array bubbler, and conformal cooling channel can reduce the cooling time by almost 50% compared to the conventional cooling channel. The increased coolant temperature in the cooling channel also affects the increase in the cooling time.

1. Introduction

In the manufacturing industry today, more than one-third of polymeric products was produced through the injection molding process [2]. The injection molding process is complicated, where the molten polymer undergoes a complex thermomechanical change, and with the improper setting of the process affects the various defects in the final product thereby increasing the amount of waste [1]. In the injection molding process, the part cooling process takes around 50% of the time from the total cycle time during the injection molding process [2]. It shows that the cooling process plays a very important role in the total cycle time in the overall injection molding process. Nowadays, many types of cooling channel designs can be used in mold manufacturing.

The common type of cooling channel design is the conventional design due to the design's easy and low cost of fabrication. However, this cooling channel design is suitable for mold that produces a flat plastic product without a curved, round, or fillet surface [1]. The conventional cooling channel also has lacks heat transfer distribution, contributing to high cooling time. To improve the performance of the conventional cooling channel, the array of baffles and bubbler cooling channel is introduced and normally is applicable in the mold that has a small region shape which has a lack cooling performance. These cooling channels can improve the performance of the heat exchanger in the injection mold and leads to more efficient and uniform control of mold temperature distribution. It is effective for use in

medium and large-sized molds with complex shapes or free-form surfaces [3]. The conformal cooling channel has been introduced with the improvement of the fabrication method. This cooling channel has four designs that are the zig-zag type, spiral type, scaffold type, and Voronoi Diagram type [4]. The conformal cooling channels can potentially improve the mold performance in terms of less warping and defect and distribution of uniform and fast cooling efficiency into the mold.
However, to produce the rectangular plastic product shape, many cooling channel designs can be used, but the wrong selection can cause an increase in the cooling time and warpage of the product, thus can contribute to the increase of mold cost fabrication and reducing the productivity in the production. Therefore, the objective of the study was to determine the effect of using different cooling channel designs on the cooling time and warpage effect with different coolant temperatures.

2. Model

To get a suitable design cooling channel for the mold that produces big rectangular plastic parts, four cooling channel designs have been modelled. It started with designing the part and gating design. This product consisting a dimension of 4.2mm x 300.6mm x 136.7mm and 2.2mm thickness and use 10 pinpoint gate as shown in Fig. 1.

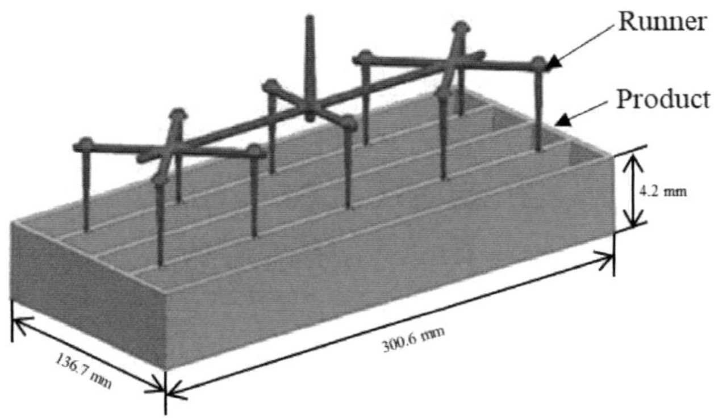

Figure 1. Rectangular plastic part with 10 pinpoint gates.

Four types of design of cooling channels have been develop and evaluated to determine the best design of cooling channel that contributes to the shortest cooling time in the injection molding process. It started with the conventional cooling channel design with are designing by following the design from Venkatesh in their study [5]. The conformal design also has been taken in the same study because this previous study has been conducted by comparing the conventional and conformal cooling channel designs in the rectangular plastic product. Besides the conventional and conformal design, two types of cooling channels have been highlighted by Tuteski that is the baffles and bubblers types [6]. This study also has developed these two designs of cooling channels to compare the effectiveness in the cooling time compared to the conformal design type. Figure 2 show the four cooling channel designs that are modelled and used in this study and the Fig. 3 show the flowchart for this analysis.

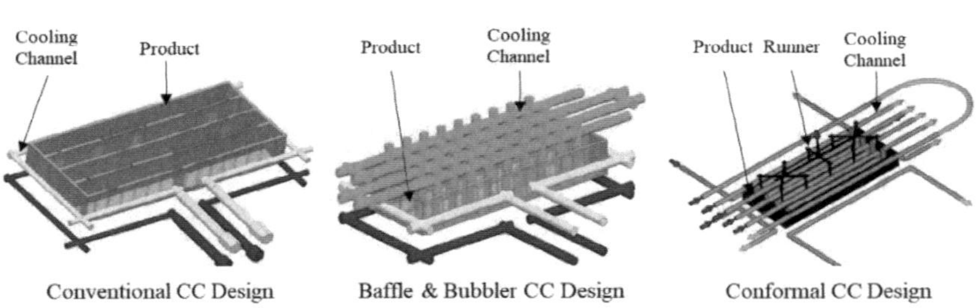

Figure 2. Design of cooling channels.

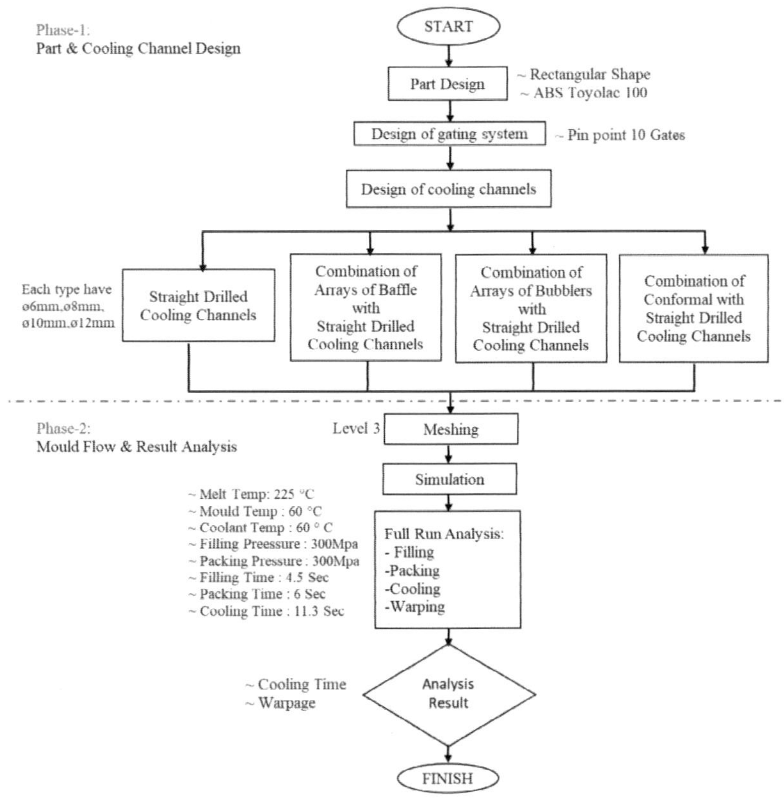

Figure 3. Flow chart of analysis.

3. Computational Analysis

A further step after designing the various types of cooling channels design is to analysed the cooling time conducted by the Moldex software. Different cooling channels have been designed with Siemens NX11.0 software and a complete CAD file was imported to the MDG file for the mold flow analysis. For the parameter set, ABS Toyolac 100 manufactured by Toray Manufacture was selected as injected material. The processing parameter for this type following the recommendation from the manufacturer which can be referred to in the MFI database is shown in Table 1 below. The flow of boundary conditions in this analysis is divided into two stages. It started with the full-scale analysis that involves filling, packing, cooling, and warping, as in the actual injection molding process. The time taken to flow the material is 4.5 seconds with the injection pressure (P) is 300MPa and the melting temperature (T) of the material is 325°C.

The second stage involves the flow of coolant in the cooling channel it started with the coolant flow in into the coolant channel (V_{in}) and flowing out at the end of the coolant channel (V_{out}) at each cooling channel as shown in Fig. 4. From this figure, for the conventional cooling channel it contains with 5 coolant flow in the cooling channel as indicate V_{1in} until V_{5in} and 5 coolant flow out started with V_{1out} until V_{5out}. The Baffle, Bubbler, and Conformal types contain 4 coolant flow in the cooling channel as indicated V_{6in} until V_{9in} and 4 coolant flow out starting with V_{6out} until V_{9out}. The coolant temperature set start with the temperature of 50°C until 80°C with increasing of 10°C for each setting with the same flow rate (Q) of 120cm³/sec for all flow in V_{in} and V_{out}. All setting is the same for each type of cooling channel design with the variable of diameter that started with a diameter of 6mm until 12mm.

IOP Conf. Series: Materials Science and Engineering 1291 (2023) 012022 doi:10.1088/1757-899X/1291/1/012022

Table 1. Process setting for ABS Toyolac 100 Material.

Process Condition	Process Setting
Melt Temperature (minimum)	190 °C
Melt Temperature (normal)	225 °C
Melt Temperature (maximum)	260 °C
Ejection Temperature	96.85 °C
Freeze Temperature	116.85 °C

Table 2. Parameter setting in CAE mode analysis.

Filling stage	Filling time (sec)	4.5
	Melt temperature (°C)	225
	Mold temperature (°C)	60
	Maximum injection pressure (MPa)	300
	Injection volume (cm³)	279.839
Packing stage	Packing time (sec)	6
	Maximum packing pressure (MPa)	300
Cooling stage	Cooling time (sec)	11.3
	Mold-open time (sec)	5
	Eject temperature (°C)	96.85
	Air temperature (°C)	25
Miscellaneous	Cycle time (sec)	26.8

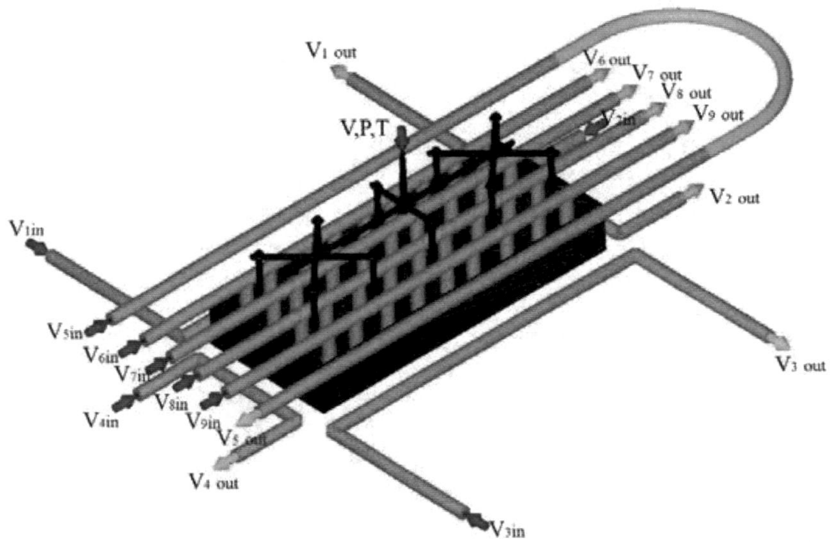

Figure 4. The example boundary condition for baffle and bubbler cooling channel.

4. Result and Discussion

Figure 5 shows the cooling time in different coolant temperatures by using the different diameters of the cooling channel and design of the cooling channel with the (a) conventional cooling channel, (b) baffle cooling channel, (c) bubbler cooling channel and (d) conformal cooling channel. Figure 6 shows the actual part temperature by using the different diameters of the cooling channel and design of the cooling channel with the (a) conventional cooling channel, (b) baffle cooling channel, (c) bubbler cooling channel, and (d) conformal cooling channel. The graph shows that the increase in the coolant temperature affects the increase in the cooling time. This increasing trend is the same for all four types of cooling channels in this analysis. This result also shows that using the other cooling channel types can reduce the cooling time by around 50% reduction compared to using the conventional type. The result between the conformal, baffle, and the bubbler is not contributed to the large difference in the cooling time.

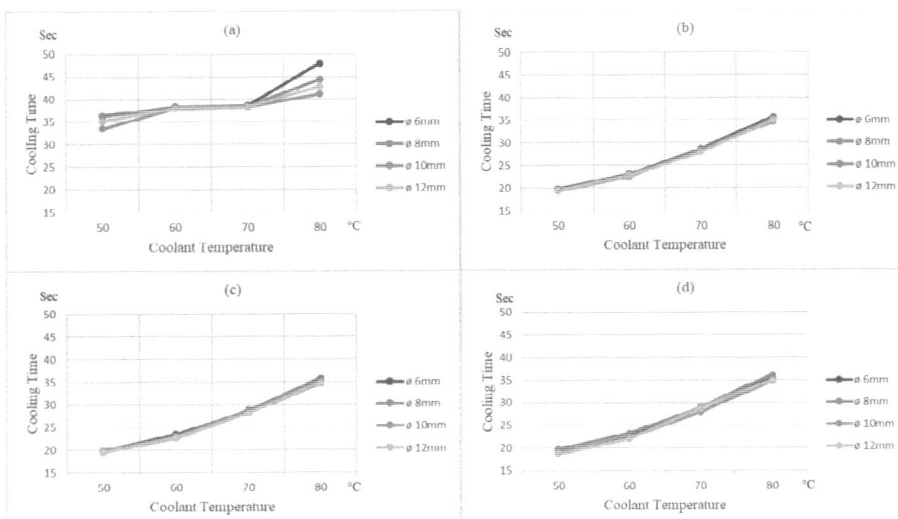

Figure 5. Cooling time result for the (a) conventional cooling channel, (b) baffle cooling channel, (c) bubbler cooling channel, and (d) conformal cooling channel.

The result of cooling time depends on the temperature of the mold cooling channel, which affected the actual part temperature for each cooling channel type with the variable of coolant temperature and diameter cooling channel. Based on this graph, shows that increasing the coolant temperature in the cooling channel affects increasing the actual temperature of the product. The increasing trend also affects the cooling time to increase. Besides that, the difference in the cooling channel design also affected the temperature in the cooling channel. From this, it shows that the conventional cooling channel design contributes to the increase of the actual part temperature almost 40% higher compared to the baffle, bubbler, and conformal cooling channel design.

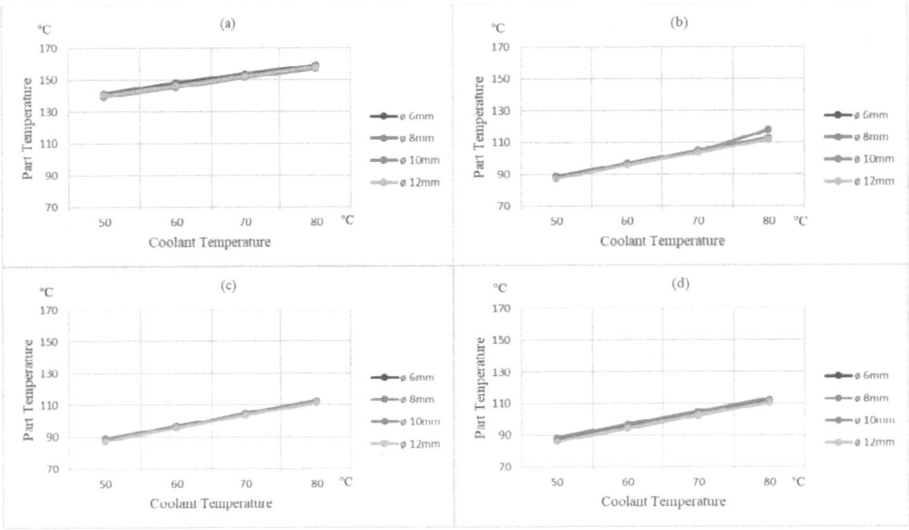

Figure 6. Actual part temperature for the (a) conventional cooling channel, (b) baffle cooling channel, (c) bubbler cooling channel, and (d) conformal cooling channel.

Figure 7 shows the warpage for four types of cooling channel designs using a diameter of 10mm with the variable of coolant temperature. From the result, it can be found that the use of a cooling channel with a type baffle, bubbler, and conformal can reduce the value of the warpage product compared to the conventional cooling channel. The result of these three cooling channels is almost similar to the warpage. Using the bubbler cooling channel type with a 50°C coolant temperature resulted in the minimum warpage compared to another set of coolant temperatures.

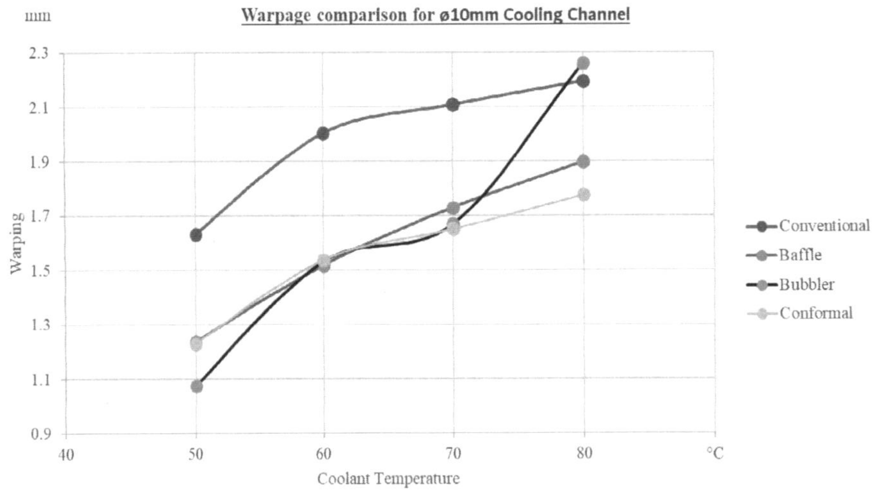

Figure 7. Total displacement for four types of cooling channels ø10mm.

This paper discussed selecting the best type of cooling channel design for the rectangular plastic product, which includes four types of cooling channels using conventional, baffle, bubbler, and conformal design. The result is focused on two factors that are important in injection molding production that focussed in productivity which focused on the cooling time result and the quality of the product which focuses on the warpage value of the product. From the simulation using software that has been done, the baffle, bubbler, and conformal cooling design can reduce the cooling time by almost 50% compared to the conventional cooling design type. Besides that, the temperature analysis that has been done shows that increasing the setting of the coolant temperature will affect the increase of the cooling time. The result for the baffle, bubbler, and conformal types does not show much difference for the cooling time and warpage result. The actual part temperature also shows that the temperature distribution of these three cooling channels is almost the same. From the warpage result, the conventional cooling channel design will contribute a high value of the warpage compared to the other cooling channel designs. From the previous study that has been done by Lu, conformal design can reduce 18% the reduction of warpage value compared to the baffle type for the circular product [7]. The difference with this study is that by using the bubbler type, the warpage is reduced almost 15% compared to the conformal type in setting coolant of 50°C temperature. In terms of low-cost fabrication cost and ease of maintenance issues, the best selection of cooling channel design in the rectangular plastic product mold is by using a baffle and bubbler cooling channel design compared to the conformal design. By study that have been done by Park, it show that the use of a conformal design can reduce 50% of the cooling time compared to the conventional type but the use of Metal 3D printing is required a high cost [8] compared to the baffle and bubbler type that only required CNC straight drilled machine to fabricate this cooling channel. For further study, it suggested analysing the use of the baffle and bubbler cooling design with different shapes from the circular design such as rectangular design, triangle design, or half circular design to compare the optimum performance in the design. Besides that, the different types of material cooling channels also can be applied for further study.

5. Conclusion

This work shows the simulation result for the cooling channel design in mold manufacturing for the big rectangular plastic product. This study shows that using an array of baffles, bubbler, and conformal designs can reduce the cooling time by almost 50% and reduce the warpage in the product. Besides that, the use of an array of baffles and bubblers contributes almost the same result compared to the conformal design. By this, manufacturing the big rectangular plastic product mold can use the array of the baffle and bubbler design, thus can reduce the manufacturing cost compare to the conformal design.

Reference

1. Feng, S., A.M. Kamat, and Y. Pei, *Design and fabrication of conformal cooling channels in molds: Review and progress updates.* International Journal of Heat and Mass Transfer, 2021. **171**.
2. Jahan, S.A., et al. *Thermo-mechanical Design Optimization of Conformal Cooling Channels using Design of Experiments Approach.* in *Procedia Manufacturing*. 2017.
3. Eiamsa-Ard, K. and K. Wannissorn, *Conformal bubbler cooling for molds by metal deposition process.* CAD Computer Aided Design, 2015. **69**: p. 126-133.
4. Park, H.S. and X.P. Dang. *Development of a Smart Plastic Injection Mold with Conformal Cooling Channels.* in *Procedia Manufacturing*. 2017.
5. Venkatesh, G., Y. Ravi Kumar, and G. Raghavendra. *Comparison of Straight Line to Conformal Cooling Channel in Injection Molding*. in *Materials Today: Proceedings*. 2017.
6. Tuteski, O. and A. Kočov, *Conformal cooling channels in injection molding tools – design considerations.* Mach. Technol. Mater., 2018. **12**: p. 445-448.
7. Lu, C.T., C.H. Chen, and S.C. Tseng. *Application of conformal cooling to reduce cooling time and warpage of a U-shaped plate.* in *AIP Conference Proceedings*. 2019.

8. Park, H.S., et al., *Design of Advanced Injection Mold to Increase Cooling Efficiency.* International Journal of Precision Engineering and Manufacturing - Green Technology, 2020. **7**(2): p. 319-328.

IOP Conf. Series: Materials Science and Engineering 1291 (2023) 012023 doi:10.1088/1757-899X/1291/1/012023

Induced Wigner crystallization in a coupled electron-hole quantum wire system at finite-temperature

Preeti Rani[1], **Kulveer Kaur**[1] **and Vinayak Garg**[1,*]

[1]Department of Physics, Punjabi University, Patiala-147 002, India

[*]E-mail: vinayak2000@gmail.com

Abstract. We explore the possibility of induced Wigner crystallization in a coupled electron-hole (*e-h*) quantum wire system at finite-temperature T by calculating the intra- and inter-wire static structure factors in the dynamic mean-field approximation of Singwi *et al* (the qSTLS theory). We find that owing to the dynamics of severely correlated particles, the *e-h* system may favour an induced phase transition from quantum liquid to the Wigner crystal state at a non-zero T and sufficiently low particle (electron/hole) number density in the close proximity of two wires. On the other hand, the static STLS approximation does not predict such phase transition due to complete neglect of the dynamics of particle correlations.

1. Introduction

A coupled quantum wire system consists of two parallel quantum nano wires placed at sufficiently small (~ nm) inter-wire distance. Depending upon the nature of charge carriers in both the wires, they are regarded as coupled electron-electron (*e-e*) or electron-hole (*e-h*) or hole-hole (*h-h*) quantum wire systems.

The coupled quantum wires have generated a great deal of theoretical [1]-[6] and experimental [7,8] curiosity over last few decades. Authors in Ref. [1-3] have predicted the charge-density-wave (CDW) instability in coupled quantum wire systems by including both the intra- and inter-wire particle correlations through the Singwi, Tosi, Land and Sjolander (STLS) approximation [9] at absolute zero. Very recently, Sharma *et al* [4] observed the CDW instability in a coupled *e-h* quantum wire system at sufficiently low temperature and particle number density using the STLS theory. On the other hand, no such instability was found in the coupled *e-e* quantum wire system at any non-zero T.

Notably, the correlations among particles are assumed to be static (i.e., time-independent) in the STLS theory. However, the dynamics of particle correlations becomes quite significant to determine the many-body properties of the system in the strongly correlated regime (i.e., low temperature, particle number density, transverse wire width and inter-wire separation), especially the Wigner crystallization. Though, Saini *et al* [5] included the dynamics of particle correlations through the quantum version of the STLS approach (the qSTLS theory [10]) to predict the CDW state in the ground-state of coupled *e-e* and *e-h* quantum wire systems, yet, assumed equal effective mass of holes and electrons (m_h^*/m_e^*=1). Nevertheless, this may not be true in a real physical situation such as *GaAs* based *e-h* quantum wire system wherein $m_h^*/m_e^* \approx 7$. Keeping this point in mind, Moudgil *et al* [6] considered the mass-asymmetry effect and employed the qSTLS approximation to study the ground-state properties of a coupled *e-h* wire system. It was deduced that the *e-h* system may favour a phase transition from quantum liquid to the CDW state or the Wigner crystallization at absolute zero in the close vicinity of two wires, and at sufficiently low particle (electron and hole) number density. The localization of holes in a coupled *e-h* system was also predicted by Steinberg and co-workers [7] at a hole density of $2 \times 10^7 m^{-1}$, electron density $6 \times 10^7 m^{-1}$ and $T = 0.25K$.

Motivated by the experimental work of Steinberg *et al*, here we explore the possibility of the formation of an induced Wigner crystal (WC) phase in a coupled *e-h* quantum wire system at a non-zero temperature by employing the qSTLS approximation.

2. Theoretical Formalism

The wave vector (q) and frequency (ω) dependent intra- and inter-wire components of density response function $\chi_{ll'}(q,\omega,\mu;T)$ for the coupled *e-h* quantum wire system at finite-T can be obtained as [11]

$$\chi_{ll'}(q,\omega;T) = \frac{\chi_l^0(q,\omega;\mu_l,T)\left[\delta_{ll'} + (-1)^{\delta_{ll'}}\,\Psi_{\bar{l}\bar{l}'}(q,\omega;T)\,\chi_{\bar{l}}^0(q,\omega;\mu_{\bar{l}},T)\right]}{\epsilon(q,\omega;T)}, \tag{1}$$

Here, l and l' are the wire indices, μ_l the chemical potential of particles in wire l, and $\bar{l}$ denotes the wire index opposite to l. Also, the dielectric function $\epsilon(q,\omega;T)$ is given as

$$\epsilon(q,\omega;T) = \epsilon_{ll'}(q,\omega;T) = \left[1 - \Psi_{ll}(q,\omega;T)\chi_l^0(q,\omega;\mu_l,T)\right]\left[1 - \Psi_{l'l'}(q,\omega;T)\chi_{l'}^0(q,\omega;\mu_{l'},T)\right] - \Psi_{ll'}^2(q,\omega;T)\chi_l^0(q,\omega;\mu_l,T)\chi_{l'}^0(q,\omega;\mu_{l'},T), \tag{2}$$

$\Psi_{ll'}(q,\omega;T) = V_{ll'}(q)[1 - G_{ll'}(q,\omega;T)]$ being the effective Coulomb interaction potential among particles with $V_{ll'}(q) = \frac{2e^2}{\varepsilon_0} K_0\left(q\sqrt{b^2 + (1-\delta_{ll'})d^2}\right)$ as the Fourier transform of bare Coulomb interaction potential. $K_0(x)$ is the zeroth-order modified Bessel's function of first kind, b the transverse width of each wire, d the inter-wire separation and ε_0 the background dielectric constant of each wire. $G_{ll'}(q,\omega;T)$ represent the dynamic local-field correction (LFC) factor at finite-T which accounts for the inclusion of dynamics of exchange-correlations among particles and in the qSTLS approximation is given as

$$G_{ll'}(q,\omega;T) = -\frac{1}{n_l}\int_{-\infty}^{\infty} \frac{dq'}{2\pi} \frac{\chi_l^0(q,q'\omega;\mu_l,T)}{\chi_l^0(q,\omega;\mu_l,T)} \frac{V_{ll'}(q')}{V_{ll'}(q)} [S_{ll'}(|q-q'|;T) - \delta_{ll'}] \tag{3}$$

Here, n_l is the particle number density (taken to be same in both the wires) and $\chi_l^0(q,q',\omega;\mu_l,T)$ is the inhomogeneous non-interacting density response function for the particles in l^{th} wire given as [12]

$$\chi_l^0(q,q',\omega;\mu_l,T) = \begin{cases} \displaystyle\int_{e^{(-\mu_l/k_BT)}}^{1} \frac{dx}{(1+x)^2}\chi_l^0(q,q',\omega;\mu_l + k_BT\ln|x|, T=0) \\ \displaystyle + \int_0^1 \frac{dx}{(1+x)^2}\chi_l^0(q,q',\omega;\mu_l - k_BT\ln|x|, T=0)\,;\ \mu_l > 0 \\ \displaystyle\int_0^{e^{(\mu_l/k_BT)}} \frac{dx}{(1+x)^2}\chi_l^0(q,q',\omega;\mu_l - k_BT\ln|x|, T=0)\,;\ \mu_l < 0. \end{cases} \tag{4}$$

which reduces to $\chi_l^0(q,\omega;\mu_l,T)$ for $q' = q$; $\chi_l^0(q,q',\omega;\mu_l,T=0)$ is the ground-state density response function of free electrons in the quantum wire analytically calculated as

$$\chi_l^0(q,\omega;\mu_l,T) = \frac{m_l^*}{\pi q}\ln\left|\frac{\omega^2 - \left(\frac{qk_{Fe}}{m_l^*} - \frac{q^2}{2m_l^*}\right)^2}{\omega^2 - \left(\frac{qk_{Fe}}{m_l^*} + \frac{q^2}{2m_l^*}\right)^2}\right| \tag{5}$$

m_l^* being the effective mass of particles in wire 'l' and k_{Fe} the Fermi wavevector of wire electrons.

In equation (3), the finite-*T* static structure factor $S_{ll'}(q;T)$ is expressed as [13]

$$S_{ll'}(q;T) = -\frac{\hbar}{\pi n_l}\int_0^\infty d\omega\, \coth\left(\frac{\hbar\omega}{2k_BT}\right) Im\chi_{ll'}(q,\omega;T) \tag{6}$$

3. Results and discussion

We use a dimensionless system of units in which the wave vector q is taken in units of the electron's Fermi wave vector k_{Fe}, wire width b and inter-wire spacing d in electron's effective Bohr radius a_{0e}^* $[= \varepsilon_0\hbar^2/(m_e^*e^2)]$, ω and μ in electron's Fermi energy E_{Fe} and T in electron's Fermi temperature T_{Fe} (i.e. $T/T_{Fe} = \tau$). We choose effective mass ratio of hole and electron to be 7 i.e., $m_h^*/m_e^*= 7$, appropriate to a *GaAs* based *e-h* wire system, and $k_{Fe} = k_{Fh}$. The number density n_l of electrons or holes in each of the wire is taken to be same and is described by a dimensionless parameter $r_s = 1/(2n_l a_{0e}^*)$; $n_l = N/L$ (total no. of electrons/holes per unit length of the wire). Throughout our calculations, we set $\hbar$, k_B and ϵ_0 equal to unity.

Further, we use the subscripts *'ee'* and *'hh'* for the presentation of intra-wire static structure factors of the electron and hole wire, respectively. Nonetheless, the subscript *'e-h'* is used for inter-wire static structure factor.

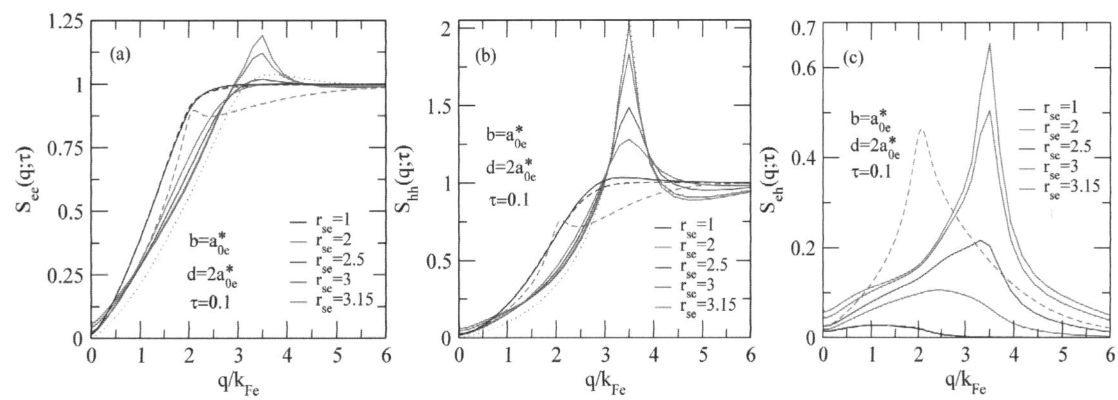

Figure 1. $S_{ee}(q;\tau)$ [in panel (a)], $S_{hh}(q;\tau)$ [in panel (b)] and $S_{eh}(q;\tau)$ [in panel (c)] of coupled *e-h* quantum wire system at $b = a_{0e}^*$, $d = 2a_{0e}^*$ and $\tau = 0.1$ and indicated r_{se}. The dashed curves are the corresponding STLS results at $r_{se} = 1$ and 3.15. The dotted curves present the static structure factor *S(q)* of a single/isolated *e*-wire [in (a)] and *h*-wire [in (b)] at $r_{se} = 3.15$.

The set of equations (1), (3) and (6) is solved numerically in a self-consistent manner for the intra- and inter-wire static structure factors $S_{ll'}(q;\tau)$ of the coupled *e-h* quantum wire system over a range of particle (electron/hole) number density (r_{se}) for a fixed value of wire width and inter-wire separation. We accepted the numerical solution when convergence in the results of $S_{ll'}(q;\tau)$ was better than 0.0001%, at each q and ω in the chosen grid of q- and ω-points.

In figure 1(a) and 1(b), we plot the intra-wire static structure factor $S_{ll}(q;\tau)$ of the coupled *e-h* system at $b=a_{0e}^*$, $d=2a_{0e}^*$, $\tau = 0.1$ and indicated r_{se} values for the *e*- and *h*-wire, respectively. It is found that particle (electron or hole) correlations grow directly with increasing coupling (r_{se}) among particles.

A significant peak starts developing in $S_{ll}(q;\tau)$ for $r_{se} \geq 2$ situated at $q \sim 3.5k_{Fe}$ (say q_0), gets stronger on further increase in r_{se} and attains maximum height at a critical r_{se} i.e., $r_{se}^c = 3.15$. However, we could not obtain the self-consistent $S_{ll'}(q;\tau)$ for $r_{se} > r_{se}^c$. The peak is stronger in $S_{hh}(q;\tau)$ [figure 1(b)] as compared to $S_{ee}(q;\tau)$ [figure 1(a)] indicating that in spite of having equal particle number density in both the wires, the holes in *h*-wire are more strongly correlated than the electrons in *e*-wire due to their heavier effective mass (we have taken m_h^*/m_e^*= 7). The emergence of very strong peak in $S_{hh}(q;\tau)$ is a precursor of the evolution of Wigner crystallization of holes in the *h*-wire at $\tau = 0.1$ and r_{se}^c= 3.15. On the other hand, this quantum effect is not much pronounced for electrons in the *e*-wire as the peak in $S_{ee}(q;\tau)$ is relatively small. We also present the structure factor [$S(q)$] of a single (isolated) *e*-wire [in figure 1 (a)] and *h*-wire [in figure 1(b)] at $r_{se} = 3.15$. It can be seen that holes in the single/isolated *h*-wire tend to localize (through peaked structure), while as, the electrons in the single/isolated *e*-wire do not show this characteristic feature as the peak is completely missing here. However, when both the *e*- and *h*-wires are coupled as in the *e-h* system, the strongly correlated holes at r_{se}^c= 3.15 tend to localize the electrons (which otherwise are in the liquid state) in the less correlated *e*-wire *via* the attractive and stronger *e-h* correlations. It is worthwhile to mention here that for $b = a_{0e}^*$, $d=2a_{0e}^*$ and $\tau = 0.1$ the qSTLS approach predicts Wigner crystallization in an isolated *e*-wire at $r_{se}^c \approx 5.6$ and in isolated *h*-wire at $r_{sh}^c \approx 4.9$. The stronger *e-h* correlations also get highlighted through a strong peak in $S_{eh}(q;\tau)$ at q_0 and r_{se}^c [see figure 1(c)].

The STLS results of $S_{ll'}(q;\tau)$ are also plotted in figure 1 at r_{se}= 1 and 3.15. We find that at $r_{se} = 1$, $S_{ll'}(q;\tau)$ remains a smooth function of q and it does not predict Wigner crystallization in either of the wires even at r_{se}^c=3.15. This is due to the fact that the STLS theory deals with particle correlations only up to a static level, whereas, the Wigner crystallization can only be observed theoretically if one includes the dynamics of correlations. Interestingly, the STLS theory also predicts a peak in $S_{ll'}(q;\tau)$ at $q \sim 2k_{Fe}$ which is more prominent in the *e-h* component than the *e-e* and/or *h-h* components.

4. Summary and conclusions

We have studied the intra- and inter-wire static structure factors of a coupled *e-h* quantum wire system at a fixed *b, d, T* and over a range of particle number density parameter r_{se} within the qSTLS approximation. It is noted that in the coupled *e-h* system, holes in the *h*-wire are more strongly correlated in comparison to electrons in the *e*-wire due to their heavier effective mass. The dynamics of particle correlations results in dominant spatial electron and hole correlations at r_{se}^c=3.15. The holes in *h*-wire of the coupled *e-h* system now prefer to be in the crystal phase at relatively lower r_{se} (r_{se}^c=3.15) as compared to that in a single/isolated *h*-wire (r_{se}^c=4.9). Interestingly, the holes localization (of *h*-wire) induces considerable Wigner crystallization in the electrons (of *e*-wire) which now takes place at much lower r_{se} (r_{se}^c=3.15) in comparison to that required in a single *e*-wire (r_{se}^c=5.6). Our this theoretical prediction is in a qualitative agreement with the experimental findings of Steinberg *et al* [7]. On the other hand, the STLS theory does not predict Wigner crystallization in the *e-h* wire system at non-zero *T* due to complete neglect of the dynamics of particle (electron or holes) correlations.

Acknowledgments

Preeti Rani is grateful to CSIR (New Delhi) for providing financial assistance in the form of JRF.

References

[1] Moudgil R K 2000 *J. Phys.: Condens. Matter* **12** 1781
[2] Thakur J S and Neilson D 1997 *Phys. Rev. B* **56** 4671
[3] Mutluay N and Tanatar B 1997 *Phys. Rev. B* **55** 6697
[4] Sharma A, Garg V, Moudgil R K 2022 *Phys. Scr.* **97** 065710
[5] Saini L K, Tankeshwar K and Moudgil R K 2004 *Phys. Rev. B* **70** 075302
[6] Moudgil R K, Garg V and Ahluwalia P K 2010 *Eur. Phys. J. B* **74** 517
[7] Steinberg H, Auslaender O M, Yacoby A, Qian J, Fiete G A, Tserkovnyak Y, Halperin B I, Baldwin K W, Pfeiffer L N and West K W 2006 *Phys. Rev. B* **73** 113307

[8] Barak G, Steinberg H, Pfeiffer L N, West K W, Glazman L, Oppen F V and Yacoby A 2010 *Nat. Phys. Lett.* **6** 489
[9] Singwi K S, Tosi M P, Land R II and Sjölander A 1968 *Phys. Rev.* **176** 589
[10] Hasegawa T and Shimizu M 1975 *J. Phys. Soc. Jpn.* **38** 965; **39** 569
[11] Verma V, Puttar D, Garg V, Moudgil R K 2023 *Phys. B: Condens. Matter* **648** 414382
[12] Maldague P F 1978 *Surf. Sci.* **73** 296
[13] Pines D and Noziers P 1966 *The Theory of Quantum Liquids* vol 1 (New York: Benjamin)

Doping effect on optical properties of pure and malachite green doped polyvinylidene fluoride films

Girja Ahirwar, Pooja Devi Sahu, P K Khare and R K Dubey

Department of Post Graduate Studies and Research in Physics and Electronics, Rani Durgavati Vishwavidyalaya, Jabalpur M.P. 482001, India

Corresponding author: Email: - [1] girjaahirwar95gmail.com
[2] pooja.rpsahu@gmail.com
[3] pkkhare.rdvv@gmail.com
[4] drrkdubey@gmail.com

Abstract: - The films of pure polyvinylidene fluoride (PVDF) and various concentration of malachite green (MG) doped have been prepared by solution cast method. The absorption spectra recorded by UV-VIS spectrophotometer 1700 Shimadzu. Absorbance versus wavelength spectra of pure and various concentration of MG doped PVDF films were observed in wavelength range 200-800 nm. With increasing malachite green in PVDF, intensity of absorbance edge and refractive index increases.

1. Introduction

Polymeric material serve insulation in transportation, air craft and automobiles, communication and medical fields [1]. A polymer is made up of many small molecules which have combined to form a single long or large molecule. The single small molecules (single unit) from which polymer is form known as monomers and the process by which the monomer molecules are link to form a big polymer is called polymerization [2]. For the study of charge storage and transportation the insulating polymers are most attractive. Such polymers also yield the best electret for practical application because of their insulating properties and high concentration of deep trapping sites [3]. In recent years, due to application in optical device, studies on the electrical and optical properties of polymers have attracted much attention. The optical properties of polymers can be suitably modified by attention of dopants / another polymer with the host matrix [4].
Polyvinylidene fluoride (PVDF) is semicrystalline, thermoplastics polymer belonging to the family of fluoropolymer due to its excellent piezoelectric properties, thermal stability and mechanical strength [5]. It has a good processability and it also possess chemical resistance property to different materials such as different acids, bases, organic solvents, oil and fat [3]. PVDF has well physical and electrical properties of a kind of many applications [6]. Its electrical properties have led to many applications as an electrical insulator, greatly as a wire coating [3]. Since its commercialization in the mid-1960, PVDF

has been used as a verity of chemical process industries sectors due to its versatility and broad attributes [7]. Glass transition temperature (T_g) of PVDF about is -38^0 C and melting temperature is 171^0 C. PVDF has low density 1.74 g/ml at 25^0 C. Structure formula of PVDF is $(CH_2\text{-}CF_2)_n$.
Malachite green (MG) is an organic compound of low molecular weight that is used as a dyestuff. MG is an Azo-dye which are the largest class of synthetic dyes.

2. Experimental

Materials and film preparation

Polyvinylidene fluoride (PVDF) in powder form (molecular weight 534,000 MW) has been used as a host polymer in the present investigation. Thin films of PVDF and malachite doped PVDF film has been prepared by using common solvent Dimethyl formamide (DMF).
Solution cast method has been used for preparation of pure and malachite green doped PVDF films. The solution was kept on magnetic stirrer at fixed temperature to get homogeneous and transparent solution. Homogeneous solution then poured onto an optically plane glass plate floating on mercury. The solvent was allowed to evaporate in an oven at desired temperature for 6 hrs to yield the desired film. This was followed by room temperature out gassing at 10^{-5} torr for further period of 12 hours remove any residual solvent. The plate was then slowly drawn out of the solution leaving a uniform film on the plate. Thus, film obtained transparent, uniformly smooth and can be easily peeled from the glass surface. The doped sample were prepared by malachite green of 0.067%, 0.117 % and 0.167 % mixing respectively in 06 gm of PVDF dissolved in 50 ml of dimethyl formamide (DMF). The doped films were green in colour, uniformly smooth and can be easily peeled from the glass surface.

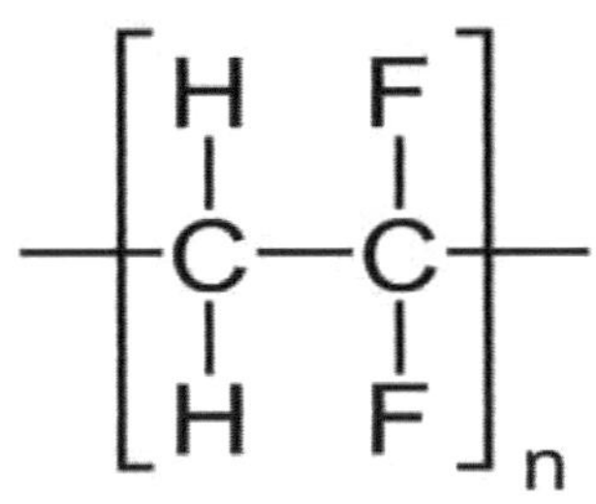

Figure 1. Sturcture formula of PVDF.

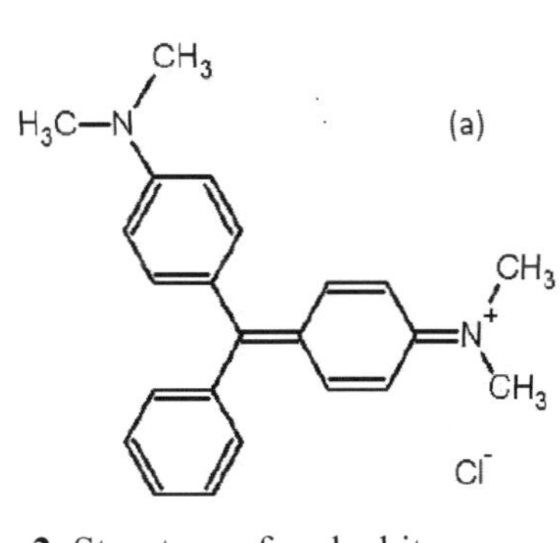

Figure 2. Structure of malachite green.

3. Results and Discussion

The absorbance spectra of pure PVDF and malachite green doped PVDF films recorded in wavelength range 200-800 nm by UV-VIS spectrophotometer 1700 Make Shimadzu. The absorption spectra for pure as well malachite green doped films of PVDF in different configuration is shown in figure 3.

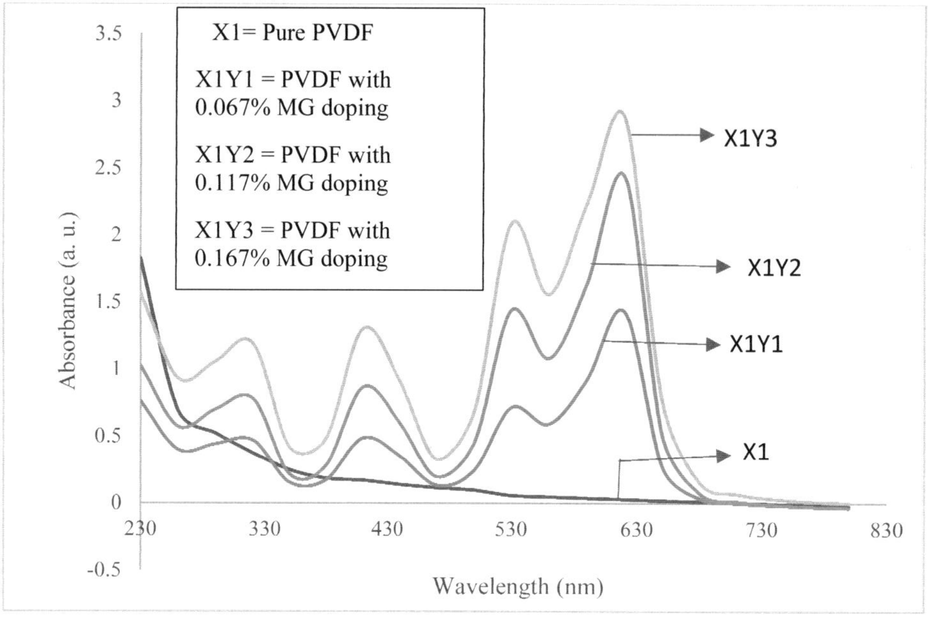

Figure 3. Absorbance versus wavelength spectra of pure and malachite green doped PVDF film.

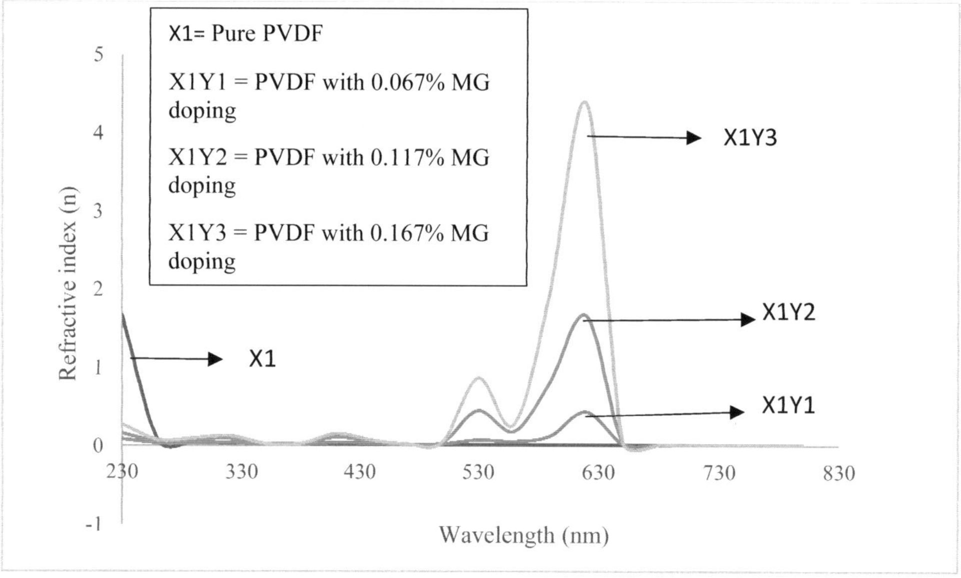

Figure 4. Spectra for refractive index versus wavelength of pure and MG doped PVDF samples.

IOP Conf. Series: Materials Science and Engineering 1291 (2023) 012024 doi:10.1088/1757-899X/1291/1/012024

For pure PVDF maximum absorption edge occur at 230 nm wavelength [8]. With increasing wavelength absorbance decreases continuously. However, addition of malachite green in PVDF gives rise to four absorption bands at wavelength around 320, 410, 530 and 620 nm for all doped samples. First absorption band observed in ultra violet region at 320 nm for 0.067 %, 0.117 % and 0.167 % doped films. Remaining three bands occurs in visible region (410-620 nm). Intensity of absorption band increases with addition of malachite green [9]. Maximum intensity of absorption band occurred for fourth band at 620 nm wavelength for all doped films. Malachite green is a cationic triphenyl methane dye [10]. As the dye is cationic, the coloured ion is charged positively [9]. The cationic nature of dye suggest that it is electron donor dye. Polymeric materials are likely to accept electrons, hence malachite green may form donor-acceptor complexes with PVDF when mixed under suitable condition [11]. Also, malachite green has a one amino group containing unshared pair of electrons which may be responsible for charge transfer complex (CTC) formation [12]. In this present study maximum absorption observed peak observed at wavelength 620 nm for 0.167 % concentration of malachite green.
Variation in refractive index for pure PVDF with different concentration of MG doped samples have been detected. Spectra of refractive index versus wavelength for pure and MG doped PVDF shown in figure 4.
Refractive index for pure and MG doped PVDF samples calculated using equation

$$n = 1/T_s + [1/(T_s - 1)]^{1/2} \qquad \text{(1)}$$

were

n = Refractive index

T_s = Percent Transmittance

For pure PVDF sample maximum refractive index occurs at 230 nm wavelength. When increasing concentration of malachite green in PVDF refractive index increases. Maximum refractive index occurs in visible range at 620 nm for all doped samples. Refractive index for pure PVDF and all MG doped samples presented in table 2.

Table 1. Absorbance versus wavelength UV spectra of pure and malachite green doped samples

Absorbance peak	Wavelength peak position (nm)	Pure PVDF (Absorbance Peak)	PVDF doped with MG		
			0.067%	**0.117%**	**0.167%**
1st	320	not found	0.468	0.779	1.191
2nd	410	not found	0.484	0.864	1.299
3rd	530	not found	0.712	1.438	2.083
4th	620	not found	1.432	2.437	2.872

Table 2. Refractive index for pure and malachite green doped PVDF samples

Sample	Wavelength (nm)	Refractive index
Pure PVDF	230	1.667
PVDF with 0.067 % malachite green	320	0.045
	410	0.046
	530	0.078
	620	0.442
PVDF with 0.117 % malachite green	320	0.095
	410	0.112
	530	0.455
	620	1.667
PVDF with 0.167 % malachite green	320	0.132
	410	0.157
	530	0.871
	620	4.375

4. Conclusion

With increasing concentration of malachite green, the intensity of absorption bands and refractive index increases. Region for change in intensity of absorption band with increasing concentration of MG in PVDF may be due to formation of charge transfer complexes.

References

1. Charles E Carraher, Jr, 2013 Introduction to polymer chemistry, Third Edition, Hand book, by Taylor & Francis Group LLC 34.
2. Gawariker, V R, Viswanathan, N V and Shreedhar J 2005 Polymer Science, New Age International, New Delhi 7-9.
3. Saxena Pooja and Shukla Prasant, 2021 Advanced Composites and Hybrid Materials **4** 8-26.
4. Rawat A, Mahavar H K, Chauhan S, Tanwar A and Singh P J 2012 Indian Journal of Pure & Applied Physics **50** 100-104.
5. Bhartti Imtiaz Noor, Banerjee M. and Bhatti Ilyas Noor, 2013 IOSR Journal of Applied Physics (IOSR-JAP) **4** 42-47.
6. Elashmawi I S, Alatawia Naifa S, Elsayed Nadia H, 2017 Result in Physics. (doi.org/10.1016/j.rinp.2017.01.022).
7. Palovcak Averrie and Pomante Jason, chemical Engineering Essentials for the CPI Professional, electronically reprinted from August 2018.
8. Ismail A M, Moammed M I and Fouad S S 2018 Journal of Molecular Structure, 1170 51-59.
9. Sahu Pooja Devi, and Khare P K, 2019 AIP Conf. Proc. 2104 020039-1-020039-4.
10. Arslan Taner, Guney Sevgi and Guney Orhan, 2019 International Journal of Environment Analytical Chemistry, Issn: 0306-7319 1-17.
11. Bhardwaj M K and Khare P K, 2012 Research Journal of Recent Sciences, **1** 238-243.
12. Nayak Rashmi J, Khare P K and Nayak J G, 2019 AIP Conf. Proc. 2162 020033-1-020022- 6.

Optical Conductivity Analysis of Green CuO Nanoparticles Using Plant Leaf Extract

B Arunkumar[1], R Meenakshi[1], S Thamarai selvi[1], N Raja[2], T Meganathan[1]

[1] Department of Science & Humanities, Sir Issac Newton College of Engineering & Technology, Nagapattinam, Tamilnadu, India-611102. (Affiliated by Anna University, Chennai)

[2] Department of Science & Humanities, Anjalai Ammal Mahalingam Engineering College, Thiruvarur, Tamilnadu, India-614403. (Affiliated by Anna University, Chennai)

Corresponding Author Mail ID: bvarun2030@gmail.com,

Abstract. The optical properties of green synthesized CuO nanoparticles such as absorption coefficient, refractive index, band gap, dielectric constant and optical conductivity were analyzed. XRD shows that upon temperature treatment of green CuO nanoparticles, the crystal size increases and thus the crystallinity increases due to the decrease in the number of crystals per unit area. Due to the high crystallinity, the absorption coefficient exhibits a high electronic shift and the refractive index exhibits good polarization of the ions. Thus the dielectric constant of the real and imaginary regions decreases as the photon energy increases. Higher temperatures exhibit higher photon energy levels. Optical conductivity has induced electron annihilation due to the absorption coefficient of increased photo energy light; Hence the optical conductivity analysis of Lantana camara capped CuO nanoparticles indicated the stability of electrical conductivity with increase in photo power.

Keywords: *Green synthesis, CuO nanoparticles, Absorption & Refractive coefficient, Dielectric Constant and optical Conductivity*

1. INTRODUCTION

Recent years have seen new trends in research speed, low cost and high energy. In this view, energies are measured quantum mechanically and the quantum field is often associated with thermodynamics and optics. The main role of optics is leading in electromagnetics, because light has dual properties like wave and particle, and it transfers information from one place to another faster, with less energy loss, less cost and more energy conversions. Optical conductivity can be measured depending on the light absorption and refractive index of the surface volume ratio of materials, because these two phenomena are related to the workability of the material, which is determined by its surface to volume ratio. The surface-to-volume ratio shows many ways to create new materials [1,2]. It plays an important role in the field of all materials and facilitates the chemical process. The surface area to volume ratio of nanoparticles has a significant effect on nanoparticle properties. Nanoparticles have a relatively large surface area compared to the same volume of matter. The radius of the sphere decreases the surface area. Here we can also conclude that when a given volume is divided into smaller pieces, the area increases. Thus as the particle size decreases, a larger fraction of atoms is found on the surface than inside [3,4]. Because the growth and catalytic chemical reactions occur at the surface, the nanomaterial is more reactive than the same mass composed of larger particles. It has also been found that materials that are stable in their bulk form react when produced in their nanoscale form [5,6]. This can improve their properties. So in this formulation, we analyze absorption coefficient, refractive index, band gap and optical conductivity of green synthesized CuO nanoparticles for electrical conductivity.

2. SYNTHESIS OF COPPER OXIDE NANOPARTICLES

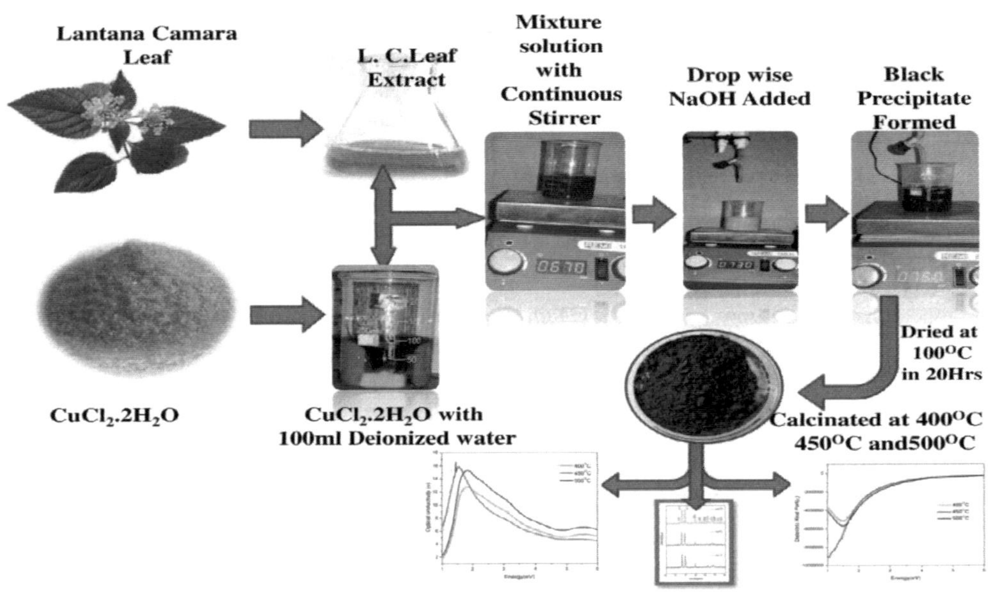

Figure 1 Schematic Diagram of Synthesis of CuO Nanoparticles

The Cleaned Lantana camara leaves chop it into small pieces and take 5g of chopped Lantana camara leaves in 250ml glass beaker and add 100ml of deionised water to it. Then it should be placed on the magnetic stirrer and stirred for 20 minutes at a temperature of 80 degrees. After 20 minutes, the extract of lantana camara leaves keep it.

Copper oxide nanoparticles were prepared by sol-gel method using Lantana camara leaf extract. Copper chloride dehydrates $CuCl_2.2H_2O$, sodium hydroxide (NaOH) and Lantana camara leaf extract are used. Take 0.2molarity copper chloride dehydrate ($CuCl_2.2H_2O$) and mix it in 100ml of deionised water. The mixed solution should be placed on a magnetic stirrer and after 10 minutes of stirred with a temperature of 80°C, 10ml of Lantana camara leaf extract should be added to it. After 15 minutes, 2M sodium hydroxide diluted in 40ml of deionised water was added drop wise to the magnetically stirred solution. When sodium hydroxide is added, the solution suddenly turns black colour and a black precipitate can be observed in the bottom of beaker. The precipitate was taken and washed twice with deionised water and the filtered. The obtained black powder was subjected to three different sintering (400°C, 450°C and 500°C) to prepare CuO nanoparticles. The prepared CuO nanoparticles samples were subjected to characterization. Figure 1 Schematic Diagram of Synthesis of CuO Nanoparticles.

3. RESULT AND DISCUSSION

3.1 Structural Analysis

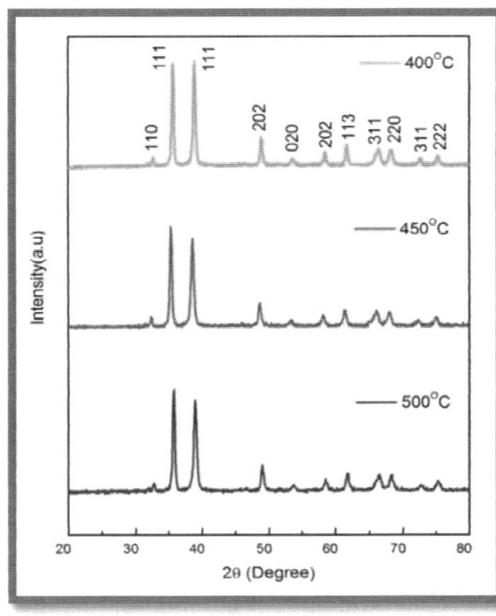

Figure 2. XRD Analysis of CuO Nanoparticles at Different Temperatures

The structural evaluation of Green synthesized CuO nanoparticles is shown in figure 2. All peaks are located at same plane and 2θ range. The figure 2 shows that the peaks are highly sharp due to high quality of crystallinity. The obtained CuO nanoparticles of quality crystallinity have indicated to correspond with optical materials [7]. These crystalline peaks are clearly matched with CuO nanoparticles, exhibits monoclinic structure and no other contamination was shown.

Table 1. Structural Parameter of Green Synthesized CuO Nanoparticles

Temperatures ^{O}C	FWHM	2θ	Crystal size(nm)	Stacking Fault	Dislocation Density $\delta \times 10^{14}$	Micro Strain (ε)
400°C	0.4506	35.7181	19.34	0.003503	2.67396	0.001872
450°C	0.3452	35.7032	25.24	0.002685	1.56946	0.001434
500°C	0.2808	35.4526	31.01	0.002192	1.03995	0.001168

The green synthesized CuO nanopaticles at different temperature treatment reveals similar structural results observed XRD analysis. The green capped CuO nanoparticles main peaks positioned at 35.7^{O} and 38.2^{O}is consistent with JCPDS 45-0937 [8]. Lantana Camara leaf capped CuO nanoparticles has modified by plant extract ions to increase the crystal size when increase temperatures. Table 1 exhibits the structural parameters like crystal size, Micro strain, stacking fault and dislocation density. The Table 1 crystal size increases with decreases stacking fault, dislocation density and Micro strain. At each temperature there is change in the surface to volume ratio due to the change in micro strain and dislocation density. The surface to volume ratio change is plotted on the stacking fault in table 1. The analysis of XRD figure1 shows that the rise of the temperature treatment of green CuO nanoparticles has leads increases crystallinity with increases crystal size and decreases magnitude of number of crystallites per unit surface area.

3.2 FTIR Spectra

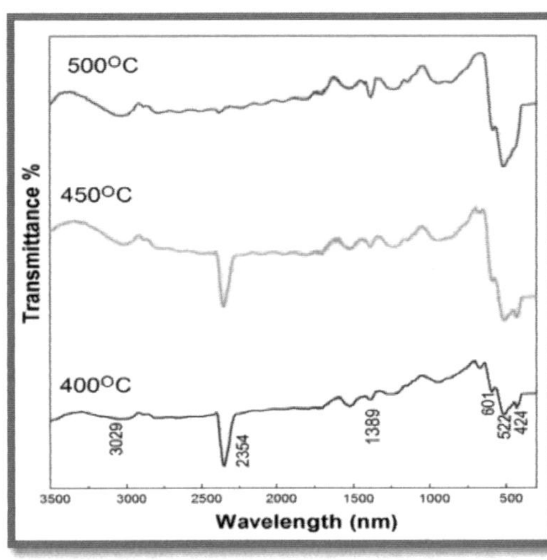

Figure 3. FTIR Spectra of Bio-synthesized CuO NPs at Different Temperatures

The FTIR spectrum of green synthesized CuO nanoparticles shown in figure 3. This FTIR spectrum revealed that peaks at 424, 522 and 601cm^{-1} has formation of CuO nanoparticles [9]. The broad peak at 3029Cm^{-1} is attributed to O-H stretching of moisture content [10]. The Peak obtained at 2354cm^{-1} has stretching vibration of O=C=O bond of carbon dioxide [11]. The peak observed at 1389cm^{-1} is bending vibration of phenol [12].

3.3 Absorption Coefficient

Absorption coefficient of green synthesizes CuO nanoparticles by UV Visible spectra as shows in the figure 4. In figure 4 shows that that lowers temperatures has high absorption and higher temperature has low absorption. But all the three temperatures of samples has higher absorption coefficient and its indicated density of electron absorb by medium of material. The absorbed by the materials is converted into heat of the material due to which electron is excited by photon energy and electron-hole fusion is formed [13]. So green synthesized CuO nanoparticles of 400oC and 500°C temperatures of higher edge absorption exhibited 1.7ev of photon energy and 450°C temperature of higher edge absorption exhibited 1.5eV. The absorption of light is reduces in higher photon energies. The Absorption coefficient is calculated by [14]

$$Absorption\ Coefficient\ \boldsymbol{\alpha} = \frac{2.303A}{d} \text{ ----------------- } 1$$

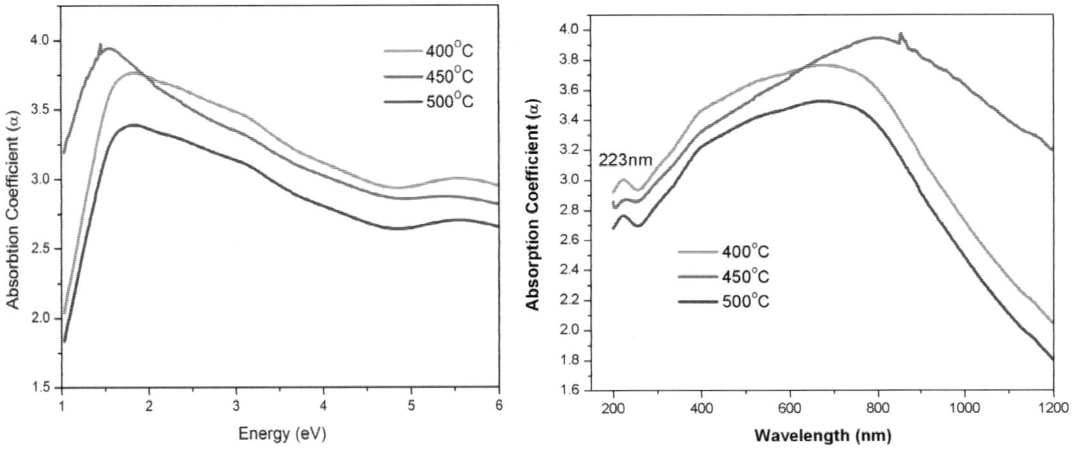

Figure 4. Absorption Coefficient of Green synthesized CuO NPs at Different Temperatures

The lantana camara capped CuO nanoparticles shows an absorption peak at 223nm its indicated nearer visible region that reveals unsaturated pair of electrons. The energy level of wavelength 223nm region π-π* which exhibits n-π*transition [11,15]. Since the wavelength of the ultraviolet region is easily scattered, the electron is excited by the photon. This type of UV absorption is associated with electronic transition in molecules from lower energy to higher energy states. The position of the atoms in a molecule that includes the orbital's involved in the transition is said to produce a transition of electron-hole combinations.

3.4. Morphological Studies

Scanning electron microscopy (SEM) images of CuO metallic nanoparticles synthesized by Sol-gel method respectively. SEM has been used to recognize the size, shape and morphology of nanoparticles. It reveals that the CuO nanoparticles are well dispersed and for the most part of spherical in shape. CuO have a standardized particle shape and size with little bit of agglomeration because they did not form a large cluster. The agglomeration can generally be confirmed by the formation of large cluster [15,16].

Figure 5. Morphological Analysis of Bio-synthesized CuO NPs at different Temperature

SEM Micrograph represents a position of Copper Oxide nanoparticles, which may be reaction in high temperatures. The crystal grain size calculated from SEM micrograph originated to assortment between 18 to 21nm. This deviation in particle size strengthens with XRD analysis. Figure 5 shows the three dimensional aggregate morphology of as synthesized material, which formed by a group of primary particles [17].

3.5 Refractive Index

Figure 6 represents the refractive index of the green synthesized CuO nanoparticles at three different temperatures. The refractive index of the photon energy plot of green synthesized CuO nanoparticles shows in figure 6. The image clearly revealed that refractive index increases with increases in photon energy. The refractive index increases with increases optical density and speed of light decreases. The optical density of a material is related to the inertial tendency of the atoms of a material to retain the absorbed energy of an electromagnetic wave in the form of vibrating electrons before giving it back as a new electromagnetic disturbance. The more optically dense a material is, slower a wave travels through the material [18].

$$\textit{Refractive Index } \mathbf{n} = \frac{\mathbf{1}}{T_s} + \sqrt{\frac{\mathbf{1}}{T_s - \mathbf{1}}} \text{ ---------------- } 2$$

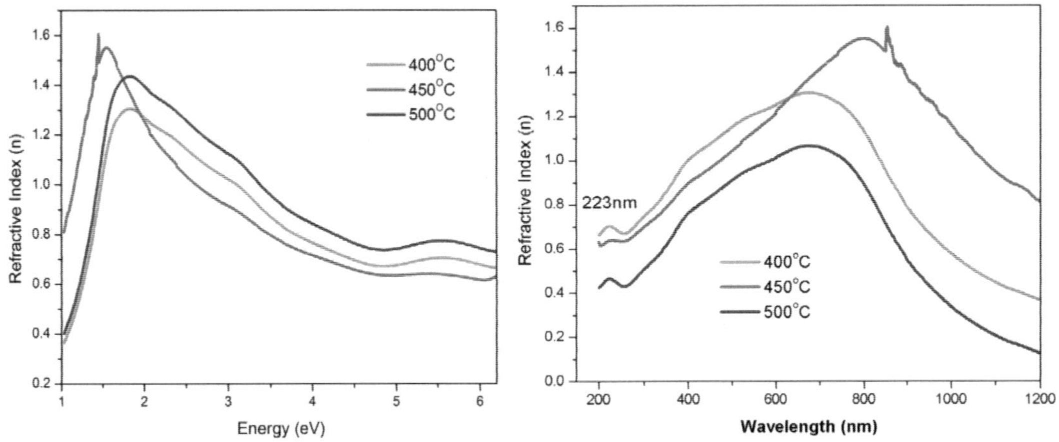

Figure 6. Refractive Index of Green synthesized CuO NPs at Different Temperatures

The figure 6 shows the variation of refractive index as a utility of wavelength, refractive index increases with increases wavelength. The edge of the refractive index is appeared at 223nm UV region and higher value of refractive index in moves in higher value of wave length. This type of refractive index of optical materials must be evaluated for integrated application of optical devices. This higher refractive index helps to electronic polarity of ions and the local field inside the material [19, 20]. The refractive index indicated to very one important feature of optical materials. The Extinction Coefficient was calculated by the formula

3.6 Extinction Coefficient

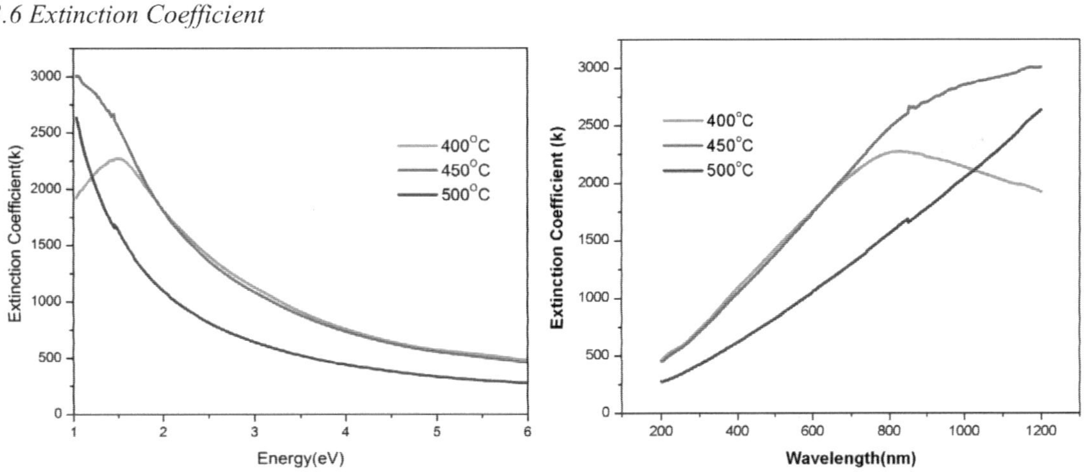

Figure 7. Extinction Coefficient of Green synthesized CuO NPs at Different Temperatures

$$\text{The Extinction Coefficient } \boldsymbol{K} = \frac{\alpha\lambda}{4\pi} \text{ ----------------- } 3$$

The extinction coefficient of green synthesized CuO nanoparticles as shown figure 7. The extinction analysis reveals that the plot of photon energy and wavelength of graph range is x10^{-3}. The high level extinction coefficient has appeared in 1.5eV and wavelength at 692nm. The plant extract capped CuO nanoparticles highest peak of extinction coefficient appeared at 1.5 eV respectively as shown in figure 5. Extinction coefficient is amplitude of decay or damping of light for optical conductivity. The absorbed light has damp in the small area and interacted with electron in nanoparticles for induced electric conductivity [21]. From the figure 5 the extinction coefficient in photon energy is increase then decrease in higher energy value and it has constant move but it comes not zero and also in wavelength has increases temperature with increase the extinction coefficient in higher wavelength. So extinction coefficient analysis of green synthesized CuO nanoparticles clearly reveals that suitable material for optical conductivity.

3.7 Band gap

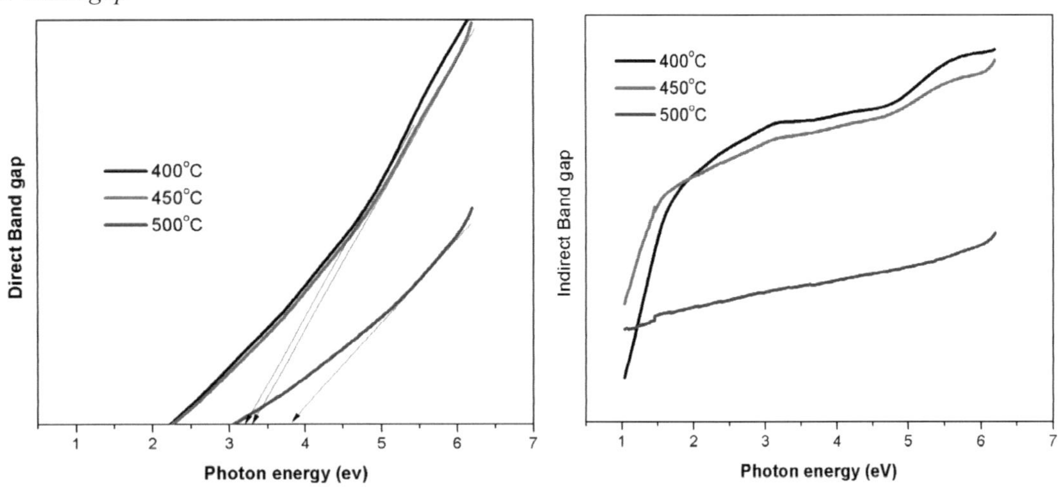

Figure 8. Band gap of Green synthesized CuO NPs at Different Temperatures

Figure 8 shows the direct and indirect band gap of green synthesized CuO nanopatrticles at different temperatures, which is express that absorption spectra 400°C of CuO green synthesized nanoparticles have slightest optical band gap energy value, it has found to be 3.19eV. The figure 8 reveals of direct band gap shows existence of high photo energy, which has been evaluated by the linear plot of direct and indirect Band gap [22,23]. The bandgap parameters are tabulated in Table 2

Table 2 Optical band gap (Eg), Urbach Energy (E_u) and optical Conductivity of Green synthesized CuO NPs at Different Temperatures

Temperature °C	E_g(eV)	E_u(eV)	σx10^{-8} S^{-1}
400°C	3.19	32.02	12.84
450°C	3.28	16.72	15.91
500°C	3.79	8.63	15.36

3.8 Optical Conductivity

Optical conductivity is a property of a material that gives a relationship between the induced current density in the material and the magnitude of induced electric field for arbitrary frequencies [24]. In this article analysed to optical conductivity of Green synthesized CuO nanoparticles by using Lantana camara leaf extract. The Optical conductivity of Green synthesized CuO nanoparticles as shown in figure 9. The Figure 9 exhibits high intensity of optical conductivity appeared at 1.7eV in photon energy and 680nm, 804nm in wavelength.

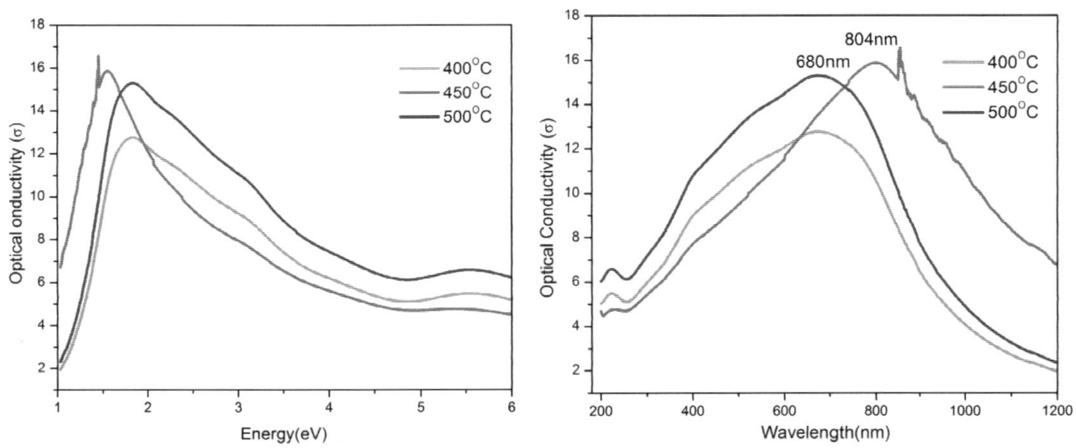

Figure 9. Optical Conductivity of Green synthesized CuO NPs at Different Temperatures

The optical conductivity found to be gradually increases against wavelength at 680nm, 804nm correlated with absorption coefficient. In view of figure 9 expressed as plant leaf doped CuO nanoparticles has absorb and damp the light in inside of material and boosted the optical behaviour and increasing the photon energy. This increased photo energy has induced electron extinction due to absorption coefficient of light [25, 26]. So that the optical conductivity analysis of Lantana camara capped CuO nanoparticles of figure 9 reveals that the assurance of the electrical conductivity on the increases in photo energy. Finally, concluded form the figure 9 of optical conductivity, the obtained green synthesized CuO nanoparticles has opted to optical devices. The table 2 reveals that the optical perameters such as optical band gap (Eg), Urbach energy (Eu) and optical conductivity.

3.9 Dielectric Constant

The dielectric constant of Green synthesized Nanoparticles as shown in figure 10. The real part of dielectric larger negative values for $500^{O}C$ than for $400^{O}C$ and $450^{O}C$ and imaginary part of dielectric constant exhibits inter band transition in ultra violet part of spectra. The real part of Dielectric constant is the factor that slows down the speed of light and thus leads to the ability of materials to electrical conductivity. Likewise, the imaginary dielectric constant acts as a measure of the material's light absorption and hence the ability to conduct electricity through dipole mobility [27].

The figure 8 shows dielectric constant of real and imaginary parts decreases with increases in photon energy. It's indicated the higher temperatures exhibit the higher photon energy level.

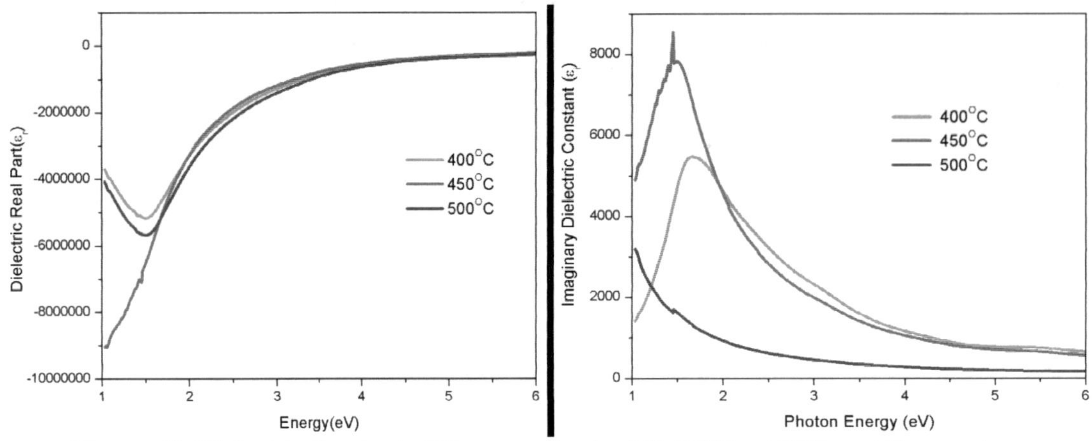

Figure 10. Dielectric Constant of Green synthesized CuO NPs at Different Temperatures

This dielectric constant analysis of figure 10 clearly provides current and hypothetical dielectric constant loss factor. In these Green chemical reacted CuO nanoparticles has exhibits the decrease in the real area of permittivity at high frequencies, the dielectric structure is composed of low-resistivity in nanoparticles, separated by poorly conducting thin nanoparticles [28]. As a result of the applied electric field, the green synthesized CuO nanoparticles obtained a local accumulation of charges, which led to interfacial polarization.

4. CONCLUSION

The optical properties of green synthesized CuO nanoparticle is analyzed likewise absorption coefficient, refractive index, band gap, dielectric Constant and optical conductivity. XRD shows that the rise of the temperature treatment of green CuO nanoparticles has leads increases crystallinity with increases crystal size and decreases magnitude of number of crystallites per unit surface area. The absorption Coefficient exhibits higher electronic transition and refractive index exhibits good polarity of ions. Dielectric constant of real and imaginary parts decreases with increases in photon energy. It's indicated the higher temperatures exhibit the higher photon energy level. Optical conductivity exhibits increased photo energy has induced electron extinction due to absorption coefficient of light; So that the optical conductivity analysis of Lantana camara capped CuO nanoparticles indicated the assurance of the electrical conductivity on the increases in photo energy.

5.REFERENCES

1. Almqvist N (1996) Fractal analysis of scanning probe microscopyimages. Surf Sci 355:221–228,- ISSN 0039-6028, https://doi.org/10.1016/0039-6028(95)01369-5.

2. Chen LB, Lu N, Xu CM, Yu HC, Wang TH (2009) Electrochemicalperformance of polycrystalline CuO nanowires as anode materialfor Li ion batteries. Electrochim Acta 54:4198–4201, https://doi.org/10.1016/0039-6028(95)01369-5.

3. B. Arunkumar, M. Jothibas, S. Johnson Jeyakumar, Blue dye degradation effect of green chemical synthesized CuO nanoparticles, Materials Today: Proceedings, Volume 66, Part 4, 2022, Pages 2207-2214, ISSN 2214-7853, https://doi.org/10.1016/j.matpr.2022.06.038.

4. Essic J, Mather R (1993) Characterization of a bulk semiconductorsband gap via near-absorption edge optical transmission experiment.Am J Phys 61:646–649, https://doi.org/10.1119/1.17173

5. Gandhi S, Subramani RHH, Ramakrishnan T, Sivabalan A, DhanalakshmiV, Nair MRG, Anbarasan R (2010) Ultrasound assisted onepot synthesis of nano-sized CuO and its nanocomposite withpoly(vinyl alcohol). J Mater Sci 45:1688–1694,- DOI:10.1007/s10853-009-4158-4

6. B. Arunkumar, S. Johnson Jeyakumar, M. Jothibas, A Strains Activity of CuO Nanoparticles using Copper Chloride Dihydrate by Sol-Gel Method,- Asian Journal of Chemistry; Vol. 31, 2019, Issue 4, pp. 886-890,https://doi.org/10.14233/ajchem.2019.21820.

7. Chiang CY, Aroh K, Ehrman SH (2012) Copper oxide nanoparticlemade by flame spray pyrolysis for photoelectrochemical watersplitting—Part I. CuO nanoparticle preparation. Int J Hydro Ener 37:4871–4879. DOI:10.1016/J.IJHYDENE.2011.09.041

8. Cho S (2013) Optical and electrical properties of CuO thin filmsdeposited at several growth temperatures by reactive RFmagnetron sputtering. Met Mater Int 19:1327–1333. **DOI:**10.1007/s12540-013-6030-y

9. B. Arunkumar, S. Johnson Jeyakumar, M. Jothibas, A sol-gel approach to the synthesis of CuO nanoparticles using Lantana camara leaf extract and their photo catalytic activity, Optik, Volume 183, 2019, Pages 698-705, ISSN 0030-4026, https://doi.org/10.1016/j.ijleo.2019.02.046.

10. J. Vasudevan, S. Johnson Jeyakumar, B. Arunkumar, M. Jothibas, A. Muthuvel, S. Vijayalakshmi, Optical and magnetic investigation of Cu doped ZnO nanoparticles synthesized by solid state method, Materials Today: Proceedings, 2021, ISSN 2214-7853, https://doi.org/10.1016/j.matpr.2020.12.429.

11. K. Rayapa Reddy, Green synthesis, morphological and optical studies of CuO nanoparticles, Journal of Molecular Structure, Volume 1150,2017,Pages 553-557, ISSN 0022-2860, https://doi.org/10.1016/j.molstruc.2017.09.005

12. B. Arunkumar, S. Johnson Jeyakumar, J. Vasudevan, M. Jothibas, A. Sindhya, Scrutiny of the magnetic properties of ZnO nanoparticles by solid state reaction method, Materials Today: Proceedings, Volume 66, Part 4, 2022, Pages 2192-2197, ISSN 2214-7853, https://doi.org/10.1016/j.matpr.2022.06.036

13. W. Avansi, L.J.Q. Maia, H.A.J.L. Mourão, C. Ribeiro, Role of crystallinity on the optical properties of Na2V6O16·3H2O nanowires, *Journal of Alloys and Compounds* (2017), DOI: 10.1016/j.jallcom.2017.10.121.

14. Wadatkar NS, Waghuley SA, Complex optical studies on conducting polyindole as-synthesized through chemical route, Egyptian Journal of Basic and Applied Sciences (2015), http://dx.doi.org/10.1016/j.ejbas.2014.12.006.

15. Green synthesis of silver nanoparticles using Azadirachta indica aqueous leaf extract Shakeel Ahmed, Saifullah, Mudasir Ahmad, Babu Lal Swami, Saiqa Ikram-doi.org/10.1016/j.jrras.2015.06.006

16. LPG sensing performance of CuO Ag2 O bimetallic oxide nanoparticles,- K.R.Nemade ,S.A.Waghuley,- doi.org/10.1016 /j.spjpm.2015.07.006

17. Applications of Surface Plasmon Resonance (SPR) for the Characterization of Nanoparticles Developed for Biomedical Purposes Mara Canovi Jacopo Lucchetti Matteo Stravalaci Francesca Re Davide Moscatelli Paolo Bigini Mario Salmona and Marco Gobbi / doi:10.3390/s121216420.

18. Synthesis and Characterization of Copper Nanoparticle using Capparis Zeylanicaleaf Extract K. Saranyaadevi, V. Subha, R. S. Ernest Ravindran, S. Renganathan - /Int.J. ChemTech Res.2014,6(10),pp 4533-4541.

19. J. Vasudevan, S. Johnson Jeyakumar, B. Arunkumar, M. Jothibas, A. Muthuvel, S. Vijayalakshmi, Optical and magnetic investigation of Cu doped ZnO nanoparticles synthesized by solid state method, Materials Today: Proceedings, 2021, ISSN 2214-7853, https://doi.org/10.1016/j.matpr.2020.12.429.

20. A.I. Arbab, On the optical conductivity, Optik, Volume 194, 2019, 163067, ISSN 0030-4026, https://doi.org/10.1016/j.ijleo.2019.163067.

21. Dhineshbabu, N.R., Rajendran, V., Nithyavathy, N. *et al.* Study of structural and optical properties of cupric oxide nanoparticles. *Appl Nanosci* **6**, 933–939 (2016). https://doi.org/10.1007/s13204-015-0499-2.

22. Riyam A. Hammoodi, Assist. Prof. Dr. Ahmed K. Abbas ,Prof. Dr .Abdul hussein K. Elttayef, Structural and optical properties of CuO thin films prepared via R.F.magnetron sputtering, International Journal of Application or Innovation in Engineering & Management, Volume 3, Issue 7(2014) 2319 – 4847.

23. Xiaojun Zhang, Dongen Zhang, Xiaomin Ni, Huagui Zheng, Optical and electrochemical properties of nanosized CuO via thermal decomposition of copper oxalate, Solid-State Electronics 52 (2008) 245–248.

24. Sohrabi, Leila, Taghavi, Amin, & Boochani, Arash (2022). The effect of Fe impurity on electronic and optical properties of graphene-like InAs: a DFT-based study. Indian Journal of Physics (Online), 96(6), 1705-1714,- http://dx.doi.org/10.1007/s12648-021-02107-z

25. Arbab, A.. (2019). Absorption coefficient and optical conductivity: a quantum perspective. TY - BOOKPY - 2019/12/16

26. Hussian, Shymaa K.. (2017). Study of Optical properties of Copper Oxide (CuO) thin film prepared by SPD technique. Al-Muthanna Journal of Pure Sciences. 4. 144-152. 10.18081/2222-4223/017-6/433-152.

27. Ch. Rayssi, S. El.Kossi, J. Dhahri and K. Khirouni,- Frequency and temperature-dependence of dielectric permittivity and electric modulus studies of the solid solution $Ca_{0.85}Er_{0.1}Ti_{1-x}Co_{4x/3}O_3$ ($0 \leq x \leq 0.1$),- (Paper) *RSC Adv.*, 2018, **8**, 17139-17150,- DOI: 10.1039/C8RA00794B.

28. Plasmonic Nanoparticles and Island Films for Solar Energy Harvesting: A Comparative Study of Cu, Al, Ag and Au Performance Ivana Fabijani´c , Vesna Janicki , Josep Ferré-Borrull , Matej Bubaš , Vesna Blažek Bregovi´c , Lluis F. Marsal and Jordi Sancho-Parramon,- Coatings 2019, 9, 382; doi:10.3390/coatings9060382.

IOP Conf. Series: Materials Science and Engineering 1291 (2023) 012026 doi:10.1088/1757-899X/1291/1/012026

Bulk metallic glass Al_2Ca from a biomedical application viewpoint: A molecular dynamics study

A. Shankar[1] and N. K. Bhatt[1]

[1]Department of Physics, Maharaja Krishnakumarsinhji Bhavnagar University, Bhavnagar – 364001, Gujarat, India

anjalirameja1993@gmail.com, anjalishankar@mkbhavuni.edu.in, nkb@mkbhavuni.edu.in

Abstract. Molecular dynamics simulations were performed to investigate the glass transition temperature (T_g) and mechanical properties of Al–Ca-based bulk metallic glass (BMG), Al_2Ca. The second-neighbour modified embedded atom method potential was used to simulate the system. We estimate the T_g between 600 to 660 K. To ensure the glassy state of the studied material, the total radial distribution function (RDF) has been calculated from 300 to 1500 K. Partial RDF for Al, Ca, and Al–Ca subsystems are also calculated. Elastic properties including bulk modulus, compressibility, Young's modulus, shear modulus, tensile yield strength and Poisson ratio are calculated. From the obtained data of elastic constants, it is concluded that Al_2Ca BMG can be the potential material to be used in the biomedical field.

1. Introduction

Zr-based bulk metallic glasses (BMGs) are routinely used for human bone implants due to their favorable elastic and non-toxic nature. Biomaterials require good fracture toughness and fatigue strength to find applications in the biomedical field. Currently, 316L SS, Ti, and Ti-6Al-4V are in-use metallic crystalline materials. They have relatively high Young's modulus and lower strength values, because of which often complications like implant failures occur [1–4]. However, recently BMG with Ca and Al as the base elements are synthesized and analyzed as a potential substitute for these biomaterials [5]. This paper reports various structural and elastic properties of Al_2Ca BMG. We show that Al_2Ca has elastic properties comparable to human bones and can find important applications in temporary implants in the human body and mimicking orthopedic replacements [2–5]. These BMGs are amorphous solid materials, and exhibit properties such as biocompatibility, and biodegradability with low density. These materials show values for Young's and shear moduli comparable to human bone, and they are osteoconductive having high corrosion and wear resistance. The motivation to choose the binary alloy Al_2Ca is two-fold: firstly, Al_2Ca alloy is already synthesized experimentally in the crystalline form, which evidences the thermodynamic stability of the composition [6,7], and secondly, Ca_xAl_y BMG can be synthesized experimentally with various compositions such as (x,y) equal to (61,39), (8,3), (13,14). This corroborates that the Al–Ca-based binary glasses, having applications in the biomedical field, are already being synthesized in a passable size [5,8].

In this study, we examine glass forming ability, glass transition temperature (T_g), various elastic moduli such as bulk modulus (B), tensile yield strength (TYS), compressibility (K), shear modulus(G), Young's modulus (E), and Poisson ratio for Al_2Ca BMG. As a routine method [9], we have employed classical molecular dynamics (MD) simulations using the modified embedded atom potential (MEAM) to account for the interaction. We show that Al_2Ca BMG has mechanical properties that are comparable to the typical values for human bone, which justifies the use of studied BMG.

The rest of the paper is organized as follows. The section-2 includes the details about the simulations followed in the study. Computed results are presented and discussed in section-3. The last section summarizes the work and draws important conclusions.

2. Simulations

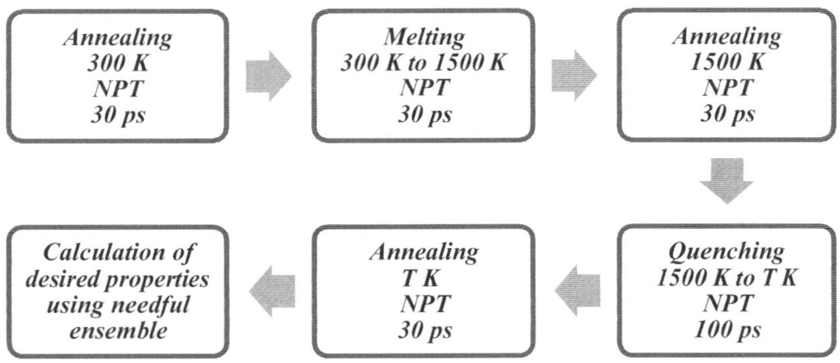

Figure 1. Flowchart describing the simulations.

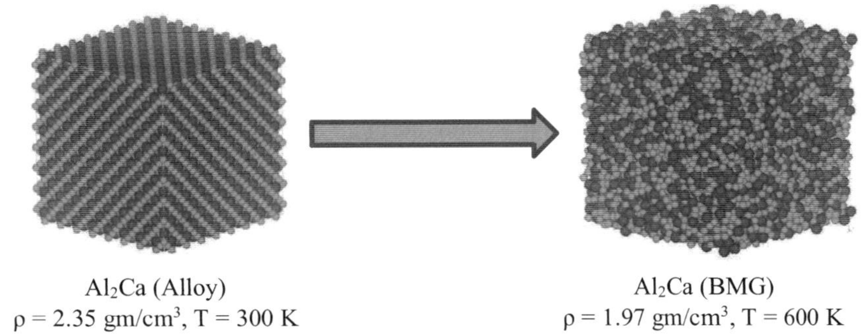

Figure 2. Alloy to the glass transition.

We performed classical MD simulation using the LAMMPS code [10] employing the second-neighbor MEAM pair interaction [11]. A cubical box containing 12288 atoms was used in the simulation with the periodic boundary to form the crystalline structure, the *C*15 crystal structure [6]. The relatively large number of atoms ensures the minimum surface effect in the simulations. The time step was kept at 0.001 ps. To obtain the T_g, the step-increment and step-decrement procedures in the temperature were followed. For each annealing step, the simulation time was optimized and finally kept at 30 ps. Whereas for each heating and cooling step, the simulation time was kept at 150 ps. Heating, annealing, and cooling were performed within the NPT ensemble. From each annealing step, the average density and temperature were extracted to plot the graph for density as a function of temperature. To simulate the glass, the system was annealed at 300 K for 30 ps in the beginning. After that, the temperature was increased to and annealed at 1500 K, each for 30 ps. Subsequently, a quenching was performed from 1500 K to the desired temperature in steps of 100 ps. Finally, the system was again annealed at the given temperature for another 30 ps. Figure 1 summarizes the simulations flowchart. The resulting glass transition from crystalline alloy is depicted in Figure 2. To derive various properties like radial distribution function (g(r)), and coordination number (CN) initially, the Al_2Ca glass was simulated using the NPT ensemble and then the properties were calculated with NVT ensemble.

The elastic constants (ECs) C_{11} and C_{12} were computed by giving a lateral strain of 2×10^9 in the z-direction, whereas the C_{44} was computed by giving a shear strain of 3 in the xy-direction. Both lateral and shear strains were given for a period of 10 ps. By finding the slope from the linear part of the stress-strain curve, we determined these ECs. We also checked the stability criteria, viz; $C_{11} > 0, C_{11} - C_{12} > 0, C_{44} > 0$, and $C_{11} + 2C_{12} > 0$ to confirm the mechanical stability of the studied BMG. These ECs are used to find other elastic and anisotropic mechanical quantities using the following equations. For cubic crystals, Voigt bounds (B_v, G_v) and Reuss bounds (B_R, G_R) can be calculated by equations (1) – (3), from which, the bulk modulus (B), the shear modulus (G), the Young's modulus (E) and Poisson ratio are obtained using equation (4) – (7):

$$B_v = B_R = (C_{11} + 2C_{12})/3 \tag{1}$$

$$G_v = (C_{11} - C_{12} + 3C_{44})/5 \tag{2}$$

$$G_R = 5(C_{11} - C_{12})C_{44}/[4C_{44} + 3(C_{11} - C_{12})] \tag{3}$$

$$B = \frac{1}{(2)}(B_v + B_R) \tag{4}$$

$$G = \frac{1}{2}(G_v + G_R) \tag{5}$$

$$E = \frac{9BG}{(3B+G)} \tag{6}$$

$$\nu = \frac{3B-2G}{2(3B+G)} \tag{7}$$

3. Results and Discussion

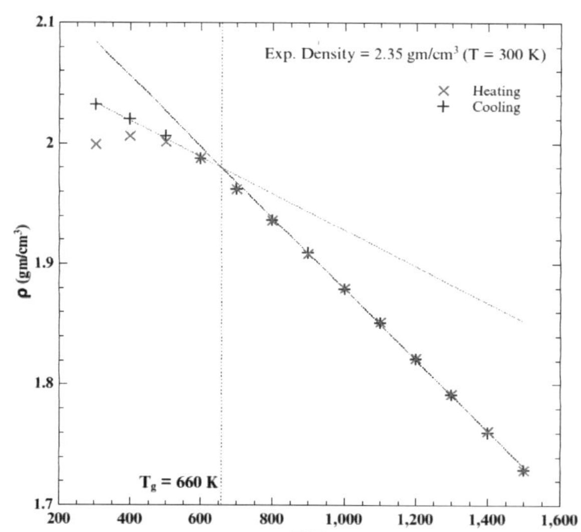

Figure 3. The glass transition temperature of Al_2Ca.

The glass transition temperature (T_g) was attained using the step-increment and step-decrement method as described in section-2. Computed results for density as a function of temperature are depicted in Figure 3. The crossing of a linear fit to heating and cooling density gives the T_g. The predicted T_g is ~660 K. For the application of BMG from the biomaterial viewpoint, we require a density compatible with human bones (1.81 gm/cm^3) [12]. We have noticed that Al_2Ca has a density of 2.35 gm/cm^3 in its alloy form but it decreases by 14% to 2.02 gm/cm^3 in the glassy state at 300 K. This indicates that Al_2Ca has a suitable density to be used as an implant in the human body. The human body temperature, in normal to hyperpyrexia cases can range from 309 to 316 K (36–43 °C). Thus, all the important properties were obtained at 300 K after the glass was formed.

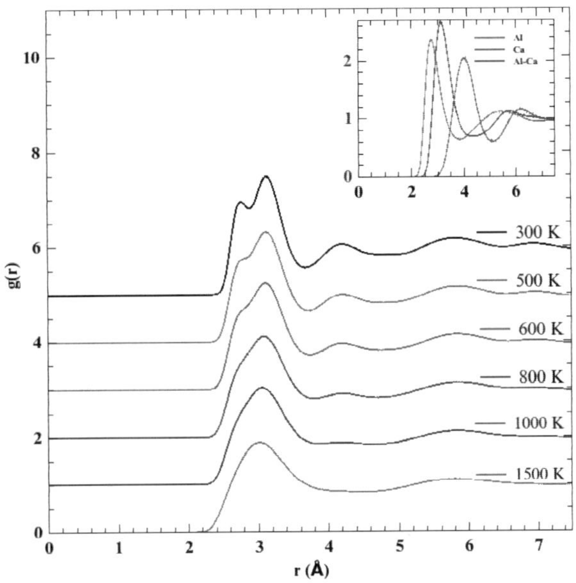

Figure 4. The radial distribution function of Al_2Ca. Inset shows the partial RDF at 1500 K.

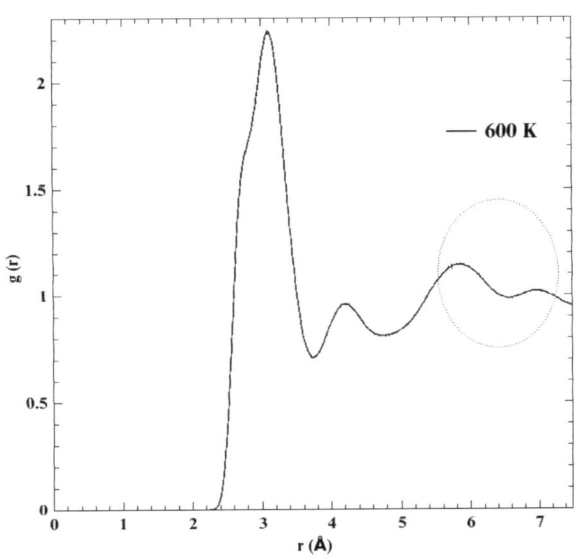

Figure 5. The radial distribution function of Al_2Ca below glass transition temperature (660K)

As a further verification, we have computed RDF (g(r)) at various temperatures, the results are shown in Figure 4. We found the following observations. (i) The slight distortion in the primary peak of g(r) is the characteristic of Al, which gradually disappears with temperatures. (ii) Inspection of Fig. 4 also indicates that at low temperatures the first dip in g(r) shows the humped structure, which corresponds to the principal peak of elemental Ca. The inset in Figure 5 shows the partial g(r) for Al, Ca, and Al–Ca subsystems at 1500 K, and confirms the peak position of elemental Ca. These observations suggest that Al_2Ca behaves as a solidus solution comprising of Al-Ca network below T_g. Which upon heating converts into a uniform liquid with a smooth principal peak and first dip in g(r). (iii) The third observation is that at temperatures 600 K and below, the signature 'shoulder' feature of amorphous material in the second peak of the g(r) is visible [13,14]. Figure 5 exemplifies this feature for a particular temperature and confirms the glassy state for Al_2Ca composition.

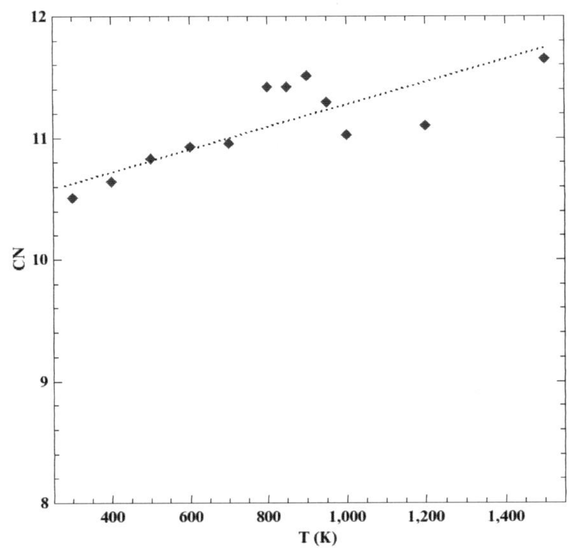

Figure 6. Coordination number as a function of temperature for Al_2Ca.

Figure 6 represents the temperature variation of coordination number. The computed results indicate that at $T < T_g$ the CN ranges between 10.5–10.9 but for temperatures $T > T_g$ it lies between 11–11.6. The range of computed CN for higher temperatures is a characteristic of a system melting from the closed-packed FCC structure. However, it is to be mentioned that the present 2NN-MEAM interaction potential is unable to predict the melting temperature correctly.

To verify the compatibility of Al_2Ca BMG with the human bone, we have studied the elastic response and mechanical strength of Al_2Ca. Figures 7 and 8 respectively show the stress-strain curve. The tensile yield strength (TYS) was obtained by giving the 20% tensile strain in the z-direction. The total strain was divided into 20 parts. Thus, the system would now experience 1% stain and then relax during each cycle, and the loop will run 20 times. Computed results for TYS were found to decrease linearly with temperature. It is respectively 12%, 10%, and 7.8% for temperatures 300, 600 (just below the T_g), and 900 K. Corresponding to human body temperature, the TYS of the Al_2Ca BMG will be ~11.8%. This value is sufficiently large to find application in implants. Figure 7 represents the TYS at 600 K. The graph indicates that until 5% of the strain the TYS for Al_2Ca BMG follows Hooks' law, between 5-10% the material undergoes the permanent deformation stage before breaking beyond 10% strain. Similarly, at lower temperatures, these values are 6-7%, 7-12% and the breaking point is >12%, respectively.

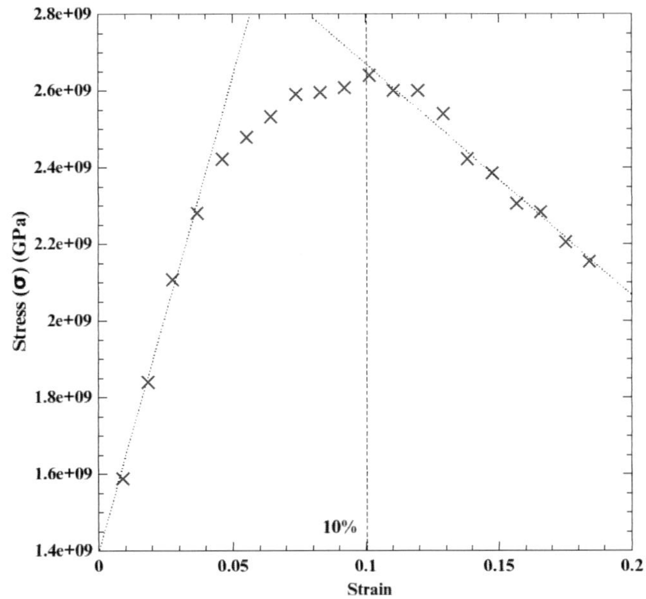

Figure 7. tensile yield Strength of Al_2Ca BMG at 600 K below the glass transition temperature (660 K).

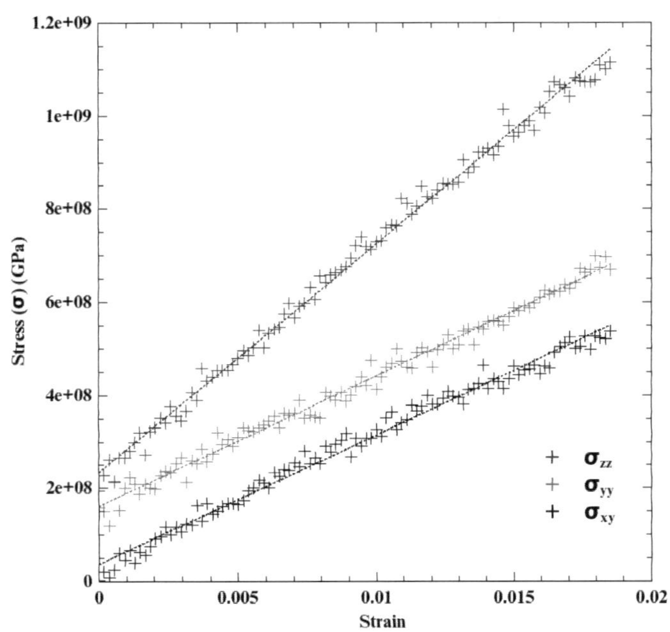

Figure 8. Stress-Strain graph for Al_2Ca BMG at 300 K.

The stress-strain curves were achieved by giving the lateral strain in the z-direction and the shear strain in the xy-direction. From the slope of the stress-strain curve shown in Figure 8, and using Eqs. (1)-(7) the ECs were calculated which are listed in Table 1. Table 1 shows the comparison of Al_2Ca

BMG and Al_2Ca alloy along with the ECs of human bones as bracketed values. Table 1 clearly shows that the ECs of Al_2Ca BMG are much lower than ECs in its alloy. This suggests that the glassy state is a soft supercooled liquid exhibiting the polymeric structure which is more useful for biomedical applications. It is also clear from Table 1 that Al_2Ca BMG has ECs comparable to human bones and therefore suitable for various implants in the body.

Table 1. Elastic moduli (in GPa) and Poisson ratio for Al_2Ca BMG and Al_2Ca alloy. Bracketed values are for femoral cortical bone in humans.

Physical Quantity	Al_2Ca BMG (Present Work)	Al_2Ca Alloy [15]
C_{11}	49.04	105.67
C_{12}	27.99	31.07
C_{44}	10.56	37.30
B	35.00	55.70
K	0.095	0.018
G	10.55 (3.3–6.0)	42.70
E	28.76 (11–21)	102.70
ν	0.36 (0.62)	---

4. Summary and Conclusion

In the present work, a classical MD has been performed to find elastic and structural properties and glass transition temperature for Al_2Ca BMG. The results for density and elastic constants in the glassy state verify the material's capability to find application in the biomedical field. Particularly, the density of Al_2Ca BMG is just 11% higher than the typical human bone, which is in better agreement than the commonly used Zr-based BMGs. Since the different compositions of Al and Ca (both stochiometric and a-stochiometric) are giving the stable glassy phase [5,8], in principle, by varying Al and Ca compositions, it is possible to tune both density and elastic properties in agreement with the human bone implant requirements.

Acknowledgments

The authors are grateful to the Department of Physics, Maharaja Krishnakumarsinhji Bhavnagar University for providing HPC computational facilities. A. Shankar is thankful to the Education Department, Government of Gujarat for providing financial support as the SHODH-ScHeme Of Developing High-quality research scholarship. Authors acknowledge the National Supercomputing Mission (NSM) for providing the computing resources of 'PARAM Ananta' at IIT Gandhinagar, which is implemented by C-DAC and supported by the Ministry of Electronics and Information Technology (MeitY) and Department of Science and Technology (DST), Government of India.

References

[1] Jamel M M, Jamel M M and Lopez H F 2022 Designing Advanced Biomedical Biodegradable Mg Alloys: A Review *Metals (Basel)* **12**

[2] Khan M M, Nemati A, Rahman Z U, Shah U H, Asgar H and Haider W 2018 Recent Advancements in Bulk Metallic Glasses and Their Applications: A Review *Critical Reviews in Solid State and Materials Sciences* **43** 233–68

[3] Li H F and Zheng Y F 2016 Recent advances in bulk metallic glasses for biomedical applications *Acta Biomater* **36** 1–20

[4] Khonik V A 2020 Recent advancements in metallic glasses *Metals (Basel)* **10** 1–3

[5] Michalik Š, Ďurišin J, Balga D, Saksl K, Ďurišin M and Drakopoulos M 2016 In situ HEXRD study of a Ca61Al39metallic glass *J Alloys Compd* **687** 188–96
[6] Zhou D W, Liu J S, Peng P, Chen L and Hu Y J 2008 A first-principles study on the structural stability of Al2Ca Al4Ca and Mg2Ca phases *Mater Lett* **62** 206–10
[7] Min X, Sun Y, Xue F, Du W and Wu D 2003 Analysis of valence electron structures (VES) of intermetallic compounds containing calcium in Mg–Al-based alloys *Mater Chem Phys* **78** 88–93
[8] Guo F Q, Poon S J and Shiflet G J 2003 CaAl-based bulk metallic glasses with high thermal stability *Appl Phys Lett* **84** 37
[9] Shankar A, Bhatt N K, Patel A B and Vyas P R 2021 Amorphous structure of binary Ca61Al39metallic glass close to transition temperature *AIP Conference Proceedings* vol 2352 (American Institute of Physics Inc.)
[10] Thompson A P, Aktulga H M, Berger R, Bolintineanu D S, Brown W M, Crozier P S, in 't Veld P J, Kohlmeyer A, Moore S G, Nguyen T D, Shan R, Stevens M J, Tranchida J, Trott C and Plimpton S J 2022 LAMMPS - a flexible simulation tool for particle-based materials modeling at the atomic, meso, and continuum scales *Comput Phys Commun* **271**
[11] Jang H S, Seol D and Lee B J 2021 Modified embedded-atom method interatomic potentials for Mg–Al–Ca and Mg–Al–Zn ternary systems *Journal of Magnesium and Alloys* **9** 317–35
[12] Rho J Y, Ashman R B and Turner C H 1993 Young's modulus of trabecular and cortical bone material: Ultrasonic and microtensile measurements *J Biomech* **26** 111–9
[13] Janssen L M C 2018 Mode-coupling theory of the glass transition: A primer *Front Phys* **6** 97
[14] Waseda Yoshio 1980 The structure of non-crystalline materials : liquids and amorphous solids 326 87-91
[15] Gao Huang Z 2021 Mechanical and thermodynamics properties of Al 2 Ca and Mg 2 Ca under various pressure: A first principle study *Materials Express* **11** 1571–8

Investigating the effect of Manufactured sand and Silica Fume on the properties of Concrete

Nandigam Bhargav Chowdary[1], Akshat Mahajan[2], Sahil Jaggi[3]

[1] School of Civil Engineering, Lovely Professional University, Phagwara, Punjab.

[2,3] School of Civil Engineering, Lovely Professional University, Phagwara, Punjab.

Email: [1] bhargavnandigam@gmail.com, [2] akshat.23424@lpu.co.in, [3] sahil.21178@lpu.co.in

Abstract: In the present era, concrete is one of the most commonly used construction materials worldwide. Consequently, the demand for natural river sand is increasing. Since the mining of natural sand has already been outlawed by the government in many regions, now it is essential to look into sustainable materials to control natural sand extraction. Hence, manufactured sand (MS) has been shown in numerous studies to be a more practical and environmentally responsible alternative to river sand (RS), which is normally used in the production of concrete. At the same time, the cement industry produces tons of greenhouse gases into the atmosphere, affecting atmospheric conditions. Now, it's time to look into a suitable replacement material for cement. To investigate the feasibility of using manufactured sand & silica fume in place of river sand and cement under normal climatic conditions fresh, mechanical and durability properties were conducted in the laboratory. Twelve samples of M30 grade cubes and cylinders at desired M sand percentages of 0, 20, 40, 60, 80, and 100 & 0% and 5% variation of silica fume with cement are included in this study as a comparison to the control mix. After 28 days of water curing, a random variation in the properties of concrete was observed in the samples. In this experimental study, SF represents Silica Fume, MS represents manufactured sand, RS represents river sand, FA represents Fine aggregates and CA represents Coarse aggregates.

Keywords: Manufactured Sand, Silica Fume, Workability, Mechanical Properties, Durability, Ph Value.

1. INTRODUCTION

Ordinary Portland Cement is the most frequently used building material in the world. Due to its application in the global expansion of the building sector, it's worth will not change in the future. Cement production requires a significant quantity of energy and raw resources. During the manufacture of cement, a significant amount of carbon dioxide (CO_2) as well as various effluents are emitted into the atmosphere [1,2,3]. The replacement of cementitious materials may reduce the production of cement and the emission of greenhouse gases into the atmosphere [4]. Silica fume is a type of admixture that is commonly used as a cement replacement material and it is produced from the wasteof ferrosilicon alloys [5]. It is a pozzolanic material that combines with Ca $(OH)_2$ and creates C-S-H, it avoids the voids or pores between concrete and enhances the strength and durability due to proper packing [5,6]. The higher per cent of silicon dioxide in silica fume is the key factor of the binding property in concrete mixes and the particle size of silica fume is less than 1 micron. However, the high surface area of silica fume absorbs more water content compared to cement material [7]. Sand is a vital raw material that is used in construction projects all over the world to create infrastructure and build homes. It is mostly used in concrete production and projects related to concrete, such as RMC plants, brickwork products, plastering, manholes, pipelines, and a range of other items [8,9]. Sand can be used as a fine aggregate in the creation of asphalt, as a filter for water and other liquids, and as a filler in manufactured items [10]. The demand for sand as a construction resource is quite high due to the expansion of the construction sector and significant infrastructure projects, which has increased the amount of illegal sand mining. Now more than ever, it is necessary to find a substitute for river sand in construction projects [11,12]. Engineers and researchers have developed concepts for reducingthe usage of river sand and for using many alternative materials instead, such as stone crusher dust, foundry sand, manufactured sand, and many other waste products [13]. Most of the research papers on manufactured sand are used as a partial replacement for concrete mixes up to M40 [7]. In the presentstudy, silica fume is used to partially replace cement and both partial and full replacements of naturalsand with manufactured sand of M30 to demonstrate the properties of concrete.

2. MATERIALS AND PROPERTIES

2.1. Cement

OPC of grade 43 was used in this experiment and purchased from *tulsi iron & cement traders, Jalandhar, Punjab* having 3.13 specific gravity [14]. The physical properties of cement are mentioned in Table 1

2.2. Silica Fume

The Silica fume used in our experimental study was produced by *KGR Agro, Ludhiana, Punjab* and the chemical properties of SF are mentioned in Table 2

2.3. Fine Aggregates

Locally available fine aggregates of grade IV were used in this experimental study. The properties of fine aggregates are shown in Table 3

2.4. Manufactured sand

Manufactured sand of average particle size 150-micron meter was used in this experimental study and was procured from the *Sri Syamala rock crushers, Hyderabad, Telangana.* The properties of the same are mentioned in Table 3

2.5. Coarse Aggregates

Coarse aggregates of a maximum size of 20 mm were used in this experiment [15]. The properties are mentioned in Table 3

Figure 1

Silica Fume

Figure 2

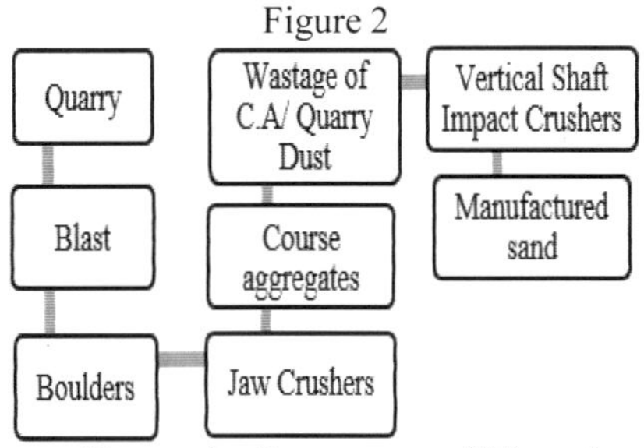

The production process of M sand

Figure 3

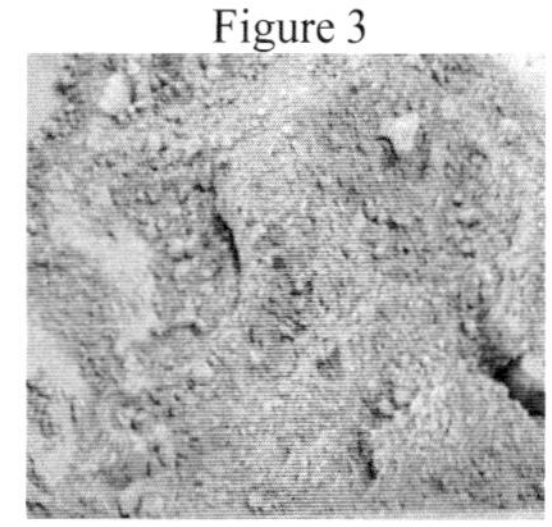

M sand

Table 1. Physical properties of cement

S. No	Test	Result
1	Specific Gravity	3.13
2	Normal Consistency (%)	28.7
3	Initial Setting (min)	58
4	Final Setting (min)	256

Table 2 Chemical composition of Silica Fume & Cement (%)

Chemicals	Cao	Sio2	Al_2O_2	Fe_2O_2	MgO	SO_3	H_2O	NaO_2	K_2O	SG
Cement	61.45	21.05	4.42	4.42	1.98	3.04	-	-	-	3.13
SF	0.25	97.14	0.25	0.25	0.56	0.12	0.6	0.26	0.58	2.21

Table 3 Physical properties of Fine & Coarse Aggregates

S. No	Test	River sand	M sand	Coarse aggregates
1	Specific Gravity	2.68	2.76	2.71
2	Water Absorption	1.1%	1.45%	1.7%
3	Fineness Modulus	2.71	2.79	2.46
4	Bulk Density	1613 kg/m^3	1715kg/m^3	1409kg/m^3

3. EXPEIMMENTAL PROCUDURE

The tests were conducted using the M sand proportions of 0, 20, 40, 60, 80, & 100% and the replacement of 5% Silica Fume of cubes & cylinders specimens of 3 were casted. These specimens were tested after 28 days of the curing period. Until the specimen failed, a constant load of 200 KN was applied using the compression testing apparatus to put the specimen in the device so that the load was distributed across the cube's opposing sides and placed the specimen in the Centre of the machine's base plate. Gently turn the movable part by hand so that it touches the specimen's top surface until the specimen fails, gradually apply the load without sudden load and constantly at a rate of 200KN and note down the peak load. The concrete specimen of dia150 mm has a height of 300 mm specimen was cast and cured according to the relevant standard, the longitudinal axis of the specimen is aligned with the platens so that it is perpendicular to the direction of the applied load and until the specimen fractures, a compressive load is applied at a rate of 14 N/mm2/min. The specimen is then set on a flat surface, a steel plate with a 150mm diameter is set on top of the specimen, and the load at which the specimen fractures is noted as the concrete's compressive strength. Until the specimen is split in half, a compressive load of 0.7 N/mm2/min is given to the steel plate [16,17,18].

The concrete's split tensile strength is measured as the load at which the specimen splits [19]. As per the RCPT, the specimen was placed between two vessels of water-saturated, 5 cm thick, 10 cm dia concrete. For 6 hours, a 60-Hz (V) applied direct current voltage was delivered to the first vessel, which contained a 3.0 per cent NaCl solution, and the second vessel, which contained a 0.3 M Na (OH)2 solution. The entire charge that was discharged & measured and this value is used to calculate how quickly chloride permeates the concrete. For saturated water absorption, the specimens were weighed before being dried in a hot air oven at 105 degrees Celsius. Until an accurate measurement of the weight differential between the 2 following measures was conducted at 24-hour intervals, the drying process was continued. Before submerging the dried samples in water, they were warmed to room temperature. The samples were taken away at regular intervals. When the surface had been dried with a clean cloth, it was weighed. This process of adjusting the weights continued until they stabilized. Using the equation provided below, the water absorption was calculated [20, 21, 22]. After the specimen was crushed, 10 g of the powdered mortar was obtained and, after being sieved through a 150-m sieve andground, diluted in 100 ml of distilled water by stirring. We measure the alkalinity of each solution with a pH meter [22,23, 24]. In this experimental study, SF represents silica fume, MS represents manufactured sand, and RS represents river sand.

IOP Conf. Series: Materials Science and Engineering 1291 (2023) 012027 doi:10.1088/1757-899X/1291/1/012027

4. MIX DESIGN

Test data of materials
Specific Gravity of Cement – 3.13

Specific Gravity of Fine Aggregates – 2.68

Specific Gravity of Coarse Aggregates – 2.71

Specific Gravity of Water – 1.0

Type of Admixture – Superplasticizer (PC)

1) Target Strength

f'ck = fck+1.65 S

f'ck – Target mean compressive strength at 28 days

S – Standard deviation

IS-10262-2019, Table 2

S value for M30 – 5N/mm^2

f'ck = 30+1.65(5) = 38.25N/mm^2

2) Water/Cement ratio

IS-456-2000, Table-3 & 5

W/C = 0.45 – 0.05 = 0.4

3) Water Content

IS-10262, Table 4

For the appropriate workability (other than 50 mm slump), the needed water content may be adjusted by around 3% for every 25 mm slump increase or reduction, or it may be determined by trial and error.

186kg – for 50 mm slump

186kg +3%+3% = 197.16kg = for 100mm slump

Let's assume 15% of water reduction

197.16kg – 10% = 167.58kg (Water Content)

4) Calculation of Cement Content

W/C ratio = Water content/Cement content

Cement content = Water content/ Water-cement ratio

= 167.58/0.4 = 418.96kg/m^3 > 320kg/m^3 (minimum cement content for M30)

Cement content = 418.96kg/m^3

5) Proportion of aggregates between coarse & Fine aggregates

IS-10262, Table 5

For 20mm coarse aggregates of W/C 0.50 – 0.62 (Zone II)

IS-10262, Table 5

W/C – 0.45

= 0.45 – 0.4 = 0.05

= 0.62+0.01 = 0.63

The volume of Coarse aggregate = 0.63

The volume of fine aggregate = 1 – 0.63 = 0.37

6) <u>Mix Calculation</u>

a) Volume of Concrete – 1m^3

b) Volume of Cement

= Mass/(SGX1000)

= 418.96/ (3.13 X 1000)

= 0.1338 M^3 X 1440

= 192.74kg (4 bags)

c) Volume of Water

= Mass/(SGX1000)

= 167.58/ (1 X 1000)

= 0.167M^3 = 167ltr

d) Volume of admixture

Mass – 1% of cement content, SG of superplasticizer – 1.2

= ((1/100) X 418.96)/ (1.2 X 1000)

= 4.1896/ 1200

= 0.00349 X 1000 = 3.49kg/m^3

e) Volume of aggregates

= 1 – (b+c+d) = 1 – (0.1338+0.167+0.0034)

= 0.6958

f) Mass of CA

= volume of all aggregate X Volume of CA X SG of CA X 1000

= 0.6958 X 0.63 X 2.71 X 1000

= 1187.93kg/m^3

g) Mass of FA

= volume of all aggregate X Volume of FAX SG of FA X 1000

= 0.6958 X 0.37 X 2.68 X 1000

= 689.95kg/m^3

Cement – 418.96kg/m^3

Water – 167.58kg/m^3

Fine aggregate – 689.95kg/m^3

Coarse aggregate – 1187.93kg/m^3

Admixture – 3.49kg/m^3

Water/ cement ratio – 0.4

Table 4 Mix proportions of all mixes (kg/m^3)

Mix	Mix designation	Cement	SF	FA	MS	CA	Water	Admixture
M1	SF0RS100MS0	418.96	0	689.95	0	1187.93	167.58	3.49
M2	SF0RS80MS20	418.96	0	551.96	137.99	1187.93	167.58	3.49
M3	SF0RS60MS40	418.96	0	413.97	275.98	1187.93	167.58	3.49
M4	SF0RS40MS60	418.96	0	275.98	413.97	1187.93	167.58	3.49
M5	SF0RS20MS80	418.96	0	137.99	551.96	1187.93	167.58	3.49
M6	SF0RS0MS100	418.96	0	0	689.95	1187.93	167.58	3.49
M7	SF5RS100MS0	398.01	20.94	689.95	0	1187.93	167.58	3.49
M8	SF5RS80MS20	398.01	20.94	551.96	137.99	1187.93	167.58	3.49
M9	SF5RS60MS40	398.01	20.94	413.97	275.98	1187.93	167.58	3.49
M10	SF5RS40MS60	398.01	20.94	275.98	413.97	1187.93	167.58	3.49
M11	SF5RS20MS80	398.01	20.94	137.99	551.96	1187.93	167.58	3.49
M12	SF5RS0MS100	398.01	20.94	0	689.95	1187.93	167.58	3.49

5. RESULT AND DISCUSSION

In concrete mixes containing only river sand and manufactured sand (M1 to M6), the increment of manufactured sand in the control mix from 0 to 100% reduces the slump value due to the high-water absorption of M sand [21]. From M1 to M6, decrement in slump value from 85mm to 73.5mm. Similarly, mixes from M7 to M12 containing the same variation of M sand with the replacement of 5% silica Fume in cement content decrease the slump value from 82mm to 70. The presence of both M sand andSilica Fume reduces the slump value of the concrete and it requires more water content at the initial stage of concrete so, a superplasticizer is suggested to maintain workability [16]. The fineness of M sand and Silica Fume fills the voids between which led to increases in the strength of the concrete. Mixes containing only M sand (M1 to M6) represent the increment in compressive strength after 28 days of curing from 33.03Mpa to 42.19Mpa [21]. Similarly, mixes incorporated with 5% Silica fume and 0 to 100%replacement of M sand (M7 to M12) increase the compressive strength of the concrete from 35.13Mpato 44.81Mpa. Compared to the control mix, the mixes containing only M sand represent the incrementin tensile strength from M1 to M6, parallelly the mixes from M7 to M12 show a greater increment in tensile strength from 3.42Mpa to 4.79Mpa. Chlorine ion permeability for the control mix (M1) showshigh penetration compared to the mixes containing M sand and Silica Fume. It represents that, the presence of M sand and SF reduces the permeability of the concrete [20]. Respective results of M1, M, M3,M4, M5, M6, M7, M8, M9, M10, M11 and M12 are 1827, 1695, 1517, 1269, 1087, 987, 1717, 1488, 1419, 1205, 964, 714. The initial stage of M sand and SF absorbs more water content but after 28 days of curing the saturated water absorption of the concrete reduces from M 1 to M6 mixes as well as M7to M12 mixes. Respective results of M1, M2, M3, M4, M5, and M6 are 2.83, 2.79, 2.72, 2.61, 2.49, 2.35 similarly, the results of M7, M8, M9, M10, M11 and M12 are 2.71, 2.56, 2.40, 2,28, 2.13, 2.01. The concrete consisting of M sand and SF reduces the pH value of the concrete. For the first six mixesfrom M1 to M6 which are only M sand variations, pH was reduced by up to 2.31% (13.40 – 13.09) andthe next six mixes (M7 to M12) which are incorporated with both M sand variation & SF content reduces even more pH value by up to 2.91% (13.37 – 12.98).

IOP Conf. Series: Materials Science and Engineering 1291 (2023) 012027 doi:10.1088/1757-899X/1291/1/012027

Table 5. Slump values (mm)

Mix	**M1**	**M2**	**M3**	**M4**	**M5**	**M6**	**M7**	**M8**	**M9**	**M10**	**M11**	**M12**
Slump Values	85	83.5	81	78	75	73.5	82	80	78	75	73	70

Table 6. Compressive & Split Tensile Strength results

Mix	**Compressive strength after 28 days of testing (Mpa)**	**Split Tensile strength after 28 days of testing (Mpa)**
M1	33.03	3.27
M2	33.88	3.51
M3	34.63	3.64
M4	37.07	3.79
M5	39.57	3.96
M6	42.19	4.19
M7	35.13	3.42
M8	36.11	3.58
M9	38.25	3.70
M10	40.11	3.88
M11	42.92	4.19
M12	44.81	4.79

Figure 4
Compressive strength testing

Figure 5
Split Tensile strength testing

Figure 6
RCPT testing

Table 7. Durability test results of all mixes

Mix	RCPT after 28 days (Columns)	Penetrability class of Cl ion	Absorption in % after 28 days of testing	Alkalinity (pH Value)
M1	1827	Low	2.83	13.40
M2	1695	Low	2.79	13.38
M3	1517	Low	2.72	13.33
M4	1269	Low	2.61	13.25
M5	1087	Low	2.49	13.17
M6	987	Very Low	2.35	13.09
M7	1717	Low	2.71	13.37
M8	1488	Low	2.56	13.32
M9	1419	Low	2.40	13.28
M10	1205	Low	2.28	13.19
M11	964	Very Low	2.13	13.11
M12	714	Very Low	2.01	12.98

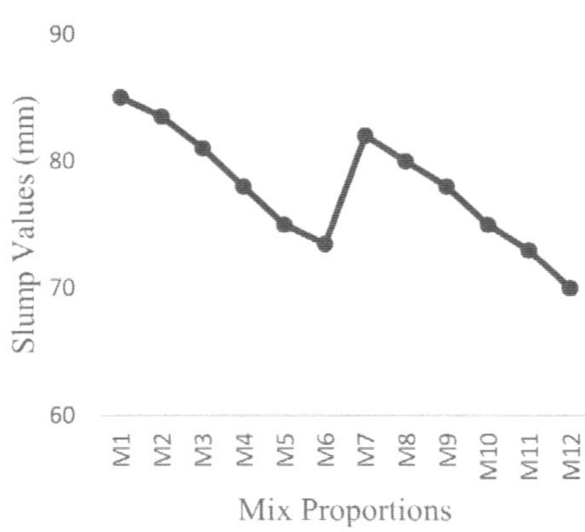

Figure 7. Variation of slump values containing various replacement % of M sand at 0 and 5% SF

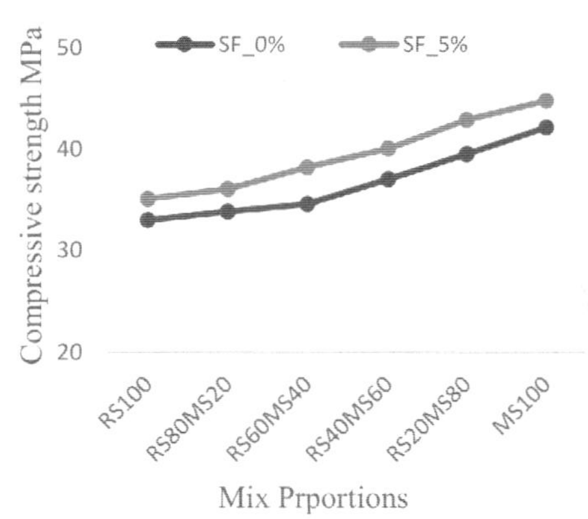

Figure 8. Variation of Compressive values of containing various replacement % of M sand at 0 and 5% SF

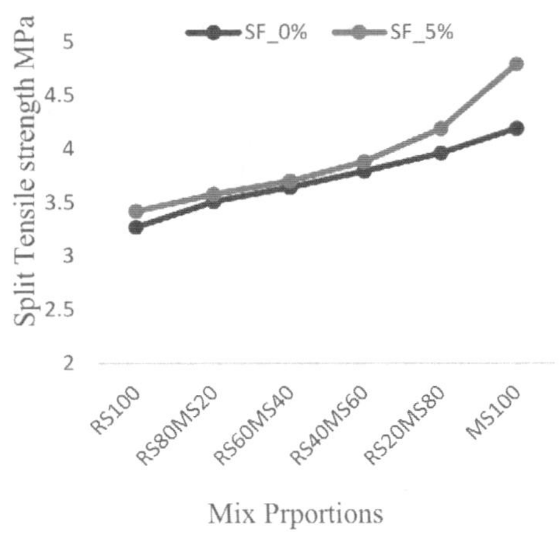

Figure 9. Variation of Split tensile values Containing various replacement % of M sand at 0 and 5% SF

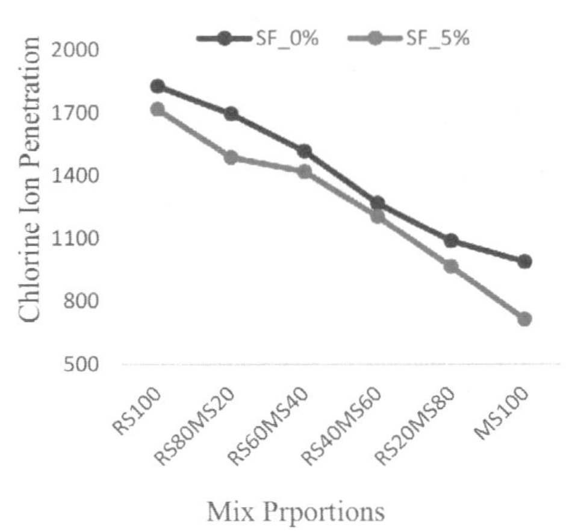

Figure 10. Variation of RCPT values containing various replacement % of M sand at 0 and 5% SF

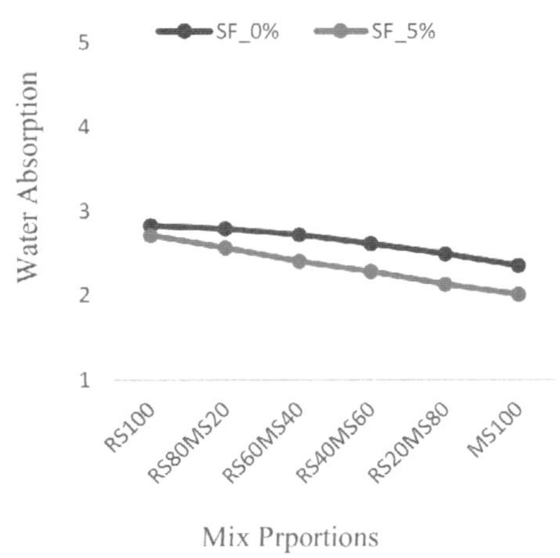

Figure 11. Variation of Saturated Water Absorption values containing various replacement % of M sand at 0 and 5% SF

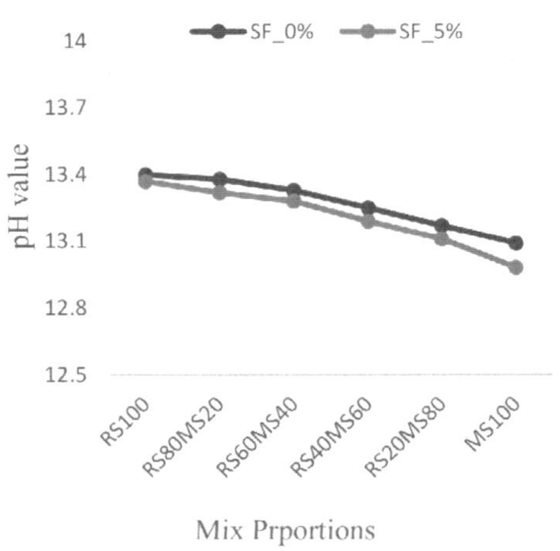

Figure 12. Variation of pH value containing various replacement % of M sand at 0 and 5% SF

6. CONCLUSIONS

The workability, strength, and durability of partial and full replacement manufactured with silica fume, the different mix proportions, were compared by Slump, Compressive, Split Tensile Strength, RCPT, Saturated Water Absorption, and Alkalinity tests to investigate the possibility of replacing natural sand with manufactured sand in concrete structures under normal environmental conditions. The findings show that:

- The increment of silica fume percentage from 0% to 5% in concrete mixes decreases the slump value of concrete by around 14%.
- Compressive & Split tensile strength of concrete increases with the increment of manufactured sand from M1 to M6 mixes by up to 21% & 20% respectively. Similarly, mixes containing both manufactured sand and silica fume increase the strength of the concrete rapidly from M7 to M12 by up to 22% & 28% respectively.
- The permeability of chlorine ions is decreased with the increment of manufactured sand, at the same time, the addition of silica fume with the same mixes reduces even more permeability compared to mixes without silica fume.
- When compared to natural sand, manufactured sand initially absorbs more water, however after 28 days of curing, superior packing of the m sand reduces the water absorption from M1 to M6 by up to 17%. On the other hand, Mixes containing silica fume reduce water absorption from M7 to M12 by up to 22.14%.
- The increase in M sand's percentage indicated a decrease in alkalinity from 13.40 to 13.09. It was discovered that concrete's alkalinity can be reduced to a certain extent by adding a higher percentage of M sand. Similarly, Silica Fume reduces even more alkalinity than normal mixes. Thus, the Silica Fume mixes contain less per cent of Cao to resist the carbonation effect.

Reference

1. Zhao, M., Dai, M., Li, J., & Li, C. (2023). Case study on the performance of pumping concrete with super-driver manufactured sand. *Case Studies in Construction Materials*, *18*(December 2022), e01850. https://doi.org/10.1016/j.cscm.2023.e01850

2. Dr.G. Velrajkumar, R. Muralikrishnan, A. Mohan, R. Bala Thirumal, P. Naveen John, Performance of GGBFS and silica fume on self-compacting geopolymer concrete using partial replacements of R-Sand, Materials Today: Proceedings, Volume 59, Part 1, 2022, Pages 909-917, ISSN 2214-7853, https://doi.org/10.1016/j.matpr.2022.01.383

3. Mo Liu, Wenting Dai, Chunling Zhong, Xue Yang, Study on mechanical properties and microstructure of manufactured sand reactive powder concrete with different curing methods, Materials Letters, Volume 335, 2023, 133818, ISSN 0167-577X, https://doi.org/10.1016/j.matlet.2023.133818
4. Sai, P. A. N., K. Chandramouli, Chaitanya, J. S. N., N. Pannirselvam, & R. Pujith. (2021). Strength Studies in Concrete with M-Sand and Silica Fume. *International Advanced Research Journal in Science, Engineering and Technology (IARJSET) ISSN (O) 2393- 8021, ISSN (P) 2394-1588*, *8*(7), 372–375. https://doi.org/10.17148/IARJSET.2021.8770
5. G. Naga Venkat, K. Chandramouli, Ezaz ahmed, V. NagendraBabu, Comparative study on mechanical properties and quality of concrete by partial replacement of cement with silica fume, metakaolin and GGBS by using M−Sand as fine aggregate, Materials Today: Proceedings, Volume 43, Part 2, 2021, Pages 1874-1878, ISSN 2214-7853, https://doi.org/10.1016/j.matpr.2020.10.819.
6. Zhang, Y., Gu, L., & Zhang, Q. (2022). The durability of manufactured sand concrete in an atmospheric acidification environment. *Case Studies in Construction Materials*,

17(October), e01613. https://doi.org/10.1016/j.cscm.2022.e01613
7. Shubha Khatri, S.S. Goliya, Shashichandra Tiwari, Utilization of crushed M−Sandy soil as an advanced sustainable construction material in reinforced concrete structures, Materials Today: Proceedings, Volume 65, Part 2, 2022, Pages 889-894, ISSN 2214-7853, https://doi.org/10.1016/j.matpr.2022.03.520.
8. N.K. Mahaa Lakshmi, C. Chella Gifta, S. Sankar Kumar, S. Stephan, Manufactured sand for the development of reinforced concrete spun pipes, Volume 64, Part 2, 2022, Pages 1116-1121, ISSN 2214-7853, https://doi.org/10.1016/j.matpr.2022.06.048.
9. Yaxin Tao, Qiang Ren, Karel Lesage, Kim Van Tittelboom, Yong Yuan, Geert De Schutter, Shape stability of 3D printable concrete with river and manufactured sand characterized by squeeze flow, Cement and Concrete Composites, Volume 133, 2022, 104674, ISSN 0958-9465, https://doi.org/10.1016/j.cemconcomp.2022.104674.
10. Mane, K. M., Joshi, A. M., Kulkarni, D. K., & Prakash, K. B. (2022). Influence of retempering on properties of concrete made with manufactured sand and industrial waste. *Cleaner Materials*, *4*(January), 100060. https://doi.org/10.1016/j.clema.2022.100060
11. Yue Li, Yunze Liu, Caiyun Jin, Jinlei Mu, Hongwen Li, Jianglin Liu, Multi-scale creep analysis of river sand and manufactured sand concrete considering the influence of ITZ, Construction and Building Materials, Volume 344, 2022, 128175, ISSN 0950-0618, https://doi.org/10.1016/j.conbuildmat.2022.128175.
12. Yingjie Xu, Hongniao Chen, Yuxiang Tang, Study on fracture parameters and fracture process zone of manufactured-sand recycled aggregate concrete, Construction and Building Materials, Volume 361, 2022, 129613, ISSN 0950-0618, https://doi.org/10.1016/j.conbuildmat.2022.129613.
13. Weiguo Shen, Jiale Wu, Xuejian Du, Zhitang Li, Dinglve Wu, Jiangtao Sun, Zhongwen Wang, Xujia Huo, Deqiang Zhao, Cleaner production of high-quality manufactured sand and ecological utilization of recycled stone powder in concrete, Journal of Cleaner Production, Volume 375, 2022, 134146, ISSN 0959-6526, https://doi.org/10.1016/j.jclepro.2022.134146.
14. Zhang, J., Li, D., & Wang, Y. (2020). Toward intelligent construction: Prediction of mechanical properties of manufactured-sand concrete using tree-based models. *Journal of Cleaner Production*, *258*, 120665. https://doi.org/10.1016/j.jclepro.2020.120665
15. Qiang Ren, Yaxin Tao, Dengwu Jiao, Geert De Schutter, Zhengwu Jiang, Rheological properties of concrete with manufactured sand: A multi level prediction, Cement and Concrete Composites, Volume 133, 2022, 104647, ISSN 0958-9465, https://doi.org/10.1016/j.cemconcomp.2022.104647.
16. Alla, S., Jayaram, M., & Asadi, S. S. (2020). An experimental investigation for replacements of river sand and cement with Robosand, fly-ash and silica fume in concrete to evaluate the influence in durability properties. 954–961. https://doi.org/10.1016/j.matpr.2020.07.340
17. Gokulnath, V., Ramesh, B., & Sivashankar, S. (2020a). Influence of M sand in self-compacting concrete with addition of steel fiber, 1026-1030. https://doi.org/10.1016/j.matpr.2019.11.270.
18. Gokulnath, V., Ramesh, B., & Sivashankar, S. (2020b). Study on M-sand in self compacting concrete with addition of steel fiber in M-25 grade-a review., 966–969. https://doi.org/10.1016/j.matpr.2019.11.229
19. Hou, W. Q., Yang, J. J., Zhang, Z. X., & Yuan, X. Q. (2021). Experimental study and application of manufactured sand self-compacting concrete in concrete-filled-steel-tube arch bridge: A case study. *Case Studies in Construction Materials*, *15*(June), e00718. https://doi.org/10.1016/j.cscm.2021.e00718
20. Kumar, S., Rai, B., Biswas, R., Samui, P., & Kim, D. (2020). Prediction of rapid chloride

permeability of self-compacting concrete using Multivariate Adaptive Regression Spline and Minimax Probability Machine Regression. *Journal of Building Engineering*, *32*, 101490. https://doi.org/10.1016/j.jobe.2020.101490.

21. Mahalakshmi, S. H. V., & Khed, V. C. (2020). Experimental study on M-sand in self-compacting concrete with and without silica fume. 1061–1065. https://doi.org/10.1016/j.matpr.2020.01.432
22. Tripathi, D., Kumar, R., Mehta, P. K., & Singh, A. (2020). Silica fume mixed concrete in an acidic environment., 1001–1005. https://doi.org/10.1016/j.matpr.2020.01.311
23. Mala self-compacting the, R., Rajagopal Sentil kumar, S. R., Prakash, A. R., Das, B. B., Chung, I. M., Kim, S. H., & Prabakaran, M. (2022). Use of Industrial Silica Sand as a Fine Aggregate in Concrete—An Explorative Study. *Buildings*, *12*(8). https://doi.org/10.3390/buildings12081273
24. Naga Venkat, G., Chandramouli, K., Ahmed, E., & Nagendra babu, V. (2020). Comparative study on mechanical properties and quality of concrete by partial replacement of cement with silica fume, metakaolin and GGBS by using M-Sand as fine aggregate., 1874–1878. https://doi.org/10.1016/j.matpr.2020.10.819

IOP Conf. Series: Materials Science and Engineering 1291 (2023) 012028 doi:10.1088/1757-899X/1291/1/012028

Rare Earth Based Iron Garnet – A Survey on Its Magnetic Properties

A Priyanshu[1], M Nath[2,*], A Bandyopadhyay[3,*]

[1,2]Department of Physics, Lovely Professional University, Phagwara 144411, Punjab, India.

[3]Department of Physics, University of Gour Banga, Malda-732103, West Bengal, India

[2,*] madhumitanath_21@yahoo.co.in

[3,*]atulbandyopadhyay@yahoo.com

Abstract. Garnet is a well-known material for a long-time by the scientific community but still today scientists are focusing on it due to the rapid application-based development of this material. When rare earth iron garnets (REIG) are formed by substituting the rare-earth ions with unfilled $4f^n$ orbitals, the magnetic properties of the iron garnets exhibit an interesting characteristic. For rare earth elements, the *4f* electrons are shielded from the crystal field as these are surrounded by *5s*, *5p*, or *5d* orbitals. That is why the exchange field between rare earth ions is much smaller than that between iron-iron and rare earth-iron. The magnetic moment of REIG will be both due to the orbital and spin moment. The magnetization of REIG at different temperatures (T) is due to the dominant contribution of different sublattices. At high and low T, the dominant sublattices are iron and rare earth sublattices respectively. The magnetic and non-magnetic ion substitution in REIG also play a very important role in deciding their magnetic property. In this review, we have tried to figure out the basic underlying physics behind the origin of remarkable magnetic behavior in REIG.

1. Introduction

Since the discovery of the fascinating properties of rare earth garnet by Forestier and Guiot-Guillain [1,2], it has become a potential candidate in various fields such as microwave applications, high-quality filters, oscillators, spintronic devices, phase shifters, and many magneto-optical devices due to its unique electromagnetic, controllable magnetic, magneto-optical, magnetic compensation, mechanical, and thermal properties [3,4]. The rare earth elements (REE) are a group of 15 elements that are located in the lanthanide series of the periodic table of elements. Scandium and yttrium are also a part of this categorization; although they are not true REEs. This is because they exhibit similar properties as that of the lanthanides and often found in the same mineral as the lanthanides [5]. Garnets are a versatile and widely distributed mineral that can be found in wide range of rock types around the world. Their diverse occurrence and properties make them a valuable resource for a range of industrial and scientific applications. A mixture of REE and iron along with the silicate mineral garnet forms rare earth-iron garnet (REIG). This composition results in unique properties like high magnetic permeability and high electrical resistance [6] etc. that can be fine tuned by altering the composition of the material. In this article, we have discussed briefly about the basic understanding of magnetism in different types of REIGs.

2. Crystal structure of REIGs

Garnet minerals share a well-known crystal composition and the generalized chemical structure of it is given by $\{X^{2+}{}_3\}[Y^{3+}{}_2](Si^{4+}{}_3)O_{12}$. In that chemical composition, "X" may be cations like Y, Gd, Lu, Tb, Tm, Er, Dy, Ho etc. and "Y" may be cations like Al, Ga, In, Fe, Mn, V or Cr. For rare earth magnetic garnet, *Fe* will be the preferred cation for Y and its chemical formula is written as $R_3Fe_5O_{12}$. Unit cell of garnet was found by Menzer in 1928 and has a body-centred cubic symmetry [7]. The unit cell consists of 8 formula units or 160 atoms per unit. Out of which 64 are cations and 96 are oxygen anions. The oxygen sub-lattice has a polyhedra combination with three sub-lattices namely tetrahedral (4-fold), octahedral (6 fold) and dodecahedral (12 fold) sites. There are 24, 16, and 24 positions in the unit cell for these sites respectively. The spin interaction between tetrahedral and octahedral sites is antiparallel, and their resultant spin interaction is antiparallel to dodecahedral sites [8]. Being large, generally, the "R" cations occupy dodecahedral sites. Fe^{3+} ions are distributed over octahedral and tetrahedral sites with different atomic positions. The configuration of cations in the tetrahedral, octahedral, and dodecahedral sites are the cubic symmetric structure of REIG along with the arrangement of different polyhedra are shown in Fig. 1(a) and 1(b) respectively.

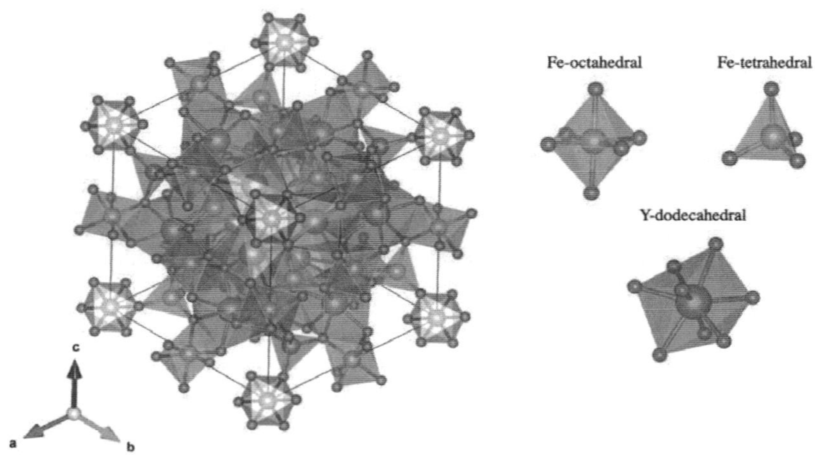

(a) (b)

Fig. 1: The cubic symmetry structure of garnet [9]; with copyright permission from the journal.

REIGs are ferrimagnetic in nature. This behaviour is a result of the exchange interaction between the Fe ions at the octahedral and tetrahedral sites mediated through oxygen anion. This interaction is crucial in deciding the magnetic properties of garnet materials, such as their magnetic ordering temperature and magnetic anisotropy. This interaction can be controlled through the use of various doping techniques, allowing for the manipulation of the crystal structure as well as the magnetic properties of the material. In the following section, we will discuss in detail about the magnetic exchange interaction of REIGs.

3. Magnetic properties of REIGs

REIG is a type of magnetic material known for its high magnetic permeability, strong magnetic anisotropy, low magnetic loss, and high-frequency properties. These properties make it a potential candidate in the field of modern technologies like microwave application due to ferromagnetic resonance [10], spintronics applications [11], wide range of optical devices [12], Spin wave logic devices [13], magnetic data storage devices [14], etc. The magnetic properties of REIG depend on the composition of the material and the arrangement of the magnetic ions. Additionally, the magnetic properties of REIG can be easily controlled through external magnetic fields, making it a versatile material for magnetic applications.

In REIGs, the magnetic ions are typically rare earth elements with partially filled *d* or *f* electron shells, which give rise to the magnetic moments. The exchange interaction between them is responsible for their alignment, so as to minimize the total energy of the material [15]. This quantum mechanical phenomenon was explained by Heisenberg in accordance with the Pauli exclusion principle. The exchange interaction between the two localized spins S_i and Sj in terms of the Hamiltonian is given by,

$$H = -2 \sum J_{i,j} (S_i . S_j) \quad (1)$$

where J_{ij} is the exchange coupling constant which is a measure of the strength of the interaction between neighbouring magnetic moments in a material. It determines the relative orientation of magnetic moments in a material. As a result, it affects the magnetic properties of the material such as its magnetic anisotropy and magnetization. The value of the exchange coupling constant is positive for ferromagnetic materials and negative for antiferromagnetic materials [16]. The exchange field between different ions in REIG in ascending order are rare earth- rare earth, rare earth – iron, and iron-iron [17]. This is due to the shielding effect of the RE *4f* electrons from the crystalline environment. But due to the strong spin-orbit coupling in this case, these materials exhibit a strong magnetic anisotropy [18]. The electrostatic interaction energy between *4f* electrons and the crystal field results in the anisotropic energy per rare earth atom. Without the application of a magnetic field, the magnetization vector aligns itself along the magnetic easy axis to minimize the energy within the material [19]. REIGs magnetic properties also depend on microstructural parameters like grain/particle size, porosity, defects and crystalline volume fraction, etc.

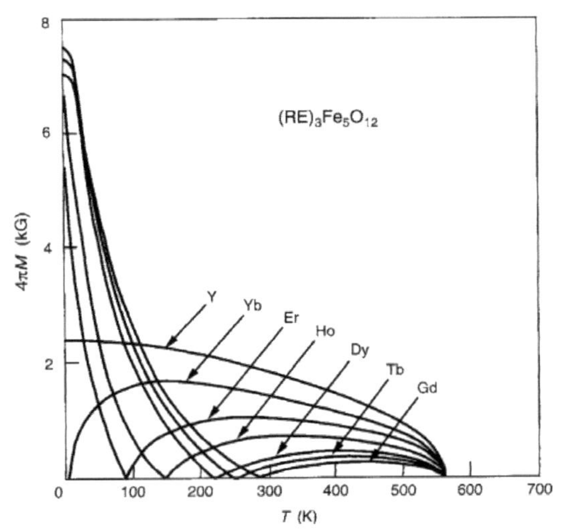

Fig. 2: Thermomagnetic curve exhibiting the magnetic compensation temperature (T_{comp}) of REIG materials [22, p. 180].

With larger grain size, the density of the sample increases which leads to the presence of better magnetic domain movement (growth of the domain, domain walls movement, and domain spin rotation [20,21]. This will result in the enhancement of magnetic permeability. Most of the garnet reveals magnetic reversal at compensation temperature (T_{comp}) which occurs due to the equality of the magnetic moments in dodecahedral sublattice with that of octahedral and tetrahedral together. This behaviour is shown in Fig.2. When the temperature is less than T_{comp}, rare earth sub-lattice dominates over the other sublattices; whereas at a temperature greater than T_{comp} iron sub-lattices are dominant. The schematic diagram for this magnetic moment dominance in the sublattice at different temperatures is explained in Fig.3(a) and 3(b). The absence of T_{comp} in some cases is due to the occupancy of dodecahedral sites by ions that are either nonmagnetic or RE with a small magnetic moment.

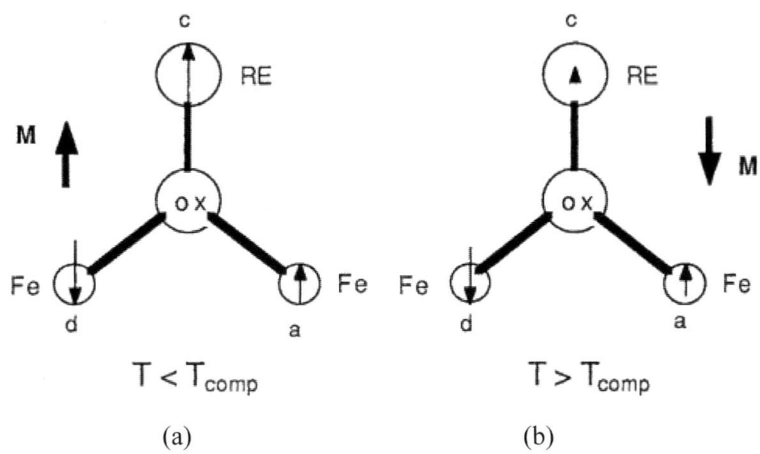

Fig. 3: The schematic diagram of magnetic moment dominance in the sublattice for the temperature (a) T< T_{comp} (b) T> T_{comp}. The sublattices a, d, and c refer to octahedral, tetrahedral and dodecahedron respectively. [22, p- 181].

In the REIG family, yttrium iron garnet (YIG), is the most explored and interesting garnet having applications as a magnetic insulator [23], magnonics [24], tunable magnetic qubit coupling [25], and many more. This extensive analysis is due to its ultralow Gilbert damping parameter [26,27] which makes it a perfect material for these applications. Apart from this, substituting Y with other RE elements like Gd [28,29], Dy [30], Tm, [31] Ho, Er, Yb [32], etc. have also attracted attention due to their complex magnetic structures that allow the alteration of the magnetic properties. In the subsequent section, we will provide a concise overview of the recent advancements made in the magnetic properties of these materials.

3.1. Magnetic properties of YIG

The magnetic properties of YIG are due to a combination of exchange interaction, magnetic anisotropy, magnetic moment, Curie temperature, magneto-optic effect, and spin wave excitations. Y^{3+} ion does not have any magnetic moment because of its electronic configuration ($4f^0$). The ferrimagnetism in YIG is due to the superexchange interaction between the $3Fe^{3+}$ ions at tetrahedral and $2Fe^{3+}$ ions at octahedral sites mediated through oxygen ions. This interaction leads to collective excitations known as spin waves or magnons. These excitations have quantized energy and can propagate through the crystal lattice, making YIG a useful material for studying spin wave physics and for developing magnon-based devices. Due to the fascinating magnetic properties (high saturation magnetization, strong magnetic anisotropy, etc.) of YIG, many researchers have tried to fine-tune its properties by performing the cationic substitution at Fe and Y sites. YIG exhibits an excellent magneto-optical effect, which means that its optical properties (such as refractive index and polarization) can be altered a magnetic field. Mao et al. [33] have substituted Ce in YIG and have shown that lower doping ($x \leq 0.2$) density can increase the saturation magnetization and remnant magnetization, while for higher doping concentration an extra phase is formed inside the sample which reduces these values.

When YIG is doped with erbium (Er), the magnetic moment is not influenced much over a wide temperature range except below 30K where the unpaired *4f* electrons are coupled antiparallelly with the *3d* electrons of iron cations [34]. Thus Er:YIG exhibits antiferromagnet-like behaviour at this lower temperature whereas above this and at room temperature it exhibits ferromagnetic behaviour. Bsoul et al. [35] have substituted Ga in Er-doped iron garnets $Er_3Fe_{5-x}Ga_xO_{12}$ ($0.0 \leq x \leq 0.8$) and obtained a significant change in the temperature-dependent behaviour of the rare earth and iron sub-lattices and a reduction of the overall magnetization above the T_{comp}. Also, they observed the decrease in Curie temperature with the increase of x. This occurs because of the substitution of magnetic ion (Fe^{3+}) by the non-magnetic ion (Ga^{3+}) that leads to decrease in the strength of super exchange interactions. This means nonmagnetic ion substitution in this case significantly reduce the magnetization over T_{comp}.

The doping of YIG with In [36] or Ni [37] or Ca [38,39] increases the magnetic moment of the octahedral sub-lattice. The substitution also makes it a charge-uncompensated material by either Fe^{4+} or Fe^{2+} formation or oxygen vacancies formation to ensure the charge neutrality of the lattice. This property makes them a potential material for device application.

3.2. Magnetic properties of Gadolinium iron garnet (GdIG)

GdIG is a ferrimagnetic insulator that exhibits a large magneto-optical activity and has low magnetic damping which means that spin waves can propagate over long distances without significant energy loss. In GdIG, the Gd^{3+} ions are in the S state with $J = 7/2$ and are magnetized parallel to the Fe^{3+} ions present in octahedral sites. Thus, when the crystalline electric field is compared with the exchange field, it can be neglected [40]. The exchange field in the lattice in ascending order is between the Gd ions, the Fe and Gd sublattice, and the two iron sublattices. The material exhibit temperature-dependent magnetization as shown in Fig. 2 and at T_{comp}, the magnetic moment of the Fe and Gd sublattices cancel each other. The compensation temperature for this material is close to room temperature (286–295 K) [41-45]. Above T_{comp}, how spontaneous magnetization may be increased, Nguyet et al. [46] suggested a mechanism for that. In this mechanism at high temperatures, Gd and Fe ions that are present on the surface layer, their spins are largely decoupled and the spins of Fe ions at the surface align themselves along the magnetic moments of the core. The magnetic anisotropy which is largely influenced by the presence of a magnetic compensation point can be fine-tuned in GdIG samples by preparing it on a different substrate and thereby varying the film strain [47]. The magnetic behaviour is also influenced by inter-particle interactions. GdIG has been extensively studied for its properties related to the Spin Seebeck effect (SSE) [48, 49]. This is a phenomenon where a temperature gradient applied to a magnetic material generates a gradient in the magnon chemical potential, which in turn drives a magnon spin current. High-energy exchange magnons, which have a large energy compared to the thermal energy, can overcome the damping in the material and travel long distances, leading to a strong SSE signal [29]. The Spin Seebeck effect in GdIG has several unique features that make it attractive for spintronics applications, energy harvesting, and thermal management.

Doping GdIG with a non-S-state ion like erbium (Er^{3+}) ion, the orbital angular momentum is quenched by the crystalline electric field. The exchange field along with the crystal electric field play a significant role in making double conical arrangement of Er^{3+} moments relative to the easy direction of the magnetization [40]. Thus, Er doping can enhance the magnetic anisotropy and saturation magnetization thereby making it an interesting material for magneto-optical and microwave devices [50]. Bismuth (Bi) substitution for gadolinium (Gd) in the GdIG crystal can lead to an increase in the Curie temperature which is attributed due to the fact that Bi^{3+} ions are larger than that of Gd^{3+}, thereby leading to the distortion of the dodecahedral site. This distortion may influence the Fe-O-Fe super exchange interactions in the structure [51]. Y^{3+} has a similar ionic radius as Gd^{3+} but a higher magnetic moment, and when it substitutes Gd^{3+} in the lattice, it affects the magnetic interactions between the Fe^{3+} ions in the structure. This can lead to changes in the magnetic anisotropy, saturation magnetization, and other magnetic properties of the material [52]. In addition to these doping, other dopants have been studied in GdIG, including cobalt (Co), calcium(Ca), cerium (Ce), Ruthenium(Ru) and manganese (Mn) etc. [53-59]. These dopants can affect the magnetic properties of GdIG in various ways, including changes in magnetic ordering, magnetic anisotropy, and magnetization.

3.3. Magnetic properties of dysprosium iron garnet (DyIG)

The main advantage of Dysprosium iron garnet (DyIG) over other magnetic materials is its high saturation magnetization, which makes it ideal for use in high-field magnets. It also has a very high

Curie temperature, making it suitable for use in high-temperature applications. It is a type of ferrimagnetic material that exhibits a large Faraday rotation effect, which is a phenomenon where the polarization plane of linearly polarized light is rotated in the presence of a magnetic field. This garnet has a particularly large Faraday rotation compared to other magnetic materials due to its high magnetic anisotropy (due to strong spin-orbit coupling) and high magnetic susceptibility. This property makes it useful in a number of applications, including in the field of magneto optics as a Faraday rotator, circulator, isolator, data storage, and magnetic field sensors [60] etc. In these devices, DyIG is used to selectively transmit or reflect light based on its polarization state, which can be controlled by an applied magnetic field. The reason of its large Faraday rotation is due to the strong interaction between the magnetic moments of the dysprosium ions and the applied magnetic field. Doping Dysprosium iron garnet (DyIG) with other elements can modify its magnetic properties. Doping DyIG by Ga, further can enhance the magnetic anisotropy resulting in a larger Faraday rotation angle [61]. Ga-doped DyIG has been used in magneto-optic devices such as optical isolators and circulators. Co doping can increase the magnetic moment of DyIG and enhance its magnetization dynamics. Co-doped DyIG has been studied for its potential use in microwave devices and magnetic sensors [60]. Al doping can reduce the magnetic anisotropy of DyIG, which can reduce its Faraday rotation angle [62]. It was shown by Rekha et al. [63] that magnetic property of DyIG (doped with Mn) can be controlled with respect to cationic distribution at different sites, the crystallite size and surface morphology of grains, porosity (that affects the magnetization process), spin–orbital interaction between the cations. Thus Mn-doped DyIG or doping DyIG with other elements can lead to interesting modifications of its magnetic properties, which can make it useful for a variety of applications in magnetics, spintronics, and magneto-optics.

Depending on the sample preparation technique and chemical composition, different garnet materials may have different values of the magnetic parameters like permeability, magnetic loss, magnetic anisotropy, saturation magnetization, Curie temperature, frequency etc. These properties are compared and tabulated in Table 1.

Table 1. Comparison of magnetic properties of different garnet materials.

Magnetic Properties	YIG	GdIG	DyIG
Saturation Magnetization	High [64]	Moderate [46]	Moderate to High [30]
Curie Temperature	High [65]	Moderate [66]	Moderate to High [67]
Magnetic Anisotropy	Strong [65]	Moderate [68]	Strong [69]
Permeability	High [70]	Moderate [71]	Moderate [30]
Magnetic Loss	Low [72]	Low [71]	Low [30]
Frequency Range	Microwave and RF [70]	Microwave and RF [73]	Microwave and RF [30]

3.4. Magnetic properties of other iron garnets

There are many other REIGs with their fascinating magnetic properties. In the small scope of the article, it is not possible to cover them all. So, here in this section we will discuss in brief about the magnetic properties of other REIGs where the rare earth elements are Ho, Sm, Er etc. Holmium iron garnet (HoIG) [74], Erbium iron garnet (ErIG) [75], Terbium iron garnet (TbIG) [76] exhibits high magnetic anisotropy, magnetic ordering, low magnetic damping making it useful in research on magnetism and magnetic materials. At the compensation temperature, they exhibit an umbrella-like magnetic structure as shown in Fig. 2. Compared to other REIGs, HoIG has large room temperature magneto-electric coupling, which is the coupling between magnetic and electric orders [77]. The magnetic domain wall pinning in HoIG leads to a complex magnetic behavior that is highly dependent on the temperature and the strength of the external magnetic field [78]. Samarium iron garnet (SmIG) exhibits a relatively weak magnetic anisotropy and does not exhibit magnetic domain wall pinning. Rather, SmIG exhibits a phenomenon called spin-reorientation, where the magnetic orientation of the material changes with changing temperature or applied magnetic field. This property makes it an interesting candidate for magnetic field sensors with high sensitivity and accuracy, magnetic storage etc [79, 80]. Thulium iron garnet (TmIG) has recently been found to exhibit spin orbit torque (SOT) effects [81]. The SOT effect is due to the strong spin-orbit coupling of the thulium ions in the material. This coupling allows the thulium ions to transfer angular momentum to the magnetic moments in the material. SOT can be used to manipulate the magnetic state of a material without the need for an external magnetic field, which could lead to more efficient and compact devices. The discovery of SOT in TmIG has potential implications for the development of spintronic devices such as magnetic memory and logic devices.

Thus from the above discussion, it is understood that this class of ferrites exhibit interesting magnetic properties that make them promising materials for various technological applications such as magnonics, quantum information processing, high entropy ceramics, biomedical applications [82-84] etc. Their magnetic properties depend on their specific composition, cationic distribution with variety of dopants, magnetocrystalline anisotropy, domain formation, spin orbit coupling etc. [85] having different magnetic and optical properties. Apart from this, the magnetic properties can also vary depending on the synthesis method, crystal structure, and sample quality.

4. Conclusion

REIGs are considered to be the important technological magnetic materials which find multiple applications in major industries like electrical, optical, magnetic, magneto-optical, microwave integrated circuits and biomedical etc. Though they are having unique magnetic properties, yet there is no unified theory which can explain their magnetic properties with different dopant and host combination. The purpose of this article is to give an outline about the basic physics behind the origin of the magnetic properties of these systems. In the limited scope of the article, it is difficult to discuss all iron garnet substituted with different rare earth elements, so we have discussed in brief about the magnetic properties of different REIGs on one platform.

5. References

[1] Forestier H and Guiot-Guillain G 1950 *Compt. Rend.* 230.

[2] Forestier H and Guiot-Guillain G 1952 *Compt. Rend.* 235.

[3] Garskaite E, Gibson K, Leleckaite A, Glaser J, Niznansky D, Kareiva A and Meyer H J 2006 *Chem. Phys.* **323(2-3)** 204.

[4] Joseyphus R J, Narayanasamy A, Nigam A K & Krishnan R 2006 *J. Magn. Magn. Mater.* **296(1)** 57.

[5] Wells W H and Wells V K. 2001 "The lanthanides, rare earth elements." Patty's Toxicology, pp 817.

[6] Nakamoto R, Xu B, Xu C, Xu H, Bellaiche L 2017 *Phys. Rev. B.* **95(2)** 024434 .

[7] Enoch, Reginal David, The magnetic domain structure in some single crystals of yttrium iron, 1973 *The magnetic domain structure in some single crystals of yttrium iron garnet.*

[8] Gilleo M A & Geller S 1958 *J. Appl. Phys.* **29(3)** 380.

[9] Redies M, Michalicek G, Bouaziz J, Terboven C, Mueller M S, Blügel S, Wortmann D 2022 *Front. Mater.* **9** 147.

[10] Akhtar M N, Sulong A B, Khan M A, Ahmad M, Murtaza G, Raza M R, Raza R, Saleem M & Kashif M 2016 *J. Magn. Magn. Mater.* **401** 425.

[11] Zanjani S M, & Onbasli M C 2019 *AIP Adv.* **9(3)** 035024.

[12] Booth R C & White E A D 1984 *J. Phys. D: Appl. Phys.* **17(3)** 579.

[13] Schneider V, Serga A A, Leven B, Hillebrands B, Stamps R L & Kostylev M P 2008 *Appl. Phys. Lett.* **92(2)** 022505.
[14] Geselbracht M J, Cappellari A M, Ellis A B, Rzeznik M A & Johnson B 1994 *J. Chem. Educ.* **71(8)** 696.
[15] Cullity B D & Graham C D, 2011 *Introduction to Magnetic Materials* (John Wiley & Sons).
[16] Blundell S 2003 *Am. J. Phys.* **71** 94.
[17] Cahaya A B 2022 *J. Magn. Magn. Mater.* **553** 169248.
[18] Skomski R and Sellmyer D J 2009 *J. Rare Earths.* **27 (4)** 675.
[19] Skomski R, Ralph H P, Oepen and Kirschner J 1998 *Phys. Rev. B.* **58** 11138.
[20] Yu R H, Basu S, Zhang Y, Parvizi-Majidi A, Xiao J Q 1999 *Appl. Phys. Lett.* **85(9)** 6655.
[21] Hapishah A N, Hamidon M N, Syazwan M M & Shafiee F N 2019 *Results Phys.* **14** 102391.
[22] Dionne G F 2009 Magnetic Oxides (New York: Springer) *pp* 180 -181.
[23] Cherepanov V, Kolokolov I and L'vov V 1993 *Phys. Rep.* **229 (3)** 81.
[24] Hauser C, Richter T, Homonnay N, Eisenschmidt C, Qaid M, Deniz H, Hesse D, Sawicki M, Ebbinghaus S G and Schmidt G 2016 *Sci. Rep.* **6(1)** 20827.
[25] Tabuchi Y, Ishino S, Noguchi A, Ishikawa T, Yamazaki R, Usami K, Nakamura Y 2015 Sci. **349 (6246)** 405.
[26] Torrão R, Alves O C, Archanjo B S, Sampaio L C, Garcia F 2022 *J. Alloys Compd.* **923** 166300.
[27] Onbasli M C, Kehlberger A, Kim D H, Jakob G, Kläui M, Chumak A B, Hillebrands B, Ross C A 2014 *APL Mater.* **2 (10)** 106102.
[28] Cramer J, Guo E J, Geprägs S, Kehlberger A, Ivanov Y P, Ganzhorn K, Coletta F Della, Althammer M, Huebl H, Gross R, Kosel J 2017 *Nano Lett.* **17 (6)** 3334.
[29] Geprägs S, Kehlberger A, Coletta F D, Qiu Z, Guo E J, Schulz T, Mix C, Meyer S, Kamra A, Althammer M, Huebl H 2016 *Nat. Commun.* **7 (1)** 10452.
[30] Bauer J J, Rosenberg E R, Kundu S, Mkhoyan K A, Quarterman P, Grutter A J, Kirby B J, Borchers J A and Ross C A 2020 *Adv. Electron. Mater.* **6 (1)** 1900820.
[31] Quindeau A, Avci C O, Liu W, Sun C, Mann M, Tang A S, Onbasli M C, Bono D, Voyles P M, Xu Y and Robinson J 2017 *Adv. Electron. Mater.* **3 (1)** 1600376.
[32] Zanjani S M and Onbaşlı M C 2020 *J. Magn. Magn. Mater.* **499** 166108.
[33] Mao T C and Jyh-Chen C 2006 *J. Magn. Magn. Mater.* **302 (1)** 74.
[34] Cho Y, Kang S, Nahm Y W, Mohamed A Y, Kim Y, Cho D Y & Cho S 2022 *ACS Omega* **7(29)** 25078.
[35] Bsoul I, Olayaan R, Lataifeh M, Mohaidat Q I & Mahmood S H 2019 *Mater. Res. Express.* **6(7)** 076114.
[36] Nazlan R, Hashim M, Ibrahim I R, Idris F M, Ismail I, Ab Rahman W N W, Abdullah N H, Zulkimi M M M and Mustaffa M S 2015 *J. Phys. Chem. Solids* **85** 1.
[37] Peña-Garcia R, Guerra Y, Santos F E P, Almeida L C & Padrón-Hernández E 2019 *J. Magn. Magn. Mater.* **492** 165650.
[38] Pardavi-Horvath M, Thavendrarajah A, Wigen P E & DeGasperis P 1993 *J. Magn. Magn. Mater.* **119 (1-2)** 193.
[39] Song Y J, Turpin G B, Bornfreund R E, Aoyama H & Wigen P E 1996 J. *Magn. Magn. Mater.* **154 (1)** 37.
[40] Mohaidat Q I, Lataifeh M, Mahmood S H, Bsoul I & Awawdeh M 2017 J. *Supercond. Nov. Magn.* **30** 2135.
[41] Wang L W, Xie L S, Xu P X & Xia K 2020 *Phys. Rev. B* **101(16)** 165137(1).
[42] Söderlind F, Selegård L, Nordblad P, Uvdal K & Käll P O 2009 *J.Sol-Gel Sc. Tech.* **49** 253.
[43] Calhoun B A, Overmeyer J & Smith W V 1957 *Phys. Rev.* **107(4)** 993.
[44] Pauthenet R 1958 *J. Appl. Phys.* **29(3)** 253.
[45] Rodić D, Tomkowicz Z, Novaković L, Szytula A & Napijalo M L 1990 *Solid state commun.***73(3)** 243.
[46] Nguyet D T T, Duong N P, Satoh T, Anh L N & Hien T D 2013 *J. Magn. Magn. Mater.* **332** 180.
[47] Holzmann C, Ullrich A, Ciubotariu O T & Albrecht M 2022 ACS *Appl. Nano Mater.* **5(1)** 1023.

[48] Yagmur A, Iguchi R, Geprägs S, Erb A, Daimon S, Saitoh E, Gross R & Uchida K 2018 *J. Phys. D: Appl. Phys.* **51(19)** 194002 .
[49] Wang L W, Xie L S, Xu P X & Xia K 2020 *Phys. Rev. B* **101(16)** 165137.
[50] Khan M I, Waqas M, Naeem M A, Hasan M S, Iqbal M, Mahmood A, Ramay S M, Al-Masry W, Abubshait S A, Abubshait H A & Mahmood Q *Ceram. Int.* **46(17)** 27318.
[51] Sharma A, Godara S K & Srivastava A K 2022 *Mater. Today: Proc.* **50** 1867.
[52] Maier-Flaig H, Geprägs S, Qiu Z, Saitoh E, Gross R, Weiler M, Huebl H, Goennenwein S T B 2017 *arXiv preprint arXiv* **1706** 08488.
[53] Llabrés J, Nicolas J and Sroussi R 1977 *Appl. Phys.* **12** 87.
[54] Torres C, Orejudo R, Francisco de C, Muñoz J M, Hernández P, Alejos O and López-Díaz L 1998 *Phys. Status Solidi (A)* **169(1)** 145.
[55] Aoki D and Shima M 2014 *Jpn. J. Appl. Phys.* **53(11)** 113001.
[56] Antonini B, Marinelli M, Milani E, Paoletti A, Paroli P, Daval J and Ferrand B 1989 *Phys. Rev. B* **39** 13442.
[57] Akhtar M N, Sulong A B, Ahmad M, Khan M A, Ali A & Islam M U 2016 *J. Alloys Compd.* **660** 486.
[58] Hansen P 1972 *Phys. Rev. B* **5(9)** 3737.
[59] Galanciak D, Grinberg M, Gryk W, Kobyakov S, Suchocki A, Boulon G and Brenier A 2005 *J. Phys. Condens. Matter* **17(46)** 7185.
[60] Tholkappiyan R & Vishista K 2015 *Appl. Surf. Sc.* **351** 1016.
[61] Yong Z, Defang S & Fuxi G 1993 *Chin. Phys. Lett.* **10 (3)** 186.
[62] Dhilip M, Saravana Kumar K, Ramesh Kumar R & Anbarasu V 2019 *J. Electron. Mater.* **48** 8243.
[63] Rekha G, Tholkappiyan R, Vishista K & Hamed F 2016 *Appl. Surf. Sc.* **385** 171.
[64] Hosseinzadeh S., Behboudnia M., Jamilpanah L., Sheikhi M.H., Mohajerani E., Tian K., Tiwari A., Elahi P. and Mohseni S.M. 2019 *J. Magn. Magn. Mater*, **476** 355.
[65] Baños-López E., Sánchez-De Jesús F., Cortés-Escobedo C.A., Barba-Pingarrón A. and Bolarín-Miró A.M. 2018 *Materials.* **11(9)** 1652.
[66] Zanatta S.C., Cótica L.F., Paesano Jr A., de Medeiros S.N., da Cunha J.B.M. and Hallouche B. 2005 *J. Am. Ceram. Soc.* **88(12)** 3316.
[67] Feldmann P., Guillot M. and Le Gall H. 1977 *IEEE Trans. Magn.* **13(5)** 1574.
[68] Lassri H., Hlil E.K., Prasad S. and Krishnan R. 2011 *J. Solid State Chem.* **184(12)** 3216.
[69] Lahoubi M., Younsi W., Soltani M.L. and Ouladdiaf B. 2010 *J. Phys. Conf. Ser.* 200(8) 082018.
[70] Tsutaoka T., Kasagi T. and Hatakeyama K. 2011 *J. Appl. Phys.* **110(5)** 053909.
[71] Nazlan R., Ibrahim I.R. and Rahim A.S.A. 2017 *Solid State Sci. Technol.* **268** 287.
[72] Burrowes C., Heinrich B., Kardasz B., Montoya E.A., Girt E., Sun Y., Song Y.Y. and Wu M. 2012 Appl. Phys. Lett. **100(9)** 092403.
[73] Fechin, P.B.A., Pereira F.M.M., Santos M.R.P., de Menezes A.S., De Oliveira R.S., Góes J.C., Cardoso L.P. and Sombra A.S.B. 2009 *J. Phys. Chem. Solid.* **70(5)** 804.
[74] Yang J, Ma S & Xu Y 2009 *J. Phys. Condens. Matter* **21(9)** 096004.
[75] Maignan A, Singh K, Simon C, Lebedev O I, Martin C, Tan H, Verbeeck J & Van Tendeloo G 2013 *J. Appl. Phys.* **113(3)** 033905.
[76] Lahoubi M & Wang M 2015 *J. Magn. Magn. Mater.* **393** 437.
[77] Selvi M M, Chakraborty D & Venkateswaran C 2017 *J. Magn. Magn. Mater.* **423** 39.
[78] Pyatakov A P, Sergeev A S, Nikolaeva E P, Kosykh T B, Nikolaev A V, Zvezdin K A E & Zvezdin A K 2015 *Phys.-Uspekhi* **58(10)** 981.
[79] Geller S & Balestrino G 1980 *Phys. Rev. B* **21(9)** 4055.
[80] Brahma M, Gaikwad V M & Ravi S 2019 *Appl. Phys. A* **125(5)** 333.
[81] Avci C O, Quindeau A, Mann M, Pai C F, Ross C A & Beach G S 2017 *Phys. Rev. B* **95(11)** 115428.
[82] Chen D, Yang Y, Chen C, Meng Y, Zhang Y & Zhang C 2023 *Ceram. Int.* **49(6)** 9862.
[83] Sharma S K and Javed Y 2020 *Springer Nature* pp 243.
[84] Chakrabarti A, Banerjee J, Chakravarty S, Samanta S, Nath M, Chattopadhyay S, Sarkar S, Banerjee S M, Chowdhury S, Dash S K, Bandyopadhyay A 2022 *J. Magn. Magn. Mater.* **563** 169957.

[85] Chattopadhyay S, Bandyopadhyay A & Nath M 2022 *Ceramic Science and Engineering* pp 165.

Analytical solution for fractional order temperature distribution in two dimensional multilayer composite circular discs

[1*] **N.J. Wange** [2]**M.N. Dandale,** [3] **Vinod Waikar**

[1] Department of Mathematics, Yeshwantrao Chavan College of Engineering, Nagpur.
**E-mail: nitin.wange02@gmail.com*
[2] Department of Mathematics, Yeshwantrao Chavan College of Engineering, Nagpur
E-mail: milinddandale@gmail.com
[3] MBA, Datta Meghe Institute of Management Studies, Nagpur
E-mail: dmimsit@yahoo.com

Abstract. This article deals with mathematical approach to discuss the time-fractional heat conduction in two dimensional multilayer circular discs. The fractional order heat conduction equation is generated by using caputo time derivative. All the layers of the two dimension multilayer circular are assume to be thermally isotropic and making a perfect thermal contact. The boundary and interface conditions are expressed by the Riemann-Liouville derivative and the combination of first and second kind boundary condition in angular direction of the circular disc and for the third kind boundary condition in the radial direction. This solution is also applicable for composite multiple layers problem with inner radius may be zero. Three multiple layer problem studied numerically and graphically....

Keywords— Composite two dimensional, circular disc, multilayer, fractional

Introduction

The study of determination of temperature distribution under the different conditions, it is a part of extensive research in the field of BVP and Thermo elasticity. In engineering practices multilayer composite spherical body play an important role. The determination of Temperature in multilayer fractional order heat conduction with internal heat source is encountered various problems. To determination of temperature in multilayer fractional order problem with heat source has a variety of engineering applications.

The analytical solution is better useful compared to other one. Bulavin and Kashcheev [3] obtained solution of single direction multilayer transient heat conduction problem with heat source by using method of orthogonal expansion and separation of variables. Y. YenerAnd M.N. Ozisik [1] discussed the solution of single dimensional multilayer unsteady heat conduction problem. N. Wange and M. Gaikwad [6] studied the behaviour of heat conduction and thermoelasicity of single dimensional multilayer hollow sphere. S. Pawar et al [8] discussed the Thermoelastic nature of composite multilayer spherical body due to asymmetric surface temperature and internal heat source. Recently Stanislaw Kukla and Urszula Siedlecka [9] discussed analytical solution of fractional order one dimensional heat conduction. S.Singh and P.K. Jain [10] Solve classical heat conduction problem of composite medium by using separation of variable method.

This investigation is to determine the transient temperature distribution for a composite multilayer circular disc method discussed by M. N. Ozisik[4], that method is applicable for heat source introducing in each layer of circular disc. Hence aim of the work is to obtain the temperature field of fractional order multilayer composite circular disc with internal heat source by using integral transform techniques and Caputo time fractional derivative. Special case is considered for our solution.

Formulation of Problem

The heat conduction equation in two dimensional fractional order n-multilayers composite circular disc with all layers are thermally isotropic and making a perfect thermal contact [4]

$$\frac{\partial^2 V_i}{\partial r^2}+\frac{1}{r}\frac{\partial V_i}{\partial r}+\frac{1}{r^2}\frac{\partial^2 V_i}{\partial \theta^2}+g_i(r,\theta,t)=\frac{1}{\alpha_i}{}_0^cD_t^\alpha V_i \quad (1)$$

Where $V_i=V_i(r,\theta,t)$

$r_0\le r\le r_n,\quad 0\le\theta\le 2\pi,\ t\ge 0$.

Boundary, interface and initial conditions

Inner boundary surface condition $(i=1)$

$$L_{in}\frac{\partial V_i(r,\theta,t)}{\partial r}+M_{in}V_i(r,\theta,t)=H_1(r,\theta) \quad (2)$$

Outer boundary surface condition $(i=n)$

$$L_{out}\frac{\partial V_i(r,\theta,t)}{\partial r}+M_{out}V_i(r,\theta,t)=H_n(r,\theta) \quad (3)$$

Periodical Boundary Conditions for $\theta=0, i=1,2,3,\dots\dots..n$.

$$\left.\begin{array}{c}V_i(r,\theta=0,t)=0\\ or\\ \frac{\partial V_i(r,\theta=0,t)}{\partial\theta}=0\end{array}\right\} \quad (4)$$

Periodical boundary conditions for $=2\pi, i=1,2,3,\dots\dots..n$.

$$\left.\begin{array}{c}V_i(r,\theta=2\pi,t)=0\\ or\\ \frac{\partial V_i(r,\theta=2\pi,t)}{\partial\theta}=0\end{array}\right\} \quad (5)$$

Interface Condition of ith layers

$$V_i(r,\theta,t)=V_{i-1}(r,\theta,t)$$

$$k_i\frac{\partial V_i(r,\theta,t)}{\partial r}=k_{i-1}\frac{\partial V_{i-1}(r,\theta,t)}{\partial r}$$

Initial condition

$$V_i(r,\theta,t=0)=\aleph_i(r,\theta) \qquad 0<\alpha\le 1 \quad (6)$$

Caputo fractional Derivative,

$${}_0^cD_t^\alpha V(t)=\begin{cases}\frac{1}{\Gamma(n-\alpha)}\int_0^t\frac{V^{(n)}(\tau)}{(t-\tau)^{1+\alpha-n}}d\tau, & n-1<\alpha<n\\ \frac{d^nV(t)}{dt^n}, & \alpha=n\end{cases} \quad (7)$$

Solution Methodology:-

By using Laplace integral transform,

$$\mathcal{L}[{}_0^cD_t^\alpha V_i(t)]=s^\alpha\mathcal{L}[V(t)]-\sum_{k=0}^{n-1}V^{(k)}(0)\,s^{\alpha-1-k},n-1<\alpha<n. \quad (8)$$

Analytical Solution in direction of θ, and θ varies from $0\le\theta\le 2\pi$

$$\overline{V}_i(r,l,t)=\int_0^{2\pi}Cos\,l(\theta-\theta')V_i(r,\theta',t)\,d\theta' \quad (9)$$

An inversion formula in direction of θ

$$V_i(r,\theta,t)=\frac{1}{\pi}\sum_{l=0}^{\infty}\overline{V}_i(r,l,t) \quad (10)$$

The integral transform according to r variable

$$\overline{\overline{V}}_i(\lambda_p,l,t)=\sum_{i=1}^{n}\frac{k_i}{\alpha_i}\int_{r_{i-1}}^{r_i}r\,R_{ilp}(\lambda_{ilp},r)\overline{V}_i(r,l,t)\,dt. \quad (11)$$

An inversion formula,
$\overline{V_l}(r,l,t) = \sum_{r=1}^{\infty} \frac{R_{ilp}(\lambda_{ilp},r)}{N_{lp}} \widetilde{\overline{V_l}}(\lambda_p,l,t)$. (12)

We have obtain the solution in r-direction as
$R_{ilp}(\lambda_{ilp},r) = a_{ilp}J_l(\lambda_{ilp},r) + b_{ilp}Y_h(\lambda_{ilp},r)$ (13)

Where the J_l be first kind Bessel function and Y_h be second kind Bessel function
Also eigen function$R_{ilp}(r)$ satisfy the orthogonality conditions.
$\sum_{i=1}^{n} \frac{k_i}{\alpha_i} \int_{r_{i-1}}^{r_i} r\, R_{ilp}(\lambda_{ilp},r)\, R_{ilq}(\lambda_{ilq},r)dr = \begin{cases} 0, & p \neq q \\ N_{lp}, & p = q \end{cases}$ (14)

By using heat flux continuity condition,
$\alpha_i \lambda_{ilp}^2 = \alpha_1 \lambda_{1lp}^2$ (15)

By using integral transform technique

$${}_0^cD_t^\alpha \widetilde{\overline{V_l}}(\lambda_p,l,t) = -\alpha_1\lambda_{1lp}^2 \widetilde{\overline{V_l}}(\lambda_p,l,t) + k_n \left[\frac{r\, R_{nlp}(\lambda_{nlp},r)}{L_{out}}\right] \int_0^{2\pi} Cos\, l(\theta-\theta')H_n(\theta',t)\, d\theta' - k_1 \left[\frac{r\, R_{1lp}(\lambda_{1lp},r)}{L_{in}}\right] \int_0^{2\pi} Cos\, l(\theta-\theta')H_1(\theta',t)\, d\theta'$$

${}_0^cD_t^\alpha \widetilde{\overline{V_l}}(\lambda_p,l,t) = -\alpha_1\lambda_{1lp}^2 \widetilde{\overline{V_l}}(\lambda_p,l,t) + Z(\lambda_p,l,t)$ (16)

Where $Z(\lambda_p,l,t) = k_n \left[\frac{r\, R_{nlp}(\lambda_{nlp},r)}{L_{out}}\right] \int_0^{2\pi} Cos\, l(\theta-\theta')H_n(\theta',t)\, d\theta' -$
$k_1 \left[\frac{r\, R_{1lp}(\lambda_{1lp},r)}{L_{in}}\right] \int_0^{2\pi} Cos\, l(\theta-\theta')H_1(\theta',t)\, d\theta'$.
(17)

$\widetilde{\overline{\aleph}}(\lambda_p,l) = \sum_{i=1}^{n} \frac{k_i}{\alpha_i} \int_{r_{i-1}}^{r_i} r\, R_{ilp}(\lambda_{ilp},r) \int_0^{2\pi} Cos\, l(\theta-\theta')\aleph_n(r,\theta')\, d\theta\, dr$ (18)

Where $L_{in} = 0\, replace \left[\frac{R_{1lp}(\lambda_{1lp},r)}{L_{in}}\right]$ $by\ -\frac{1}{M_{in}}\frac{dR_{1lp}(\lambda_{1lp},r)}{dr}$
Where $L_{out} = 0\, replace \left[\frac{R_{nlp}(\lambda_{nlp},r)}{L_{out}}\right]$ $by\ -\frac{1}{M_{out}}\frac{dR_{nlp}(\lambda_{nlp},r)}{dr}$
By using fractional order differential equation Laplace transform formula,

$$s^\alpha \hat{\widetilde{\overline{V}}}(\lambda_p,l,s) - s^{\alpha-1}\widetilde{\overline{f}}(\lambda_p,l) = -\alpha_1\lambda_{1lp}^2 \hat{\widetilde{\overline{V}}}(\lambda_p,l,s) + \hat{z}(\lambda_p,l,s)$$

$$(s^\alpha + \alpha_1\lambda_{1lp}^2)\hat{\widetilde{\overline{V}}}(\lambda_p,l,s) = s^{\alpha-1}\widetilde{\overline{\aleph}}(\lambda_p,t) + \hat{z}(\lambda_p,l,s)$$

$\hat{\widetilde{\overline{V}}}(\lambda_p,l,s) = \frac{s^{\alpha-1}\widetilde{\overline{\aleph}}(\lambda_p,t)+\hat{z}(\lambda_p,l,s)}{s^\alpha+\alpha_1\lambda_{1lp}^2}$ (19)

Where cap Λ denot is used for respect s time t laplace transform
$\hat{\widetilde{\overline{V}}}(\lambda_p,l,s) = \int_0^\infty e^{-st}\, \widetilde{\overline{V}}(\lambda_p,l,t)dt$ (20)

$\hat{z}(\lambda_p,l,s) = \int_0^\infty e^{-st} z\,(\lambda_p,l,s)dt$ (21)
By using the Mittag-Leffler function in two parameter.
$\int_0^\infty e^{-st} t^{\alpha k+\beta-1} E_{\alpha,\beta}^{(k)}(\pm a t^\alpha)dt = \frac{k!\, s^{\alpha-\beta}}{(s^\alpha \mp a)^{k+1}}$ (22)
We get,
$\widetilde{\overline{V}}(\lambda_p,l,t) = E_{\alpha,1}(-\alpha_1\lambda_{1lp}^2 t^\alpha)\widetilde{\overline{\aleph}}(\lambda_p,l) + \int_0^t z(\lambda_p,l,t-\tau)U_{lp}(\tau)d\tau$ (23)

Where $U_{lp}(t) = \mathcal{L}^{-1}\left[\frac{1}{s^{\alpha}+\alpha_1\lambda_{1lp}^2}\right] = t^{\alpha-1}E_{\alpha,\alpha}\left(-\alpha_1\lambda_{1lp}^2 t^{\alpha}\right)$

$$E_{\alpha,\beta}(Q) = \sum_{k=0}^{\infty}\frac{Q^k}{\overline{|\alpha k+\beta}}$$

$$E_{\alpha,\beta}^{(k)}(Q) = \frac{d^k}{dz^k}\left[E_{\alpha,\beta}(z)\right]$$

$$V_i(r,\theta,t) = \frac{1}{\pi}\sum_{l=0}^{\infty}\sum_{p=1}^{\infty}\frac{R_{ilp}(\lambda_{ilp},r)}{N_{lp}}\left\{E_{\alpha,1}\left(-\alpha_1\lambda_{1lp}^2 t^{\alpha}\right)\tilde{\tilde{\aleph}}(\lambda_p,l) + \int_0^t z(\lambda_p,l,t-\tau)U_{lp}(\tau)d\tau\right\} \quad (24)$$

Where $\alpha = 1$,

$$V_i(r,\theta,t) = \frac{1}{\pi}\sum_{l=0}^{\infty}\sum_{p=1}^{\infty}\frac{R_{ilp}(\lambda_{ilp},r)}{N_{lp}}e^{-\alpha_1\lambda_{1lp}^2 t}\left\{\tilde{\tilde{\aleph}}(\lambda_p,l) + \int_0^t z(\lambda_p,l,t-\tau)e^{\alpha_1\lambda_{1lp}^2\tau}d\tau\right\} \quad (25)$$

Case study:-

We consider three layer (i=1,2,3) semicircular annulus region $r_0 \le r \le r_3$ and $0 \le \theta \le \pi$. Also consider $H_1(r,\theta) = 0,\ H_2(r,\theta) = 0$ For time t > 0, $\aleph_i(r,\theta) = 0$

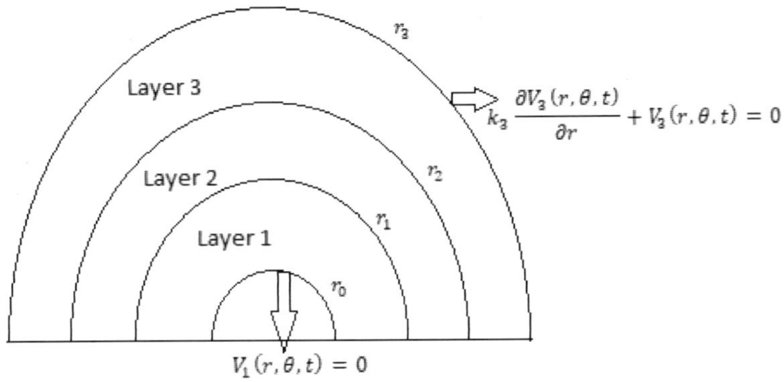

Fig.1 Three multilayer composite semicircular disc

Dimensionless parameter values as

$L_{in} = 0, L_{out} = k_3$ and $M_{in} = 1, M_{out} = 1, {}^{k_2}/_{k_1} = 2, {}^{k_3}/_{k_1} = 4, {}^{\alpha_2}/_{\alpha_1} = 4, {}^{\alpha_3}/_{\alpha_1} = 9, {}^{r_1}/_{r_0} = 2,$

${}^{r_2}/_{r_0} = 4$

${}^{r_3}/_{r_0} = 6$

Heat source term

$g_i(r,\theta,t) = \frac{g_0}{k_i}\frac{\delta(r-r_{i0})}{r}\theta^2(\pi-\theta)^2 t^{\varepsilon}$ at $0 \le \theta \le \pi$

We obtain the solution for i=1,2,3.

$$V_i(r,\theta,t) = \frac{g_0}{\pi}\sum_{l=0}^{\infty}\sum_{p=1}^{\infty}\frac{R_{ilp}(\lambda_{ilp},r)}{N_{lp}}\left(\frac{1-E_{\alpha}(-\alpha_1\lambda_{1lp}^2 t^{\alpha})}{\alpha_1\lambda_{1lp}^2}\right)\int_0^{\pi}\theta'^2(\pi-\theta')^2\cos l(\theta - \theta')d\theta'\sum_{i=1}^{3}R_{ilp}(\lambda_{ilp},r_{i0})$$

Numerical results and Discussion

For numerical computation we consider three layers semicircular annulus region where inner layer is copper $r_0 \le r \le r_1$, middle layer is zinc $r_1 \le r \le r_2$ and outer layer is aluminium.

Thermal properties of copper, zinc and aluminium at room temperature as

Thermal conductivity [W/cmC] copper is 3.86, zinc is 1.13 and aluminium is 2.04.

Thermal diffusivity [cm^2/s] copper is 1.11, zinc is 1.12 and aluminium is 0.97.

Table for eigenvalues λ_{1lp} by using mathematica software and heat flux continuity equation

λ_{1lp}	1	2	3	4	5	6	7	8	9
l=0	1	1.8459	3.0175	4.2309	5.3179	6.4619	7.7221	8.9035	9.9999
$l = 1$	1.0721	1.9642	3.0975	4.2870	5.3630	6.4985	7.7661	8.9287	10.0345
$l = 2$	1.5351	2.2750	3.2808	4.4770	5.4800	6.6081	7.8370	9.0130	10.100
$l = 3$	2.0785	2.7250	3.5930	4.7130	5.6830	6.7742	7.9931	9.1285	10.200
$l = 4$	2.6420	3.2504	4.0001	5.0630	5.9371	7.0268	8.2010	9.2969	10.3818
$l = 5$	3.20749	3.8181	4.4002	5.4820	6.3065	7.3286	8.4575	9.5301	10.5851
$l = 6$	3.7721	4.4040	4.8360	5.9590	6.6815	7.6726	8.7905	9.7655	10.8295
$l = 7$	4.3343	4.9935	5.3030	6.4729	7.1181	8.050	9.1345	10.061	11.111
$l = 8$	4.8929	5.5795	5.7940	7.0120	7.6030	8.4761	9.5210	10.3815	11.4477
$l = 9$	5.4469	6.1578	6.3040	7.5670	8.1211	8.6461	9.9635	10.7745	11.800

Fig.2 show the behaviour of temperature against r in three different layers for different values of t=0.1, $\theta = \frac{\pi}{2}$ and $\beta = 0$ it seen that the temperature at inner layer is suddenly increasing and then decreasing, in middle layer temperature slowly increasing and slowly decreasing, in the outer layer temperature slowly decreasing.

Fig.3 show the behaviour of temperature against r in three different layers for different values of t=0.5, $\theta = \frac{\pi}{2}$ and $\beta = 1$ it seen that the temperature at inner layer is suddenly increasing and then decreasing, in middle layer temperature slowly increasing and slowly decreasing, in the outer layer temperature slowly decreasing.

Fig.4 show the behaviour of temperature against time t for taking $\alpha = 1, \alpha = 0.8, \alpha = 0.5$ with $\beta = 0$ as initially time varies suddenly temperature increases and then after time t=1.5 very slowly temperature increases.

Fig.5 show the behaviour of temperature against time t for taking $\alpha = 1, \alpha = 0.8, \alpha = 0.5$ with $\beta =$ 1as time varies temperature increases.

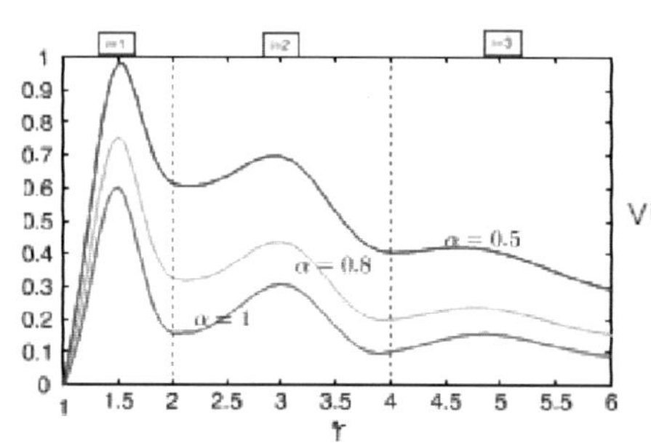

Fig.2 the behavior of $V_i(r, \theta, t)$ against r for $\alpha = 1, \alpha = 0.8, \alpha = 0.5$

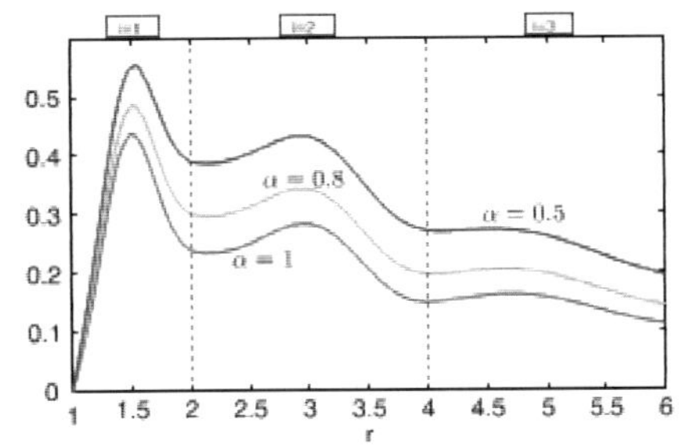

Fig.3 the behavior of $V_i(r,\theta,t)$ against r for taking $\alpha = 1, \alpha = 0.8, \alpha = 0.5$

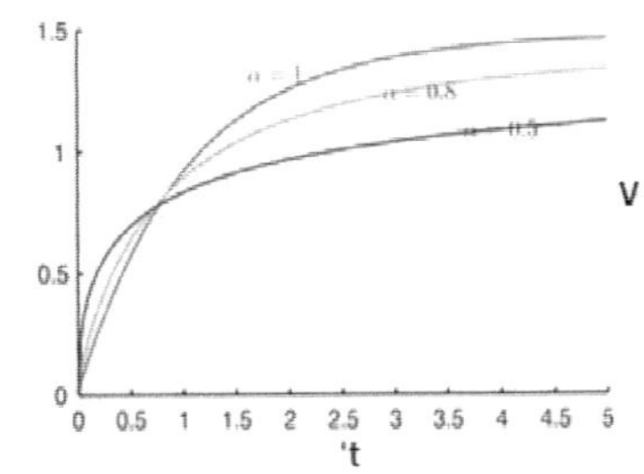

Fig.4 the behaviour of $V_i(r,\theta,t)$ against t for taking $\alpha = 1, \alpha = 0.8, \alpha = 0.5$

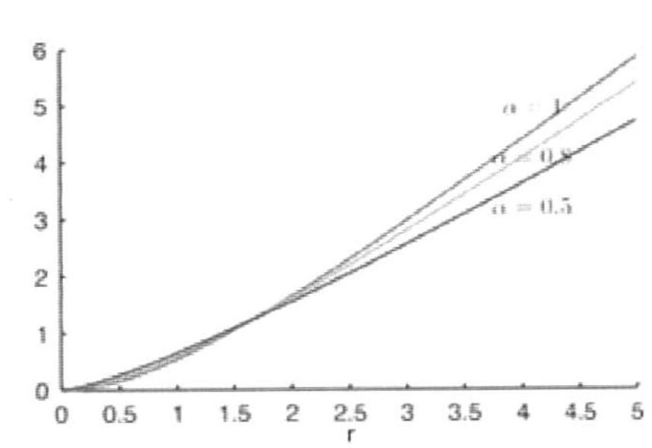

Fig.5 the behavior of $V_i(r,\theta,t)$ against t for taking $\alpha = 1, \alpha = 0.8, \alpha = 0.5$

Conclusion

In this article, by using Laplace transform & the finite integral transform technique obtained analytical solutions for this fractional partial differential equation in a two dimensional multilayer composite circular disc. By the use of caputo derivative for fractional order derivative is defined. Three layer problem studied numerically and graphically and result were discussed.

References

[1] Y. Yener, and M. N. Ozisik, On the solution of unsteady heat conduction in multiregion media with time dependent heat transfer coefficient, Proceedings of the 5th International Heat Transfer Conference,Tokyo, pp 188-192,(1974).

[2] H. S.Carslaw andJ. C Jaeger, Conduction of Heat in Solids, Oxford Clarendon, (1986).

[3] P.E. Bulavin, and V.M. Kascheev, Solution of the non-homogeneous heat conduction equation for multilayer bodies, Int.Chemical Engineering, Vol.1(5),pp.112-115 (1965).

[4] M. N. Ozisik, Heat Conduction, Wiley and sons, Inc,2nd Ed,(1993).

[5] 5.N.Noda, R. B. Hetnarski, and Y.Tanigawa, ,Themal Stresses, 2nd Ed. Taylor and Francis, New York, (2003).

[6] N.J.Wange, M N Gaikwad, and S P Pawar, Solution for One-Dimensional Transient Heat Conduction And Thermoelasticity in a Multilayer Hollow Sphere, International Journal of Mathematics and it's applications, 6(2–B), 315–327, (2018).

[7] Valery V. Vasiliev and Evgeny V. Morozov, Advanced Mechanics of Composite Materials, Elsevier, (2007).

[8] S P Pawar, N J Wange and M N Gaikwad, Thermoelastic Response to surface Temperature asymmetry and Internal Heat Generation Parameter in a Composite Spherical Body, International Journal of Mathematical Archive,9(3), 215-224, (2018).

[9] Stanislaw Kukla and UrszulaSiedlecka, Heat conduction problem in a two layered hollow cylinder by using the Green's function method, J of Applied Mathematics and Computational Mechanics, 12(2), pp 45-50, (2013).

[10]S. Singh, P. K. Jain and Rizwan-uddin, Int.J.Therm. Sci. 47, 261, (2008).

PLC-based Automated Aqua-Hydroponics System

Selvalakshmi S[1], Bharath Singh J[2*], Priyadharshini S[3], Sushmitha S[4]

[1&3]Assistant Professor, Department of Electronics and Instrumentation Engineering, R.M.K. Engineering College, Tamil Nadu.
[2]Associate Professor, Department of Computer Science and Engineering, Kalasalingam Academy of Research and Education, Tamil Nadu.
[4]Assistant Professor, Department of Electronics and Communication Engineering, Alpha College of Engineering, Chennai.

E.mail: j.bharathsingh@gmail.com

Abstract. Aquaponics refers to the growth of fish and plants in a single platform. As the demand for land increases due to urbanization, the growing of crops and other vegetables with less soil area is needed. Growing plants and aquatic animals in the same environment will save the usage of soil and the consumption of water levels will also be minimized. To automate the Aquaponics environment and to monitor the plant growth from the seed level to the fruiting level, a system is designed. This proposed method will help the user to monitor the water quality parameters like pH, humidity, and temperature and intimates the user about the current level, and alarms the user for any overshooting. The proposed system is designed, fabricated, and tested in laboratory conditions. The plant growth is monitored from the starting seed stage to the final fruiting stage. The essential nutrients required for plant growth are monitored and the discrepancy is satisfied accordingly. The bacteria and the microorganisms present in the soil decompose the plant waste into food that aquatic animals consume. The remains and excreta of the aquatic animals are sediments and given to plants as nutrition. Due to proper plumbing, water consumption is reduced. Different plants require different nutrient levels for their growth. The nutrient levels are listed and the test values are loaded into the microprocessor according to the plant selected. Here spinach is selected. The pH level, temperature level, and humidity levels of the soil and the water are monitored and the values are updated to the user using the web interface. From the web page, the user can get the parameter values and a comparison can be made accordingly.

1. Introduction

Aquaponics was coined in the early 1970s and became famous over the years. It was first adopted in countries where the land space is not sufficient to progress in cultivation. Aquaponics is a combination of aquaculture and hydroponics. Aquaculture refers to the growing of aquatic animals and harvesting them. Hydroponics is the technique of growing plants in soil-less conditions but rich in nutrients. In recent years hydroponics has gained the most popular due to the non-availability of land. We have to search for new agricultural techniques to serve the overpopulated world [12]. In order to feed the unnourished and poor people all over the world, the techniques like hydroponics are implemented. In general, Aquaponics is a technique to grow aquatic animals like fish and do agriculture at the same time. Agriculture means we can grow paddy as well as vegetables on our rooftops. The cost of Aquaponics is very less as we serve only the fish and the remains of fish are served as the input for agriculture. The fish excreta is rich in nutrients. This avoids the usage of fertilizers for our plants. The water usage and land usage in Aquaponics are very less when compared to aquaculture and agriculture separately. As we have avoided pesticides and other harmful fertilizers, the yield from the plants gives us healthier and organic foods. The measurement of soil quality parameters is also very important in agriculture. The yield will be more when the suitable plantation is done according to the type of soil. The soil quality is analyzed with various methods. Earlier, it was done

directly by the farmers by visiting the lands and inspecting them. But this method is not accurate. Then the physical method is used to find the soil depth and the types of pores. The biological method is used to find the microorganisms and bacteria present in the soil. Currently, we are using chemical methods to find the soil pH, salt content, chemicals, and composition present in the soil. This chemical method provides accurate results and it is a commonly used method [8].
The main components of Aquaponics are a rearing tank. This tank holds the fish. Here the fish is grown. Below the tank, there will be a settling bin for collecting the excreta from the fish and other fine materials. Then bio filter portion will help in converting ammonia into nitrates which the plants consume. The sump system helps to collect the cleaned water from the plants and then drive them back to the fish tanks. The water pumping and draining are taken care of by the DC motors. All the tanks are connected internally with the help of a plumbing system. The pipe materials are selected according to cost and quality. PVC pipes are commonly used because they are easily available and have moderate quality. The vinyl is of low quality but it is very cheaper. The steel and copper pipes are of high quality but it is very costly [4,6]. The comparison between hydroponics, Aeroponics, and Aquaponics shows that the saving of water, soil, and fertilizers are very less in Aquaponics. Aeroponics is the technique of growing plants in the air. The comparison is listed in the table and illustrated as a graph [3].

Table 1: Comparison of different techniques

Parameters	Hydroponics	Aeroponics	Aquaponics
Water Consumption savings	80	85	90
Fertilizer savings	55	85	90
Production	75	100	125

The comparison shows the advantages of Aquaponics when compared with Aeroponics and Hydroponics. Water consumption is more in Aquaponics and the usage of fertilizers is also less. Productivity shows a drastic improvement in terms of quantity and quality. The output is fertilizer and pesticide free.

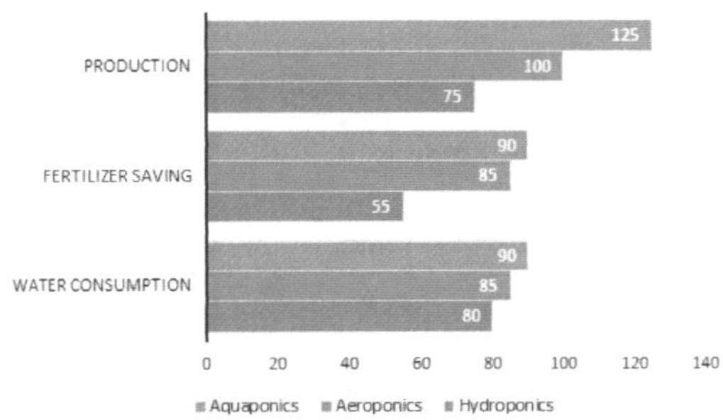

Figure 1: Comparison of different agriculture methods

The plants and fishes that are commonly grown in Aquaponics environments are listed. The common fish grown is Tilapia. This fish is inexpensive and do not have great taste. This fish needs less maintenance and it is consumed by more people. Any leafy vegetables like lettuce, and cauliflower and herbs like basil can be grown in the Aquaponics environment [7,10].
We have worked on this research paper in the following way. The introduction about Aquaponics and the importance of soil are discussed in the introduction part. The advancements in the research area are studied by conducting a detailed literature review and the difficulties of most of the authors faced are summarized. The proposed methodology is studied and the extensions are implemented. The hardware components are handpicked and the test circuit is drawn and simulated. The threshold values of the different parameters used are fixed after studying the soil parameters. Then the test circuit is transferred as a proposed model with hardware implementation along with the software programming. The interaction between the hardware and the web is carried out and the user is alerted to the values present as a graphical representation.

2. Literature Survey

Slamet W [15] has developed a system to monitor the radial growth of guava fruit using IoT. A reflective tape is wound on the guava fruit. The optoelectronic sensor senses the changes in the bar code of the reflective tape. The data collected is monitored and real-time analysis is done. The developed system can measure the radial growth of the fruit with a 2mm error and the success rate is 97.54% Mi-Hwa Song [11] has developed a smart gardening system that is able to auto-monitor the humidity and temperature of surroundings, air quality, and soil moisture. He has used Node MCU in the IoT platform to collect the data from the sensor. By analyzing the data the system is capable of initializing automatic irrigation of the soil when the soil moisture is low.

Ranjit Raut [13] has analyzed the specific relationship between Physio-chemical parameters, soil micronutrients, cropping patterns, and pH. From the analysis, he has found that the pH will affect the physical and chemical properties of the soil. So continuous irrigation, canal water, and the use of fertilizer have an adverse effect on the soil parameters. Shahana Firdous et.al [14] have analyzed 10 soil parameters statistically. Analysis of variance (ANOVA), cluster analysis (CA), and principal component analysis (PCA) was performed to evaluate correlation in soil quality parameters on spatiotemporal and vertical scales. This analysis has greatly reduced the sampling sites and helped in water management also.

A B Kalinin [1] has developed a system to eliminate soil compaction. He used digital measurement of parameters so that the soil state is monitored up to the depth of 60cm. The data obtained are used to select rational settings for sub-soil to fulfill the required indicators of the quality of tillage while minimizing energy consumption. Jason J. Danaher [9] examined the growth of plants and other aquatic animals by monitoring the water quality parameters. Abigail J. Lynch & Cooke S.J [2,5] informed the importance of fisheries in social and other related areas. He posted that the conservation of resources in the aquaculture field will yield good results.

3. Proposed Methodology

Due to overpopulation and the non-availability of the land surface for agriculture, the proposed model is fabricated. This proposed model is tested on the rooftop and the results are satisfactory. This model can be extended to any land area and the system will adapt accordingly. The monitoring of soil parameters and the service accordingly will help the user to get detained from the system.

3.1. Problem Identification

As the earth is insufficient of land due to buildings, industries, and residential complexes, an optimal solution to grow plants without soil and conservation of water, land, and fertilizers is proposed. Since Aquaponics is an emerging solution for the above-said issues, advancement in the proposed area is required. Moreover, the monitoring of soil parameters along with seed growth monitoring is not available in the market.

3.2. Proposed Model

The model has two modules. The first one is the control side, which has a processor along with the sensors. The sensor values along with the programming control the processor. The other side is the plant/fish side, where the parameters of the soil and water are monitored using sensors fixed. The whole setup values can be monitored by updating the values in the website using the Internet of Things.

3.2.1 Essential components of the proposed Method

The essential components are classified into two categories. Living and nonliving components. The living components consist of plants, fish, and bacteria. The other components are sensors, processors, connecting wires, pipes, fish tanks, and a soil bed. The living components are classified as fish and plants.

The fishes like Tilapia are commonly grown in the Aquaponics environment. It is a freshwater fish and it is commonly found in rivers, lakes, streams, and ponds. The reason for growing this particular fish is that it can tolerate any pH change, temperature change, and levels like ammonia and oxygen. Even if the levels are either up or down, this fish will grow for about 9 months and no harm will occur. The commonly grown plants in the Aquaponics environment are creepers (Watermelon, Cucumber, Bitter Gourd, and Ridge Gourd), Underground (Radish, Garlic Chives, and Carrot), Herbs (Celery, Rosemary, Parsley, Oregano, Thyme, and Sage) and plants (Brinjal, Capsicum, Green Chilli, Lady's Finger and Tomato) The specifications of the materials used are presented in Table 2.

Table 2: Specifications of the components

Components	Specifications	Remarks
Plastic Tray	51 x 27 cm	Rectangle
Piping	3 mm x 5 mm	Silicone
Fish Tank	48x21x28 cm	Glass
Layers	Clay, coco, net, Foodgrade	Growing area
Nutrition tank	Nitrogen, Phosphorous, and Potassium	
Processor	Supporting NodeMCU	
Sensors	Soil moisture, pH, Humidity	
Relays	4 channel	
motor	DC submersible	

3.2.2 Proposed model Block diagram and specifications

The proposed work is divided into two blocks. The main block is the control side which has sensors, processors, motors, light control, and IoT display interfacing. The raw seeds before germination are kept in coco-peat cups. These cups are used in gardening as a medium alternative to soil. It is produced from coconut fibers. The waiting period will be started until the seeds start germination. Once the waiting period got over, the germinated seeds are planted in the grow bed. The second block in our proposed system is the continuous monitoring of the germinated seeds. The seeds need enough water and nutrients for their growth. The water level needed is monitored by the moisture sensor and the nutrient levels are monitored by the pH sensor.

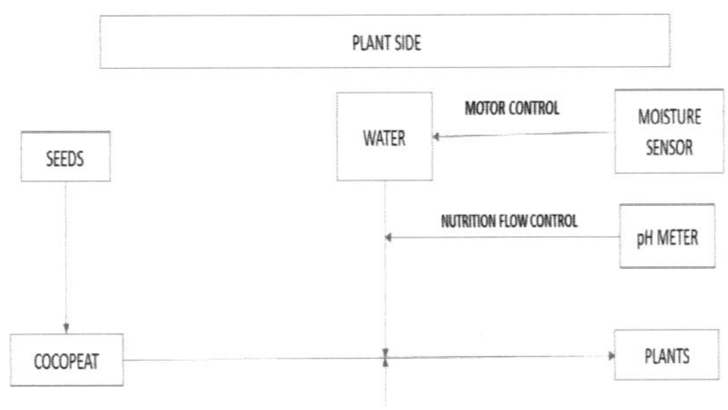

Figure 2: Block diagram – Plant Side

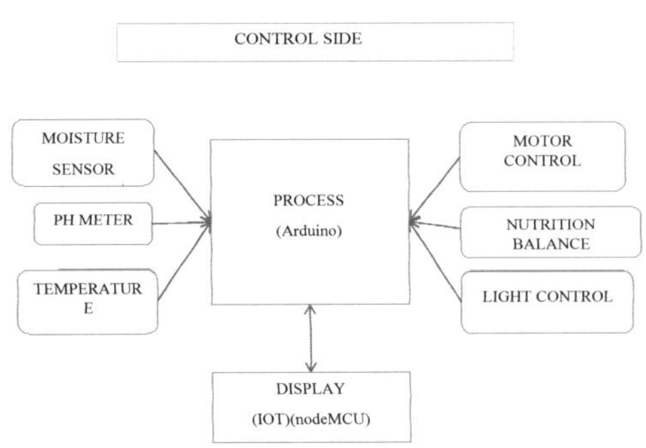

Figure 3: Block diagram – Control Side

This cycle of monitoring the temperature, pH, and humidity will be continuous and the measurements are transferred to the user through IoT. The proposed model is implemented by carrying out the workflow. Initially, the process is studied in detail. Then the necessary materials are collected and the specifications are noted. Next, the designing happened in two dimensional. Since the proposed model has both aquaculture and hydroponics, the fish tank and plant growing bed are fabricated using the materials collected. Then the connections are provided. PVC connections and the electronics side connections to the processor are done. Finally, the observations are noted and the analysis is done on the calculated values. The necessary parameters are displayed in the display device with the help of IoT.

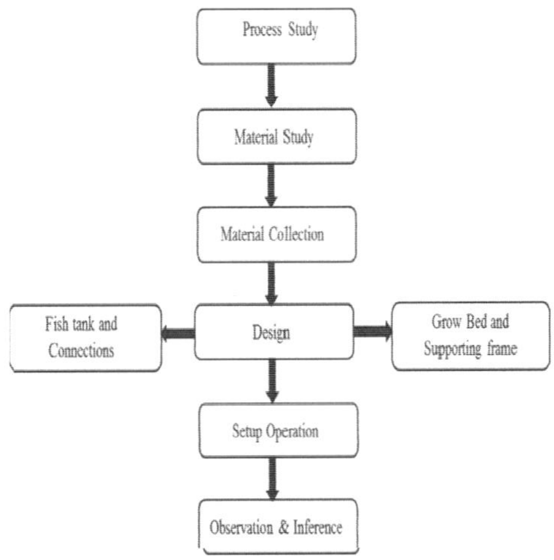

Figure 4: Workflow diagram of the proposed model

3.2.3 Hardware and Software design

The parameter measuring sensors used are pH sensor, Humidity sensor and moisture sensor. The processor used is Arduino Mega and fan is used as a temperature controller. Four subimersible motors are connected to the processor using the 8 channel relay module. The relay circuit helps in connecting the interfaces like motors, bulbs to the processor and the interfaces can be controlled by the processor through the relays.

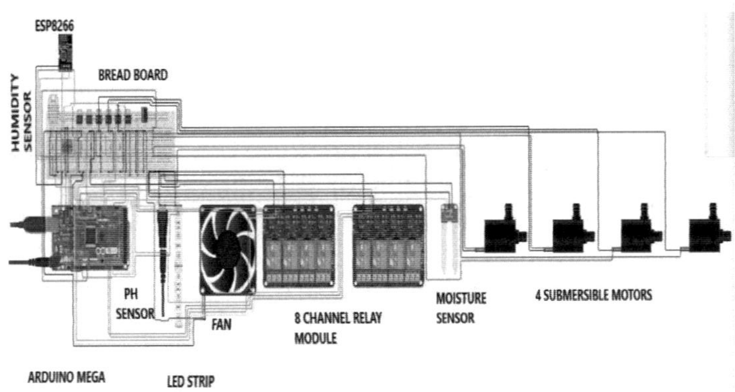

Figure 5: The hardware design of the proposed model

The other components like the fish tank and the plant growth bed are designed as per requirements. The Delta PLC programming software is used for the simulation of the proposed system. This software is preferred as it can be used for large-scale applications. The PLC ladder diagram is drawn for easy understanding of the process.

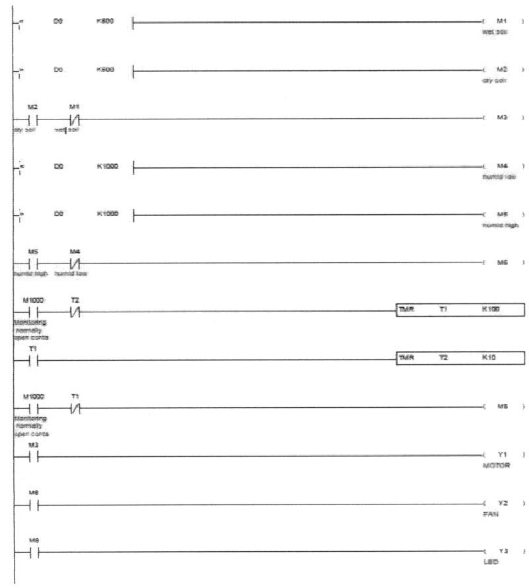

Figure 6: PLC Ladder Logic Program

3.2.4 Target values of the parameters

The water quality parameters mainly monitored in Aquaponics are dissolved oxygen, pH, temperature, hardness, and the dissolved nitrogen components. The growing cycle of any plant has five phases. The amount of water and other nutrient requirements will vary accordingly. The five phases of the plant life cycle are germination to a sapling (P1), vegetative growth beginning (P2), advanced stage of growth (P3), flowering stage (P4), and fruit formation (P5).

Table 3: Nutrient level requirements

Plant Varieties	pH Zone	Nutrient level for each growth stage NPK (ml/L)				
		P1	P2	P3	P4	P5
Coriander	6.0-7.0	0.5	1	1.5	2	2.5
Mint	5.5-6.5	0.5	1	2	2.5	3
Sprouts	6.0-7.0	1	2	3	4	5
Ladies Finger	6.0-7.0	0.5	1	2	2.5	3
Tomato	5.5-6.5	1	2	3	4	5.5
Spinach	**6.0-7.0**	**0.5**	**1**	**1.5**	**2**	**3**
Peppers	5.5-6.5	0.5	1	2	2.5	3
Onions	6.0-7.0	0.5	1	1.5	2	2.5
Capsicum	6.0-7.0	0.5	1	2	2.5	3
Beans	5.5-6.5	1	2	3	4	5.5

The pH levels and the requirement of nutrient levels will vary for plant varieties. We have used spinach as the plant variety for the proposed model. The ph value must be maintained from 6-7 and the nutrient level increases for different phases. Growing tomato in an Aquaponics environment is difficult because the nutrient requirements are very high. The fixed target values are monitored by the sensors and the deviation is corrected and intimated to the user by the web interface.

3.2.5 Experimental setup of the proposed system

After the initial designing of hardware and software is completed, the experimental setup is fabricated accordingly. As the proposed system involves setting up the fish tank and the plant grow bed simultaneously, great care must be taken. The piping system is carefully designed and the water leakages must be checked as the leakage may damage the wiring connections.

Figure 7: Plant growth bed setup

Figure 8: Hydroponics Fish tank setup

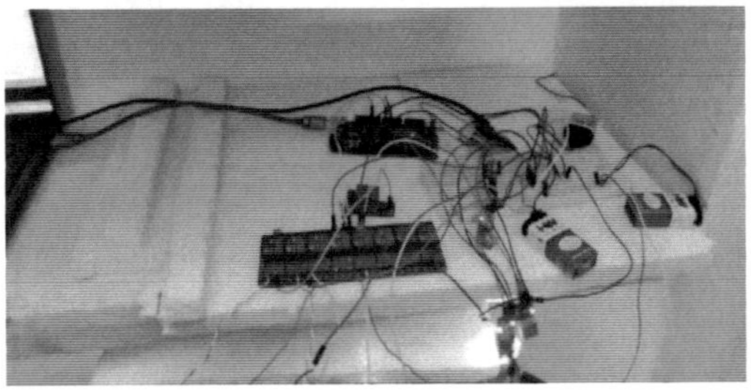

Figure 9: The hardware setup of the proposed system

The proposed system has three stages. The first stage is the preparation of the growth bed. The plant germination stage is monitored separately and the plants are shifted to the growth bed after germination. The growth bed is filled with clay pebbles. The second stage is hydroponics. The fish tank is taken for prototype and the parameters are monitored accordingly. The final stage is automation. The essential parameters are monitored using the sensors and the control of the parameter target values are done by the processor and the user will be provided with real-time values using the Internet of Things.

Figure 10: Final experimental prototype of the proposed system

4. Inferences from the results

The final prototype is designed and fabricated. The test values are loaded into the microprocessor using coding. The web interface is initiated and the values of pH and humidity will be plotted as a graph. The graph is provided with values on the y-axis and time on the x-axis. The target levels can be fixed and the user is alarmed by the faulty values. The target values of pH for our growing specimen is 6-7.

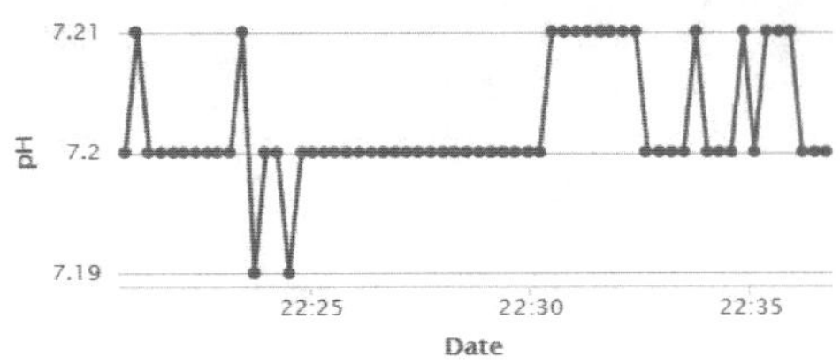

Figure 11: pH Measurement

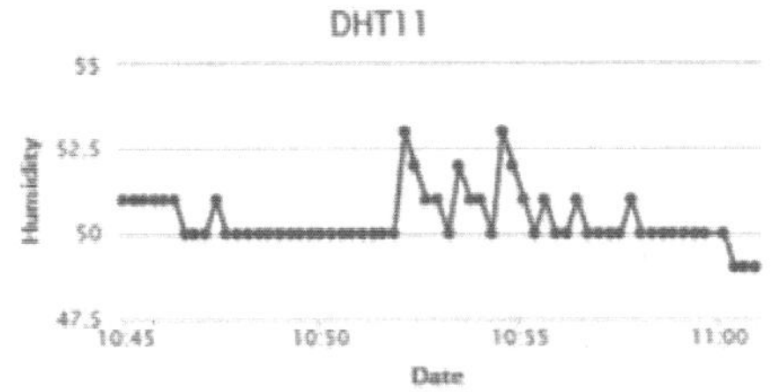

Figure 12: Humidity Measurement

5. Conclusion and future scope

The agenda of the designed system is to promote Aquaponics in an automated scheme. Since Aquaponics is an emerging area in the overpopulated world, automation and monitoring of the parameters which is necessary for the growth of plants and aquatic animals will provide good yields. The proposed work helps the user to monitor the growth of plants from the seed stage to the fruiting stage. Moreover, the main parameters of water like pH, temperature, and humidity are continuously monitored and displayed to the user through the web interface. The target values of the parameters can be updated in the microcontroller and the user will be alarmed when the values get overshot.

6. Acknowledgement

We would like to thank the management of R.M.K. Engineering College and Kalasalingam Academy of Research and Education for providing the required resources.

7. Compliance with Ethical Standards

7.1 Funding

The authors declare that they have no conflicts of interest to report regarding the present study.

7.2 Ethical Conduct

This chapter does not contain any studies with human participants or animals performed by any of the authors

8. References

[1] A B Kalinin, I Z Teplinsky2021, V A Ruzhev, V A Kalinina and V E Gerasimova, Methods and means of digital measurement of soil parameters and conditions of functioning of tillage machines for deep loosening of soil, *IOP Publishing*, 2021.

[2] Abigail J. Lynch, Steven J. Cooke, Andrew M. Deines, Shannon D. Bower, David B. Bunnell 2016, The social, economic, and environmental importance of inland fish and fisheries, *Environmental Reviews*, Volume 24, Number 2, June 2016.

[3] Ali AlShrouf 2017, Hydroponics, Aeroponic and Aquaponic as Compared with Conventional Farming, Environmental Science, *American Scientific Research Journal for Engineering, Technology, and Sciences*, January 2017

[4] Beecher LE 2021, Aquaponics: system layout and components. *Clemson (SC): Clemson Cooperative Extension, Land-Grant*, 2021. LGP 1123.

[5] Cooke S.J., Paukert C., and Hogan Z. 2012. Endangered river fish: factors hindering conservation and restoration. *Endang. Species Res.* 17: 179–191.

[6] D. Allen Pattillo1, 2017, An Overview of Aquaponic Systems: *Aquaculture Components Iowa State University Extension and Outreach Technical Bulletin*, October 2017

[7] Espinosa-Moya, Azucena; Álvarez-González, Alfonso; Albertos-Alpuche, Pedro; Guzmán-Mendoza, 2018, Growth and development of herbaceous plants in aquaponic systems, *Acta Universitaria*, vol. 28, no. 2, 2018.

[8] Hamid kheyrodin, 2014, Important of soil quality and soil agriculture indicators, *Academia Journal of Agricultural Research* 2(11): 231-238, November 2014.

[9] Jason J. Danaher, Charlie R. Shultz, James E. Rakocy 2011, Effect of a Parabolic Screen Filter on Water Quality and Production of Nile Tilapia (Oreochromis niloticus) and Water Spinach (Ipomoea aquatica) in a Recirculating Raft Aquaponic System, *International Journal of Recirculating Aquaculture* 12(1).

[10] KK Gloger, K.C., J.E. Rakocy, J.B. Cotner, D.S. Bailey, W.M. Cole and K.A. Shultz. 1995, Waste treatment capacity of raft hydroponics in a closed recirculating fish culture system. *World Aquaculture Society, Book of Abstracts*: 126-127.

[11] Mi-Hwa Song, 2020, A Study on IoT-based Real-Time Plants Growth Monitoring for Smart Garden, *International Journal of Internet, Broadcasting and Communication*, Vol.12 No.1 130-136, 2020

[12] Overpopulation: a threat to sustainable agriculture and food security in developing countries? *A Review*, April 2017, DOI: 10.13140/RG.2.2.20613.04325

[13] Ranjit Raut, Prashant Harale and Anil Kurhe 2020, Studies on Soil Quality Parameters in Relations to Cropping Patterns, Micronutrients and pH from Goagalgaon area in Ahmednagar District of Maharashtra, India, *International Journal For Innovative Research In Multidisciplinary Field*, Volume - 6, Issue - 9, Sept – 2020.

[14] Shahana Firdous & Shaheen Begum & Azra Yasmin 2016, Assessment of soil quality parameters using multivariate analysis in the Rawal Lake watershed , *Research gate*, DOI 10.1007/s10661-016-5527-5, Aug 2016.

[15] Slamet W, M Irham N and Sutan M S A, 2018, IoT-based Growth Monitoring System of Guava Fruits, *IOP Publishing*, doi:10.1088/1755-1315/147/1/012048, 2018

Impact of materials on performance of Vine Robot

Archit Borkar[1,*], Shubham Bene[2], Mahesha Gonal[3] and B.S. Manohar Shankar[4]

[1-4]Department of Mechanical Engineering, Goa College of Engineering, Farmagudi, Ponda, Goa 403401, India.
*Corresponding author: *architborkar9@gmail.com*

Abstract. Vine Robots are soft continuum robots designed with low-cost fabrication and for the navigation of difficult environments. Due to their movement patterns resembling those of natural vines, these robots also known as "everting vine robots" grow by pressure-driven eversion. This paper presents the impact of different materials on the performance of Vine Robot. Three of the most easily accessible materials namely Transparent LDPE (125 Microns), Black LDPE (150 Microns) and Tarpaulin HDPE (250 Microns) were acquired and utilized in fabricating the body of Vine Robot. Under the application of these materials, difference in various properties of the robot were studied. By systematic study, the suitability of one material over the other was determined in specific environmental scenarios.

1. Introduction

Vine Robot is a subset of Soft Robotics, these robots are called so because of their similarity to plants (vine) in their growth-trailing behaviour. Due to their growth-based movement, vine robots are well suited for navigation and exploration in cluttered environments[1]. Non-destructive exploration of small areas remain difficult for existing robot designs in a number of possible robotic applications, including inspection, medicine, search and rescue, and archaeology. Vine robots can potentially fill this need for robots able to move in highly constrained environments[2].

The technology has gone through a lot of innovations and upgrades over the year and extensive research work is still being conducted in the field of vine robots based on the control systems, material, construction, modelling, and sensors being used. Researchers fabricated the robot using soft lithography whose simplicity allowed it to iterate the design rapidly[3].

Researchers fabricated a soft robot body driven by a pneumatic pump[4] and used it to evaluate the rate of lengthening by changing the internal pressure and derived a relationship between the two. Control chambers were used to selectively permit one side of the robot body to extend in comparison to the other side in order to accomplish active direction control. Researchers have also demonstrated advancements on control systems, material, modelling, and sensors[5]. Researchers have further investigated the retraction of robot body along with a model predicting when controlled retraction is and is not possible, and present a design to aid in controlled retraction[6].

While unique designs could call for labor-intensive and complicated manufacturing procedures, the majority of everting vine robots were built in a very short amount of time. Everting vine robots were created in the most basic form by sealing one end of a tube and reversing this sealed end inside the body[7]. Researchers have also been successful in developing a general model form, The Lockhart-Ortega equation, which relates the rate of volume change to the difference between the turgor pressure and yield pressure. A modification of this viscoplastic model of plant cell extension described by Green relates pressure to a linear extension rate as shown in equation (1).

$$r = \varphi\{P - Y\}^n \tag{1}$$

where r is the linear extension rate, Y is the yield pressure below which no extension occurs, P is the internal pressure, φ is the extensibility, and n is a power term close to unity. The model describes a monotonically increasing relationship between pressure and extension rate with an offset[8].

This study investigates the use of various conventional materials in fabricating the body of Vine Robot. These materials are inextensible enough to produce eversion as opposed to radial expansion upon pressurization and that are both fluid impermeable and sealable, such that a closed pressure vessel can be developed. By making the use of materials like LDPE and HDPE sheets, the impact on the performance of Vine Robot was studied by plotting quantities such as Average Displacement in time, Diameter of body tube, Pressure input to the robot and Lengthening Rate.

2. Materials and Methods

For the effective functioning of the Vine Robot, a base station was fabricated to hold all the circuit components of the robot along with the confined body of the robot. The body was coiled around the spool connected to a DC Motor. Microcontroller system was used to manage Pressure input as well as the motor to cause eversion and inversion of the Vine Robot. In addition to this, series pouches were sealed along the sides of the body which allowed the steering of the Robot body due to inflation on one side with respect to the second side.

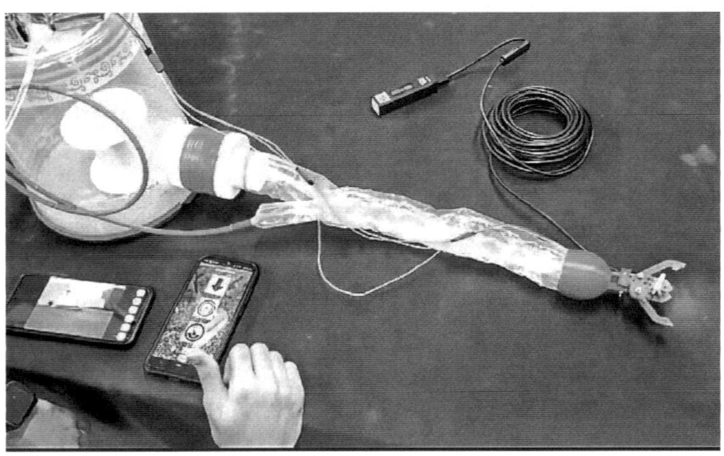

Figure 1. Fabricated Vine Robot

Initially, transparent LDPE (125 Microns) was utilized in making the body of the robot. The material was sealed using conventional plastic heat sealer, and a long 2m tube was created. An air compressor rated 1/3HP and 1440RPM was utilized along with a pressure regulator to specify a particular pressure level. By varying the pressure input to the robot, the distance travelled within a constant time was measured thus achieving the extension speed. The table was then plotted in order to relate the results to the General Model Form or Lockhart Ortega Equation.

For Polyethylene Y=13.3kPa, $\varphi = 0.75\text{m/s-kPa}^{-n}$, n= 1.1 and Extension speed(r) = $\varphi\{P - Y\}^n$.

Table 1. Pressure v/s Extension Speed

Pressure (kPa)	Extension Speed (m/s)
27.6	13.9
41.4	27.4
55.2	45.6

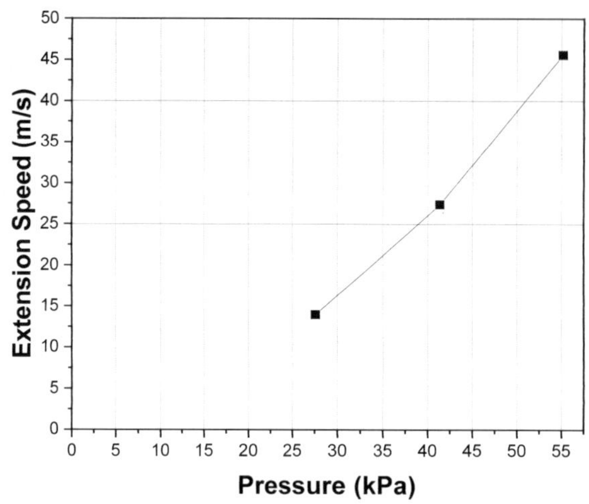

Figure 2. Experimental plot for General Model Form

As shown in 'figure 2' the results obtained in the experiment were found satisfying the Lockhart Ortega Equation as plotted by researchers in past.

Same material was utilized with three different Diameters to measure Average Displacement travelled in each case. A similar experiment was conducted by changing the Pressure input to the robot and Average Displacement was noted down in each case. Lengthening rate was also plotted against Diameter of the body. The same experiment was repeated with two other materials, Black LDPE (150 Microns) and Tarpaulin HDPE (250 Microns). Using the obtained results, the performance of the Vine Robot was studied due to the influence of different materials.

3. Results and Discussions

The values measured in the experiment conducted are tabulated below, and are plotted on a graph to understand the difference in performance of the Vine Robot.

Three tubes with different diameters were sealed out of all the three materials. A constant pressure was supplied each time to observe the Average Displacement of the tip of the Robot at a constant time. Constant Pressure = 27.6 kPa (4 psi), Constant time = 5 Seconds and conversion of Diameter 3 in = 0.08 m.

Table 2. Average Displacement v/s Diameter in Transparent LDPE

Diameter (m)	Average Displacement (m)	Initial Time to inflate (sec)
0.08	0.05	1.5
0.09	0.02	2.5
0.10	0.02	4.5

Table 3. Average Displacement v/s Diameter in Black LDPE

Diameter (m)	Average Displacement (m)	Initial Time to inflate (sec)
0.08	0.03	1.5
0.09	0.03	2.5
0.10	0.01	5

Table 4. Average Displacement v/s Diameter in Tarpaulin HDPE

Diameter (m)	Average Displacement (m)	Initial Time to inflate (sec)
0.08	0.03	4
0.09	0.02	4.5
0.10	0.01	6.5

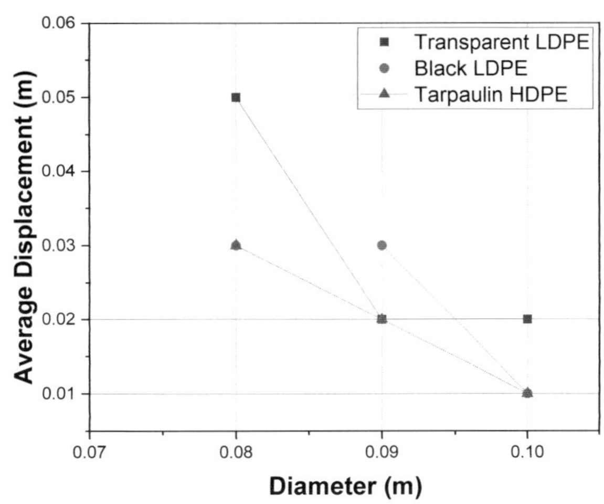

Figure 3. Average Displacement v/s Diameter

For the next set of readings, a constant diameter of the tube was selected for all three materials and corresponding Average Displacement was measured for a constant time.

Constant Tube Diameter = 0.1 m (4 in), constant time = 5 Seconds and conversion of Pressure 4 psi = 27.6 kPa.

Table 5. Average Displacement v/s Pressure in Transparent LDPE

Pressure (kPa)	Average Displacement (m)	Initial Time to inflate (sec)
27.6	0.018	4.5
41.4	0.02	3.5
55	0.025	2.5

Table 6. Average Displacement v/s Pressure in Black LDPE

Pressure (kPa)	Average Displacement (m)	Initial Time to inflate (sec)
27.6	0.01	5
41.4	0.025	5.5
55	0.03	3

Table 7. Average Displacement v/s Pressure in Tarpaulin HDPE

Pressure (kPa)	Average Displacement (m)	Initial Time to inflate (sec)
27.6	0.015	6.25
41.4	0.018	4.75
55	0.02	4

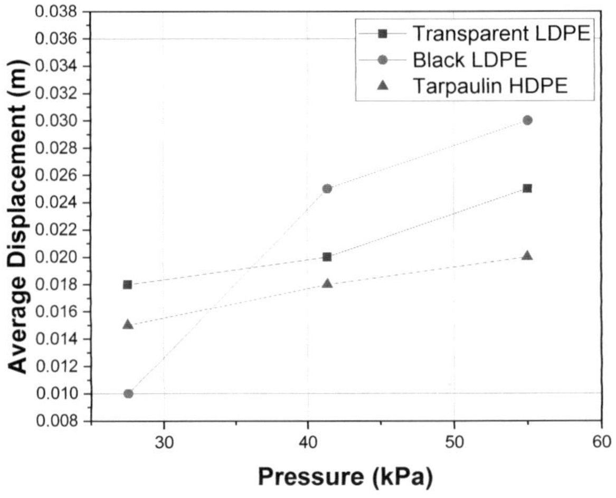

Figure 4. Average Displacement v/s Pressure

To study the change in lengthening rate of the robot, air pressure of 27.6 kPa (4 psi) was maintained and the Average Displacement was observed in each case of tube diameter for all three materials. Using the Average Displacement, lengthening rate was determined using the equation (2):

Lengthening rate = Average Displacement / (Total time − Initial time to inflate) (2)

Table 8. Lengthening Rate v/s Diameter in Transparent LDPE

Diameter (m)	Initial Time to Inflate (sec)	Average Displacement (m)	Lengthening Rate (m/sec)
0.08	1.5	0.055	0.011
0.09	2.5	0.0225	0.0045
0.10	4.5	0.0175	0.0035

Table 9. Lengthening Rate v/s Diameter in Black LDPE

Diameter (m)	Initial Time to Inflate (sec)	Average Displacement (m)	Lengthening Rate (m/sec)
0.08	1.5	0.03	0.006
0.09	2.5	0.03	0.006
0.10	5	0.01	0.002

Table 10. Lengthening Rate v/s Diameter in Tarpaulin HDPE

Diameter (m)	Initial Time to Inflate (sec)	Average Displacement (m)	Lengthening Rate (m/sec)
0.08	4	0.025	0.005
0.09	4.5	0.0225	0.0045
0.10	6.25	0.015	0.003

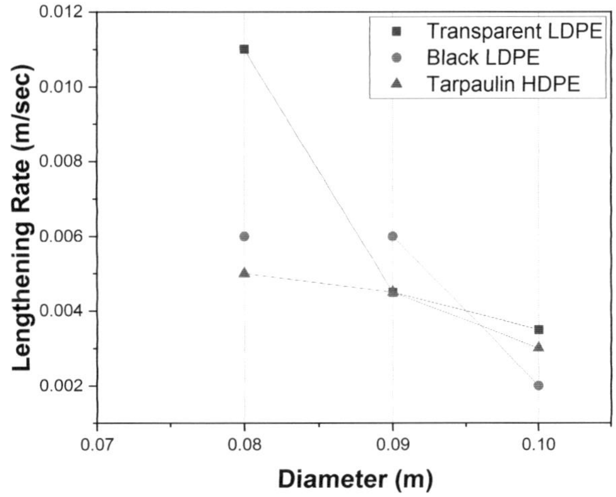

Figure 5. Lengthening Rate v/s Diameter

From the 'figure 4', it was evident that a larger diameter of the robot body slowed down the displacement achieved by the robot because of the larger volume of air to cover thus slowing down the speed of travel. In contrast, with increase in Pressure, the speed was seen to be increased as seen in 'Figure 5'. Lengthening rate being directly proportional to the Average displacement, gave a similar plot to Average displacement v/s Diameter. However, by observing all the graphs, it can be noted that Tarpaulin HDPE material shows a very small change in its speed with change in diameter and pressure thus making it a suitable option in case of stable travel. Furthermore, the strength of the material was seen to be protecting the material in sharp and rough terrains. However, for the customised applications of the robot, transparent LDPE material was seen to be most useful as it was found to be the easiest to prototype among the three.

4. Conclusions

From the observations and results it is concluded that-

1. In applications where higher displacement is required at a given pressure, the robot having smaller inflated diameter has to be used since the displacement decreases as the diameter increases. In such a situation, lighter material like transparent LDPE can also be prioritized.
2. In applications where the space of the robot is fixed that is the inflated diameter of the robot is fixed, in such a scenario to achieve higher displacement, either a higher internal pressure can be used or lighter material can be employed.
3. Lengthening rate is defined as the change in length per unit change in time. It is observed that at a given pressure as the diameter increases the lengthening rate decreases.
4. From the plots of lengthening rate v/s diameter it was observed that LDPE material delivers a higher lengthening rate in comparison to the other two materials. However, in rough terrains, LDPE got punctured due its thin and light structure. In rough terrains, HDPE material was seen to be performing well due to its thin yet strong structure.

5. References

[1] Coad, Margaret M., et al. "Vine robots: Design, teleoperation, and deployment for navigation and exploration." IEEE Robotics Automation Magazine 27.3 (2019): 120-132.
[2] El-Hussieny, Haitham, et al. "Development and evaluation of an intuitive flexible interface for teleoperating soft growing robots." 2018 IEEE/RSJ International Conference on Intelligent Robots and Systems (IROS). IEEE, 2018.
[3] Shepherd, Robert F., et al. "Multigait soft robot." Proceedings of the national academy of sciences 108.51 (2011): 20400-20403.
[4] Hawkes, Elliot W., et al. "A soft robot that navigates its environment through growth." Science Robotics 2.8 (2017): eaan3028.
[5] Walker, James, et al. "Soft robotics: A review of recent developments of pneumatic soft actuators." Actuators. Vol. 9. No. 1. MDPI, 2020.
[6] Coad, Margaret M., et al. "Retraction of soft growing robots without buckling." IEEE Robotics and Automation Letters 5.2 (2020): 2115-2122.
[7] Blumenschein, Laura H., et al. "Design, modeling, control, and application of everting vine robots." Frontiers in Robotics and AI 7 (2020): 548266.
[8] Blumenschein, Laura H., Allison M. Okamura, and Elliot W. Hawkes. "Modeling of Bioinspired Apical Extension in a soft robot." Conference on Biomimetic and Biohybrid Systems. Springer, Cham, 2017.

Case study of electrical, structural and optical properties of CdSe Quantum Dots using DFT based computational tools.

Abyaz Abid

Department of Mechanical and Production Engineering, Islamic University of Technology, Board Bazar, Gazipur 1704, Bangladesh

Corresponding author: abyazabid@iut-dhaka.edu

Abstract.
Recent advancement in the field of nano-structured electronic devices has finely tuned up various progenition and manufacturing process of Quantum Dots that can directly harvest energy from solar power. Duly acknowledging different types of procuring processes, the paper primarily focuses on eight different CdSe (Cadmium Selenide) Quantum DOT samples which are initially synthesized in the material testing lab from the available precursor. Structural and electrical properties of these synthesized nano materials are inspected using DFT (Density function theory) based commercial tool - ESPRESSO. Later on, different photo luminous spectroscopic methods are used to understand the optical characteristics of the CdSe based nanomaterials. Light scattering method is applied to comprehend greater insights about the same sampled nano materials based on different parameters such as – polydispersity index (PI), Stokes shift analysis, zeta potential etc. Finally, the most equipped CdSe Quantum DOT sample is chosen for future purpose and a sensitive case study covering diverse advantages, disadvantages and future challenges of QDOTs is carried out.

Keywords: Quantum Dots, Nanomaterials, Cadmium Selenide, Density function theory, Zeta potential

1. Introduction

The materials that possess nanostructure are of immense interest due to their potential to bridge the chasm between the macroscopic and atomic scales, opening up entirely new application possibilities, particularly in electronics, optoelectronics, and biology [1]–[3]. Apart from these industries, nanomaterials are also being extensively used in agricultural production, feed materials and in food additives to meet the ever increasing demand of human consumption [4]. Nanotechnology has been centred on Quantum Dots, nanoscale semiconductor particles with exceptional optical, electrical, and photonic properties. This field of engineering is relatively new. Since their creation, quantum dots have been used and researched for a wide range of applications, including cancer detection, drug delivery, therapy, a replacement for organic dyes, and also as effective fluorescents because of their versatility in terms of size and applications. [5]. QDs are zero-dimensional in comparison to the bulk, and the density of states (DOS) of QDs contains discrete quantized energies as a result of the limited number of electrons. Despite being zero-dimensional, quantum mechanics views it as a box; the size of the box is crucial. Sometimes a single electronic charge in a Qdot acts to repel the addition of another charge, causing an I-V curve DOS. The radius of the Qdots is reciprocal to the step size of the staircase. Due to its optoelectronic and optical properties, we concentrate on the compound semiconductor-based nanostructured material CdSe and its multimodal applications in this lab. [6], [7].

In 1932, Rocksby used x-ray diffraction (XRD) to identify that the colours were caused by crystallites of CdS and CdSe [8]. In 1982, Efros and Efros proposed that by altering the size or stoichiometry of CdS_xSe_{1-x} [9]. Rosetti et al. explained the colour change of colloidal semiconductor solutions in 1991[10]. Several distinct methods of synthesis were developed during this time period [11]–[14]. A new nano structured QD WO_3 was synthesized by novel pulsed leaser ablation method and the morphological investigation was carried out in [15]. Colloidal dispersion method was used to synthesize InP, a very narrowly sized quantum dot made from the precursor of chloroindium oxalate complex solution $P(SiMe_3)_3$ by Olga I and et al [16]. Ethylene diamine tetra acetic acid (EDTA) was introduced as a facilitator to produce PbS quantum dot without any aggregation of Pb^{2+} - solving an agelong problem in the field of integrating nano structured quantum dots [17], [18].

Density function theory (DFT), which was developed after the discovery of quantum mechanics in the latter 20th century, has emerged as a remarkable tool for investigating the optical, electrical, structural, and mechanical properties of nanostructured materials. Although the Schroedinger's equation solution for a single electron is the fundamental idea behind DFT-based tools, ongoing improvements and sophisticated approximations from numerous predecessors have made it reliable to investigate the aforementioned properties of any system [19], [20]. A density function-based simulation tool is significant because it can foretell a variety of behaviours and properties of a system made up of parallel layers of different materials, even when the environment is constantly changing, and that too without the input of any experimental data. The fundamental principle of this DFT-based modelling is that, rather than focusing on a single electron, electron density of any system is taken into account, which can represent all of the system's ground state properties, including energy, diffusivity, etc. Numerous projects have used a customised DFT-based method up to this point [21], [22]. Shiqi Zhou and co-workers have proposed a new use for this DFT-based simulation technique, in which he used classical density function theory to bridge the gap between uniform and non-uniform systems by demonstrating alignment of the first order and second order direct correlation functions. [23]. By developing a thorough understanding between time-dependent and classical density function theorems, different luminescent properties, such as UV spectra, IR spectra, etc. are investigated for metal-organic framework (MOF-5) [24][25][26]. Svetlana V. Kilina and colleagues have proposed a novel modification of the fundamental electron-photon interaction to improve the efficiency of CdSe and PbSe QD based solar cells by generating multiple bottleneck relaxations between electronic and vibrational frequencies. [27].

Energy harvesting nanoparticles has greatly increased due to the desire to comprehend many fundamental Qdot properties and commercialization efforts.[28]–[30]. In this lab, commercially available precursors in solution were used to create the robust and well-studied colloidal nanoscale material known as cadmium selenide quantum dots (CdSe QDs) under inert conditions at high temperatures. In this project, the structure-electrical properties of one of the well-known nanomaterials which has a hexagonal Crystal System, were studied using DFT-based software called Quantum ESPRESSO. Previous material science projects provided the background information on CdSe, and the outcomes of this project will aid in the computational modelling of devices based on quantum dots. To describe the optical characteristics of these nanocrystals, spectroscopy techniques such as UV-visible and photoluminescence were employed. These nanocrystals were cleaned up, and then dynamic light scattering techniques were used to characterise them. A quantum dot sample was finally made for use in applications that harvest and emit light.

2. Methodology

- **Synthesis of CdSe nanoparticles for hydrogen evolution reaction**
  - I. Selenium precursor was prepared using 60 mg of selenium powder; 4.0 mL of 1-octadecene and 1.0 mL of Trioctylphosphine.
  - II. Cadmium precursor was prepared in a 25 mL three-neck flask using 134 mg of Cadmium oxide, 8 mL of 1-octadecene (90%), and 1.20 mL oleic acid. These mixtures were stirred in a vacuum environment for 5 minutes at 1500 rpm.
  - III. After 5 minutes in a vacuum, nitrogen flow was again started and the whole apparatus was wrapped in glass wool to insulate.
- **Synthesis of CdSe quantum dots**
  - i. 3 mL of toluene was taken in a set of 11 cultivation tubes.
  - ii. After cadmium precursor solution reached 225 °C, waited till solution starts to turn clear.
  - iii. 5 mL of selenium precursor was rapidly injected.
  - iv. 8 samples of aliquots (at the reaction time of 8s, 16 s, 32 s, 1 min 30 s, 2 min 30 s, 3 min, 4 min, 5 min) were extracted over the course of the 5 min reaction. The final aliquot was taken when the temperature finally falls 150 °C. And the set of aliquots was observed with the UV-lamp.
- **Purification**
  - I. Approximately 15-20 mL of CdSe in octadecene was carefully poured into a plastic 50 mL centrifuge tube when the reaction mixture reaches 100 °C.
  - II. 7- 10 mL of CdSe dots was transferred to a second centrifuge tube. And 15 mL of ethyl acetate were added to each centrifuge tube. Centrifuge tubes were filled to the 45 mL mark with methanol to precipitate the quantum dots. Following this, the tubes were centrifuged for 10 min at max RPM to crash out the particles.
  - III. The organic waste was Poured off and resuspended the quantum dots in about 3 mL of toluene. Following that, ~ 7.5 mL of ethyl acetate was added. The tubes were filled to 30ml with methanol and centrifuged to crash out particles.
  - IV. Repeated step 3 for 3 times.
  - V. Resuspend the particles in 10 mL of toluene and carefully pour each 10 mL sample of QDs into its own scintillation vial.
- **Characterization**
  - i. The collected samples were characterized using UV-Vis and DLS.

3. Results

Optical, electrical and structural properties of the synthesized CdSe QD are enlightened following two methods – Photoluminescence (PL) analysis method and Dynamic Light Scattering (DLS) set-up. A non-contact and non-destructive material probing system such as PL analysis captures the luminescence spectrum generated from CdSe QD as it falls to the ground state from excited state. To understand the structural properties and size variation of the QD in the solution, DLS method is followed and the outcomes are presented in terms of their comparative zeta potential and dispersity index.

3.1 Absorbance and photoluminescence (PL) analysis

Photoluminescence analysis is done to understand the opto-electrical properties of the integrated quantum dots. These characteristics are represented for eight different samples with respect to their invigilation time and in terms of their absorbance of light irradiated from a laser beam and emission upon the cessation of it.

3.1.1 Absorbance vs Wavelength plot

When a number of luminescent spectra are reflected off the solution, the quantum dot absorbance phenomenon occurs. The QD sample elevates to a higher energy level after absorbing the photon. The absorbance of the synthetic QD is examined eight times during this demonstration in order to fully grasp its absorbance potential. The absorbance of the quantum dot for various wavelengths at their corresponding times is shown in Fig. 1. Even though the overall absorbance of the QD initially rises with the passage of time, a general decaying trend of absorbance can be seen for any specific wavelength (between 550 nm and 625 nm) at its specific time. This merely asserts that the CdSe QD sample is better equipped to absorb energy at shorter wavelengths than it is at longer wavelengths. According to Fig. 1, the wavelength (320 nm) exhibits a much higher absorption (4.5) when exposed to light for longer than 5 minutes. Initially, it exhibits an absorbance of about 0.50 for its initial 8 second exposure to light. It is easier to argue that the CdSe quantum dot exhibits higher absorption when its exposure time to light is increased because the other six time-variant samples drift similarly. On the other hand, for any particular time-dependent observed sample, the absorbance falls exponentially as the wavelength increases, even though it initially manifests a higher absorption value (apart from some absorbance peaks). For instance, a sample that was exposed to light for 4 minutes had a higher absorbance of 3.75 at the 320 nm initial wavelength compared to 0.75 at the 525 nm initial wavelength. This alone, however, does not change the fact that the timely observed quantum dot samples displayed absorption peaks occasionally as the wavelength increased, and that the number, frequency, and size of these absorbance peaks are sensitive to the length of time the samples are exposed to light.

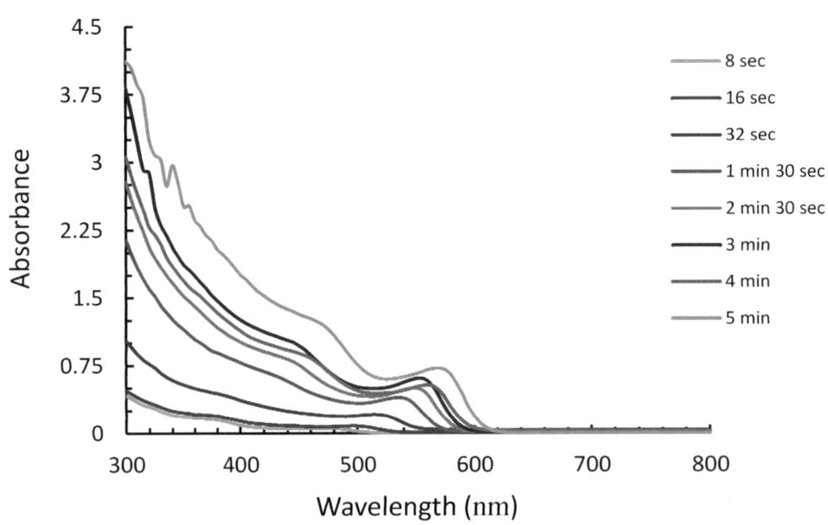

Fig 1: Absorbance vs Wavelength(nm)

3.1.2 Normalized emission spectra vs wavelength plot

The CdSe QD sample emits proton after being given energy and raised to an excited level, but only before returning to the ground state. The characteristics of this proton emission are shown in Fig. 2, and the intensity of the emission has been classified according to how much light the sample was exposed to. It's interesting to note that the emission peaks on the intensity curve of the samples that were examined at 1, 2, 3, 4, and 5 minutes are at a higher wavelength than those on the initial samples, which were collected at 8 and 16 seconds. The sample that was collected after the initial 32 seconds of exposure had a value of 700 au within the range of 500 nm to 525 nm wavelength, and the sample that was examined after 5 minutes of exposure had the lowest intensity of the emission. A potential aspect of the intensity vs. wavelength plot shows that it is either better to harvest energy from the synthesised QD sample with little or no manifestation to light for samples with lower wavelengths or with greater manifestation to light for samples with higher wavelengths.

The normalised plot of Fig. 3, in which the intensity is normalised under the usual scale of maximum intensity (700 a.u), provides a clearer understanding of the CdSe QD sample's intensity. The bench mark for the earlier inferred time-wavelength-emission incident is provided by this plot. The later samples corresponded to their higher wavelengths, whereas the earlier samples managed to find the maximum intensity at relatively lesser wavelengths.

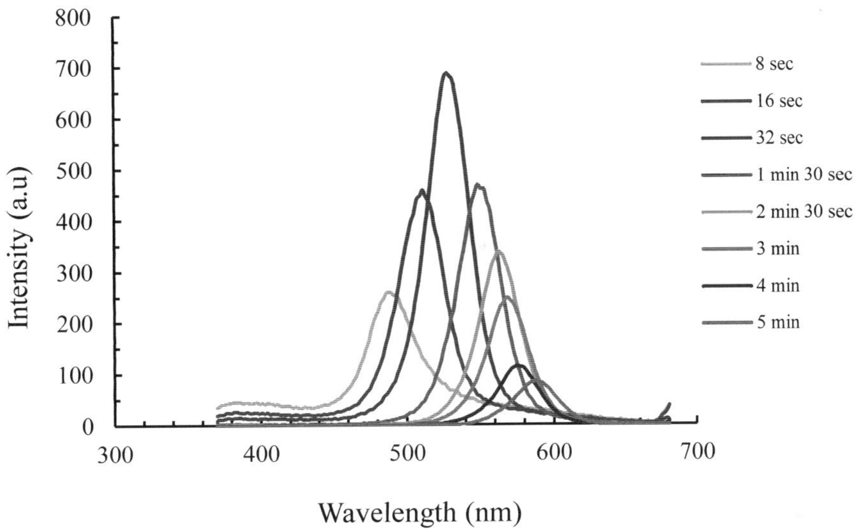

Fig 2: Intensity(au) vs Wavelength plot (Not normalized)

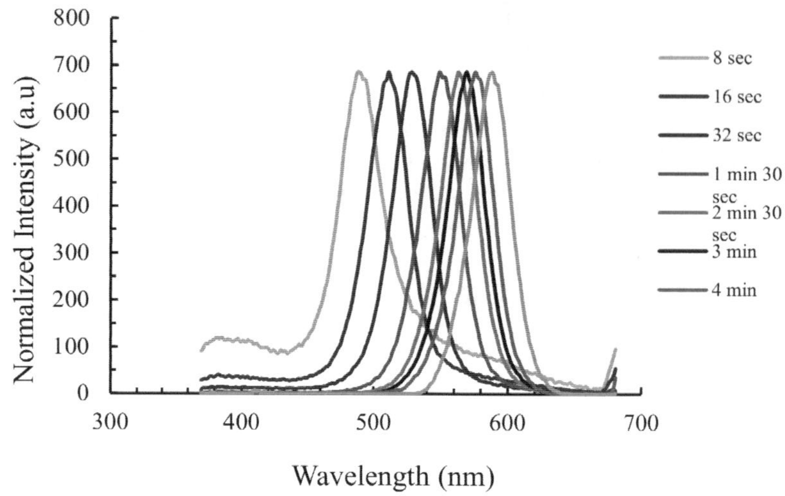

Fig 3: Intensity (a.u) vs Wavelength plot (Normalized)

3.1.3. Absorbance maxima vs emission maxima plot

The correlation in Fig. 4 indicates that the emission wavelength and absorbance wavelength are very close to one another. In other words, a particle's energy release and absorption are almost equal, suggesting that less energy is lost during the vibrational relaxation phase.

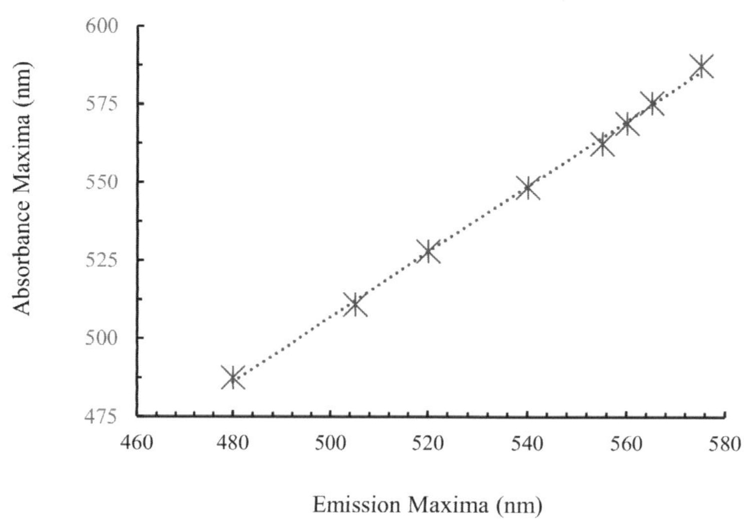

Fig 4: Absorbance Maxima (nm) vs Emission Maxima (nm) plot.

3.1.4 The energy of the Stokes shift analysis

The Stokes shift predicts the spectral position of the maximum absorption and emission band, making it a crucial parameter for comprehending the molecular spectroscopic behaviour of QDs. The wavelength difference between the absorption and emission peak maxima for a given transition is known as the Stokes shift (SS) [25]. The Stokes shift value represents the sample's relaxation time from the initial state to the excited state. The stokes shift of the manufactured CdSe quantum dot is presented in terms of wavenumber under the current study, and the value is discovered to be 2659.89

cm-1. The calculation was performed using the observed values of 14492.75 cm-1 for emission and 17152.65 cm-1 for absorption. The wavelength in cm-1 is simply expressed as a wavenumber. Only the local maxima of the absorbance curve were taken into account in Fig 5. When this curve was plotted, the data to the left of the local maxima was ignored. Following that, the trajectory of the absorbance curve was predicted using higher-order polynomial forecasting. The primary loss mechanism that significantly reduces the efficiency of power conversion is reabsorption of emitted photons. To put it another way, the photon that was released will be easily reabsorbed in the emission layer. The absorption and emission spectra do not overlap very much or at all, though, when the stock shift is greater. As a result, very few photons will be reabsorbed, leaving us with a large number of photons that can be used to generate power. Therefore, solar applications benefit from a greater stock shift.

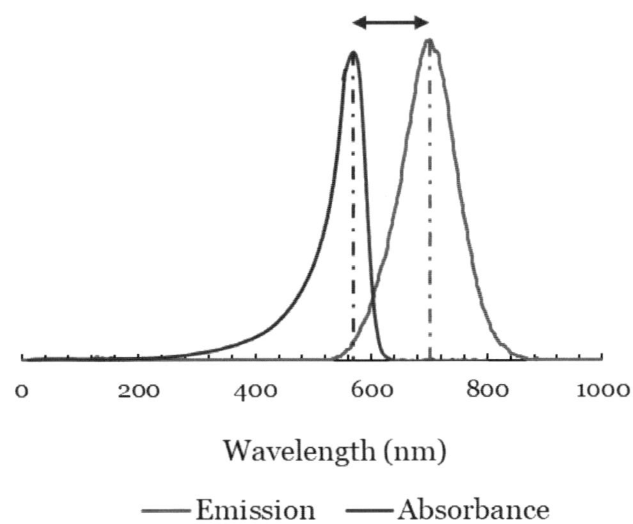

Fig 5: Stokes shift of the wavelength

3.1.5 Gaussian Curve and Jacobian Conversion Analysis

The Jacobian conversion process is a forerunner method to deal with any ambiguity during the data manipulation process, whereas the Gaussian curve is an excellent distributive method to understand the behaviour of real valued randomly and independently generated data. To fully comprehend the electro-optical phenomena of the synthesised CdSe QD, both of these approaches are used.

3.1.5.1. Determination of the FWHM of the Gaussian Curve

The width of the distribution at a level that is only half the maximum ordinate of the peak is known as the full width half maximum (FWHM). The duration of pulse waveforms is one phenomenon that is described by FWHM. The spectral width of optical communication sources and the resolution of spectrometers are examples of a signal from a baseline value to a higher or lower value, followed by a rapid return to the baseline value. It can be seen in Fig. 6 that the sample examined after 2.30 minutes and 1.30 minutes has the sudden peaks for the FWHM duration. The intensity of the quantum dot generated emissions is plotted in terms of 1000 times of its original e.V.

IOP Conf. Series: Materials Science and Engineering 1291 (2023) 012032 doi:10.1088/1757-899X/1291/1/012032

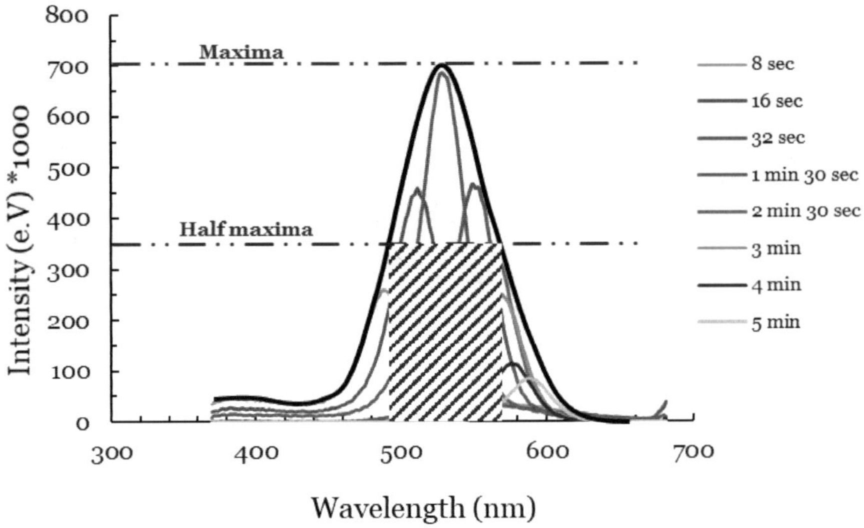

Fig 6: Fitting the PL spectrum to the Gaussian Curve.

3.1.5.2. Jacobian conversion

Spectra are typically expressed in wavelength units, but expressing data as a function of energy provides a more comprehensive view of the physical universe. Nonetheless, the inverse relationship between the two variables poses difficulties and has occasionally led to peculiar conclusions, such as the notion that the human eye's spectral response complements the solar spectrum. To assure enlightening scientific research, it is essential that spectral data be presented accurately. Photoluminescence (PL) spectroscopy is a widely used experimental method for the study of photochemistry and photo physics. Quantitatively analysing PL intensity and spectral profiles requires contemplating more than intensity versus wavelength. Nanomaterials engineered for luminescent devices typically exhibit a broad spectral window with multiple emissive sources, resulting in emission over a broad spectrum of wavelengths. Prior to quantitative analysis, it is essential to resolve such ambiguities in PL spectra, which are frequently observed in nanocrystals of semiconductors. In addition, in order to account for these ambiguities and wavelength conversions, signal values must be scaled by (h*c/E2) and the Jacobian conversion, depicted in Figure 7, is conducted. This conversion not only modifies the geometry of the line, but also accounts for the wavelength's direct effect on energy. Therefore, aliquots emitting in green rather than red would have the same FWHM in terms of wavelength but differ in energy. As indicated in the first clause of the Jacobian conversion, employing energy units such as eV provides a more comprehensive representation of the physical world.

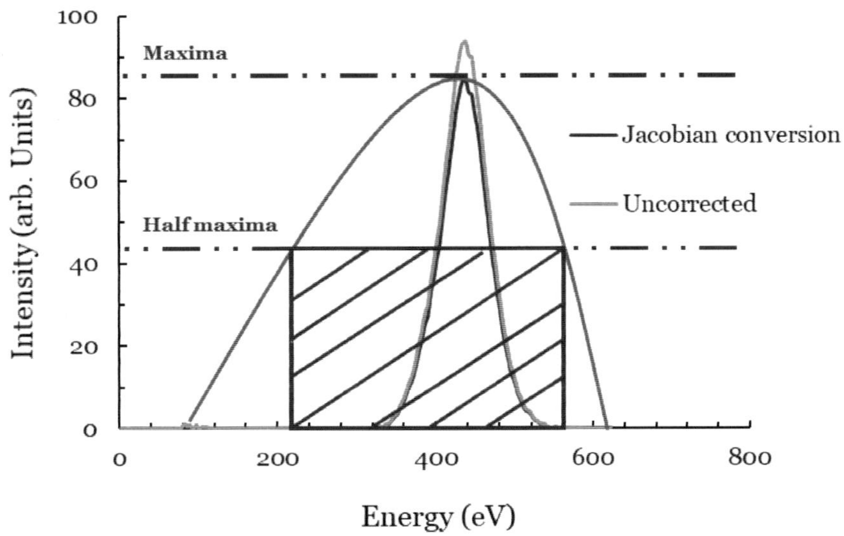

Fig 7: Jacobian converted and uncorrected absorption and fluorescence spectra for CdSe nanocrystals

3.1.5.3. The size calculation of the CdSe quantum dots

The study provides the relationship between the diameter of the CdSe quantum dot and the wavelength.[31].

Here the wavelength (λ) of the CdSe quantum dots at the final time point is 583.5nm. So, the diameter of the particles can be calculated to be 4.0375 nm.

3.2. Evaluate DLS data

The photon correlation spectroscopy (PCS), also referred to as the dynamic light scattering (DLS) method, is a crucial tool for comprehending how nanoparticles or macromolecules in solution diffuse. The size and shape of the molecules or samples are taken into consideration when calculating the hydrodynamic radii of the micro molecules after the diffusion coefficient has been determined. Under this subsection, the size-shape comparison of the examined sample of CdSe QDs and the DLS corresponded data are presented.

3.2.1. Zeta potential

Zeta potential is the name for the force that prevents negatively charged particles from interacting with one another. The amount of negative charge in a solution or the amount carried by particles can both be calculated using the zeta potential. These negatively charged particles can form micro flocks and eventually macro flocks by adding a charge neutralizer to overcome this repelling force.
Figure 8 displays the zeta potentials of quantum dots that have been pre- and post-ligand exchanged. Pre-ligand exchanged quantum dots, as shown in the figure, have a zeta potential of 0.170 mV, while post-ligand exchanged quantum dots have a zeta potential of -26.5 mV. The negatively charged CdSe-S2- that is present in the solution as a result of the ligand exchange also causes the zeta potential to be negative, indicating the attraction between these negatively charged ions. And this repellent force has a 26.5mV strength.

IOP Conf. Series: Materials Science and Engineering 1291 (2023) 012032 doi:10.1088/1757-899X/1291/1/012032

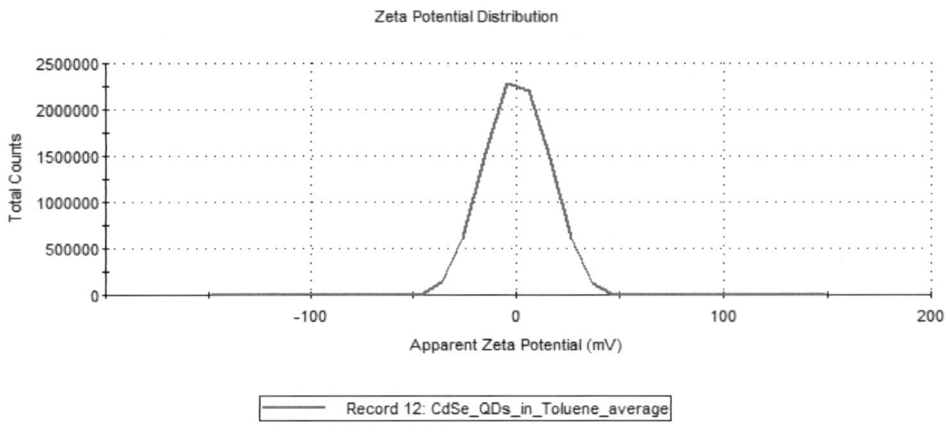

Fig 8: Zeta potential distribution - Pre Lingard-Zeta potential distribution

3.2.2. Comparison of the size measurement of the quantum dot samples

The pre-ligand exchanged quantum dots' size, as determined by DLS data, is approximately 9.429 nm. which is larger than the size that we estimated using the reference's equation [26]. Given that the DLS experiment was carried out three weeks after the quantum dots were created, the size difference could be the result of the quantum dots being combined between the two experiments. As an alternative, the higher-order term's neglection in the equation's standard error could be to blame for this variation. Furthermore, the size analysis of pre-ligand exchange quantum dots shows that there are roughly 30% of 9.429 nm-sized quantum dots in the solution. The standard deviation of the sample is 1.812 nm. This indicates that between 5.805nm and 13.053nm is the diameter range for about 95% of the particles. The graph of the ligand exchange size analysis has three peaks; the longest peak is at 4601 nm, the shortest peak is at 63.25 nm, and the two largest peaks are at 306.9 nm and 63.25 nm, respectively. In addition, according to %Number, 6.8% of nanoparticles have a diameter of 306.9 nm, 3% have a diameter of 4601 nm, and 8% have a diameter of 63.25 nm. The results of the post-Lingard size analysis show that the increase in particle size is caused by the presence of CdSe-S2- particles rather than by the presence of CdSe in the toluene solution before the Lingard exchange.

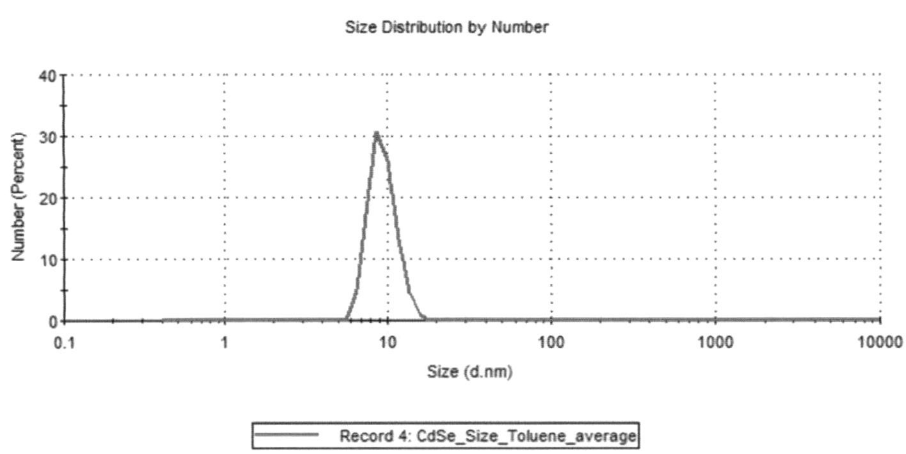

Fig 9: Size distribution by %Number - Pre Lingard size distribution by %Number.

4. Discussion

Apart from the successful manufacturing of these highly potent QDs, an aftermath and deliberate impact of different parameters on the integration process are discussed in this paper. Also, the influence these synthesized nanostructured particles will have in near future compared to some heavy and large-scale energy extraction devices is another motivation leveraged under this section. These potentialities of the newly synthesized CdSe QDs are journaled in three different sections to enlighten the reader more persuasively and discretely.

References

[1] D. Bera, S. C. Kuiry, and S. Seal, "Synthesis of nanostructured materials using template-assisted electrodeposition," *JOM*, vol. 56, no. 1, pp. 49–53, Jan. 2004, doi: 10.1007/s11837-004-0273-5.

[2] S. V. N. T. Kuchibhatla, A. S. Karakoti, D. Bera, and S. Seal, "One dimensional nanostructured materials," *Prog. Mater. Sci.*, vol. 52, no. 5, pp. 699–913, 2007, doi: 10.1016/j.pmatsci.2006.08.001.

[3] A. Henglein, "Small-particle research: physicochemical properties of extremely small colloidal metal and semiconductor particles," *Chem. Rev.*, vol. 89, no. 8, pp. 1861–1873, Dec. 1989, doi: 10.1021/cr00098a010.

[4] R. J. B. Peters *et al.*, "Nanomaterials for products and application in agriculture, feed and food," *Trends Food Sci. Technol.*, vol. 54, pp. 155–164, 2016, doi: 10.1016/j.tifs.2016.06.008.

[5] L. Qi and X. Gao, "Emerging application of quantum dots for drug delivery and therapy," *Expert Opin. Drug Deliv.*, vol. 5, no. 3, pp. 263–267, 2008, doi: 10.1517/17425247.5.3.263.

[6] A. P. Alivisatos, "Perspectives on the Physical Chemistry of Semiconductor Nanocrystals," *J. Phys. Chem.*, vol. 100, no. 31, pp. 13226–13239, Jan. 1996, doi: 10.1021/jp9535506.

[7] P. Walter *et al.*, "Early use of PbS nanotechnology for an ancient hair dyeing formula," *Nano Lett.*, vol. 6, no. 10, pp. 2215–2219, 2006, doi: 10.1021/nl061493u.

[8] H. P. Rocksby, "Color of selenium ruby glasses," *J. Soc. Glas. Technol.*, vol. 16, no. 171, pp. 171–181, 1932.

[9] A. L. Efros and A. L. Efros, "INTERBAND ABSORPTION OF LIGHT IN A SEMICONDUCTOR SPHERE.," *Sov. physics. Semicond.*, vol. 16, no. 7, pp. 772–775, 1982.

[10] R. Rossetti, J. L. Ellison, J. M. Gibson, and L. E. Brus, "Size effects in the excited electronic states of small colloidal CdS crystallites," *J. Chem. Phys.*, vol. 80, no. 9, pp. 4464–4469, May 1984, doi: 10.1063/1.447228.

[11] F. Grieser, D. N. Furlong, D. Scoberg, I. Ichinose, N. Kimizuka, and T. Kunitake, "Size-quantised semiconductor cadmium chalcogenide particles in Langmuir–Blodgett films," *J. Chem. Soc. Faraday Trans.*, vol. 88, no. 15, pp. 2207–2214, Jan. 1992, doi: 10.1039/FT9928802207.

[12] C. B. Murray, D. J. Norris, and M. G. Bawendi, "Synthesis and characterization of nearly monodisperse CdE (E = sulfur, selenium, tellurium) semiconductor nanocrystallites," *undefined*, vol. 115, no. 19, pp. 8706–8715, Sep. 1993, doi: 10.1021/JA00072A025.

[13] A. R. Kortan *et al.*, "Nucleation and growth of CdSe and ZnS quantum crystallite seeds, and vice versa, in inverse micelle media," *J. Am. Chem. Soc.*, vol. 112:4, no. 4, pp. 1327–1332, Feb. 1990, doi: 10.1021/JA00160A005.

[14] Y. Spanhel, M. Haase, H. Weller, and A. Hengleiir, "Photochemistry of colloidal semiconductors. 20. Surface modification and stability of strong luminescing CdS particles," *undefined*, vol. 109, no. 19, pp. 5649–5655, Sep. 1987, doi: 10.1021/JA00253A015.

[15] A. M. Mostafa, S. A. Yousef, W. H. Eisa, M. A. Ewaida, and E. A. Al-Ashkar, "WO 3 quantum dot: Synthesis, characterization and catalytic activity," *J. Mol. Struct.*, vol. 1185, pp. 351–356, 2019, doi: 10.1016/j.molstruc.2019.03.007.

[16] O. I. Mićić, C. J. Curtis, K. M. Jones, J. R. Sprague, and A. J. Nozik, "Synthesis and characterization of InP quantum dots," *J. Phys. Chem.*, vol. 98, no. 19, pp. 4966–4969, 1994, doi: 10.1021/j100070a004.

[17] R. K. Giri, S. H. Chaki, A. J. Khimani, S. R. Patel, and M. P. Deshpande, "Thermal investigation of nanospheres and nanowhiskers of CuInS2," *Eur. Phys. J. Plus 2021 1363*, vol. 136, no. 3, pp. 1–17, Mar. 2021, doi: 10.1140/EPJP/S13360-021-01241-1.

[18] A. B. Hirpara *et al.*, "Biological investigation of sonochemically synthesized CZTS nanoparticles," *Appl. Surf. Sci. Adv.*, vol. 12, p. 100338, Dec. 2022, doi: 10.1016/J.APSADV.2022.100338.

[19] A. J. Khimani, S. H. Chaki, R. K. Giri, R. R. Meena, R. M. Kannaujiya, and M. P. Deshpande, "Thermal exploration of sonochemically achieved SnS2 nanoparticles: Elemental, structural, and morphological investigations of TG residual SnS2," *Chem. Thermodyn. Therm. Anal.*, vol. 9, p. 100104, Mar. 2023, doi: 10.1016/J.CTTA.2023.100104.

[20] R. M. Kannaujiya *et al.*, "Mechanistic insights into antibacterial and anti-biofilm activities against multidrug-resistant microbes of SnTe nanospheres synthesized by sonochemical method," *Chem. Phys. Impact*, vol. 6, p. 100219, Jun. 2023, doi: 10.1016/J.CHPHI.2023.100219.

[21] Z. S. Kachhia *et al.*, "Thermal decomposition study of cadmium telluride (CdTe)," *Mater. Today Proc.*, Mar. 2023, doi: 10.1016/J.MATPR.2023.02.240.

[22] R. K. Giri, S. H. Chaki, A. J. Khimani, and M. P. Deshpande, "Mechanistic insights into transport properties of chemical vapour transport grown CuInS2 single crystal," *J. Alloys Compd.*, vol. 959, p. 170487, Oct. 2023, doi: 10.1016/J.JALLCOM.2023.170487.

[23] S. Zhou and E. Ruckenstein, "A density functional theory based on the universality of the free energy density functional," *J. Chem. Phys.*, vol. 112, no. 18, pp. 8079–8082, 2000, doi: 10.1063/1.481407.

[24] M. Ji, X. Lan, Z. Han, C. Hao, and J. Qiu, "Luminescent properties of metal-organic framework MOF-5: Relativistic time-dependent density functional theory investigations," *Inorg. Chem.*, vol. 51, no. 22, pp. 12389–12394, 2012, doi: 10.1021/ic301771b.

[25] A. Abid and A. A. Bhuiyan, "A finite element analysis approach to design and optimize the static structural impact on a skateboard," *Mater. Today Proc.*, vol. 60, no. xxxx, pp. 2171–2187, 2022, doi: 10.1016/j.matpr.2022.02.424.

[26] A. Abid and M. T. Sarowar, "Heat Transfer, Thermal Stress and Failure Inspection of a Gas Turbine Compressor Stator Blade Made of Five Different Conventional Superalloys and Ultra-High-Temperature Ceramic Material: A Direct Numerical Investigation," *J. Fail. Anal. Prev.*, vol. 22, no. 3, pp. 878–898, 2022, doi: 10.1007/s11668-022-01413-w.

[27] S. V. Kilina, D. S. Kilin, and O. V. Prezhdo, "Breaking the phonon bottleneck in PbSe and CdSe quantum dots: Time-domain density functional theory of charge carrier relaxation," *ACS Nano*, vol. 3, no. 1, pp. 93–99, 2009, doi: 10.1021/nn800674n.

[28] M. Sundaram, S. A. Chalmers, P. F. Hopkins, and A. C. Gossard, "New quantum structures.," *Science*, vol. 254, no. 5036, pp. 1326–1335, Nov. 1991, doi: 10.1126/SCIENCE.254.5036.1326.

[29] R. Cingolani and K. Ploog, "Frequency and density dependent radiative recombination processes in III V semiconductor quantum wells and superlattices," *AdPhy*, vol. 40, no. 5, pp. 535–623, Jan. 1991, doi: 10.1080/00018739100101522.

[30] A. P. Alivisatos, "Semiconductor Clusters, Nanocrystals, and Quantum Dots," *Science (80-.).*, vol. 271, no. 5251, pp. 933–937, Feb. 1996, doi: 10.1126/SCIENCE.271.5251.933.

[31] J. Jasieniak, L. Smith, J. van Embden, P. Mulvaney, and M. Califano, "Re-examination of the Size-Dependent Absorption Properties of CdSe Quantum Dots," *J. Phys. Chem. C*, vol. 113, no. 45, pp. 19468–19474, Nov. 2009, doi: 10.1021/jp906827m.

Effect of Electroless Nickel Coating on Wear Rate of EN8 Steel

S Ghosh[1], B Bose, S Gaji, A V Lakra, S Debnath, S Ghosh

Department of Mechanical Egineering, JIS College of Engineering, Kalyani, India

Abstract. Electroless Nickel coating has emerged as a major breakthrough in the development of novel materials with excellent tribological properties. In the present work Ni-P coating has been performed over EN8 steel specimens for an hour. Mechanical and tribological character was determined for the coated samples. Tribological tests were performed for various loads and the wear behavior of coated steel has been analyzed. The coated samples were found to be harder with improved frictional characteristics like lesser wear under similar working conditions.

Keywords: Coating; Tribology; Electroless; Coating; Nickel; Wear.

1. Introduction

A Number of design applications involves wear and tear process, where parts or joints are subjected to relative motion. Wear can be considered as an undesired removal of material due to mechanical action which cannot be eliminated completely. However, in most of the cases alloys and composites are developed which possess low friction and low wear properties under dry sliding conditions against smooth metallic counterparts [1]. EN8 is an unalloyed medium carbon steel which is a popular choice in industry from the automotive trade to wider general engineering applications. Electroless Nickel-Phosphorus plating process is an industrial chemical plating process where a uniform layer of Nickel-Phosphorus alloy is deposited on the surface of the solid substrate. As the name suggests the electroless plating 4 does not require the passing of electric current [2]. EN coatings have a wide range of industrial applications including fluid handling valves & cylinders, medical equipment, owing to their excellent mechanical, electrical, corrosion and wear resistance properties. The electroless bath consist of a solution of metal ions, complex agents, reducing agents and stabilizers operating in a specific metal ion concentration, temperature and pH ranges [3]. The superior quality deposition improves the mechanical and microstructural properties of base metal. With a refined microstructural arrangement on the top surface, the cladding material exhibits outstanding properties like withstanding high temperature, impact loads and high wear resistance.

Various Electroless Nickel plating processes like Ni-P, Ni-B and pure Ni coating can be achieved based on the reducing agent used in the coating bath. Electroless plating process involving Hypophosphite as the reducing agent has received commercial success because of various factors such as low cost, ease of control and good corrosion resistance. Even though, electroless Ni-P deposits give satisfactory performance in many cases, enhancing their performance to suit different end uses warrants further development [4]. Many literatures depict a fact that electroless coating process is to impart smoothness and hardness to the surface for application areas requiring excellent tribological characteristics. The extent of friction and wear is highly responsible for limiting the life of several important links and joints in automotive and aerospace industry. The friction behavior of Electroless Ni-P coated materials can be modified through various surface treatments and incorporation of particles [5]. To improve the quality of deposition, complex agents such as Citrates, Phosphate and Succinate have been used in several experimental works reported so far [6, 7, 8]. However, there is very less reports on the effect of loading conditions on the coating behavior in tribological perspectives. Therefore, more experiments are needed to grow confidence on the behavior of coated medium carbon steels. In the present work electroless Ni-P coating has been performed on EN8 steel specimens to assess the mechanical and tribological behavior of coated steel samples under dry sliding conditions. The tribological characteristics has been experimentally analyzed for various applied loads on the specimens.

[1] Corresponding author. Tel.: +917003772535
E-mail address: sandip.ghosh@jiscollege.ac.in

2. Methodology

Electroless coating can be considered as a 'self-reduction' process, which is primarily based upon the auto-catalytic reduction process of metal ions in an aqueous solution which contains sodium hypophosphite, NaH_2PO_2 as a chemical reducing agent [9]. The main function of stabilizer is to prevent the decomposition of plating bath solution. Sodium acetate and Sodium hydroxide are used as buffer which control the pH and to obtain a uniform coating. Two types of baths have been mainly used for depositing Ni-B, Ni-P alloy which are acidic and alkaline baths depending on the reducing agent used. The commonly used Reducing agent of EN coating are Sodium hypophosphite for Ni-P Deposition and Sodium borohydride for Ni-B Deposition. Hypophosphite baths are the most common type of electroless nickel bath because of their higher deposition rates, increased stability and greater simplicity of bath control. The various bath components and their function are described in Table 1.

Table 1. Components of Coating Bath

Component	Function
Nickel ion	Source of nickel
Reducing Agent	Source of electron
Buffers	Controls pH
Accelerators	Activate reducing agent
Complex agent	Stabilize the solution

3. Experimental Procedure

3.1. Coating Deposition

Electroless Ni-P coating is deposited on mild steel of two types: square and cylindrical. Square specimens are used for microstructural characterization related studies while the cylindrical ones are used for the friction and wear tests. The basic experimental setup is depicted in Fig. 1(a) and 1(b).

a b

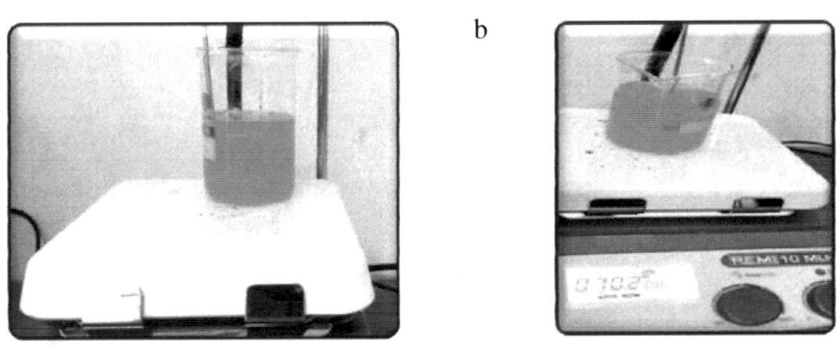

Figure 1. (a) Electroless Ni-P Coating Bath with magnetic stirrer; (b) Solution heating arrangement.

Initially, specimens are mechanically cleaned to remove foreign matter or any kind of corrosion product. After that, remaining organic particles of steel specimen are cleaned off by rinsing with distilled water. Then the specimen is submerged in con. HCL for 15 seconds. After that the specimen is submerged in Palladium Chloride for 15 seconds at 55°C. Palladium chloride activation is given to each of the samples prior to the deposition to have a good initiation of the coating onto the mild steel substrate. After that the temperature of the heater is set to 82°C and the deposition of substrate material is to be done. The deposition was carried out for 1 hour at a stretch before the bath shows signs of decomposition.

3.2. Measurement of frictional parameters

Friction and wear characteristics of the EN-coated specimens are studied with pin on disk wear testing apparatus. The wear tests are performed against Coated cylindrical samples under dry, non-lubricated condition at room temperature. The experimental instrument includes a rotating specimen with disk shape and a cylindrical stationary specimen. The stationary pin specimen holder is attached to a lever arm that has a pivot. Addition of weights produces a test force proportional to the mass of the weight applied. The loading lever's deflection corresponding to the friction force of the specimen plate is measured by a displacement sensor. The friction force and vertical displacement of pin are automatically measured in the pin-on-disk machine using a load cell and a linear variable differential transformer. A variable speed motor rotates the disk at different speeds and is mounted in such a manner that its vibration does not affect the test. In the present work the normal load applied on the disk through the pin are 20N, 30N, 40N, 10N respectively. No lubrication was applied during the tests to estimate the maximum wear rates at minimal or no lubrication conditions. The friction and wear tests are done setting the track diameter of the disk is 80 mm with a rotating speed of 60 RPM in room temperature. Each test was performed for 10 minutes of duration. The friction force values are acquired periodically along the experiment through a data acquisition system into a computer for further processing. Figure 2(a) shows the entire experimental setup while figure 2(b) illustrates the pin on disk arrangement.

(a) *(b)*

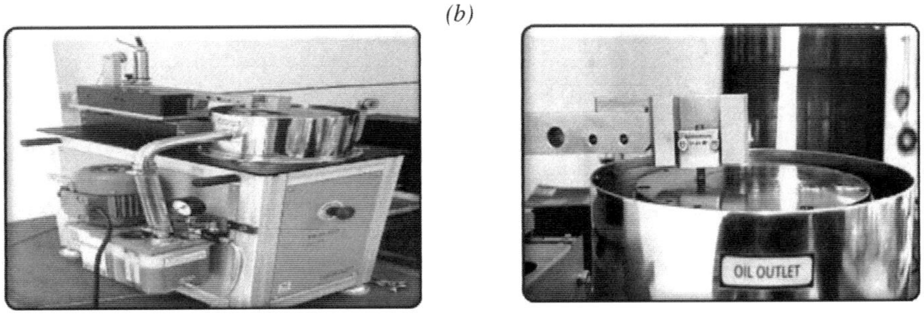

Figure 2. (a) Tribometer Setup with loading arrangements; (b) Pin setup on rotating disk.

4. Result and Discussion

(a) *(b)*

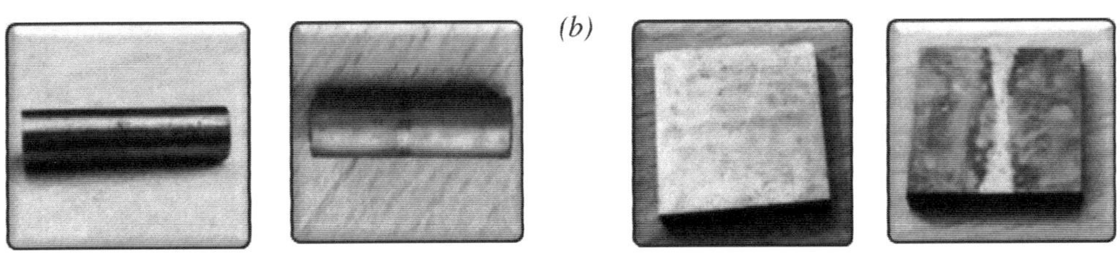

Figure 3. Pictures of test samples before and after coating (a) Cylindrical Sample; (b) Square Sample.

Figure 3 shows the pictures of test samples before and after coating. As the samples are dipped into the solution with cotton threads, there are certain marks of the thread where the coating has deposited. The hardness of the samples was tested with the applied load of 60 N before the coating. After the coating the microhardness was measured for the coated samples. The comparison as reported in Table 2 shows that the coating process has improved the hardness of EN8 steel specimens. Similar results have been reported in several literatures [11, 12] and it has been found that hardness and wear resistance can be increased with heat treatment.

Table 2. Effect of Coating on Hardness

Sample No.	Steel Substrate hardness	Coated sample hardness
1	42.5	44
2	42.2	46
3	42	47
Average	42.23	45.66

Table 3 shows comparison of results for the variation Coefficient of friction at different applied loads at track rotational speed of 60 RPM. It also shows the variation of wear rate of electroless Ni-P coated samples before any without any specific heat treatment. From the results it can be noticed that the Coefficient of friction is decreased and the wear is increased. Fig. 4 illustrates the variation of Coefficient of friction with the applied load. It can be noticed that is Coefficient of friction consistently decreased with the increasing load up to 30 N load. However, there has been a slight increase in Coefficient of friction when 40 N load was applied as observed in several experimental works reported in the literature [3,13]. Micro-structural properties and crystallization behavior of EN deposits highly depend upon the phosphorus content in the coating which attributes to their response to varying loads.

Table 3. Experimental findings from tribotest of cylindrical samples

Sl.	Load (N)	Disc speed (RPM)	Coefficient of friction	Wear (micron)
1	10	60	0.81	9.75
2	20	60	0.59	158.19
3	30	60	0.46	221.24
4	40	60	0.56	496.58

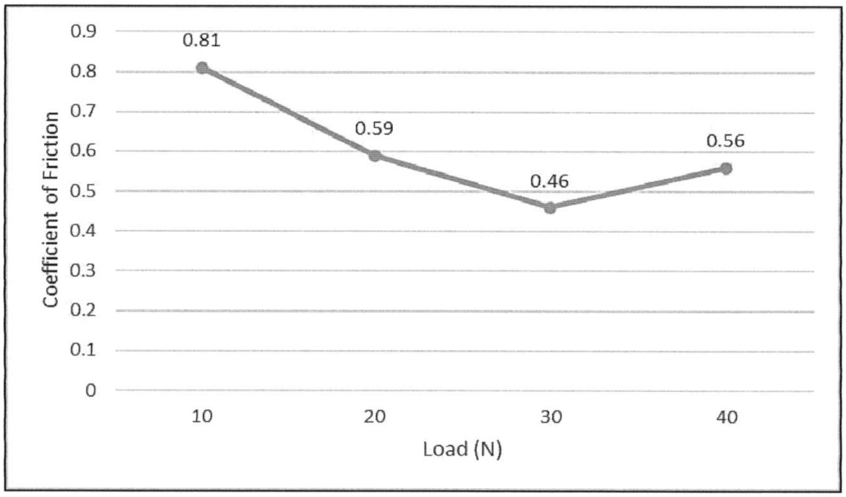

Figure 4. Effect of load on coefficient of friction of coated EN8 steel specimen

Electroless nickel coatings are susceptible to major adhesive wear under dry sliding contact, because the low interfacial and high surface free energies of this material combination produce highly compatible surfaces. In figure 5, the measured rate of wear from the data acquisition system has been plotted with the applied load to check the wear resistance capabilities of coated steel. It is observed that the wear is increased with increased application of normal load in a decreasing rate till 30 N. However, there is a sudden rise in wear when the load reaches 40 N as it measures 496.58 micron. It indicates the change in wear resistance over a particular load for the current thickness of coating. Such occurrence can be attributed by the presence of wear debris on the rotating disc which have affected the wear results for the particular set of experiment. When compared with uncladded EN8 samples, the coated samples always show better wear resistance as found in the present work and also reported in several literature [14]. The solution to the severe wear can be achieved through a controlled heat treatment by inducing crystallization of amorphous nickel to Ni_3P [15]. Ni-P coated samples tested under brine environment [16], also confirms the present trend of frictional parameters reported in the present work.

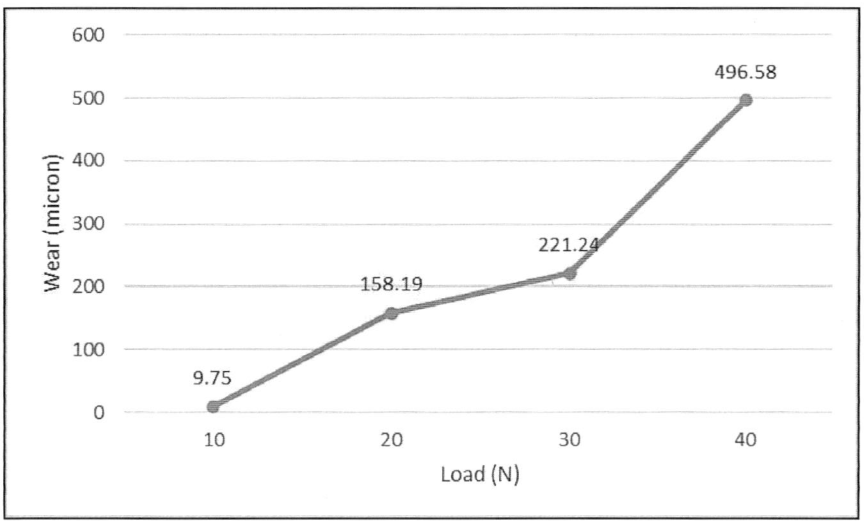

Figure. 5 Wear characteristics of coated EN8 steel samples

5. Conclusion

The present work demonstrates electroless Ni–P coating process on EN8 steel substrate. The mechanical and tribological behavior of the coating has been analyzed in the perspective of its applications. The property changes offered by Ni-P coatings has proved the way for their use in numerous industrial applications. The Hardness of the coated sample were found higher when compared with noncoated samples. The Wear and Friction characteristics were determined for varying loading conditions. It was found that on average the coefficient of friction decreases with increasing load. Electroless nickel in the as-deposited state exhibits severe adhesive wear in dry sliding contact. The rate of wear was observed to be proportional with applied load in the tribometer testing. Further developments in this field will secure a prominent place for these coatings in the surface engineering of metals and alloys.

Acknowledgements

The work has been carried out under an AICTE funded research project in the project lab of Mechanical Engineering Department of JIS College of Engineering.

References

[1] Tunalioglu MS and Tuc B 2014 *Wear* **309**(1–2) pp. 208–215.
[2] Chowdhury MA and Nuruzzaman DM 2013 *Tribol Ind* **35**(1) pp. 42–50.
[3] Taheri R., Oguocha I.N.A. and Yannacopoulos S 2001 *Wear* **249**(5–6) pp.389–396.
[4] Balaguraju J.N., Sankara Narayanan T.S.N. and Seshadrie S.K. 2003 *Applied Electrochemistry* **33** pp. 807–816.
[5] Sahoo P. and Das S. K. 2011 *Materials Design* **32**(4) pp.1760-1775.
[6] Masoumi F, Ghasemi RH, Ziaei AA and Shahriari D 2012 *The International Journal of Advanced Manufacturing Technology* **62** pp.1063–1070.
[7] Agarwala RC and Agarwala V 2003 *Sadhana* **28** pp. 475–493.
[8] Saha M., Bakshi S., Mandal S., Biswas P. and Ghosh S. 2022 *Materials Today Proceedings* **66**-9 pp. 3890-3896.
[9] Godse RS, Gawande SH and Keste AA 2016 *J Bio Tribo Corros* **2**:3; doi:10.1007/s40735-016-0034-3
[10] Alirezaei S, Monirvaghefi SM, Salehi M and Saatchi A 2007 *Wear* **262**(7–8) pp. 978-985.
[11] Keong KG, Sha W and Malinov S 2003 *Surface and Coatings Technology* **168**(2–3) pp. 263-274.
[12] Sribalajia M., Rahmana O.S.A., Lahab T., Keshri A.K. 2016 *Materials Cheistry and Physics* **177** pp. 220-228.
[13] Charoo MS, Wani MF, Hanief M, Chetani A and Rather MA 2017 *Advanced Materials Proceedings* **2**(7) pp. 445-449.
[14] Gawne DT and Ma U 1987 *Wear* **120**(2) pp. 125-149.
[15] Bhaumik S, Mukherjee M, Sarkar P, Nayek A and Paleu V 2020 *Metals* **10**, 958; doi:10.3390/met10070958.
[16] Panja B, Das SK and Sahoo P 2014 *J. Inst. Eng. India Ser. D* **95**, pp. 153–159. doi.org/10.1007/s40033-014-0041-9

Determination of Transmission Error and Mesh Stiffness in Spur Gear Pair using CAD-FEM Integration Approach

V K Karma [1], G Maheshwari [2]

[1,2] Mechanical Engineering Department, Institute of Engineering & Technology, Devi Ahilya University, Indore, Madhya Pradesh, 452001, India

vkarma@ietdavv.edu.in

Abstract. Unwanted sound and sensations are key sources of excitations in gear drives. They developed because of Transmission Error (TE). TE is a consequence of geometric or engineering flaws like profile/pitch error etc. or distortions like spall, indentation, crack, rubs etc. arising due to the diverse load conditions. These distortions and various gear parameters like number of teeth, contact ratio, module, pressure angle etc. have huge effects on TE and produce changes in mesh stiffness. This demands to investigate the effects of these parameters on TE and mesh stiffness for the avoidance or early finding of causes of malfunctions in the transmission system using computer models using CAD-CAE (FEM) software. In this paper the TE and mesh stiffness is determined using CAD-FEM integration approach and the impact of some of the gear parameters, tooth tip radius and geometric flaw- pitch error is analyzed. The modeling of spur gears & assembly to form gear pair is done in CAD software SOLIDWORKS. The gear pair formed is then introduced into FEM software Ansys for simulation using transient structural analysis method to find angular deformations and then TE and mesh stiffness. Several charts are drawn amongst rotation angle of one of the gears (driver gear) and TE/mesh stiffness for different gear parameters. The work demonstrated the use of CAD-CAE integration approach and point out that all these parameters affect TE and mesh stiffness in a meaningful way.

Key words: SOLIDWORKS, Pitch Error, Tooth Tip Radius, Tooth Geometry Error, Deformations, Ansys

1. Introduction

Gears are the backbone of the transmission system, and spur gears are the most frequently used elements in it. They are also exposed to distortions because of various engineering/geometric flaws or altering load situations. These distortions are responsible for the development of unsolicited sound and excitations resulting in power or torque loss. To explore the effects of these on power transmission systems such as gear drives, several methods ranging from theoretical or mathematical to experimental or computer simulations are frequently applied successfully. The work presented here shows the use of a CAD and FEM software integration approach developed by [1] for the calculation of mesh stiffness to find the TE and mesh stiffness to analyse the effects of various gear parameters, tooth tip radius and pitch error on TE and mesh stiffness. Pitch error is the difference in theoretically correct pitch and obtained pitch. In the presented work pitch error is created purposely by boosting the theoretically correct tooth thickness to 1% & 2%.

Using FEA software, Bartosova et al. [2] parametrically modelled spur gear to assess the effects of load fluctuation, centre distance, and tooth variations on static transmission error (STE). By modifying the parameters, parametric modelling allows for the creation of a variety of gear models. Benacha et al. [3] discuss the effectiveness and accuracy of a multibody technique based on the Lagrange formulation to simulate gear TE in spur and helical gears. The STE is calculated without making any assumptions regarding the placement and orientation of the contact lines. Here, tooth and gear body flexibility, as well as tooth profile variations, are considered. The STE, dynamic transmission error (DTE), and time-varying mesh stiffness (TVMS) of spur gears with varied backlash at different coefficients of friction are predicted by Ambaye et al. [4] using ABAQUS software using a plane strain analysis. For spur gears with profile change under non nominal or variable load circumstances, a simple analytical model for meshing stiffness, load sharing ratio, and TE is described by Pleguezuelos et al. [5]. Improved analytical mesh stiffness and quasi static transmission error (QSTE) models are suggested and tested using the results of a finite element approach that includes tooth profile variation and wear both. The findings reveal that both types of deviations have a considerable impact on the mesh stiffness as well as QSTE, but wear effect is more than tooth profile deviation [6]. With validation from encoder readings acquired from a dry gear wear test and under varied operating circumstances, Zhang et al [7] showed the robustness of the suggested approach in getting absolute TE measurements and the efficacy of the DC component of TE as a wear indicator. Czako et al [8, 9] used the FEA approach to identify STE in helical gears. They looked at how contact ratio and tooth changes affected TE. They employed stress-strain quasi-static contact analysis as their FEA approach.

Randall et al [10] tested and measured TE on an input shaft over a wide range of speeds and torques using single-stage gear box equipment. The TE is further classified into geometric transmission error (GTE), i.e., deviation of the combined profile from the correct profile; STE, caused by elastic displacement under low load; and DTE, deformations caused by speed and load variations. Chang et al. [11] predicted TE by devising a novel approach that takes into account manufacturing, assembly, tooth deflection, and tooth profile alteration faults. They also looked at the backlash and contact ratio over time. Duan et al. [12] calculated TE by taking into account the flexibility of various gearbox components, such as the housing wall, shaft, bearing, and so on. They conducted various tests and compared the findings to theoretical predictions. When a huge dataset is produced, statistical analysis is a useful alternative. Lee et al. [13] worked on developing an algorithm for determining TE in spur gear with lots of errors in deformations. They used statistical analysis to do this. Kohler et al [14] performed several studies on gears with pitch error and concluded that pitch error has a significant impact on TE. Using semi-analytical and non-hertzian contact analysis approaches, Bruzzone et al. [15] computed STE, load sharing characteristics, and pressure distribution on a 2D gear model. Using a 2D CAD model in the FEA programme Ansys.

Kiekbusch et al [16] devised a method for determining combined torsional mesh stiffness. They double-checked the findings of a parametrically created 3D model using Ansys. By creating an analytical relationship from prior methodologies, Sanchez et al. [17] investigated the impacts of dental profile alteration factors such as quantity, length, shape, and others on TE and load sharing ratio. Dogan et al. [18] provided a technique for determining the gear system's echo zones based on different design characteristics. They looked at the impact of mesh stiffness, contact ratio, TE, and other gear factors on spur gear dynamics. Nina Sainte-Marie [19] research work focuses on the creation of a numerical model for predicting the dynamic behaviour of gear systems with many gear stages and various types of gears. The equations of motion are built using time-varying functions that describe mesh excitations, such as mesh stiffness functions, quasi-static transmission error under load, and kinematic (or no-load) transmission error. The combined effect of tooth pitch errors and load on gear transmission dynamic behaviour is also investigated. The relationship between dynamic transmission errors (DTE) and mesh force/root stress dynamic factors is then investigated using an expanded three-dimensional model and a reduced torsional version. For classic spur gear transmissions, Chen et al.

[20] presented a revolutionary high-order phasing gear to lessen mesh stiffness variation. The gear pair is used to study the contact line, and the phasing parameters are optimised to provide almost no mesh stiffness fluctuation. To reconcile the kinematics and centrifugal fields of spur gears, Zheng et al. [21] created an analytical mesh stiffness model that included tooth contact and centrifugal expansions. In comparison to previous models, FEM, and experiments, quasi-static and dynamic solutions exhibit acceptable accuracy and discernible centrifugal stiffening at high-speed conditions. Gear wear increases vibration and noise level and has an impact on meshing excitations like mesh stiffness and gearbox error. Shen et al. [22] suggested a geometric model to assess how gear wear affects mesh stiffness and USTE. According to experimental findings, gear wear only modifies the amplitude of the meshing frequencies and their sidebands rather than the spectrum's overall structure. He and Yang [23] offer a unique gear thermal stiffness model to investigate the static and dynamic properties of gear system vibration. For time-varying thermal mesh stiffness and gear thermal mesh stiffness, it derives general analytical formulations. A new nonlinear dynamic model for a spur gear system that takes into account dynamic force increment and velocity-dependent mesh stiffness was proposed by Liu et al. [24]. The effect of velocity-dependent mesh stiffness and dynamic force increment on the dynamic properties of the spur gear system is depicted and quantified using numerical data.

It has been learned from the literature that the researchers are using an analytical, experimental, or numerical methods to measure or forecast the effects of different displacements on TE and mesh stiffness. To calculate/predict TE and mesh stiffness in gears, numerous CAD/CAE software for modelling, analysis, and simulation are utilised on a regular basis. For these computations, several researchers use an individual methodology or a mix of methodologies. But still, there is a lot of scope for the use of simple and easy to understand procedures to find the TE and mesh stiffness by considering deformations due to geometric/manufacturing/loading conditions using CAD and CAE (FEM) software to benefit numerous researchers. Also, a very few of the researchers attempted pitch error, gear parameters and tooth tip radius together effects on TE and mesh stiffness.

The 3D modelling and assembly of different spur gears are done in SOLIDWORK's software, and these gear pairs of various parameters are then introduced into CAE/FEM software Ansys for the analysis purpose. From these analysis results numerous graphs are drawn amongst TE/mesh stiffness and rotation angle or mesh angle of driver/driven gear.

2. Modelling and Assembly of Spur Gear

2.1. Gear Input Parameters

The main gear parameters required for gears are module, number of teeth and pressure angle. Other gear parameters such as addendum circle, dedendum circle, base circle, pitch circle diameter, tooth thickness etc. are calculated from these parameters based on the type of tooth proportion system. The tooth form is involute curve and is drawn using equation driven curve option. These relations are represented in table 1 [25-26].

2.2. Gear Modeling

The healthy (correct) and faulty (defective) gears of various parameters are modelled based on the procedure explained in [25-26]. Fig. 1 shows the various curves of the gear and modelled gear respectively.

2.3. Gear Assembly (Gear Pair Formation)

After completion of geometric modelling of different gears, now the various combinations (pairs) of gears are formed. The assembly module of SOLIDWORKS software is open and gear 1 is inserted and then gear 2 is inserted. To make the assembly different mate (positions two components relative to one another) options coincident, distance and mechanical gear mate etc. are used to form the pairs. Fig. 2 show assembly of one of the pairs and various mates used.

Table 1. Parameters and relations used for drawing different curves.

1. Module (m)	11. Pitch Circle Diameter (PCD)= Z * m
2. Number of Teeth (Z)	12. Addendum Circle Diameter (ACD) = PCD+ 2*m
3. Pressure Angle (Φ)	13. Dedendum Circle Diameter (DCD) = PCD - 2*1.25*m
4. Addendum= 1 * m	14. Base Circle Diameter (BCD) = PCD*Cos(Φ)
5. Dedendum= 1.25 * m	15. Fillet= Clearance * cos (Φ) * tan ((90 + Φ)/2)
6. Working Depth= 2 * m	16. Pitch Angle (PA) = 360°/z
7. Whole Depth= 2.25 * m	17. Alpha (α) = sqr ((PCD) ^2 – (BCD) ^2)/BCD*180 /pi - PA
8. Clearance= 0.25 * m	18. Beta (β) = (PA/4 + α)
9. Tooth Thickness (t)= pi /2* m	19. Phi = sqr ((ACD/2) ^2 - (BCD/2) ^2)/(BCD/2)
10. Circular Pitch (p)= pi * m	

20. xt_{left} = BCD/2*(cos (-t-($beta_{left}$-β) *pi/180) -t *sin (-t-($beta_{left}$- β) *pi/180)) 21. yt_{left} = BCD/2*(sin (-t-($beta_{left}$- β) *pi/180) +t*cos (-t-($beta_{left}$- β) *pi/180)) 22. $t1_{left}$=0; $t2_{left}$=Phi; $beta_{left}$ = 270°	(Left Involute curve equation parameters)
23. xt_{right} = BCD/2*(cos (t+($beta_{right}$-β) *pi/180) +t*sin (t+($beta_{right}$- β) *pi/180)) 24. yt_{right} = BCD/2*(sin (t+($beta_{right}$-β) *pi/180) -t*cos (t+($beta_{right}$- β) *pi/180)) 25. $t1_{right}$=0; $t2_{right}$=Phi; $beta_{right}$ = 90°	(Right Involute curve equation parameters)

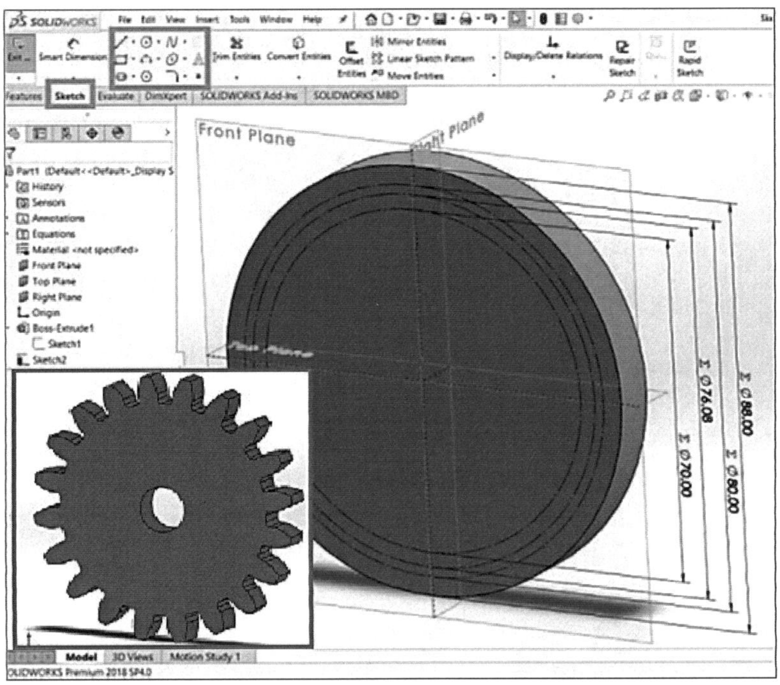

Fig. 1. Various Curves of Gear and Modelled Gear.

3. Determination of TE and Mesh Stiffness using Ansys Workbench

In an uncorrected drive, TE is the consequence of the output gear departing from its theoretical position and the actual position held by it. Geometric imperfections, manufacturing flaws, tooth deformations, lubrication, and other factors can all contribute to TE in gear drives. The presence of TE in gear drives results in a grating point in the noise and vibration waves. Pitch Error (PE) is a type of geometric inaccuracy in teeth. It's the distinction between real and theoretically precise pitch. PE is created in this study by boosting the tooth thickness to 1% and 2% of the theoretically correct thickness.

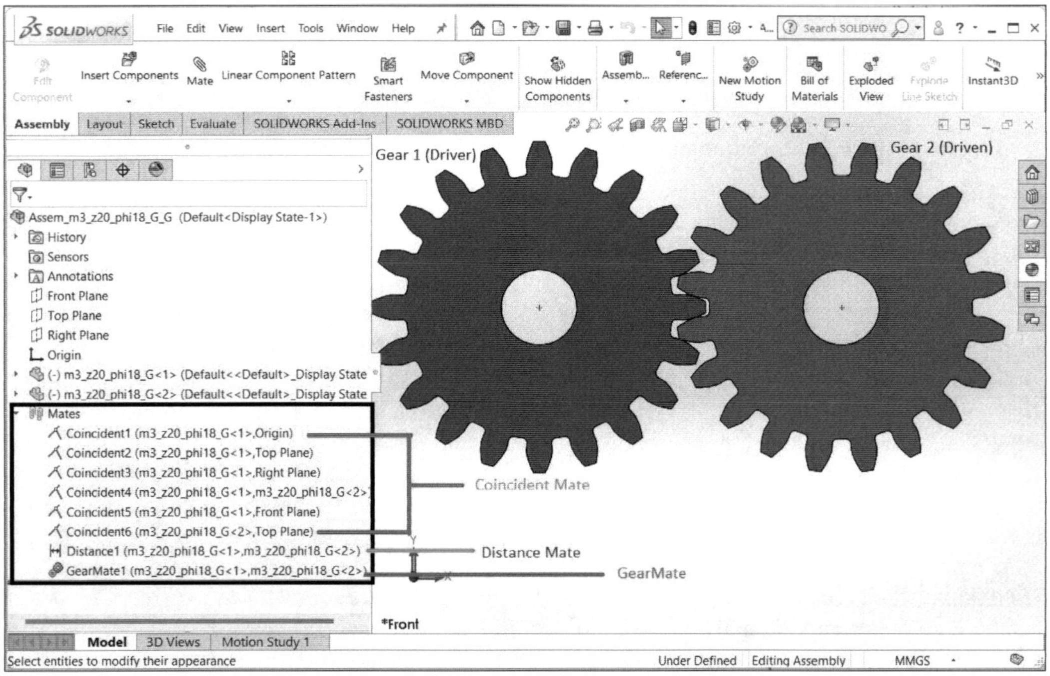

Fig. 2. Spur Gear pair and different mate options.

The angular and linear representations of TE are given by (1) and (2), which are evaluated at the base circle or pitch circle. In the current work TE is calculated at the base circle. The mesh stiffness is then represented in torsional and linear way as given by (3) and (4).

$$TE\ (radian) = \theta_{gear2} - \left({R_{bgear1}}/{R_{bgear2}} \right) \theta_{gear1} \tag{1}$$

$$TE\ (\mu m) = R_{bgear2}\theta_{gear2} - R_{bgear1}\theta_{gear1} \tag{2}$$

$$K_{Torsional} = \frac{T}{|TE(radian)|} \tag{3}$$

$$K_{Linear} = \frac{K_{Torsional}}{R_{bgear1}^2} = \frac{K_{Torsional}}{R_{bgear2}^2} \tag{4}$$

where θ_{gear1} & θ_{gear2} are the angle of rotations, R_{bgear1} & R_{bgear2} are base circle radius of driver gear 1 & driven gear 2, respectively. T represent resistive or braking torque and $K_{Torsional}$ and K_{Linear} represent torsional mesh stiffness (Nm/radian) & time varying mesh stiffness (N/m) respectively [1, 3-4]. The most convenient representation of mesh stiffness in literature is the time varying mesh stiffness (TVMS) representation, so in this work it is used for analysis. The technique provided by [1] is utilised for calculating TE and mesh stiffness. All the spur gears necessary for the investigation are parametrically designed in the CAD software SOLIDWORKS.

By modifying various gear properties, parametric modelling allows for the creation of many gears. After that, the modelled gears are put together to make different gear pairs. The distinct gear pairs are healthy gear pairs, which contain both healthy gears, and faulty gear pairs, which contain one healthy gear and another defective gear (1% & 2% PE). This is already explained in section 2. The specifications for these gears and gear pairs are shown in table 2. There are 24 gears total, and 24 gear pairs (assemblies) are created.

Ansys workbench is chosen owing to its wide ability to solve any complicated linked physics computations in a simplified way in a single environment. Any finite element analysis/method involves three major processes, known as preprocessing, processing, and postprocessing.

Table 2. Gear parameters used in presented work.

Parameters	Healthy Gear Pair (0% PE)	Faulty Gear Pair (1% PE)	Faulty Gear Pair (2% PE)
Module, m (mm)		3, 4, 5	
No. of Teeth, Z_{gear1}/Z_{gear2}		18/18, 20/20, 24/24, 30/30	
Pressure Angle, Φ (°)		18, 20, 25	
Face Width, b (mm)		10	
Modulus of Elasticity, E_{gear1}, E_{gear2} (N/mm^2)		210,000	
Poisson's ratio, ν		0.3	
Tooth tip radius (mm)		0.5, 1.0, 1.5	

3.1. Preprocessing

In this step, the type of analysis system, geometry, material properties, surface contact definition, mesh generation, boundary conditions, and different analysis settings are performed. These are explained as below.

1) Analysis System

There are different analysis modules/systems for solving various applications in Ansys Workbench. Here, a transient structural analysis is used for the analysis work. Fig.3 show the Workbench containing various analysis systems and a user interface for transient structural analysis.

2) Geometry and Material Properties

Ansys has an interlinked user interface for SOLIDWORKS. Any file containing 3D models can be directly opened in the Ansys mechanical environment while maintaining all design details. The gear pair is imported into the Ansys Workbench environment and material is assigned to each of the gears in the gear pair (See Fig. 3). The material assigned is structural steel, whose modulus of elasticity is listed in table 1.

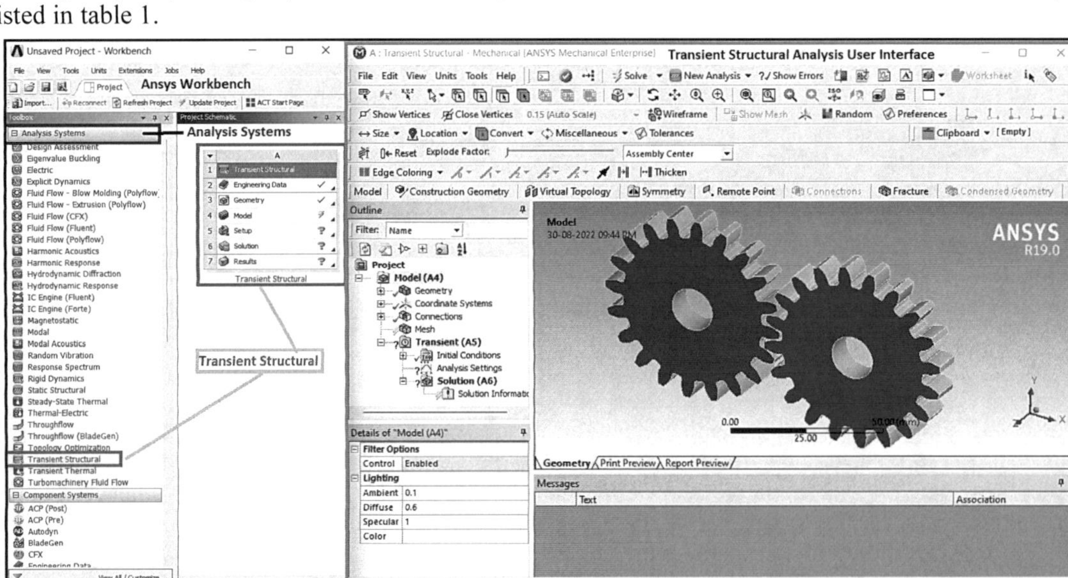

Fig. 3. Ansys Workbench Environment and Transient Structural Analysis user interface.

3) Defining Contact Regions

The contact zone between gear is chosen as frictionless with 0.1 mm penetration tolerance value and normal stiffness factor as 1. To achieve convergence, initial contact is established using "Adjust to Touch" interface treatment. Contact stiffness sensitivity is reduced by applying "Augmented Lagrange formulation". This is represented in Fig. 4.

4) Mesh Generation

Once the contact region is defined then element formation is done. The SOLID185 and CONTA170 element types are used. Because FEM takes a long time to compute, so few pairs of teeth are fine meshed to speed up the process. The meshed gears and tiny meshed teeth are shown in Fig. 3.

5) Defining Boundary Conditions

Body ground revolute joints are attached at the hub of the driver and driven gears. All other degrees of freedom are fixed, and the gear pair is allowed to rotate along its axis of rotation only. The driver gear is given a constant very low rotational velocity equivalent to 1 mesh cycle angle and the driven gear is given a braking torque. (Fig.5.).

6) Analysis Setting

To disable the inertial effects, the time integration feature is switched off. However, the large deflection should be on. For quasi-static analysis, a system damping value of 0.1 is utilised. Auto Time Stepping can be used to speed up the computing process if necessary and the sub-steps are selected in such a way that 1 sub-step equals 1° rotation angle. Fig. 6 show the analysis settings.

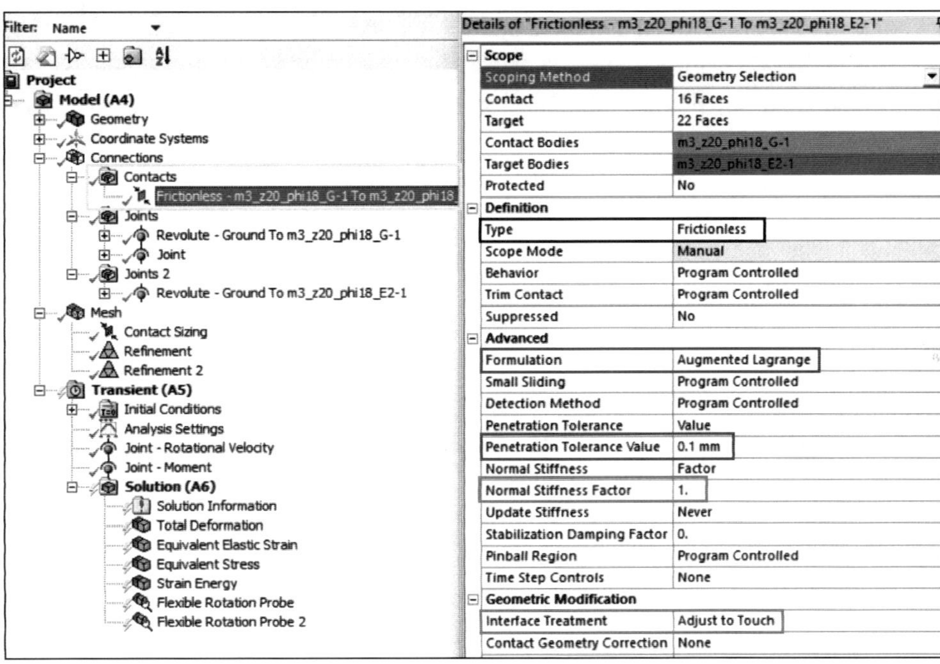

Fig. 4. Defining Contact Regions in Ansys.

3.2. *Processing (Analysis)*

After completing pre-processing stage, the non-linear loaded tooth contact analysis is performed.

3.3. *Post Processing (Result Solutions)*

The relative rotation of the driver gear and driven gear is measured with the help of joint probes attached to the two gears of the pair. Fig. 7 shows the joint probes. The Quasi-static Algorithm is utilised to compute the TE and TVMS [1]. The simulation data produced is entered into (1), (3) and (4) to get the TE and TVMS corresponding to the rotational angle.

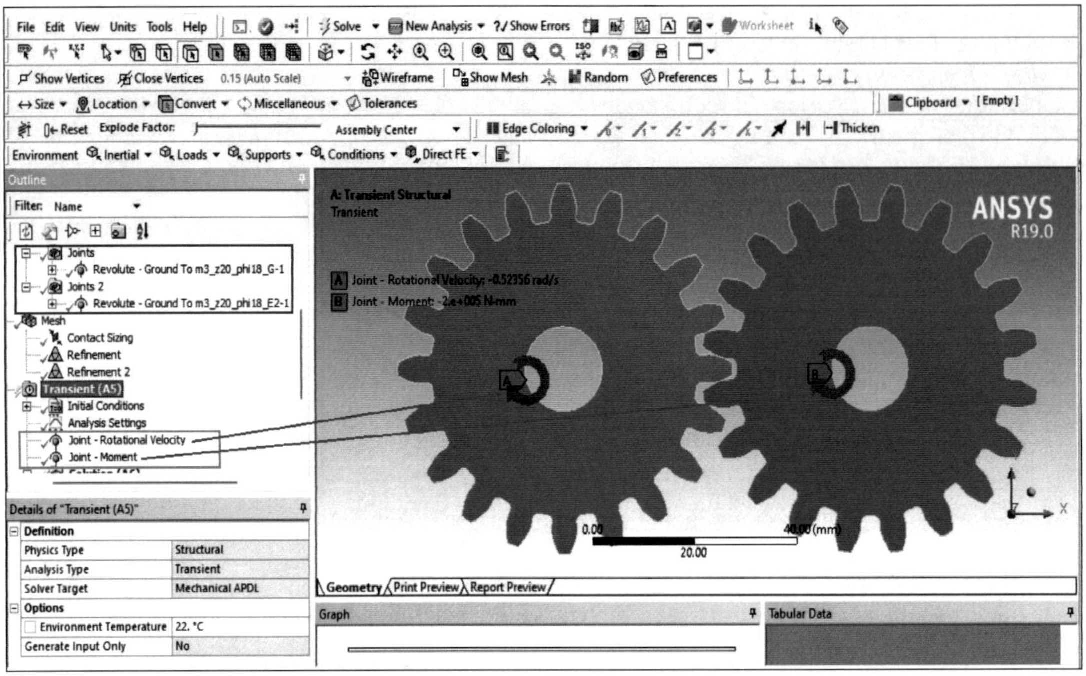

Fig. 5. Defining Boundary Conditions.

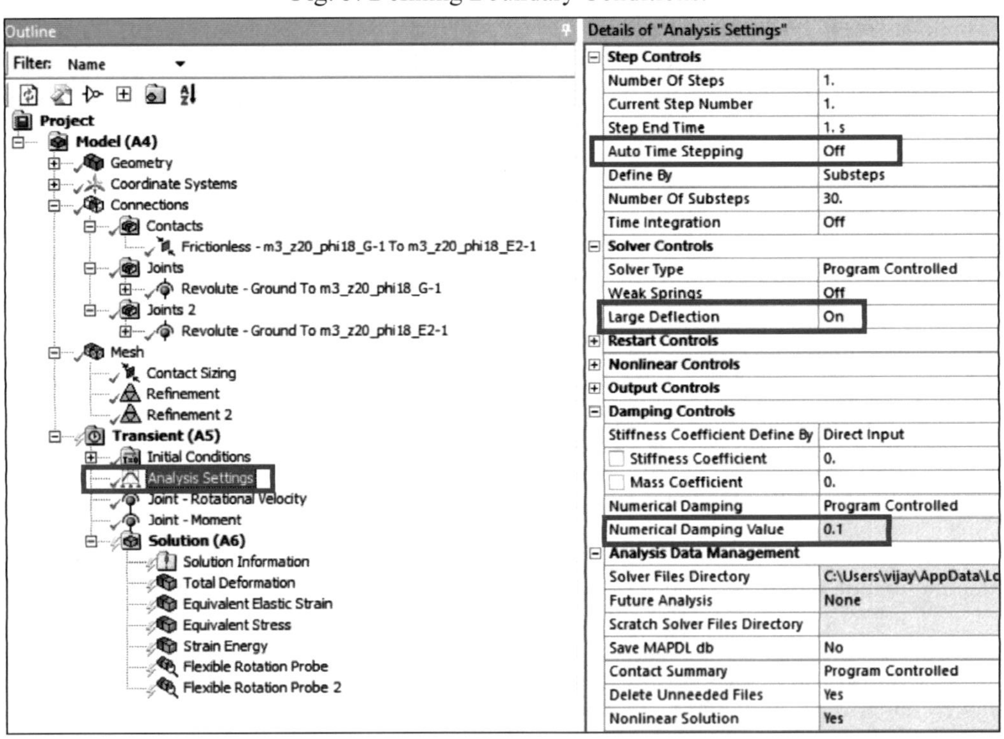

Fig. 6. Details of Analysis Setting.

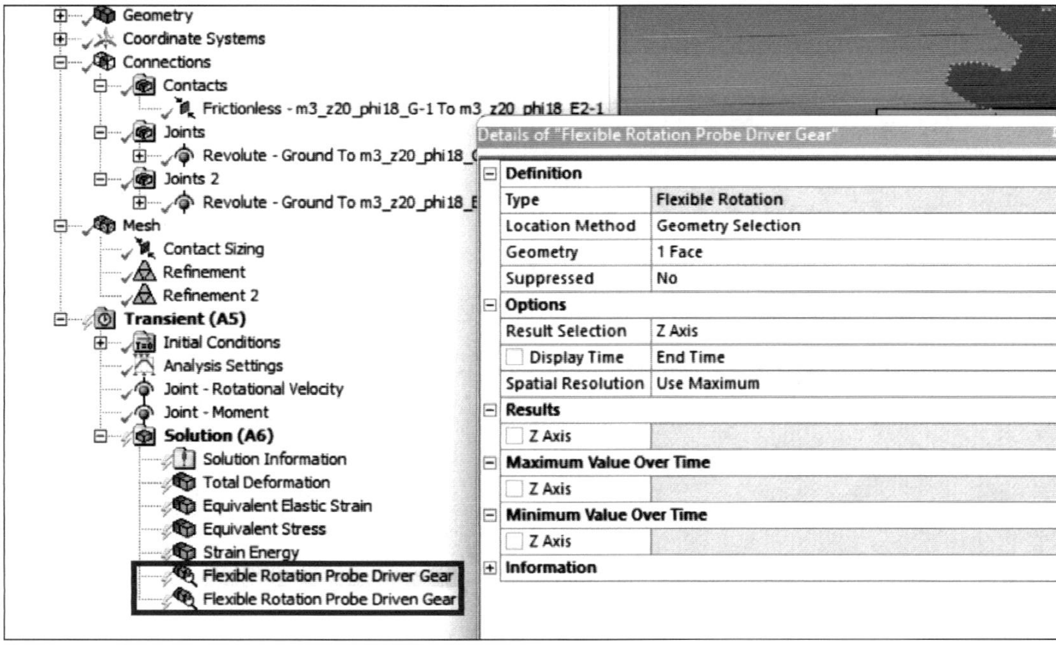

Fig. 7. Result of Solution (using probes to measure angular deformations).

4. Numerical Illustration and Discussion

The gears used in the work are listed in table 2. Now to use FEM to find the TE and mesh stiffness, import one of the healthy gear pairs, say module 3 mm, number of teeth 18, pressure angle 18°, into Ansys workbench first. The method employed is transient structural analysis. The SOLID185 and CONTA170 finite elements are then used to mesh this pair, resulting in 339308 elements and 682373 nodes. The contact pair is frictionless, with 0.1 mm of penetration tolerance and a touch interface treatment. Because FEM takes a long time to compute, 4 pairs of teeth are fine meshed to speed up the process. The meshed gear pair and tiny meshed teeth are shown in Fig 3. All other degrees of freedom are fixed, and the gear pair is allowed to rotate along its axis of rotation. The driver gear is one, while the driven gear is the other. The driver gear is given a constant rotational velocity of 0.5254 rad/sec, which is equivalent to 1 mesh cycle angle and the sub-steps are selected in such a way that 1 sub-step equals 1° rotation angle, while the driven gear is given a braking torque of 200 Nm. To obtain relative angular positions, the flexible rotation probes are connected to both the driver and driven gear. After a successful simulation/analysis, the varied angular locations acquired using probes on the driver and driven gear are input into (1) - (4) to provide TE and mesh stiffness. Following that, the analysis is carried out in the same manner for all other gear pairs as listed in table 2 and various plots of the TE and mesh stiffness curves for pressure angle, number of teeth, module, pitch error and tooth tip radius variations as a function of driver gear rotations are obtained (Figs. 8-15). The maximum and minimum values of TE and mesh stiffness i.e., TVMS are represented in table 3 and 4 respectively. The minimum value of TE occurs when double tooth contact takes place, and the maximum value occurs when single tooth contact takes place during the meshing of gear pairs while maximum value of mesh stiffness occurs in double tooth contact zone and minimum value occurs in single tooth contact zone.

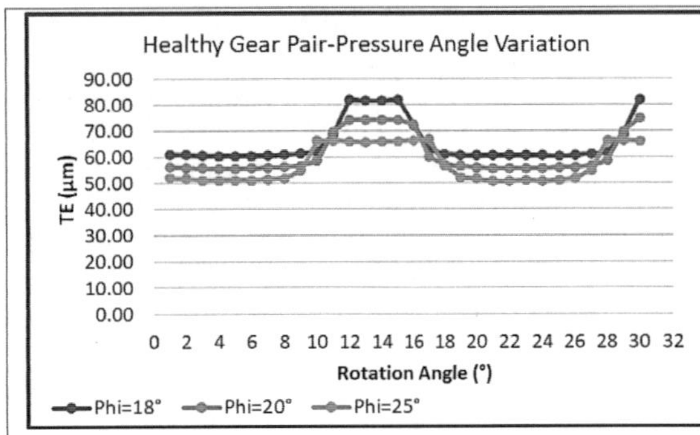

Fig. 8. (a) Healthy Gear Pair (m=3, $Z_{gear1}=Z_{gear2}=20$, $\Phi=18°$, 20°, 25°) - Influence of Pressure Angle on TE (µm).

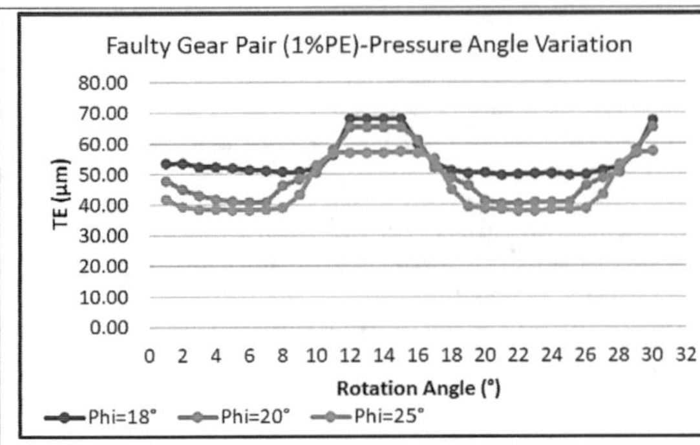

Fig. 8. (b) Faulty Gear Pair 1% PE (m=3, $Z_{gear1}=Z_{gear2}=20$, $\Phi=18°$, 20°, 25°) - Influence of Pressure Angle on TE (µm).

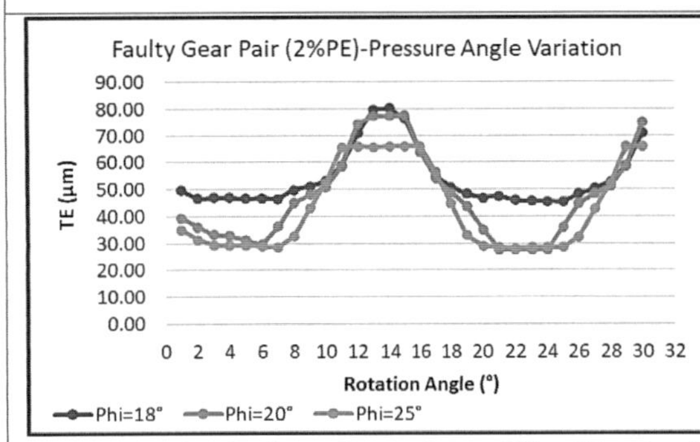

Fig. 8. (c) Faulty Gear Pair 2% PE (m=3, Zgear1=Zgear2=20, $\Phi=18°$, 20°, 25°) - Influence of Pressure Angle on TE (µm).

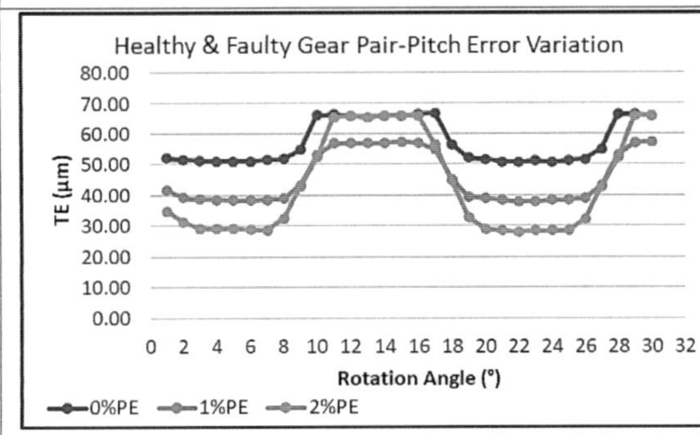

Fig. 8. (d) Healthy & Faulty Gear Pair (m=3, $Z_{gear1}=Z_{gear2}=20$, $\Phi=18°$) - Influence of Pitch Error on TE (µm).

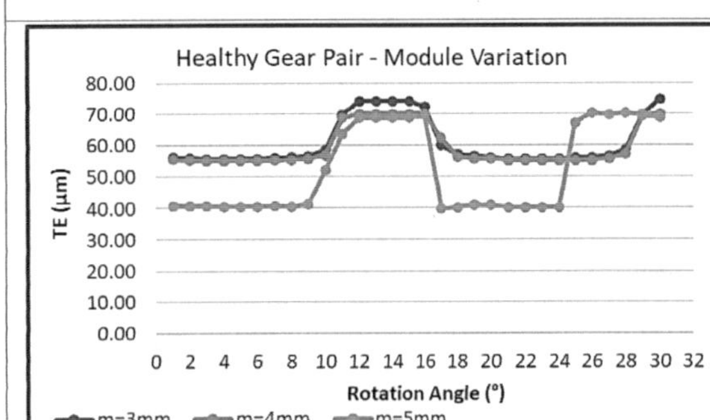

Fig. 9. (a) Healthy Gear Pair (m=3,4, 5 Zgear1=Zgear2=20, $\Phi=18°$) - Influence of Module on TE (µm).

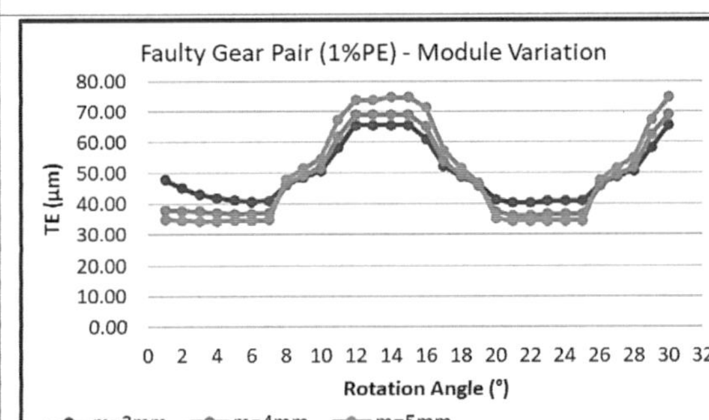

Fig. 9. (b) Faulty Gear Pair 1% PE (m=3, 4, 5 Zgear1=Zgear2=20,$\Phi=18°$)-Influence of Module on TE (µm)

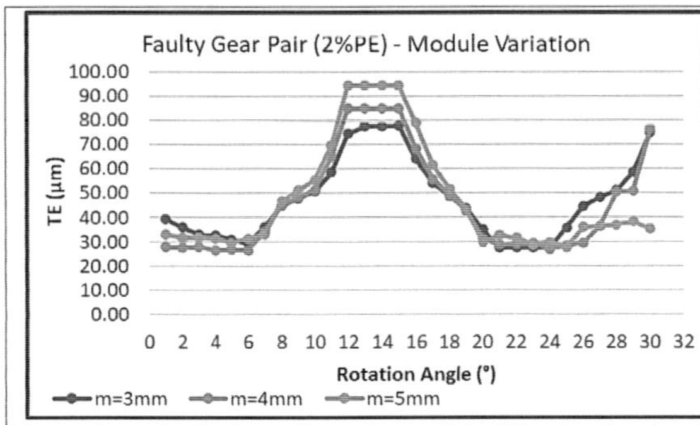

Fig. 9. (c) Faulty Gear Pair 2% PE (m=3, 4, 5 Zgear1=Zgear2=20, Φ=18°) - Influence of Module on TE (µm).

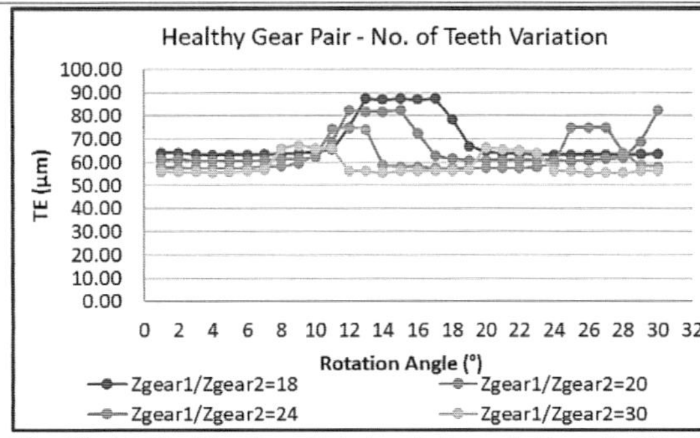

Fig. 10. (a) Healthy Gear Pair (m=3, Zgear1=Zgear2=18, 20, 24, 30, Φ=18°) - Influence of No. of Teeth on TE (µm).

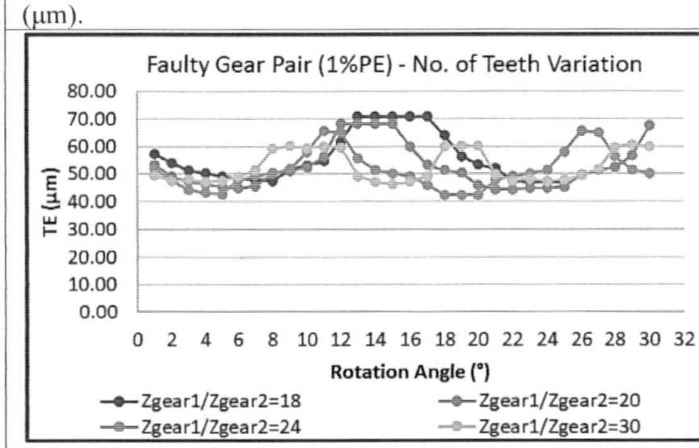

Fig. 10. (b) Faulty Gear Pair 1% PE (m=3, Zgear1=Zgear2=18, 20, 24, 30, Φ=18°) - Influence of No. of Teeth on TE (µm).

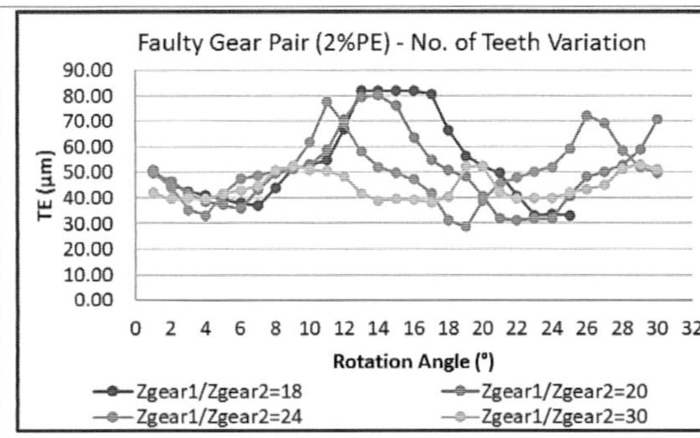

Fig. 10. (c) Faulty Gear Pair 2% PE (m=3, Zgear1=Zgear2=18, 20, 24, 30, Φ=18°) - Influence of No. of Teeth on TE (µm).

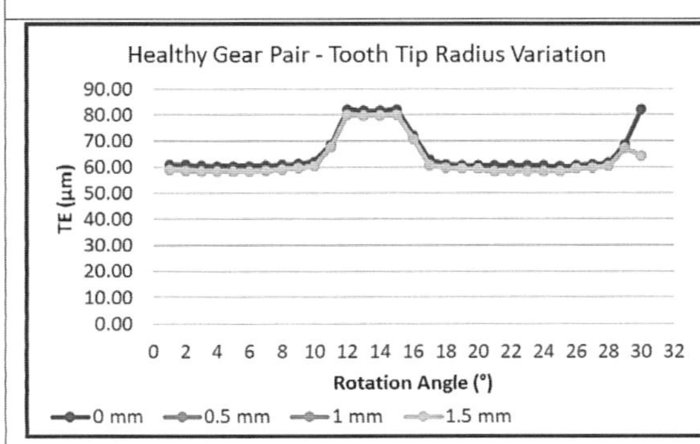

Fig. 11. (a) Healthy Gear Pair (m=3, $Z_{gear1}=Z_{gear2}=20$, Φ=18°) - Influence of Tooth Tip Radius on TE (µm).

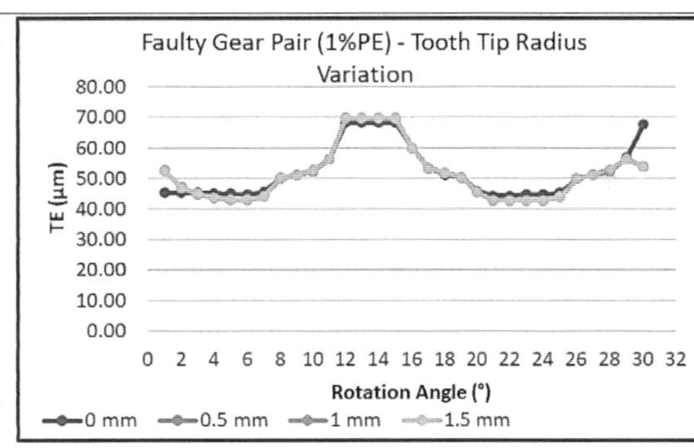

Fig. 11. (b) Faulty Gear Pair 1% PE (m=3, Zgear1=Zgear2=18, 20, 24, 30, Φ=18°) - Influence of Tooth Tip Radius on TE (µm).

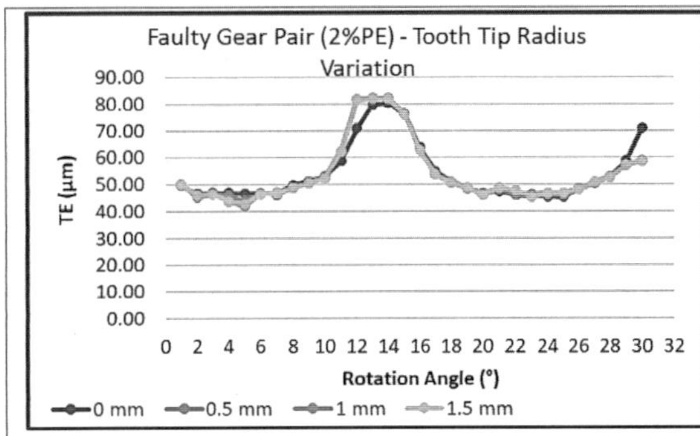

Fig. 11. (c) Faulty Gear Pair 2% PE (m=3, Zgear1=Zgear2=20, Φ=18°) - Influence of Tooth Tip Radius on TE (µm).

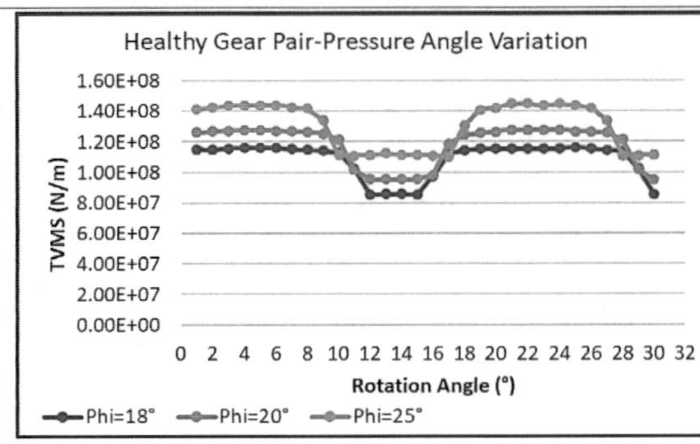

Fig. 12. (a) Healthy Gear Pair (m=3, Zgear1=Zgear2=20, Φ=18°, 20°, 25°) - Influence of Pressure Angle on mesh stiffness (N/m).

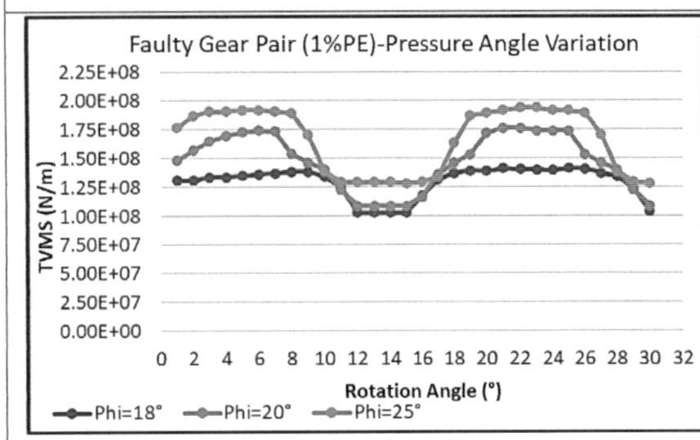

Fig. 12. (b) Faulty Gear Pair 1% PE (m=3, Z_{gear1}=Z_{gear2}=20, Φ=18°, 20°, 25°) - Influence of Pressure Angle on mesh stiffness (N/m).

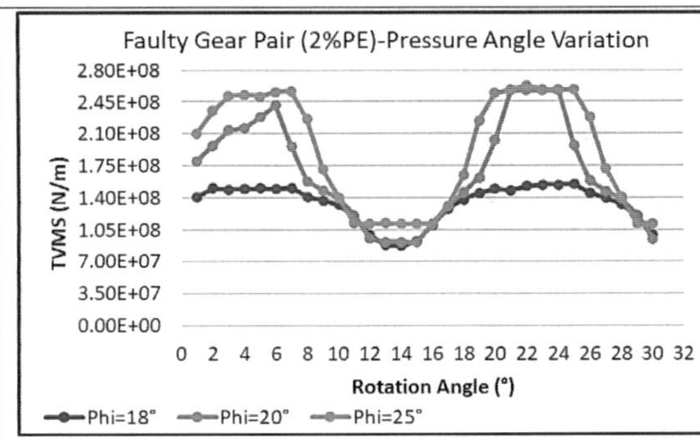

Figure 12. (c) Faulty Gear Pair 2% PE (m=3, Z_{gear1}=Z_{gear2}=20, Φ=18°, 20°, 25°) - Influence of Pressure Angle on mesh stiffness (N/m).

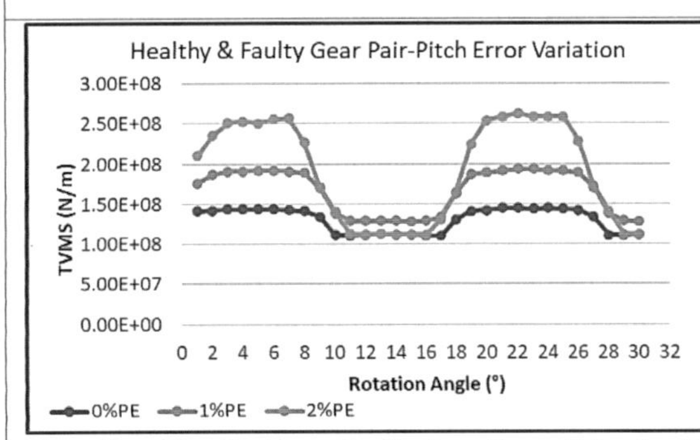

Fig. 12. (d) Healthy & Faulty Gear Pair (m=3, Zgear1=Zgear2=20, Φ=18°) - Influence of Pitch Error on mesh stiffness (N/m).

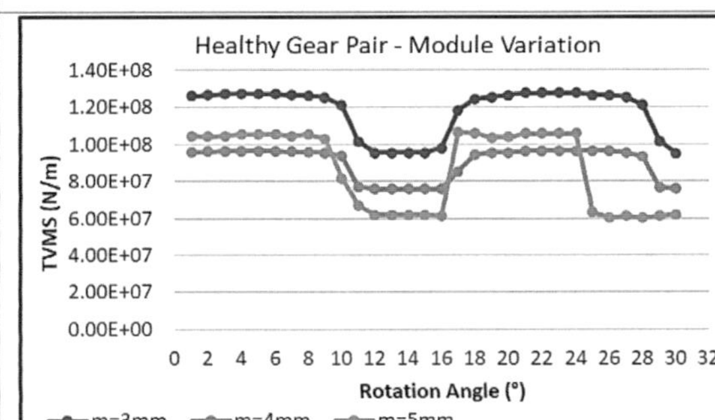

Fig. 13. (a) Healthy Gear Pair (m=3,4, 5 Zgear1=Zgear2=20, Φ=18°) - Influence of Module on mesh stiffness (N/m).

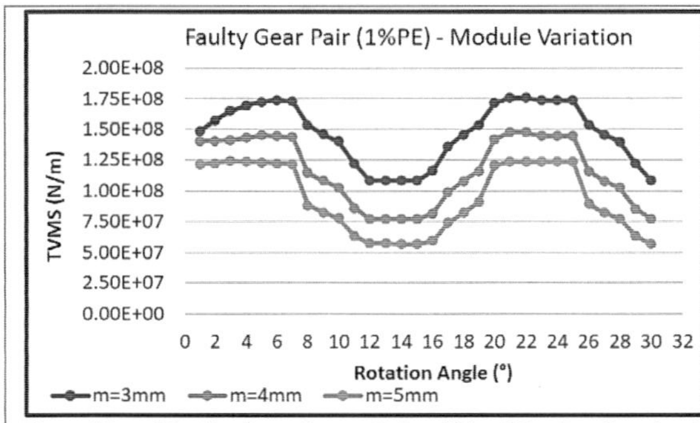

Fig. 13. (b) Faulty Gear Pair 2% PE (m=3, 4, 5 Zgear1=Zgear2=20, Φ=18°) - Influence of Module on mesh stiffness (N/m).

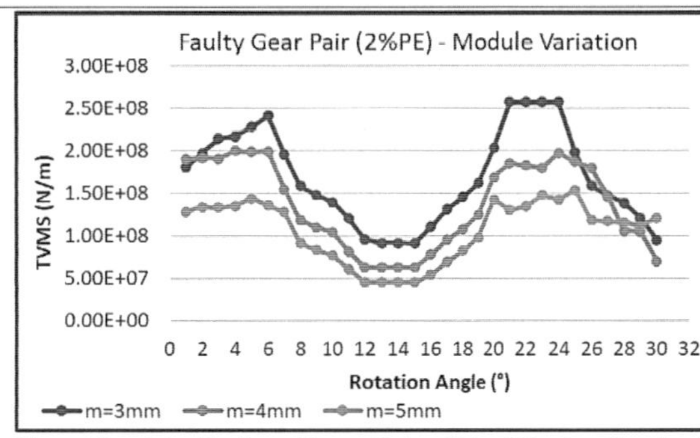

Fig. 13. (c) Faulty Gear Pair 2% PE (m=3, 4, 5 Zgear1=Zgear2=20, Φ=18°) - Influence of Module on mesh stiffness (N/m).

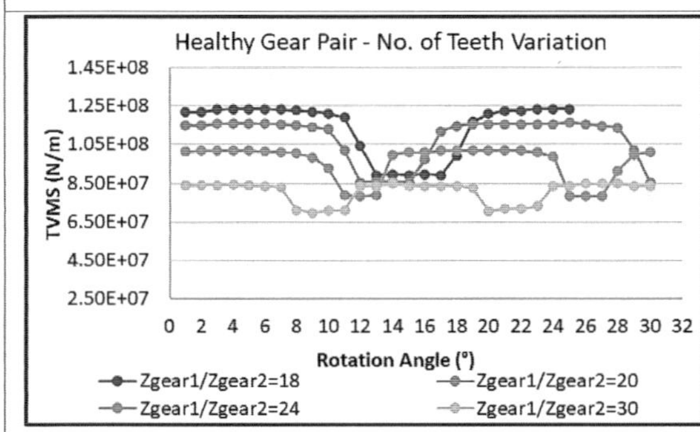

Fig. 14. (a) Healthy Gear Pair (m=3, Zgear1=Zgear2=18, 20, 24, 30, Φ=18°) - Influence of No. of Teeth on mesh stiffness (N/m).

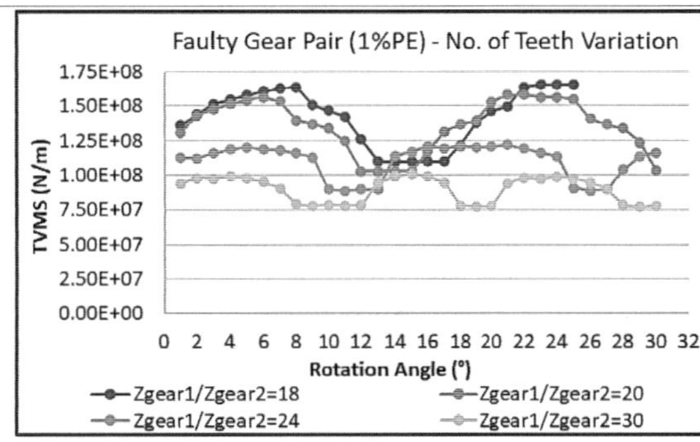

Fig. 14. (b) Faulty Gear Pair 1% PE (m=3, $Z_{gear1}=Z_{gear2}=18$, 20, 24, 30, Φ=18°) - Influence of No. of Teeth on mesh stiffness (N/m).

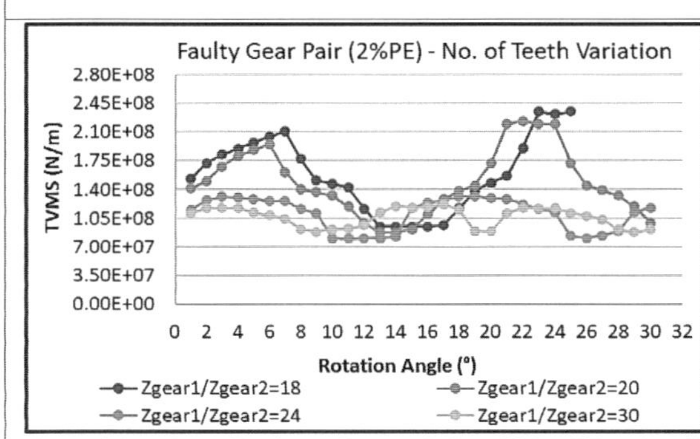

Fig. 14. (c) Faulty Gear Pair 2% PE (m=3, Zgear1=Zgear2=18, 20, 24, 30, Φ=18°) - Influence of No. of Teeth on mesh stiffness (N/m).

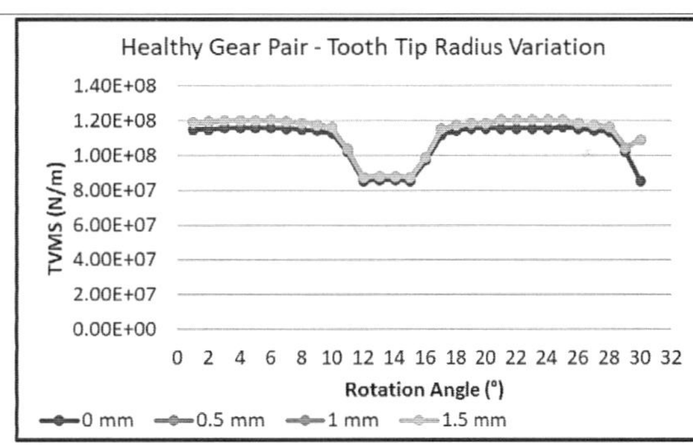

Fig. 15. (a) Healthy Gear Pair (m=3, $Z_{gear1}=Z_{gear2}=20$, Φ=18°) - Influence of Tooth Tip Radius on mesh stiffness (N/m).

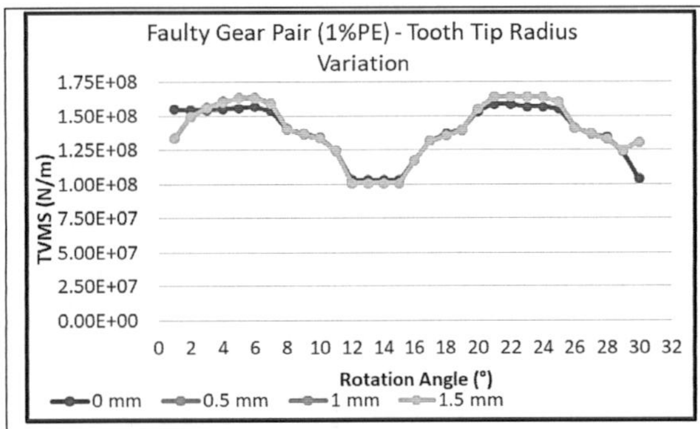

Fig. 15. (b) Faulty Gear Pair 1% PE (m=3, Zgear1=Zgear2=18, 20, 24, 30, Φ=18°) - Influence of Tooth Tip Radius on mesh stiffness (N/m).

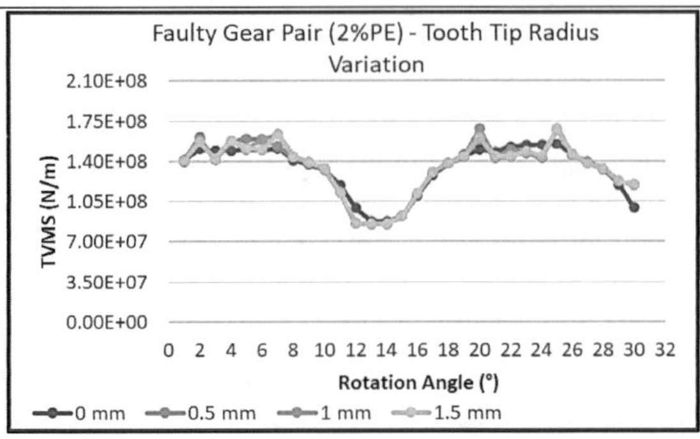

Fig. 15. (c) Faulty Gear Pair 2% PE (m=3, $Z_{gear1}=Z_{gear2}=20$, Φ=18°) - Influence of Tooth Tip Radius on mesh stiffness (N/m).

Table 3. Maximum and minimum values of TE (μm).

Parameters Variation	Healthy Gear Pair (0%PE)		Faulty Gear Pair (1% PE)		Faulty Gear Pair (2% PE)	
	Maximum	Minimum	Maximum	Minimum	Maximum	Minimum
Module = 3 mm	74.78	55.59	65.43	40.34	77.73	27.55
Module = 4 mm	70.19	55.10	68.88	36.08	84.62	26.63
Module = 5 mm	70.52	39.93	74.62	34.28	94.30	27.72
Pressure Angle = 18°	82.16	60.25	68.22	49.64	80.17	45.26
Pressure Angle = 20°	74.78	55.59	65.43	40.34	77.73	27.55
Pressure Angle = 25°	66.91	50.77	57.42	37.96	65.96	28.00
No. of Teeth = 18/18	87.39	63.10	70.81	47.05	82.01	33.16
No. of Teeth = 20/20	82.16	60.25	68.22	44.32	80.17	31.37
No. of Teeth = 24/24	74.69	57.36	65.73	42.42	77.68	28.68
No. of Teeth = 30/30	67.07	55.27	60.57	46.31	53.18	38.24
Tooth Tip Radius =0 mm	82.16	60.25	68.22	44.32	80.17	45.26
Tooth Tip Radius =0.5 mm	80.17	58.26	69.71	42.82	82.16	43.92
Tooth Tip Radius =1.0 mm	80.17	58.26	69.71	42.82	82.05	41.81
Tooth Tip Radius =1.5 mm	80.17	58.26	69.71	42.82	81.95	42.71

Table 4. Maximum and minimum values of mesh stiffness (N/m).

Parameters Variation	Healthy Gear Pair (0% PE)		Faulty Gear Pair (1% PE)		Faulty Gear Pair (2% PE)	
	Maximum	Minimum	Maximum	Minimum	Maximum	Minimum
Module = 3 mm	1.28E+08	9.49E+07	1.76E+08	1.08E+08	2.58E+08	9.13E+07
Module = 4 mm	9.66E+07	7.58E+07	1.47E+08	7.73E+07	2.00E+08	6.29E+07
Module = 5 mm	1.07E+08	6.04E+07	1.24E+08	5.70E+07	1.54E+08	4.51E+07
Pressure Angle = 18°	1.16E+08	8.53E+07	1.41E+08	1.03E+08	1.55E+08	8.74E+07
Pressure Angle = 20°	1.28E+08	9.49E+07	1.76E+08	1.08E+08	2.58E+08	9.13E+07
Pressure Angle = 25°	1.45E+08	1.10E+08	1.94E+08	1.28E+08	2.63E+08	1.12E+08
No. of Teeth = 18/18	1.23E+08	8.91E+07	1.66E+08	1.10E+08	2.35E+08	9.50E+07
No. of Teeth = 20/20	1.16E+08	8.53E+07	1.58E+08	1.03E+08	2.23E+08	8.74E+07
No. of Teeth = 24/24	1.02E+08	7.82E+07	1.22E+08	8.89E+07	1.32E+08	8.05E+07
No. of Teeth = 30/30	8.46E+07	6.97E+07	1.01E+08	7.71E+07	1.22E+08	8.79E+07
Tooth Tip Radius =0 mm	1.16E+08	8.53E+07	1.58E+08	1.03E+08	1.55E+08	8.74E+07
Tooth Tip Radius =0.5 mm	1.20E+08	8.74E+07	1.64E+08	1.01E+08	1.68E+08	8.53E+07
Tooth Tip Radius =1.0 mm	1.20E+08	8.74E+07	1.64E+08	1.01E+08	1.68E+08	8.54E+07
Tooth Tip Radius =1.5 mm	1.20E+08	8.74E+07	1.64E+08	1.01E+08	1.68E+08	8.55E+07

4.1 Influence of module on TE and Mesh Stiffness

Figs. 9a-9c and 13a-13c depict graphs related to the variation of module effect on TE and mesh stiffness, respectively. Tables 3 and 4 represent the maximum and minimum values of TE and mesh stiffness, respectively, for different parameters. When the module changes from 3 mm to 5 mm, there is a little variation in TE in the case of a healthy gear pair. TE increases for gear pairs with pitch errors. It shows some variation in mesh stiffness in the case of a healthy gear pair as well as a gear pair with pitch errors. This can be seen from the maximum and minimum values given in tables 3 and 4 respectively.

4.2 Influence of number of teeth on TE and Mesh Stiffness

In case of number of teeth variation, it is observed from Figs. 10a-10c, 14a-14c and Tables 3-4, there is some variation observed in TE and mesh stiffness. There is some decrease in TE and mesh stiffness observe when number of teeth varied from 18 to 30. This observation is noted for both healthy and faulty gear pairs.

4.3 Influence of pressure angle on TE and Mesh Stiffness

The pressure angle variation effect on TE and mesh stiffness is depicted in Figs. 8a-8c and 12a-12c, while the maximum and minimum values are shown in Tables 3-4, respectively. From this information, it is seen that as the pressure angle increases from 18° to 25°, there is a decrement in TE, but the variation is very small or negligible in both healthy and faulty gear pairs. The mesh stiffness shows an increasing trend. As the pressure angle increases, it causes tooth to become thin at the tip and thick at the bottom portion, making the gear strong at the lower portion.

4.4 Influence of pitch error on TE and Mesh Stiffness

Due to pitch error variation, it is noted that there is an increase in TE values and an increase in mesh stiffness. Since the pitch error is created by increasing the tooth thickness by 1% and 2% of the original value, the tooth becomes thick at the lower portion and increase the single tooth contact zone. (Figs. 8d, 12d and Tables 3-4).

4.5 Influence of tooth tip radius on TE and Mesh Stiffness

The tip radius is created by removing a little quantity of material around the tip of the gear tooth. Because tip relief is commonly utilized to minimize transmission errors, dynamic loads, vibration, and noise, it's crucial to look at how tip radius affects the TE and TVMS. The effect of tip radius on the TE and TVMS is seen in Figs. 11a-11c and Figs. 15a-15c respectively. In general, changing the tooth tip radius has only a little impact on the TE and TVMS's maximum and lowest values. However, due to the larger tooth tip radius, the two-tooth contact zone is enlarged while the one tooth contact region is reduced.

5. Conclusions

By adapting and using the CAD and FEM integrated approach for analysing the TE and mesh stiffness, the presented work investigated the impacts of the gear parameters module, teeth number, pressure angle, tooth tip radius, and geometric precision pitch error on TE and mesh stiffness. SOLIDWORKS is used to create the 3D models of gear pairs, which are then seamlessly loaded into the finite element system as the ANSYS workbench platform offers direct, associative, bidirectional interfaces with all major CAD packages. Figs. 8 to 15 show the results of Ansys simulations for various gear specifications plotted against the driver gear's rotation angle on the horizontal axis and the TE/mesh stiffness on the vertical axis. The observations from the work are as follows.

- The adapted method for determining the TE and mesh stiffness using the CAD-FEM integration approach is successfully applied in the case of gears with pitch error, tooth tip radius variation, and other gear parameter variations.
- Gear parameters module, pressure angle, and teeth quantity affects TE and mesh stiffness. Pressure angle variation has a very little impact on TE, but it increases the mesh stiffness. Number of teeth affects TE mesh stiffness. Lower teeth gear pair have more value of TE and mesh stiffness. Module variation in case of faulty gears has some impact on TE and reduction in mesh stiffness while healthy gear pair saw very less or negligible impact on TE, but mesh stiffness variation observed.
- In case of geometry flaw – pitch error impact on TE and mesh stiffness, pitch error variation causes TE fluctuations. There is some increase in mesh stiffness noted.
- There is some reduction in TE and mesh stiffness observed when tooth tip radius changes from 0 to 0.5 mm and after that its impact is negligible.

The integration approach adapted and used in the current work may be applied for the analysis of other deformations such as crack, spalling, pitting, etc and for gears other than spur gear.

Acknowledgment

The authors thanks Dr. Sanjiv Tokekar, Director IET DAVV for providing the research facilities in the institute and Dr. Ashesh Tiwari, Head Mechanical Engineering Department for motivating to do the research work.

References

[1] Zhan, J., Fard, M., Jazar, R. (2017), A CAD-FEM-QSA integration technique for determining the time-varying meshing stiffness of gear pairs, Measurement vol. 100 pp. 139-149 (2017). DOI: http://dx.doi.org/10.1016/j.measurement.2016.12.056

[2] Bartosova, D., Otipka, V., Rehak, K., (2018), Determination of transmission error in spur gear by numerical approach, Vibroengineering Procidia, September 2018, vol. 19 pp. 284-288 (2018). DOI: https://doi.org/10.21595/vp.2018.20239

[3] Y Benaïcha, Joël Perret-Liaudet, J-D Beley, E Rigaud, F Thouverez. (2021), On a flexible multibody modelling approach using FE-based contact formulation for describing gear transmission error. 2021. hal-03178317 DOI: https://hal.archives-ouvertes.fr/hal-03178317

[4] Ambaye, G. A., Lemu, H. G., (2021), Effect of backlash on transmission error and time varying mesh stiffness, IWAMA 2020 LNEE 737 pp. 18-28 (2021). DOI: https://doi.org/10.1007/978-981-33-6318-2_3

[5] Pleguezuelos, M., Sánchez, M. B., Pedrero J. I., (2021), Analytical model for meshing stiffness, load sharing, and transmission error for spur gears with profile modification under non-nominal load conditions, Applied Mathematical Modelling 97 (2021) 344–365. DOI: https://doi.org/10.1016/j.apm.2021.03.051

[6] Chen, z., Ji, P., (2020), Research on the variation of mesh stiffness and transmission error for spur gear with tooth profile modification and wear fault, Engineering Failure Analysis 122 (2021) 105184. DOI: https://doi.org/10.1016/j.engfailanal.2020.105184

[7] Zhang, Y. C., Wade A. S., Borghesani, P., Randall, B. R., Peng, z., (2021), Absolute transmission error: A simple new tool for assessing gear wear, Mechanism and Machine Theory vol. 146 107070 (2021). DOI: https://doi.org/10.1016/j.ymssp.2020.107070

[8] Czakó, A., Řehák, K., Prokop, A., (2020), Determination of static transmission error at helical gear, 26th International Conference, Engineering Mechanics 2020, Brno, Czech Republic, November 24-25, 2020, pp. 118-121 (2020). DOI: 10.21495/5896-3-118

[9] Czakó, A., Řehák, K., Prokop, A., Ranjan, V., (2020), Determination of static transmission error of helical gears using finite element analysis, Journal of Measurement in Engineering vol. 8 issue 4 pp. 167-181 (2020). DOI: https://doi.org/10.21595/jme.2020.21825

[10] Randall, R., Chin, z. Y., Smith, W., Borghesani, P., (2019), Measurement and use of transmission error for diagnostics of gears, In Surveillance, Vishno and AVE conferences, Lyon, France, (2019).

[11] Chang L., Wan-kai, S., Cura, F. M., Mura, A., (2020), A novel method to predict static transmission error for spur gear pair based on accuracy grade, J. Cent. South Univ. vol. 27 pp. 3334−3349 (2020). DOI: https://doi.org/10.1007/s11771-020-4550-4

[12] Duan, T., Wei, J., zhang, A., Xu, z., Lim, T. C., (2021), Transmission error investigation of gearbox using rigid-flexible coupling dynamic model: Theoritical analysis and experiments, Mechanism and Machine Theory vol. 157 104213 (2021). DOI: https://doi.org/10.1016/j.mechmachtheory.2020.104213

[13] Lee, J.-H., Choi, H.-S., Sohn, J.-H., Lee, G.-H., Park, D.-I., Kim, J.-G., (2021), Statistical Analysis for Transmission Error of Gear System with Mechanical and Thermal Deformation Uncertainties, Appl. Sci. vol. 11 6582 (2021). DOI: https://doi.org/10.3390/app11146582

[14] Kohler, H., Regan, R., (1985), The Derivation of Gear Transmission Error from Pitch Error Records, Proceedings of the Institution of Mechanical Engineers, part C: Journal of

Mechanical Engineering Sciences, 1985, vol. 199, pp. 195-201 (1985). DOI: https://doi.org/10.1243/PIME_PROC_1985_199_114_02
[15] Bruzzone, F., Maggi, T., Marcellini, C., Rosso, C.., (2021), 2D nonlinear and non-Hertzian gear teeth deflection model for static transmission error calculation, Mechanism and Machine Theory vol. 166 104471 (2021). DOI: https://doi.org/10.1016/j.mechmachtheory.2021.104471
[16] Kiekbusch, T., Sappok, D., Sauer, B., Howard, I., (2011), Calculation of the Combined Torsional Mesh Stiffness of Spur Gears with Two and Three Dimensional Parametrical FE Models, Strojniški vestnik - Journal of Mechanical Engineering vol. 57 (2011) 11, pp. 810-818 (2011). DOI: 10.5545/sv-jme.2010.248
[17] Sánchez, M. B., Pleguezuelos, M., Pedrero, J. I., (2019), Influence of Profile modifications on meshing stiffness, load sharing ratio and transmission error of involute spur gears, Mechanism and Machine Theory vol. 139 pp. 506-525 (2019)
[18] Doğan, O., Karpat, F., Kopmaz, O. *et al.* (2020), Influences of gear design parameters on dynamic tooth loads and time-varying mesh stiffness of involute spur gears, *Sādhanā* **45,** 258 (2020). DOI: https://doi.org/10.1007/s12046-020-01488-x
[19] Nina Sainte-Marie, (2019), A transmission-error-based gear dynamic model: Applications to single- and multi-mesh transmissions. Mechanics [physics.med-ph]. Université de Lyon, 2016. English. NNT: 2016LYSEI133. tel-02067109
[20] Chen, X., Wei, J., Zhang, J., Zhang, C., Wang, C. L., Xu, Z., Gao, H., Zhang, A., & Yu, G. (2022, April). A novel method to reduce the fluctuation of mesh stiffness by high-order phasing gear sets: Theoretical analysis and experiment. Journal of Sound and Vibration, 524, 116752. https://doi.org/10.1016/j.jsv.2022.116752
[21] Xingyuan Zheng, Wenjun Luo, Yumei Hu, Zao He, Sheng Wang, (2022), Analytical approach to mesh stiffness modeling of high-speed spur gears, International Journal of Mechanical Sciences, Volume 224, 2022,107318,ISSN 0020-7403,https://doi.org/10.1016/j.ijmecsci.2022.107318
[22] Shen, Z., Yang, L., Qiao, B., Luo, W., Chen, X., & Yan, R. (2022, March). Mesh relationship modeling and dynamic characteristic analysis of external spur gears with gear wear. Frontiers of Mechanical Engineering, 17(1). https://doi.org/10.1007/s11465-021-0665-z
[23] He, Y., & Yang, H. (2023, March 11). Analytical Model of Gear Thermal Stiffness Based on the Potential Energy Method. Applied Sciences, 13(6), 3599. https://doi.org/10.3390/app13063599
[24] Liu, G., Liu, F., Ma, T., Jiang, H., Yang, C., & Zhang, J. (2023, June 11). Dynamic analysis of spur gears system with dynamic force increment and velocity-dependent mesh stiffness, Nonlinear Dynamics 2023. https://doi.org/10.1007/s11071-023-08603-7
[25] Gonzalo Gonzalez Rey, and Ing. Gerardo Brianza Gordillo, "Parametric Geometric Modelling of a Spur Gear Using SOLIDWORKS", (2017). https://gearsolutions.com/features/parametric-geometric-modeling-of-a-spur-gear-using-solidworks/
[26] Tom, "Accurate Involute Gears in SOLIDWORKS", (2015). https://thebloughs.net/involute-gear-generation-in-solidworks/

IOP Conf. Series: Materials Science and Engineering 1291 (2023) 012035 doi:10.1088/1757-899X/1291/1/012035

Experimental Investigation of SPIF Processfor Conical Frustum

Hiteshkumar Patel[1] and Kapil Surani[1]

[1] Department of Mechanical Engineering, Gujarat Power Engineering and Research Institute,Gujarat, India. (*hit2424@gmail.com*)

Abstract: In sheet metal industries, single point incremental forming (SPIF) is a popular technology for quick prototyping and small series manufacturing techniques. In producing specific components with complicated geometries, the technique has become known for its flexibility, effectiveness, and simplicity of usage. The objective of this research work is to carry out the numerical investigation of the SPIF process to study the effect of different parameters on the formability for truncated cone. Al 1060 grade of aluminium is used for this study which is commonly used for kitchenware, reflectors, electronic components etc; where strength is not a major concern. Three dimensional models were developed for different set of parameters using commercial finite element software ABAQUS 6.13with Dynamic/Explicit solver using solid mechanics model. Taguchi analysis has been carried out to find the effect of parameters (Tool rotational speed, Feed, Step size, Tool diameter and Wall angle) on thickness reduction. Each parameter was varied for three levels and L9 orthogonal array was used. The ANOVA tool has been used to summarize the contribution of each variable. Results have beencompared with reported experimental results for validation. The process modelling is concluded with abrief overview of approaches for improving part quality.

Keywords: Single point incremental forming (SPIF), Flexibility, Solid mechanics model, Taguhi Analysis, ANOVA, and Process modelling

1. Introduction:

The creation of flexible manufacturing methods that can easily maintain the difficult product proficiency in the market is currently in growing market. A relatively new technique of moulding is called incremental sheet forming (ISF). Its flexibility is the key for increasing use of the process in the industries that can conform to the needs of current scenario. ISF holds a competitive advantage for fabricating low volume functional sheet metal products in an economic and efficient way. Complex shapes can be manufactured using this method and the with less set up cost, less manufacturing time, without use of dies and higher degree of flexibility. Mason evaluated the actual estimation of the popular present ISF in 1978 for use with small batches of customized parts. The idea suggests using a singular spherical tool with three axes of numerical control. The development of technology, more particularly the appearance of numerically controlled machines, made it possible to use this method. Later, Mason's work was carried in Japan, leading to additional inventions [1]. This has been proved to be an efficient process for small batch production and customized products in sheet metal forming.

Tomaž Pepelnjak et al compared finite element (FE) analysis to experimental validation of a progressively built truncated pyramid. The emphasis was on the potential simplicity of the FE process modelling and its impact on the dependability of the obtained findings, particularly on the geometric precision of the component and the bottom pillowing effect. The FE modelling of the SPIF process was carried out using the programme ABAQUS [2]. Slim Bouzidi et al, used the SPIF technique on AZ31B Magnesium alloy. A comparison of the gathered responses from portions acquired without holes and those obtained from perforated plates revealed inconsistencies [3]. M. Sbayti et al. developed hot single point incremental forming (HSPIF) procedures to increase the shape complexity and formability of the Ti6Al4V alloy. They used four contemporary meta-heuristic algorithms for met modeling: Multiverse (MVO), Moth-flame (MFO), Harris Hawk (HHO), and Marine Predictor (MPA). The computational findings show that the strategies under consideration are quite competitive in geometry optimization. Among these four optimizers, MPA is the best in estimating the pillow defect in HSPIF rapidly [4]. Sherwan Mohammed Najm et al. used Single Point Incremental Forming (SPIF) to perform several forming experiments under various forming circumstances. The arithmetical mean roughness (Ra) and ten-point mean roughness (Rz) of an AlMn1Mg1 sheet were predicted using an Artificial Neural Network (ANN). The findings demonstrated that ANN with a single output provides an optimal R-Square [5]. Adham E. Ragab et al. investigated the impact of four process factors (tool diameter, feed rate, step size, and sheet thickness) on the final product properties. They use principle component analysis (PCA) to find the correlations between answers [6]. Weining Li et al.investigated heating-assisted SPIF systems for high-strength alloy sheets in order to overcome current constraints. They investigated heating system methods, tools, tool path design, lubricants, and macro- and micro-numerical assessments [7].

Single Point Incremental Forming (SPIF) is a kind of AISF and a novel technique for highly localized deformation by the gradual feed of a generic tool following a predefined and programmed path with the help of a CNC machine leading to the final geometry of the parts shown in Figure 1. Unlike its close forming processes, shear forming and spinning, ISF is able to form complex asymmetric geometries due to computer assisted forming [8].

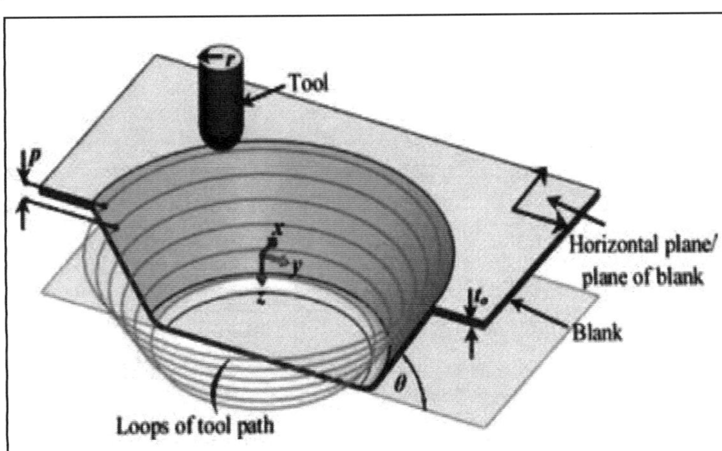

Figure 1. Tool progressions with path

ISF is a sheet material formation method built on additive manufacturing process concepts. Slices cut horizontally deform the sheet portion locally. The NC technique is used to move the locus of the forming tool (also known as the tool path) in these slices that are built to the work piece.

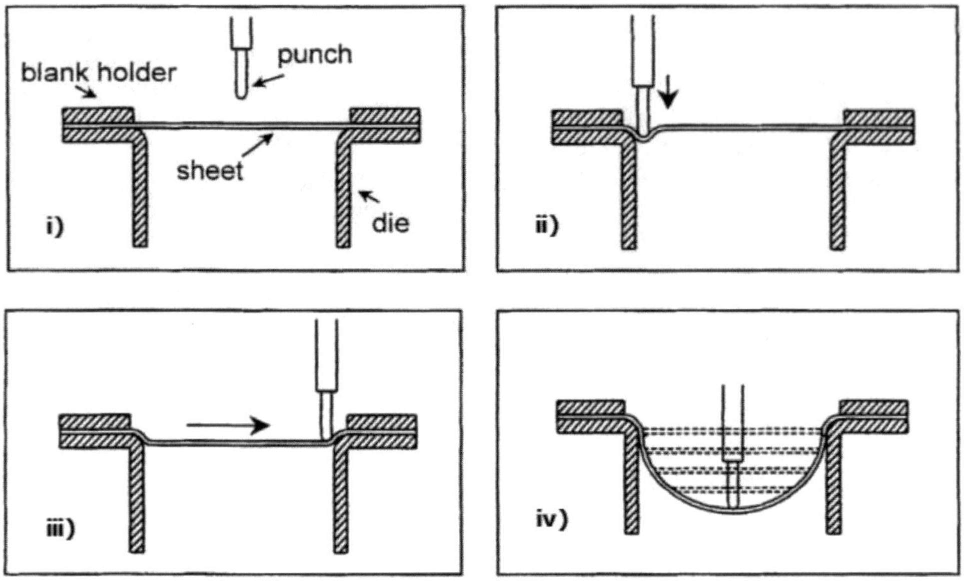

Figure 2 Schematic diagram of ISF

Using CAPP techniques, the tool path is directly generated from the CAD model of the finished object. The forming instrument has an end that is hemispherical. The edges of the metal sheet are rigidly fixedon a straightforward frame as it travels along the tool path [9]. The ISF process's entire main steps are shown in Figure 2.

i. The steel plate is securely clamped to the structure.
ii. The forming tool descends and makes touch with the sheet.
iii. The tool moves in accordance with the initial tool path, which has the end product's geometric shape.
iv. The tool keeps performing these actions until it reaches the conclusion of its path.

In the incremental sheet metal forming, generally two types of tool are used to form the sheet: Water Jet (WJ) and solid tools such as punches. Solid tool generally hemispherical and flat headed. Compared to hemispherical end tools, flat end tools can offer greater contour precision and formability. Additionally, especially in comparison to hemispherical end tools, flat end tools needed comparatively less forming force [10]. The Forming Limit Curve in SPIF is roughly 2.7 times higher than standard Forming Limit Curves (FLC) than traditional processes, such as deep drawing and stamping [11]. The large through thickness shear or the irregular strain path caused by cyclic plastic deformation are the two factors responsible for this rise in formability [12]. In spiral tool paths give better surface finish in comparison to stepped contours and no entry and exit marks are observed. However, spiral paths are typically not allowed by CAM programmes. The easier approach to defining spiral path is via set of coordinates. When moving at faster velocities, surface flaws like sheet waviness get worse [13]. Surface roughness also increases with speed. For the FEM modelling of the SPIF procedure, use ABAQUS/Standard (implicit solver). The authors enhanced the aluminium of the plastic deformation zone in the SPIF process using a FE sub-modelling method. The outcome demonstrated that there are differences in the contact pressure distribution between the sheet and the shaping instrument. The contact could typically be divided into two sections, according to a comparison of the distribution of the contact pressure under various operating circumstances. To reduce processing time in their research, a cone model with a 40° pie is used for modelling [14].

2. Experimental Work

To raise the calibre of products and procedures, the Taguchi technique is applied. When a better degree of performance is consistently attained, improved quality follows. By choosing the ideal mix of design elements, the greatest efficiency is achieved. Making the product/process resistant to the impact of uncontrollable factors allows for uniformity in performance. In the Taguchi method, the best design is chosen using design-of-experiment principles, and performance constancy is attained by running the trial condition while taking noise factors into consideration. To explore their influence on the final result, certain factors, including tool design, methods, and substance parameters, have been chosen.

Table 1. Values of Factors

Parameters	Level 1	Level 2	Level 3
Tool Rotational speed (RPM)	750	1000	1250
Feed (mm/min)	75	100	125
Step size (mm)	0.25	0.5	0.75
Wall angle	55	65	75

The selection of parameters in accordance with formability, are selected. Four parameters (Tool rotational speed, Feed, Step size and Wall angle) have been selected with an objective to observe the effect of these process parameters on product geometry and percentage thinning. Four levels of factors were selected for DOE. The factors and there levels used in present study are listed in Table 1. Constant parameters for the experiment design are listed in Table 2.

Table 2. Constant parameters

Parameters	Value
Initial Cone Diameter	44 mm
Cone Height	12 mm
Tool Diameter	6 mm
Sheet Thickness	0.91 mm

The trials can now be chosen and their methodology created using the variables and levels identified during the ideation session. The Taguchi technique is founded on conducting evaluations or experimentation to evaluate the sensitivity of response variables to a set of control parameters by taking into consideration trials in "Orthogonal Array (OA)" with a goal to achieve the ideal setup of process parameter. An optimal collection of well-balanced (minimum) trials is provided by orthogonal arrays.

IOP Conf. Series: Materials Science and Engineering 1291 (2023) 012035 doi:10.1088/1757-899X/1291/1/012035

Table 3. L9 Orthogonal Arrays for defined factors and levels

Sr. Number	Rotational Speed (RPM)	Feed (mm/min)	Step Size (mm)	Wall Angle (degree)
1	750	75	0.25	55
2	750	100	0.5	65
3	750	125	0.75	75
4	1000	75	0.75	65
5	1000	100	0.25	75
6	1000	125	0.5	55
7	1250	75	0.5	75
8	1250	100	0.75	55
9	1250	125	0.25	65

In the present study orthogonal array was designed using five factors with four levels. Here we choose L_9 orthogonal array was selected. Table 3 shows the L_9 orthogonal array for defined factors and their levels. Percentage thickness reduction and profile error in the simulated results was compared with practical output in order to validate the model. Further effect of process parameters over thickness reduction are analysed using "Analysis of Variance (ANNOVA)" tool.

3. Modelling and Simulation

Modelling is representation of graphical data into the model of actual part with the help of aluminium software. Modelling is at the heart of modern engineering practices. In this study efforts are made to perform FEM analysis of SPIF of AA1060 for conical frustum. The commercial FEA software ABAQUS 6.13 is used with explicit aluminium. Aluminium alloys have a wide range of application from general purpose to industry specific works. It possesses good mechanical strength, excellent weight to volume ratio, machinability and weldability. Aluminium / Aluminium 1060 alloy is a low strength and pure Aluminium / Aluminium alloy with low strength and good corrosion resistance characteristic, basically used for manufacturing of general purpose items where strength is not a major concern like utensils and electrical appliances. Economical cost and ease of machining favour it for the selection.

'Hemispherical Punch' is created as 3D Analytical rigid revolved shell object, 3D model of which is shown in Figure 3 (a). Different radius is taken for the geometry of Hemispherical Punch for different simulation runs as per DOE. Sheet metal blank is considered to be '3D deformable shell planar' object with a dimension of 90 mm × 90 mm. (Figure 3 b).

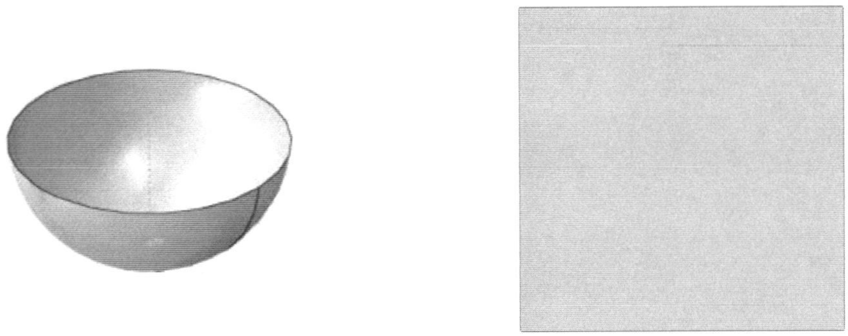

(a) (b)

Figure 3. 3D Model of parts (a) Hemispherical Punch ; (b) Sheet metal Blank

No material has been required to define for the deforming tool as it is considered to be an analytical rigid object. Sheet metal is assigned with the material property of AA1060 asper Table 4.

Table 4. Material Properties [15]

Properties	**Value**
Material Name	AA1060
Density	2.68×10^{-5} N/mm^3
Elastic Properties	Isotropic
	Young's Modulus = 78 Gpa
	Poisson's Ratio = 0.33
Plastic Properties	Hardening – Isotropic
	Yield Stress = 75 Mpa
	Plastic Strain = 0
Rate Dependent	Power Law
	Multiplier = 328
	Exponent = 0.19

Part geometries are assembled according to the actual configuration in single coordinate system (Figure 4). Tool radius compensation is considered while positioning the tool over the sheet. Tool Radius Compensation is given as per the equation 1.

$$\partial = r\, X \left(\frac{1}{\sin\alpha} - \frac{1}{\tan\alpha}\right) \tag{1}$$

Where;
∂ = tool radius compensation
r = radius of the tool
α = wall angle

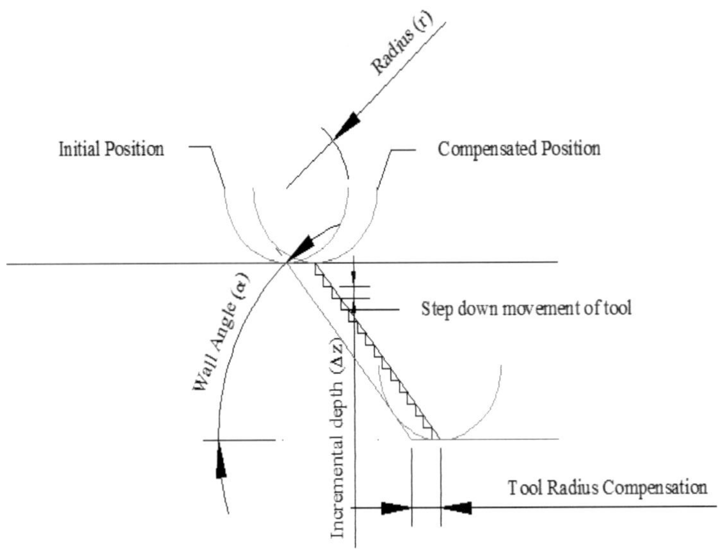

Figure 4. Tool radius compensation

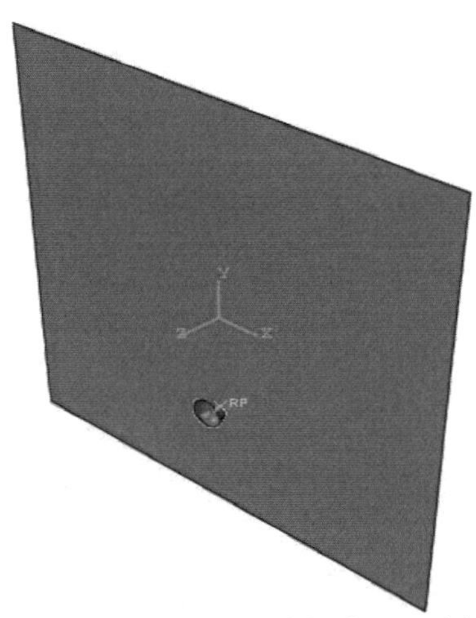

Figure 5. 3D Model of assembly

After creating assembly of tool and blank, the surface to surface type interaction between the outer surface of the hemispherical tool, taken as master surface, and the top surface of the sheet metal blank, taken as slave surface is assigned (Figure 4). The contact property taken is tangential behaviour type, directionality taken is isotropic and friction coefficient between the two surfaces is taken as 0.09 [16]. The Figure 5 shows the view of the assembly with contact definition. 3D model of final assembly is shown in Figure 6.

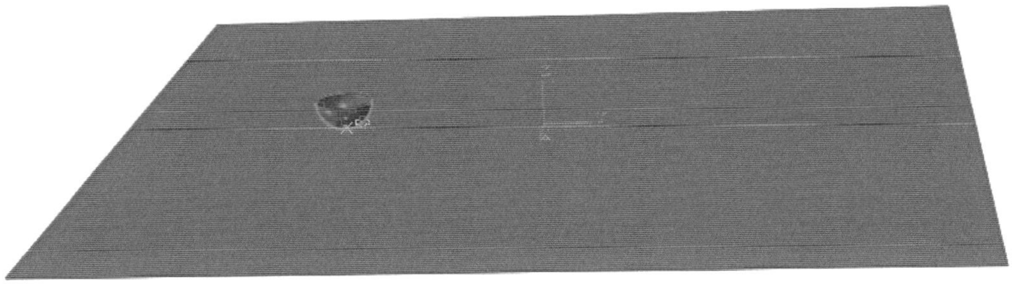

Figure 6. Assembly showing interaction between tool and blank

The Sheet blank was fixed at its four edges, to hold the sheet at its place as it is done by blank holderin experimental tests. The mid surface area of the sheet is not supported and is kept free to deform freely as shown in Figure 7.

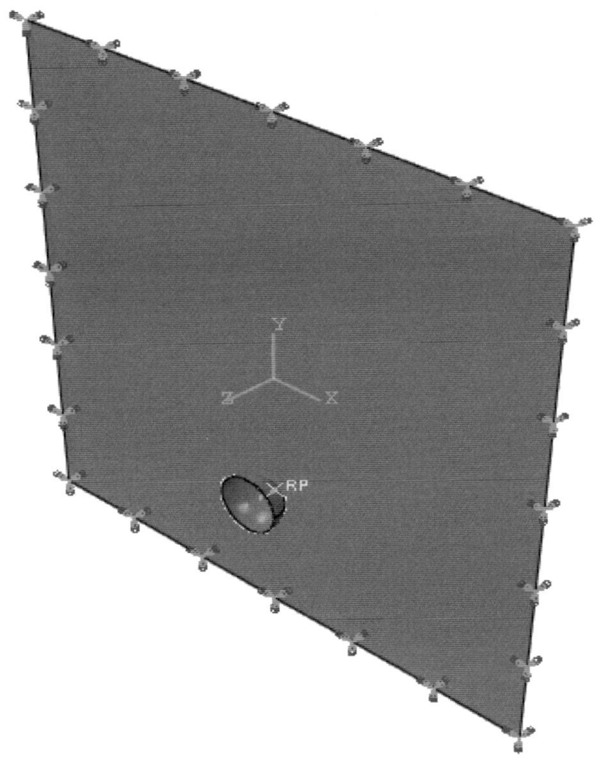

Figure 7. 3D Model of assembly with Encastred edge

The load is defined in terms of velocity assigned to it. First tool movers downward following the wall angle and then planar movement takes place to form the circle. For the circular movement of tool amplitudes with smooth steps are incorporated for planar velocities in X and Y in terms of time and velocity as described by the equation 2 and 3 for each of the circular contour.

$t = cos\theta$ (For amplitude along X axis) (2)

$t = sin\theta$ (for amplitude along Y axis) (3)

IOP Conf. Series: Materials Science and Engineering 1291 (2023) 012035 doi:10.1088/1757-899X/1291/1/012035

Where;

t = smallest time step for the circle.

θ = incremental angle for circle Example of steps is shown is Figure 8.

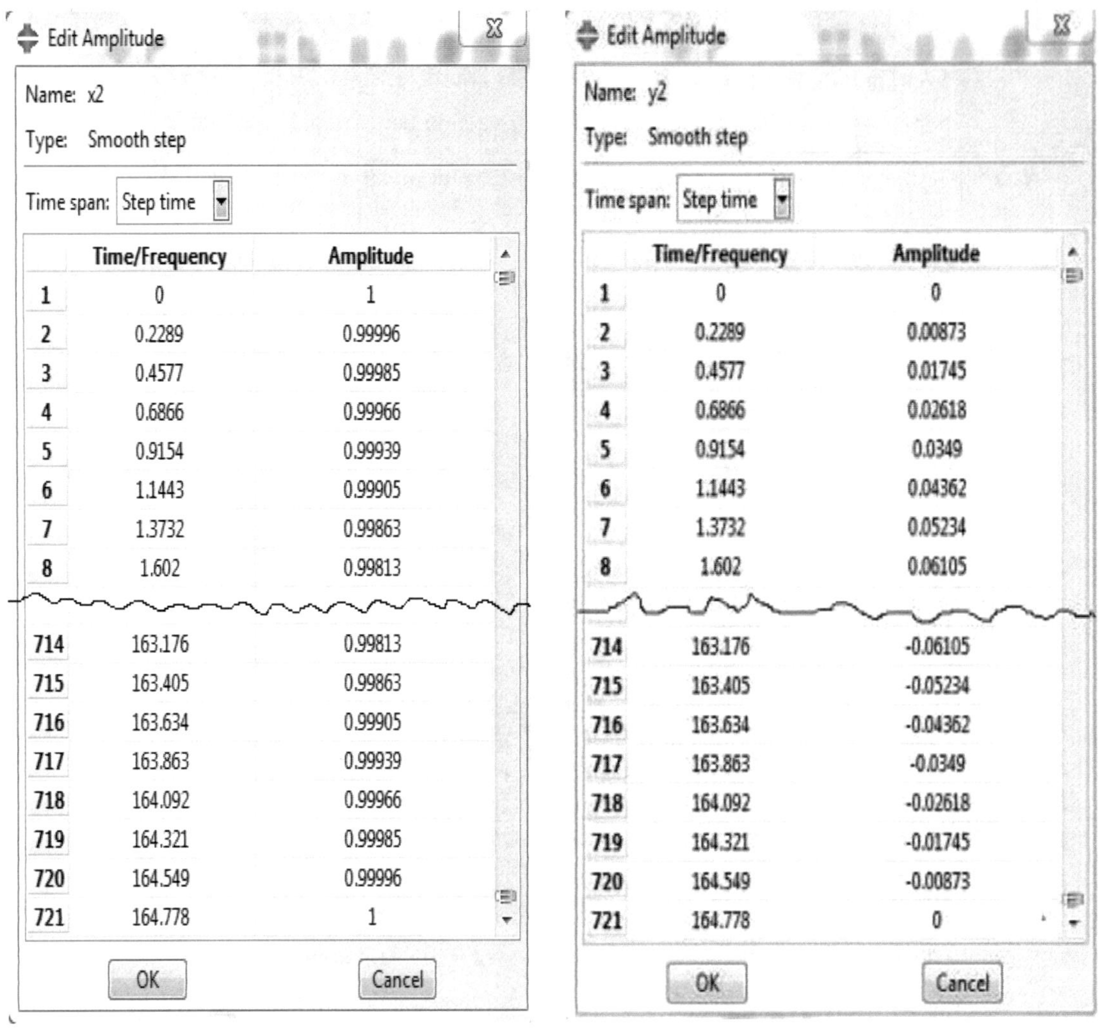

	Time/Frequency	Amplitude
1	0	1
2	0.2289	0.99996
3	0.4577	0.99985
4	0.6866	0.99966
5	0.9154	0.99939
6	1.1443	0.99905
7	1.3732	0.99863
8	1.602	0.99813
714	163.176	0.99813
715	163.405	0.99863
716	163.634	0.99905
717	163.863	0.99939
718	164.092	0.99966
719	164.321	0.99985
720	164.549	0.99996
721	164.778	1

	Time/Frequency	Amplitude
1	0	0
2	0.2289	0.00873
3	0.4577	0.01745
4	0.6866	0.02618
5	0.9154	0.0349
6	1.1443	0.04362
7	1.3732	0.05234
8	1.602	0.06105
714	163.176	-0.06105
715	163.405	-0.05234
716	163.634	-0.04362
717	163.863	-0.0349
718	164.092	-0.02618
719	164.321	-0.01745
720	164.549	-0.00873
721	164.778	0

(a) Amplitude along X axis **(b)Amplitude along Y axis**

Figure 8: Amplitude definition for time step 164.778 second

4. Result and Discussion

Simulated model was further processed for results and analysis. Visualizations for different outputlike Sheet-thickness distribution (STH), Equivalent plastic strain (PEEQ), Von-mises stress (S, Mises) etc were obtained (figure 9 a,b,c)

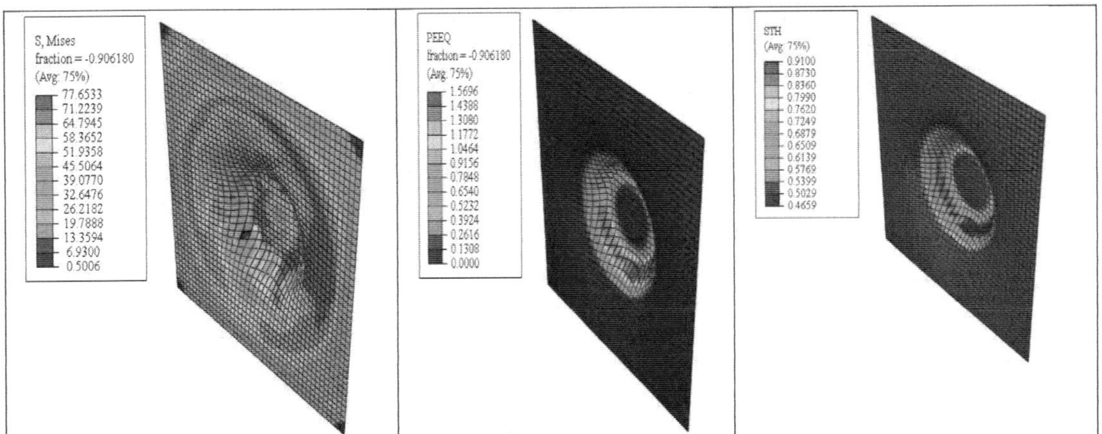

(a) **S. Mises Stress** (b) **Plastic Equivalent Strain** (c) **Sheet Thickness Distribution**

Figure 9. Output Visualizations

A radial path is created by selecting the nodes (Figure 10). Field output data is plotted against thispath for analysis

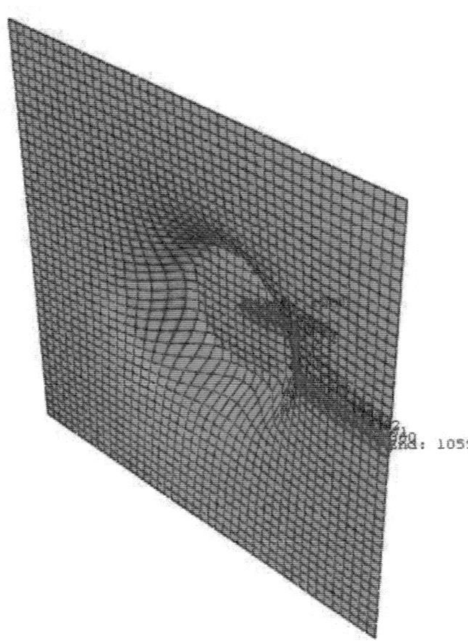

Figure 10. Path selection for output data

Product geometry of simulated cone is compared with ideal geometrical profile (Table 5) and deviations are observed (Figure 11). Deviation from the ideal geometry is observed in the cone profile. Material behaviour and tool path have significant influence over the geometrical profile. However backing plates help to lower this deviation considerably. During simulations element size functions as an important factor for profile accuracy but simultaneously it costs the simulation time.

Table 5. Comparison of simulated profile with Ideal geometry

(Wall angle 55°, Step size 0.75 mm, Feed 100 mm/min, Rotational speed 1250 RPM)

X (mm)	Simulation z (mm)	Ideal geometry z (mm)	Error
0	-11.607	-12.000	-0.393
2	-11.605	-12.000	-0.395
4	-11.602	-12.000	-0.398
6	-11.610	-12.000	-0.390
8	-11.624	-12.000	-0.376
10	-11.839	-10.649	1.190
12	-12.281	-9.011	3.270
14	-11.480	-7.372	4.108
16	-8.445	-5.734	2.711
18	-5.776	-4.096	1.680
20	-4.243	-2.457	1.786
22	-3.246	-0.819	2.427
24	-2.595	0.000	2.595
26	-2.114	0.000	2.114
28	-1.747	0.000	1.747
30	-1.450	0.000	1.450
32	-1.195	0.000	1.195
34	-0.971	0.000	0.971
36	-0.768	0.000	0.768
38	-0.581	0.000	0.581
40	-0.405	0.000	0.405
42	-0.236	0.000	0.236
44	-0.075	0.000	0.075
46	0.010	0.000	-0.010

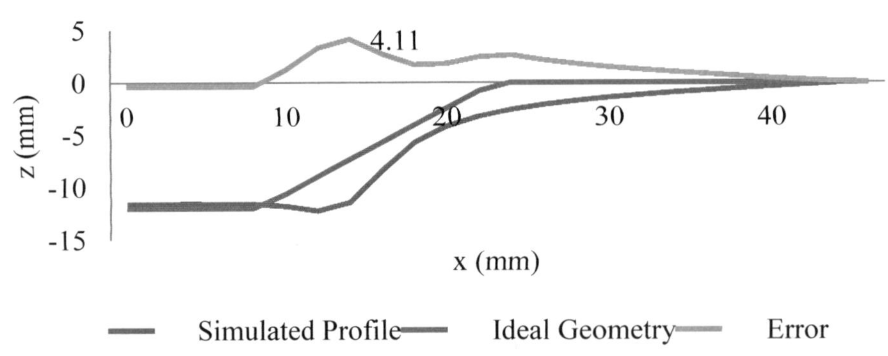

Figure 11. Comparison of simulated profile with Ideal geometry (Wall angle 55°, Step size 0.75mm, Feed 100 mm/min, Rotational speed 1250 RPM)

IOP Conf. Series: Materials Science and Engineering 1291 (2023) 012035 doi:10.1088/1757-899X/1291/1/012035

From the visualization of plastic equivalent strain (Figure 12), it is observed that strain distribution is not same in the deformation zone. Maximum strain is observed along in middle zone of the cone wall.

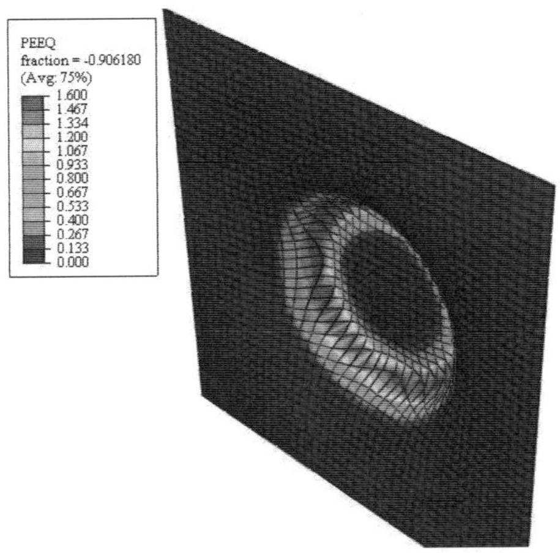

Figure 12. Visualization of Plastic Equivalent Strain (Illustrative)

5. Conclusion

One of the most researched shapes in the area of incremental sheet formation is the conical frustum. Basic modelling setup (element type, meshing, steps, boundary conditions, and schemes (dynamic/explicit)) are reviewed and explored through practice on a FE simulation of a cone. Additionally, emphasis has been placed particularly on geometrical precision, thickness strain, and thickness reduction in the study of modelling results. Solver time for different models varied with feed and number of contour path. FE modelling for conical frustum was successfully done with ABAQUS 6.13. Deviation of simulated profile from ideal geometry was observed. Spring back, Tool path, absence of back plate and element size are speculated reasons for the cause. Thickness strain is found changing drastically in deforming zone. Maximum strain is observed in the middle zone of the deformed wall. Change in strain rate is also observed with element position. Percent thickness reduction data of simulation runs for designed orthogonal array were compared with experimental results and error was found in range of 6.42% -22.74%. Reduction in element size and mass scaling factor can improve the output with significant cost over solver time. The ability to minimize the number of measurements necessary to preserve the quality of the final SPIF output was the key contribution of the research given in this study. The findings reported in this research provide valuable insight into the link between the various properties of SPIF products. By applying the same approach to other materials, product shape, and process parameters, expertise in this field may be enhanced.

6. Future Scope

It is critical to understand the process limiting variables in industrial micro forming implementations. The limiting issues might be related to the micro forming's unique characteristics. Many studies are being conducted to enhance the micro forming process, such as

- Development of forming boundary conditions in incremental forming process
- Formability enhancement with Perform design
- Technique for warm incremental sheet metal forming

- Creation of computing tools based on the Finite Element Method
- Quick plastic's deformation computation, optimization of processes

7. References

1. W. C. Emmens, "Formability", Springer Briefs in Applied Sciences and Technology, DOI: 10.1007/978-3-642-21904-7_3, Wilko C. Emmens 2011, pp-5.
2. Pepelnjak, T.; Sevšek, L.; Lužanin, O.; Milutinovi´c, M. Finite Element Simplifications and Simulation Reliability in Single Point Incremental Forming. Materials 2022, 15, 3707. https://doi.org/10.3390/ma15103707
3. Bouzidi, S., Ayadi, M. & Boulila, A. Feasibility Study of the SPIF Process Applied to Perforated Sheet Metals. Arab J Sci Eng 47, 9225–9252 (2022). https://doi.org/10.1007/s13369-022-06570-6
4. M. Sbayti, A. Ghiotti, R. Bahloul, H. BelhadjSalah, S. Bruschi, Effective strategies of metamodeling and optimization of hot incremental sheet forming process of Ti6Al4Vartificial hip joint component, Journal of Computational Science 2022, https://doi.org/10.1016/j.jocs.2022.101595.
5. Najm, S.M., Paniti, I. Predict the Effects of Forming Tool Characteristics on Surface Roughness of Aluminum Foil Components Formed by SPIF Using ANN and SVR. Int. J. Precis. Eng. Manuf. 22, 13–26 (2021). https://doi.org/10.1007/s12541-020-00434-5
6. Ragab, A.E. Analyzing SPIF Product Characteristics Using Full Factorial Design-Integrated PCA Approach. Processes 2023, 11, 1254. https://doi.org/10.3390/pr11041254
7. Li, W., Attallah, M.M. & Essa, K. Heat-assisted incremental sheet forming for high-strength materials — a review. Int J Adv Manuf Technol 124, 2011–2036 (2023). https://doi.org/10.1007/s00170-022-10561-0
8. Hussain, G., Al-Ghamdi, K. A., Khalatbari, H., Iqbal, a., & Hashemipour, M. (2014). "Forming Parameters and Forming Defects in Incremental Forming Process: Part B". Materials and Manufacturing Processes, 29(4), 454–460.
9. Kim, T. J., & Yang, D. Y. (2000). "Improvement of formability for the incremental sheet metal forming process", 42, 1271–1286
10. Ziran, X., Gao, L., Hussain, G., & Cui, Z. (2010). "The performance of flat end and hemispherical end tools in single-point incremental forming.
11. Bramley, A. N., Jeswiet, J., Micari, F., Duflou, J., & Allwood, J. (2005). "Asymmetric single point incremental forming of sheet metal", CIRP Annals of Manufacturing Technology, 54, 623-649.
12. Jeswiet, J., & Young, D. (2005). "Forming limit diagrams for single-point incremental forming of aluminium sheet". Proceedings of the Institution of Mechanical Engineers, Part B: Journal of Engineering Manufacture, 219(4), 359–364.
13. Hagan, E., & Jeswiet, J. (2004). "Analysis of surface roughness for parts formed by computer numerical controlled incremental forming". Proceedings of the Institution of Mechanical Engineers, Part B: Journal of Engineering Manufacture, 218(10), 1307–1312.
14. Eyckens, P., Van Bael, A., Aerens, R., Duflou, J., & Van Houtte, P. (2008). "Small-scale finite element modelling of the plastic deformation zone in the incremental forming process". International Journal of Material Forming, 1(1), 1159-1162
15. Desai, B. V., (2016). "Parametric investigations during Single Point Incremental Sheet Metal Forming of AA1060 & AA1200-H14". PhD Thesis, Mechanical Engineering, Sardar Vallabhbhai Institute of Technology
16. Arfa, H., Bahloul, R., & Bel Hadj Salah, H. (2013). "Finite element modelling and experimental investigation of single point incremental forming process of 14luminium sheets: influence of process parameters on punch force monitoring and on mechanical and geometrical quality of parts". International journal of material forming, 6(4), 483-510.

Certification of welder for tungsten inert gas welding in unmanned aerial vehicle systems

N Subramanya[1], A R Satheesh Kumar[1], Visakh D G[2], Balasubramanya K[3]

[1] Scientist, Aeronautical Development Establishment (ADE), Defence R&D Organisation, Bangalore, India
[2] Technician, Aeronautical Development Establishment (ADE), Defence R&D Organisation, Bangalore, India
[3] Technical Officer, Aeronautical Development Establishment (ADE), Defence R&D Organisation, Bangalore, Indi
subramanya-n.ade@gov.in

Abstract. Aluminum and its alloys is one of the extensively used non-ferrous metal in aerospace engineering applications due to its excellent properties like higher strength to weight ratio, non-magnetic property, high thermal conductivity, good machinability, recyclability, and excellent formability. However, welding of aluminum and its alloys has always been a challenge due to drastic reduction in strength near heat affected zone (HAZ). Due to softness, ductility, lower melting point of aluminum and its alloys impose additional challenges to weld aluminum alloy components. Due to this, the quality and quantum of skill set needed to weld aluminum alloy joints for flight critical components is much more than that of welding of ferrous materials. The qualification and certification of welder is one of the prime requirements in welding applications especially in aerospace industry. This paper mainly discusses the various aspects involved in qualifying and certifying a welder for tungsten inert gas (TIG) welding process of aluminum alloys, which is treated as one of the special processes. The welder certification is a written verification that weld joints produced by welder meets the prescribed standard of welder performance. It is to verify the ability of an individual to execute a qualified welding procedure specification to produce a sound weld. Aluminum alloy 6xxx series is considered for base metal for sample preparation and weld joint assessment studies, as it is one of the prevalently used aluminum alloy material in unmanned aerial vehicle (UAV) applications. Weld samples are prepared as per internationally practiced standards and welding procedure specification (WPS) is recorded. Various mechanical tests carried out on weld samples during certification process are discussed in the paper. Weld joints are subjected to tensile testing, bending test and macro test. The certifying agencies are involved at every stage to ensure consistency of welder performance.

Keywords: Welder Certification; Welding; Unmanned Aerial Vehicle (UAV); Manufacturing

1. Introduction to Welding

Metal joining by welding process has been extensively used in almost all sectors of manufacturing industries. Welding has become very dependable, efficient method in aerospace industry in recent years. Due to its cost effectiveness, ease of maintenance, lesser weight penalty, structural soundness advantages, welds joints are replacing riveted joints and bolted joints since world war-2 and hence, become one of the favorite for military, ship building and aerospace industry. Welding is typically defined as joining o two similar or dissimilar metals which are melted at the joining interface and usually a filler material is added at the molten metal pool with application of pressure and with / without application of heat. Welded joints are permanent, portable, and most economical m method of joining metals in terms of fabrication cost and material usage. Based on type of materials to be welded, type of shielding gas use, the manner in which heat and pressure are applied, welding techniques are classified as liquid state (fusion) welding and solid state welding. Liquid state welding melts the metal (by external heating) at the interface and uses an external filler rod to fill the molten metal pool. In solid state welding surfaces to be joined are brought into close proximity by heating the material (without melting the interface) and applying normal pressure. Arc welding, gas welding and resistance welding are few types of arc welding process. Forge welding, friction welding, explosion welding, diffusion welding and ultrasonic welding are few types of solid state welding method. However, dissimilar metals can also be welded by using advanced welding techniques like laser welding, electron beam welding, Plasma Arc welding etc. Due to minimum distortion, high impact strength, high speed, high quality of weld, low maintenance cost and strong weld joints, further arc welding is majorly classified as Shielded Metal Arc Welding (SMAW), Gas Metal Arc Welding (GMAW) (Also called Metal Inert Gas welding (MIG)), Shielded Metal Arc Welding (SMAW), Flux Coated Arc Welding (FCAW), Gas Tungsten Arc Welding (GTAW) (also called Tungsten Inert Gas welding (TIG)).

2. Welding of Aluminum and its Alloys

The existence of aluminum was first discovered by Sir Humphrey Davy in 19th century [1]. Usage of aluminum in aircraft structures became popular and extensive during the second world war. Aluminum and its alloys is one of the extensively used non-ferrous metals in aerospace engineering applications due to its excellent properties like higher strength to weight ratio, non-magnetic property, high thermal conductivity, good machinability, recyclability, and excellent formability. However, welding of aluminum and its alloys has always been a challenge due to drastic reduction in strength near heat affected zone (HAZ). Due to softness, ductility, lower melting point of aluminum and its alloys impose additional challenges to weld aluminum alloy components. Due to this, the quality and quantum of skill set needed to weld aluminum alloy joints for flight critical components is much more than that of welding of ferrous materials. Aluminum alloys are classified based on the key ingredients like copper, silicon, manganese, magnesium etc. Figure 1 shows the overall classification of aluminum alloy [2]. Hardenability, machinability, and weld-ability of aluminum alloys are driven by the key constituent alloy component. Non-hardenable alloy groups are those materials whose strength cannot be increased through heat treatment, in contrast to that of hardenable alloys which are capable of increasing strength through heat treatment. Percentage usage of aluminum alloys used in some of the commercial airlines is shown in figure 2 [3]. Welding of aluminum and its alloys is a critical and one of the challenging areas in the domain of welding engineering. Aluminum welding is a discipline that needs to be understood and learned. Correctly applied, welders find that an aluminum welded product is quicker to manufacture than an equivalent performing steel welding and offers about 40 % mass advantage [4].

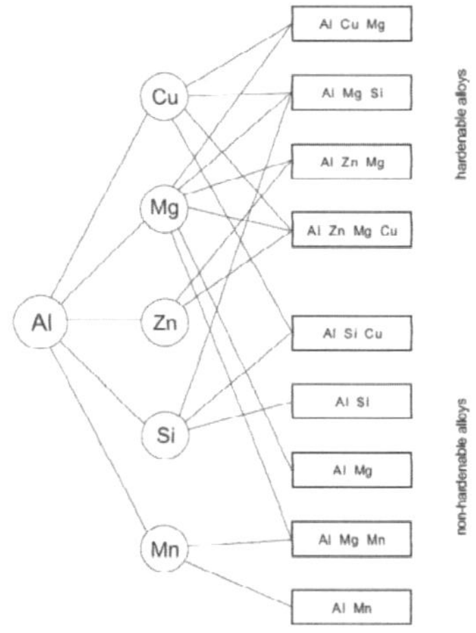

Figure 1. Classification of Aluminum Alloys

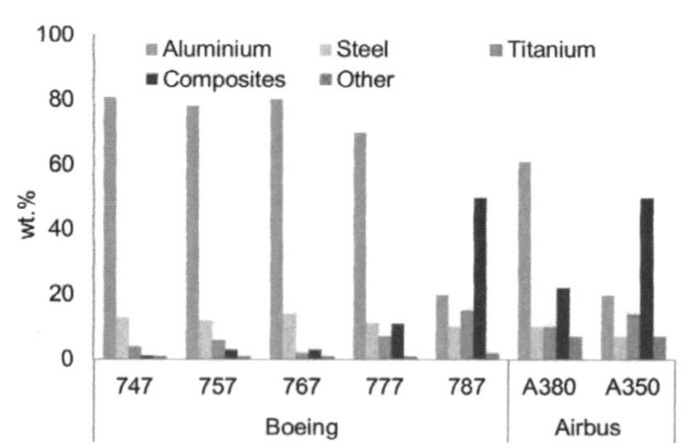

Figure 2. % by wt. usage of aluminum for aircrafts

3. Qualification of Weld Joints and Welder Certification

In welding terminology, weld joint qualification refers to qualification of joints envisaged by the designer to evaluate its fitness to use for the intended function [5,6,18]. It is a design parameter. During weld joint qualification studies samples are prepared as per internationally approved industry practices, Welding Procedure Specification (WPS) and PQR (Procedure Qualification Records) [19] prepared, and joints are subjected to various mechanical and metallurgical testing to assess the quality of weld joint [7]. However, Welder qualification (or certification) is a manufacturing aspect. Here welder or welding operator is qualified to operate particular type of welding technique for specific kinds of applications. It is an index representing the ability of the operator to generate high quality defect free welds. In other words, a qualified welder or welding operator is the one who meets ASME Section IX requirements [5]. The qualification and certification of welder is one of the prime requirements in welding applications especially in aerospace industry.

This paper mainly discusses the various aspects involved in qualifying and certifying a welder for Tungsten Inert Gas (TIG) welding process of aluminum alloys, which is treated as one of the special processes [20,21]. The welder certification is a written verification that weld joints produced by welder meets the prescribed standard of welder performance. It is to verify the ability of an individual to execute a qualified welding procedure specification to produce a sound weld. Aluminum alloy 6xxx series is considered for base metal for sample preparation and weld joint assessment studies, as it is one of the prevalently used aluminum alloy material in unmanned aerial vehicle (UAV) applications [22]. Weld samples are prepared as per internationally practiced standards [5] and welding procedure specification (WPS) is recorded. Various mechanical tests carried out on weld samples during certification process are discussed in the paper. Weld joints are subjected to tensile testing, bending test and macro test. The certifying agencies are involved at every stage to ensure consistency of welder performance.

4. Test Sample Preparation by Welder Under Certification

3mm thick silicon and manganese based aluminum alloy in 6xxx series is considered for base metal for sample preparation. Standard samples are prepared under suitable inspection supervision as per the standard [9, 10]. The dimensions of the tensile test specimen, bending test specimen and macro test specimen are depicted in figure 3. Test specimen details are recorded in WPS document. After welding, the surface of the welds shall be sufficiently free from coarse ripples, grooves, overlaps, abrupt ridges and valleys. The welded test coupons shall be subjected to visual inspection, liquid dye penetrant inspection, and radiographic examination as per American Society for Mechanical Engineers (ASME) Boiler and Pressure Vessel Code (BPVC) [5, 8] is referred. Welded test samples are shown at figure 4.

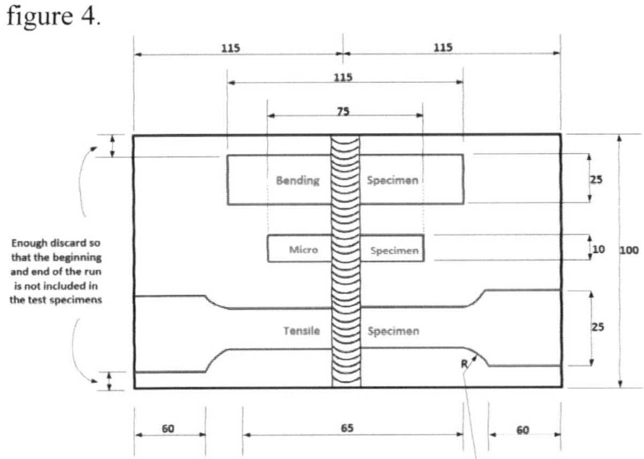

Figure 3. Test specimen dimensional details

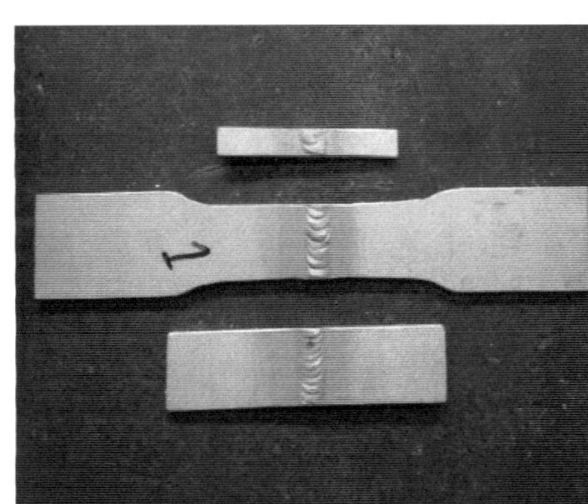

Figure 4. Welded Test samples

Initial Raw material was tested and conformed to specifications using handheld portable XRF material

IOP Conf. Series: Materials Science and Engineering 1291 (2023) 012036 doi:10.1088/1757-899X/1291/1/012036

analyzer machine [11, 12, 13]. Tungsten Inert Gas welding is carried out using the TIG welding machine (Make : Miller Dynasty 400 model), which has a unique auto-line technology. Non-consumable tungsten electrode [14] was used along with recommended AWS ER 5356 filler rod [16] and argon shielding gases as per MIL-A-18455C [15] were used to generate high quality weld specimens. Alloy deposit of ER 5356 filer rod gives good resistance to sea water corrosion, good anodizing property [4,24]. Typical Mechanical properties of ER 5356 filler rod are shown in table 1. All the activities were done in presence of certifying agency. The reinforcement and penetration of weld bead should be machined so that the whole welded coupon shall be of same thickness as that of base metal, 3mm.

Table 1. Properties of TIG 404 AWS ER 5356 Filler Rod

Parameter	UTS	YS	Elongation (L=4D)	Melting Point
TIG 404 AWS ER 5356	21-26 kgf/mm^2	8-10 kgf/mm^2	15-25%	562-633^0 C

5. Experimental Investigation

The welded test samples were subjected to mechanical destructive test namely tensile test and 180 degree bend test. Samples were also subjected to a macroscopic test to evidence quality of weld throughout the thickness.

Tensile testing of welded specimen : The tensile test specimen were tested in direct tension in Universal Testing Machine (UTM) (Machine : MTS 810 Material Test System). The test specimen mounted on the UTM machine is shown in figure 5. 0.2 % proof stress, ultimate tensile stress (UTS), percentage elongation and failure location were recorded in PQR.

180^0 Bend test : The test specimen is to be bend so that the weld is along the axis of the bend (as shown in figure 6) and is bent round a former of the required radius with the weld face on the outer side (that is to say, the base of the 'V' is to be on the inner side of the specimen after bending). To facilitate close contact with former, bend test specimens may be dressed on both sides by filing, grinding or machining until the weld metal is in level with the parent metal. The edges of the samples in the vicinity of the bend should be given a reasonable radius. The specimen must be bent through 1800 over a radius equal to five times the nominal thickness of the parent metal (As shown in figure 7). The bending strength and position of the break are recorded

Macro examination : Specimen for the examination to be cut from the middle of the sample as shown in figure 8. The cross section of the sample to be examined utilizing micro-scope. In case if no facilities ae available for mounting the micro-section for polishing, a convenient method is to bend the two ends of the micro sample to form a 'U' or 'V' section. Specimen shall be examined in etched and unetched condition. Width of the wed, height of the weld, root height are measured and recorded

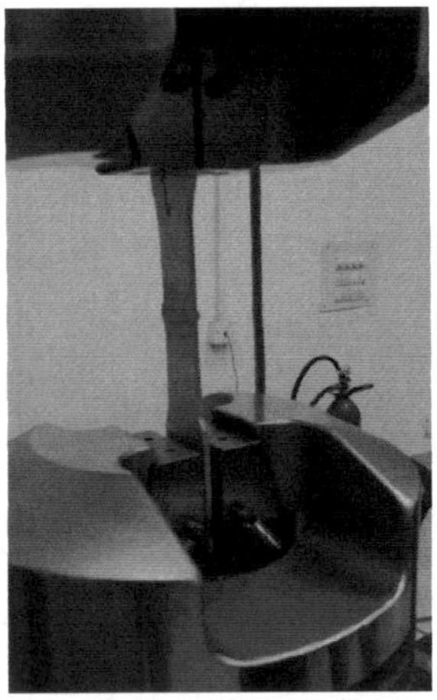

Figure 5. Tensile specimen on UT machine

Figure 6. Test set for 180^0 bending test

Figure 7. 180^0 degree bend test axis

Figure 8. Specimen prepared for Macro-analysis

6. Results and Analysis

The tensile testing of the specimen confirmed that the failure occurred at the heat affected zone (HAZ) [23] as expected and is shown in figure 9 and 10. In welded aluminum alloy materials, weld bead is stronger than base material and the failure is expected to occur at HAZ region. This infers that the weld samples prepared by the welder is of sound quality. The bend test resulted in failure of the specimen at the weld bead and the same is showed in figure 10 and 12. The macro examination revealed no cracks, porosity, indicated good quality weld.

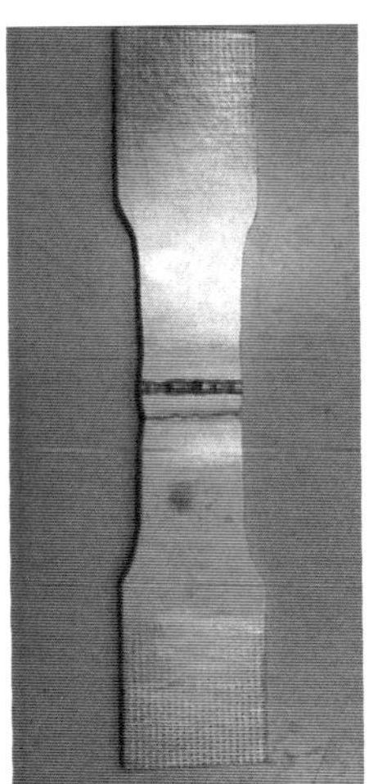

Figure 9. Fractured tensile specimen

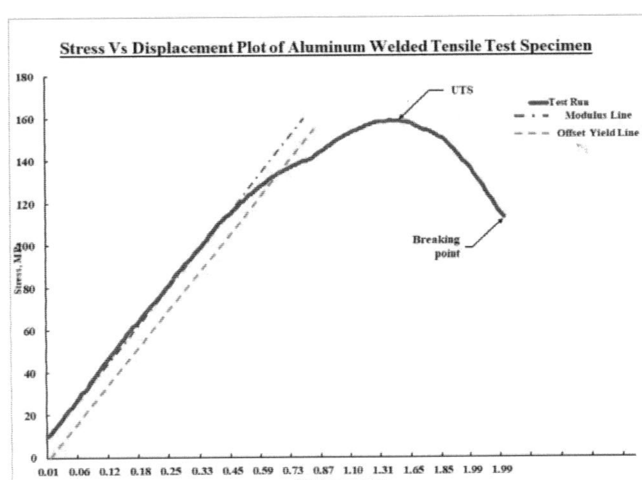

Figure 10. Stress-strain plot of tensile testing

7. Conclusions

The welder qualification for TIG welding of aluminum alloys for aerospace applications is satisfactorily carried out. Weld samples are prepared as per the internationally approved standards, and subjected to various tests to evaluate the fitness of the welder to produce sound welds with a particular machine type and welding parameters. Test results are satisfactorily and failure locations indicated that the weld joints were prepared as per the designers requirements. It is observed that the skill set required to TIG weld aluminum and its alloys is high due to low melting temperature, HAZ and reduction in strength of the material. The welding filler rod, number of passes, purging parameters and most importantly operator skill dictates the weld quality. Defect free good quality weld joints ensure that the aircraft structures are manufactured as per designer requirements of zero defect. The weld specimens prepared by welder met all the requirements towards welder qualifications. The inspection agency issued a certificate for the qualified TIG welder. However, in general, welder qualifies for a particular type of material combination, machine and welding parameters. The welder examination body (WEB) recommends that the qualification certificate needs to renewed at a frequency of one year. A typical sample of welder qualification certificate format is shown in figure 13 [17].

Figure 11. Fractured bending test specimen

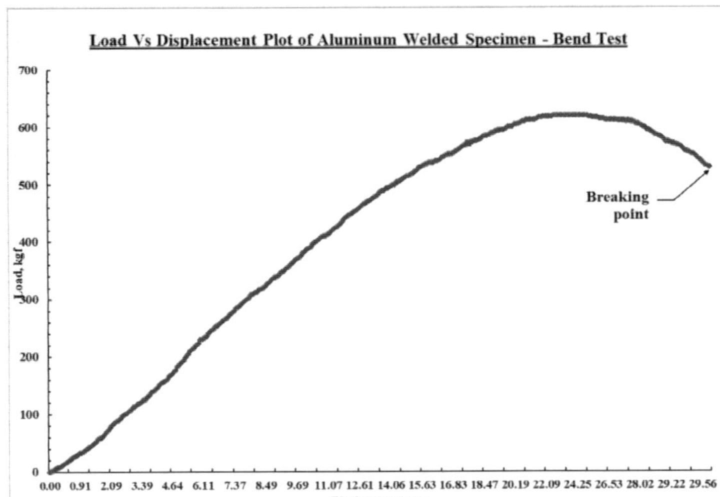

Figure 12. Bending Test plot showing

Welder's name:		Date of birth:	Photograph
Cert. No.		Sex.	
Identification No.			
Employer's name and address			
WPS/pWPS No.			
Date of initial approval			
This is to certify that the welder has passed the qualification test [/and re-validation record audit] according to the rules of [the Society], and is qualified to undertake welding operation specified in range of qualification of this certificate.			
Items	Test piece	Range of qualification	
Welding process			
Base metal			
Filler metal type			
Plate thickness			
Type of welded joint			
Welding position			
Revalidation method	In accordance with 6.2.1 a) ☐ b) ☐ c) ☐		
Other details			

This certificate is issued at______ [place], and valid until______ [DD/MM/YYYY].

Signature/seal of examiner______ Issued on______ [DD/MM/YYYY].

	Report No. to be reviewed	Date of report	Signature of Employee	Date of signature
1				
2				
3				
4				
5				
6				

Figure 13. Sample welder qualification certificate format issued by WEB [17]

8. Acknowledgements
Authors would like to thank the Director of their establishment for providing facility to carry out the research work and for giving permission to present this technical paper. Authors are also grateful to their Technical Paper Review Committee for critically reviewing the paper and providing valuable suggestions.

9. References
[1] Gene Mathers, *The welding of aluminum and its alloys*, **2002**, Woodhead Publishing Limited, Cambridge, England, CRC Press

[2] J Gilbert Kaufman, *Introduction to aluminum alloys and tempers*, **Nov 2000**, ASM International

[3] Muhammad Aamir , Khaled Giasin, Majid Tolouei-Rad et al. *A Review: Drilling performance and hole quality of aluminum alloys for aerospace applications*, Journal of material research and technology, Vol **9**, Issue 6, **Nov-Dec 2020**, Pages 12484-12500

[4] The welding of aluminum MIG and TIG fusion– pocket guide, aluminum federation of southern africa, 2nd edition, **2004**

[5] ASME Boiler and Pressure Vessel Code : *Qualification standard for welding and brazing procedures, welders, brazers, and welding and brazing operators*, section 9, **2004 July**

[6] IS 16003 : **2012**, *Specification and qualification of welding procedures for metallic material*

[7] Christi V Jacob, Subramanya N, Md. Shahid Quamar et al. *Fabrication and Qualification studies of aluminum fuel tank for an experimental UAV*, Materials Today: Proceedings 5 (**2018**) 13514–13520

[8] ASME Boiler and Pressure Vessel Code: *Non Destructive Examination*, Section 5, **2004**

[9] ASTM A 370, *The standard test methods and definitions for mechanical testing of metals,*

[10] DTD 901F, *Cleaning and preparation of metal surfaces*, **Nov 1964**

[11] Peter Brouwer, *Theory of XRF, Getting acquainted with principles*, PANalytical B V Publications, **2010**

[12] ASTM E 34, *Standard test methods for chemical analysis of aluminium and its alloys*, **1998**

[13] ASTM E 101, *Spectroscopic analysis of aluminium and aluminium alloys*, **1983**

[14] ISO 6848 : 2015, *Arc eeldin and cutting – Non consumable tungsten electrodes*

[15] MIL-A-18455C, *Military specifications for Argon gas of high purity*, **1986**

[16] AMS 4190, *Aluminum alloy welding rods*, **Jan 2008**

[17] IACS, W32, *Qualification scheme for welders of hull structural steels*, Revision 1, **Sep 2020**

[18] M Tavodova, Michaela Hnilicova et al. *Quality Evaluation of welding joints by different methods,* FME Transactions, 816-824, **Aug 2020**

[19] Sachin Bhardwaj and R M Chandima Ratnayake. *Welding procedure qualification record (WPQR) for welds fabricated at proximity.* The international journal of advanced manufacturing technology, **120**, 4477-4491, **2022**

[20] Tadak Singh Kirade, Purushottam Sahu. *Multi-objective optimization of TIG welding Aa 6061 T6alloy using resposne surface methodology (RSM).* International research journal of engineering and technology (IRJET), Volume 9, Issue 12, **Dec 2020**

[21] M D Sameer, P Karunakar, T Prashant kumar et al**. *Influences of TIG welding parameters on tensile behaviour of aluminium AA6061alloy V groove joint,*** Journal of material science and mechanikcal engineering, ISSN 2393-9109, Volume 2, Number 4, pp 317-320, **April 2015**

[22] Md Rasidi Ibrahim, Muhamad Firdaus Azman, Ahmad Hamdan Ariffin et al. *Overview of unmannes aerial vehicles (UAV) parts material in recent applications*. Structural Integrity and monitoring of composite materials, pp 179-189, **28 January 2023**

[23] N Nazemi, F Ghrib, Jerry Sokolowski et al. *The HAZ in aluminium welding revisited.* 3rd speciality conference on engineering mechanics and materials. Montreal, QB, Canada, **June 2013**

[24] Badr Kamoon, Salam Abeed Dahe, Zuhair H. Obaid et al. *Microstructural Evaluation for TIG Welding of AA6061-T6 using ER5356 Filler Metal; in Postweld and as-Weld Treatment.* Journal of Mechanical Engineering Research and Developments, Vol. 43, No. 2, pp. 154-164, **2020**

Improving the vortex tube parameters using CFD analysis for sustainable manufacturing

S Tamil Selvan[1], K Venkataraman[1], K Arun Balasubramanian[2] and G Venkatesan[2]

[1]Department of Mechanical Engineering, Rathinam Technical Campus, Coimbatore, Tamil Nadu, India
[2]Department of Mechanical Engineering, Sethu Institute of Technology, Virudhunagar, Tamil Nadu, India

E-mail: tamilau08@gmail.com

Abstract. Sustainable development will provide innovative methods to lessen manufacturing's negativeenvironmental effects. Due to the rising desire for a more environmentally conscious populace, sustainability is becoming increasingly crucial in today's world. Without using power or refrigerant, vortex tubes are a green method of spot cooling. Vortex tubes don't need to be maintained and are maintenance free. Through this effort, the tube's parameters will be optimized to raise the cooling temperature. Via CFD study, different vortex tubes with different lengths. Through CFD simulation, the ideal parameters for achieving the lowest cold gas temperature are discovered, including the ideal length of the tube to diameter of the tube ratios, freezing air collection tube diameter (dc), and parameters. By examining the behaviour of vortex tubes with various profiles, length-to-diameter ratios, icy orifice diameters, also a hot end areas using a computerized fluid dynamic tool, it will be possible to demonstratethat they are an alternative and long-lasting instrument for spot cooling in machining operations.

1. Introduction

In order to cool the cutting edge during the machining operation, the study discussed in this paper is designing a viable clean and renewable tool vortex tube also used to cool the circuit board of a fully automated machine. The vortex tube of Exair brand is shown in figure 1.Sustainable construction referred to by the term making of industrially produced item by utilizing pollution-free techniques, preserve fossil fuels and a variety of sources, and are both financially sensible as well as worker safety., Additionally to battle of market, revenue growth, and yield` care for the environment and sustainable developments referred collectively as the three pillars of sustainability.

Figure 1. Vortex tube of the EXAIR Medieval style with a 15 SCFM vortex generator

In order to provide pinpoint chilling, vortex tubes are a viable instrument. Rudolf Hilsch, a German engineer, refined it after French scientist Ranque created it in 1933. Having no working parts, a vortex tube is a simplest gadget. In the vortex tube, cold and warm air under increased pressure separated. The vortex tube is made up of orifices, swirling chambers, inlet nozzle, and central tubes, as well as a control valve and an orifice. High pressure air enters through tangential nozzles in the vortex tube, where it expands and reaches high angular velocities, creating turbulence. With two exit points, a hot end and a cold side, each with a divergent outlet, a vortex tube. The regulating valve is situated on the hot side. At the warm exhaust a control valve manages the rate of mass flow. It's challenging to understand the swirling tube's energy splitting phenomenon. Swirling flow with greater rotational velocities provide a stronger and more effective energy separation effect, which is connected to a power exchange among the two zones that form within the vortex tube.

Table 1. Vortex tubes Size

Parameters	Values
Inner tube diameter, D	10.10 mm
Diameter of the cooler outlet dc	4.57mm
Length / Diameter ratio	11.35
The multitude of nozzles	6
Width of the circumferential inflow tonal	1.42 mm
Entrance altitude of a vortex	0.98mm
Gap at hot end exit	0.31mm

According to Eiamsa-ard [1], the ideal range for cold mass friction (c) is 30% to 40% for maximum cooling temperatures. but so far It is unclear how the energy separates in the vortex tube. The typical and non-invasive flow imaging approaches are challenged by the recirculation nature and high velocity whirling flow described by Aljuwayhel et al [2]. A preliminary quantitative tube design technique can be accomplished to attain the important characteristics of the main tube, such as length, diameter, and diameter of the cold exit, based on a basic description of the counter flow structure in a restricted thin tube according to Xiangji Guo et al [3] research. Yunpeng Xue et al[4], show a clear connection between how hot and cold streams are created and how the tube's vortex changes. The temperature drops at the cold side due to the pressure differential in the anterior vortex tube, The heated end's temperature goes up as a result of the inter flow arrangement in the primary tube's periphery. The dimensional parameter of the EXAIR Medieval style with a 15 SCFM vortex tube is given in the Table 1.

2. Vortex tube geometry

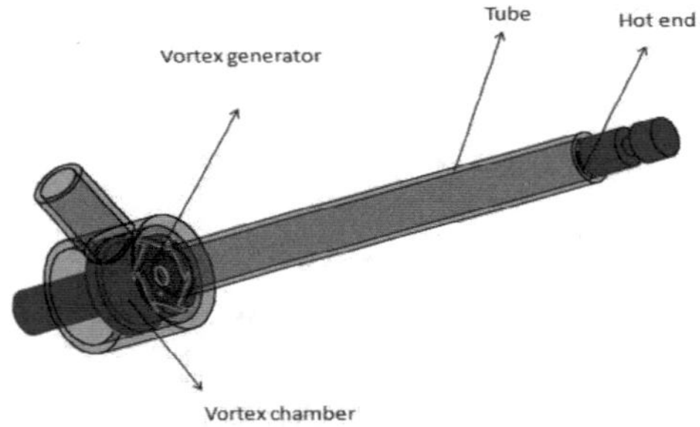

Figure 2. Vortex tube design

The circular tube, vortex chamber, vortex generator, barrier, hot side check valve, and six tangential bays make up this whirling pipe. Utilized to regulate the mass movement at the chilly end. By use of the hot side control valve. The mathematical characteristics of the vortex tube, we look at factors including its length, pipe diameters, nozzle size and hot junction air escape gap. figure 2 show the vortex tube design in solid works.

The CFD study makes use of working tubes of various lengths. The primary tube's diameter is 10.10mm. According to research by Nader Pour Mahmoud et al. [5], increase of temperature separation by increase of nozzle numbers. The six tangential nozzles are intended to produce considerable turbulence. The cold opening diameter is 4.56mm. the nozzles' flow width is 1.42mm, and the height of the notch is 0.98mm. For analysis, L/D radio 14.21, 12.35, 11.37, 10.39, 9.31, 8.33, 7.35, 6.37, and 5.38 are taken. This CFD model makes use of L/D radio 5.39, which in our situation provides the optimum energy separation. Additionally, several L/D ratios (1.98, 1.63, 2.32, 2.05, 3.42) were examined, with 1.643 providing the best results. The fluid path of the existing dimension will be shown in figure 3 and vortex generator which creates the turbulent flow is shown in figure 4.

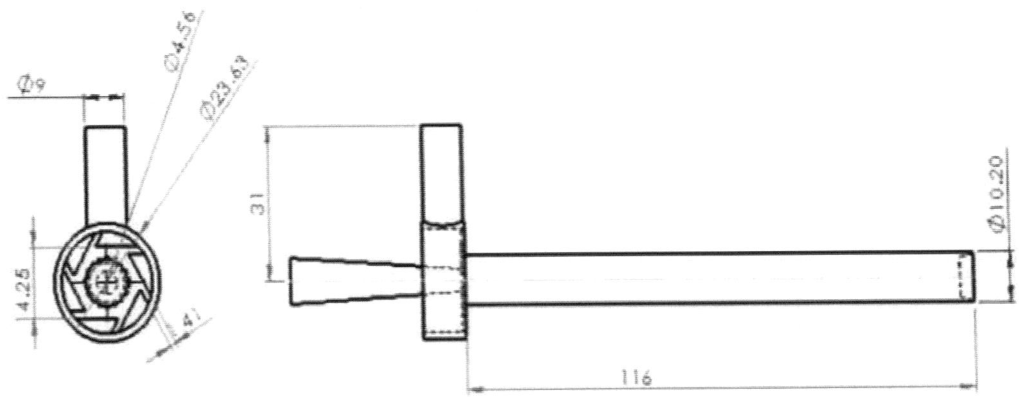

Figure 3. Vortex tube fluid path geometry

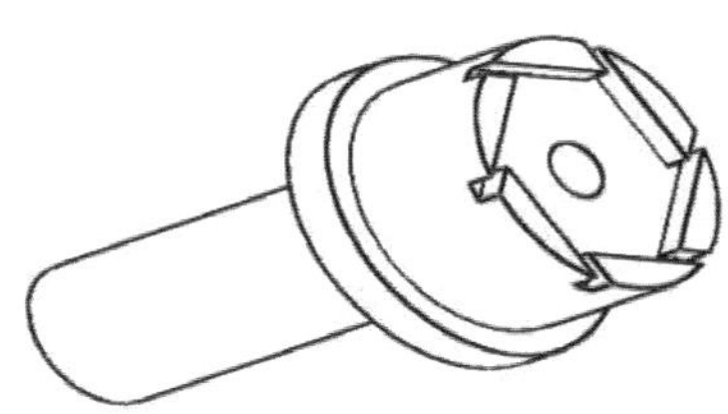

Figure 4. Vortex generator

3. CFD Simulation

The Ansys ICEM CFD is used to construct the volume mesh for our component design. It makes use of the common k-epsilon turbulence model. Skye et al. [6] learn about the RNG k-epsilon turbulence model, Yet the outcomes would not match the observational evidence, for this simulation, as well, walls are taken into account as the friction less limit constraint, thus the Reynolds stress equation also cannot converge. The forced and free vortex zones develop within the pipe due to the velocity in the axial direction. Tria components are used in the ICEM mesh generation tool to texture the three - dimensional object.

Ansys Fluent 14.5 is the volume-based solver used in the current work. In order to solve the vortex tube issue, the solver employed a three-dimensional stable, compressible, pressure-based setup. To record the energy and heat spread, the energy equation was activated. There is a K-epsilon model. In order to calculate the wall friction on a fluid, a two-equation model can be utilized. By using the viscous heating function, it is possible to find the heat loss between the layers. As a material body, air is employed. The ideal gas equation is utilized since it is believed that the flow is compressible. There is no connection between the surrounding and the energy separation realm. The air temperature is provided as an input in the model along with the definitions of the vortex tube's cold and hot outlets as well as its entrance as a pressure inlet and output as a pressure outlet. We suppose that the wall is an adiabatic state. The mass, momentum, and energy equation was solved using the fluent programme.

Mass balance equation

$$\frac{\partial x}{\partial t} + \nabla.(\rho u) = 0 \quad (1)$$

Momentum balance equation

$$\frac{\partial(\rho u)}{\partial t} + \nabla.(\rho u u) = -\nabla p + \mu \nabla^2 u \quad (2)$$

Energy balance equation

$$\rho C_p \left[\frac{\partial T}{\partial t} + \nabla.T(u)\right] = k\nabla^2 T \quad (3)$$

Ideal gas equation

$$\frac{p}{\rho} = RT \quad (4)$$

The input values for the intake pressure and inlet ambient temperature are 5.5 bar and 27°c, respectively. Both the hot outlet and the cold outlet have their pressures tuned to the environment. Due to the twofold whirling flow arrangement, the vortex tube separates its energy. The cold end's negative pressure, which is necessary for reverse flow, is formed as a result of pressure differences in the whirling part, hot and cold outlet flow.

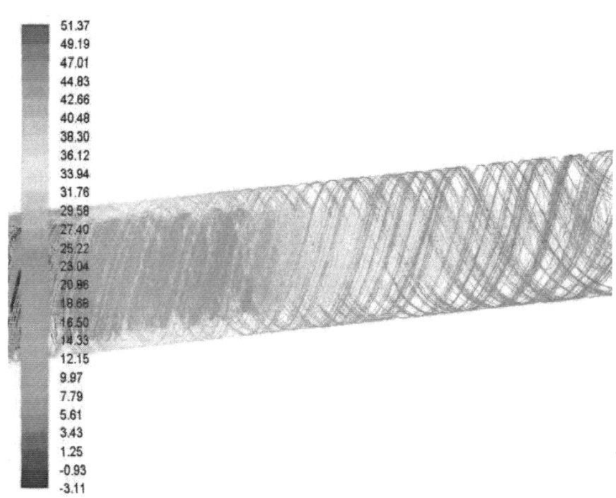

In T°

Figure 5. Separation of energy in the vortex tube L/D=5.38

The separation of two phases demonstrates a reliable energy transfer among systems, with motion serving as thermal energy in the internal core.

Different gas properties have been linked by research to the influence of the temperature separation. P.Ambedkar et al.[7] so in this project air is taken as a medium for the sustainable cooling system.

4. Change in design

Although convergent type vortex inlet is employed for length and diameter ratio 5.38 and 2.33 mm orifice radio, the cold end temperature is not improved. Due to wall friction, convergent nozzle lowers entrance velocity. Figure 6 show the vortex inlet with converging inlet nozzle.

Figure 6. Vortex inlet that is converging

5. Simulation result

Different length and diameter ratio and the cold end out is listed in the Table 2 This result shows that out temperature at cold end is minimum for 5.38 L/D ratio. Energy Separation in the cold end and hot end is shown in Figure 7.

Table 2. Cold side temperature and thus length and diameter ratio

L/D ratio	Cold end temperature in°c
5.38	2.33
6.38	2.86
7.36	3.38
8.34	3.43
9.32	4.94
10.38	5.65
11.36	3.55
12.33	7.28
14.22	6.88

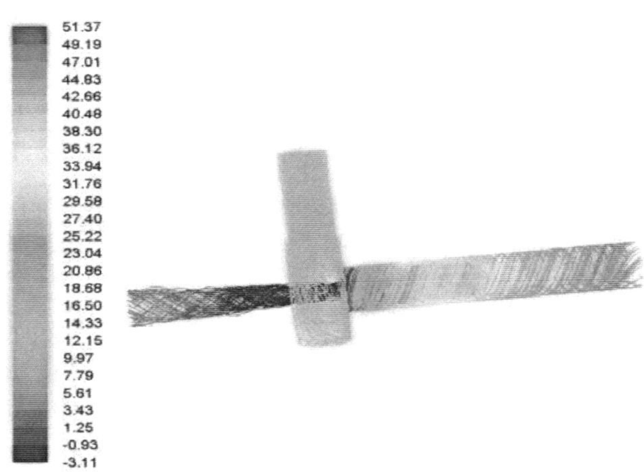

Path line T°c for L/D=5.39

Figure 7. Line of energy separation between the cold and hot sides

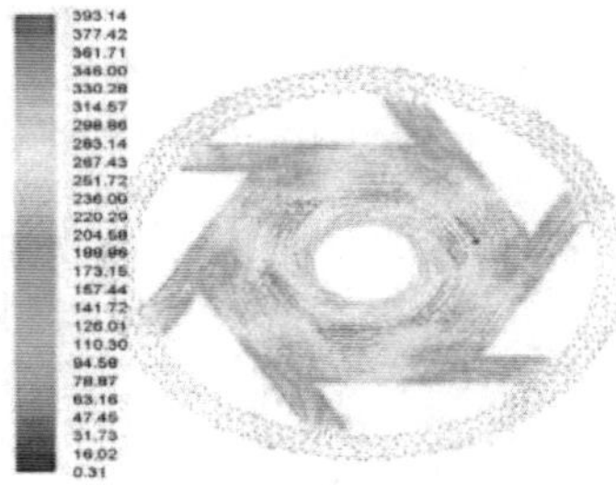

Figure. 8 velocity vector of L/D at vortex path=5.38

Velocity at different position of the nozzle is shown in figure 8 for length and diameter ratio of 5.38.

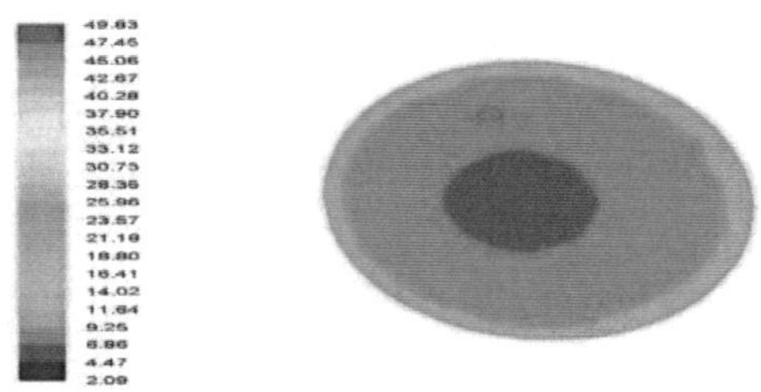

Figure 9. L/D 14.22 Depiction of the outside heat in °C

Figure 10. L/D 5.38 Contoured cold side heat in °C

Different cold end temperature at actual and optimized values are shown in figure 9 and figure 10. The best cold end temperature is achieved when the spacing between two input nozzles and the warm end is 0.41 mm for an Length and diameter ratio of 5.38 and an size of the orifice is R=2.23.

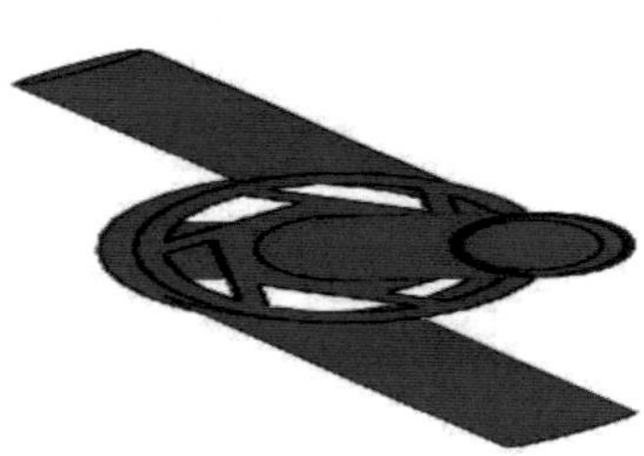

Figure 11. Two input nozzles

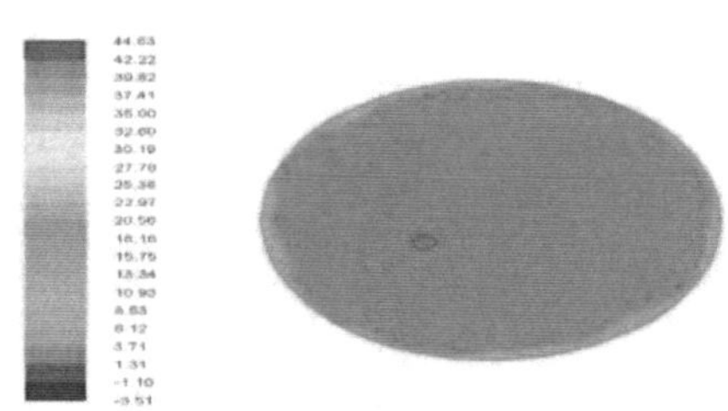

Figure 12. L/D 5.38 Cold side heat ranges in degrees Celsius for both holes.

Energy separation is improved when two nozzles is used for the L/D ratio 5.38 figure 11 and figure 12.enery separation for the above ratio is shown in figure 13 for the optimum result.

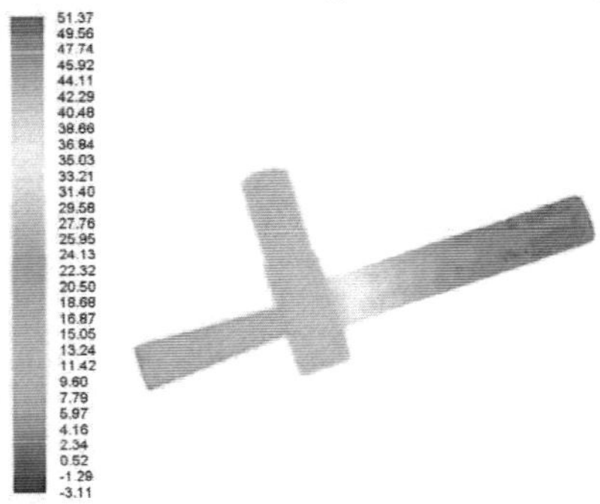

Figure 13. Temperature Contours in° C

6. Conclusion

The optimal ratio for the vortex tube length and diameter is found to be 5.39 based on the numerical results of the L/D 19.7, 14.5, 11.76, 11.37, 10.39, 9.31, 8.33, 7.35, and 6.37. At the hot end, 48 degrees Celsius, and at the cold end, 2.37 degrees Celsius were measured length and diameter ratio below the5.39 increase the temperature at cold end. The converging 6 number of tangential inlet also produce high vortex compared to previous single and double inlet.

References

[1] S Eiamsa-Ard, C Thianpong, P Promvonge Experimental investigation of heat transfer and flow friction in a circular tube fitted with regularly spaced twisted tape elements, International Communications in Heat and Mass Transfer, Volume 33, Issue 10, December 2006, Pages 1225-1233, ISSN 0735-1933,

[2] N.F. Aljuwayhel, G.F. Nellis, S.A. Klein, Parametric and internal study of the vortex tube using a CFD model, International Journal of Refrigeration, Volume 28, Issue 3, 2005, Pages 442-450, ISSN 0140-7007,

[3] Xiangji Guo, Bo Zhang, Analysis of the unsteady heat and mass transfer processes in a Ranque–Hilsch vortex tube: Tube optimization criteria, International Journal of Heat and Mass Transfer,Volume 127, Part C,2018,Pages 68-79,ISSN 0017-9310,

[4] Yunpeng Xue, Maziar Arjomandi, Richard Kelso, Experimental study of the thermal separation in a vortex tube, Experimental Thermal and Fluid Science, Volume 46, 2013, Pages 175-182,ISSN 0894-1777.

[5] Nader Pourmahmoud, Amir Hassanzadeh, Omid Moutaby,Numerical analysis of the effect of helical nozzles gap on the cooling capacity of Ranque–Hilsch vortex tube, International Journal of Refrigeration, Volume 35, Issue 5,2012,Pages 1473-1483,ISSN 0140-7007

[6] H.M. Skye, G.F. Nellis, S.A. Klein, Comparison of CFD analysis to empirical data in a commercial vortex tube, International Journal of Refrigeration, Volume 29, Issue 1,2006, Pages 71-80, ISSN 0140-7007,

[7] P. Ambedkar, T. Dutta, CFD simulation and thermodynamic analysis of energy separation in vortex tube using different inert gases at different inlet pressures and cold mass fractions, Energy, Volume 263, Part B,2023, 125797, ISSN 360-5442,

Organic pesticide from non-conventional natural resources

T Bhattacharjee[1], A A Kotame[1*], A Mahajan[1], A Ghorpade[1], A Kothadia[1], A Mohammadi[1] and A Bhandarkar[1]

[1]Department of Chemical Engineering, Vishwakarma Institute of Technology, Pune 411047, India

[*]Corresponding author. Tel.: 9420318312, E-mail address: aniket.kotame20@vit.edu

Abstract. The new varieties of organic pesticides and their application as a substitute for chemical pesticides are the main topics of this paper. Due to the residue left on the vegetables,fruits, etc. which we consume on a daily basis, many of the contemporary, chemical pesticides being utilised today are reducing the soil fertility and negatively damaging human health. Study has shown that natural oils also work as pesticides but the emulsion of oil and water is big task so for that We need chemical based emulsifier so to avoid chemicals we are studying on various fruits and vegetable extracts which can be used as emulsifiers. According to the data of year 2020-21 Maharashtra leads with the 13243 MT of usage of pesticides. Orange peel extract contain chemicals such as saponins and flavonoids, which have the property of emulsifier for the emulsion of oil and water. After studying various papers, we found that orange peel extract can definitely work as emulsifier for oil and water emulsion. In this paper we got to know the presence of saponins with the help of froth test (test for saponins), test for flavonoids and the UV-Vis absorbance spectrum of the orange peel. In this we got the yield of orange peel extract using two different solvents. Paper tells us about the different analysis performed with the help of the parameters obtained for finding the presence of emulsifying properties present in orange peel.

1. Introduction

Organic pesticides are formed from the substance which are natural ingredients [1]. Formulation of essential oils can work as pesticide on the crop to cure the diseases like different pest, fungus, bacteria, etc [2]. Due to excessive use of chemical pesticides according to WHO (World Health Organisation) Years can pass before some of the older, less expensive pesticides are removed from soil and water. Although many of these chemicals are no longer utilised in agriculture in wealthy nations, they are still prevalent in many underdeveloped nations.

Pesticides have the potential to be hazardous to people and can have both short-term and long-term health consequences. Studies indicate that orange peels are abundant in bioactive compounds and are by products of fruit processing. India produces approximately 25 lakh tons of orange annually, which results in the production of a significant amount of orange by product waste, such as peels. Punjab, Madhya Pradesh, Andhra Pradesh, Maharashtra, Rajasthan, Assam, and Karnataka are India's primary orange-producing states. The orange peel builds up in large quantities during the production of orange juice and other orange products, posing a threat to the environment.

Neem Oil is a holistic pesticide due to the nature of its active components. This means that it is effective at all stages of the insect's life cycle. Azadirachtin, a key element in neem oil, functions as a natural insecticide in organic farming. As a result, it's critical to figure out where these peels can be used. Numerous phytochemicals and nutrients can be found in orange peels; Consequently, they may be useful

in numerous food and drug products [3]. So for the alternative for chemical pesticides we can use organic pesticides which are derived from natural ingredients. Formulation of essential oils can work as pesticide on the crop to cure the diseases like different pest, fungus, bacteria, etc. Orange peel extract can work as a stabilizer in the oil-water emulsion which can form organic pesticide. Later on the extract of fruits or vegetables which contain saponins and flavonoids can work as stabilizers [3].
The agricultural land in India is very huge. Agriculture is the only industry on which farmers depend. According to the most recent figures, India's agricultural sector generates 20% of the country's total GDP. The total amount of chemical pesticides consumed in India in 2021–2022 was 58720 MT. Recent years have seen a rise in environmental concerns from both experts and the general population due to the overuse of pesticides. Pesticides are thought to be used to treat crops with about 2.5 million tonnes of pesticides annually, resulting in $100 billion in global economic losses.
This is due to two factors: pesticides' high toxicity and nonbiodegradability, as well as the health-threatening residues found in crops, water supplies, and soil. To address the issue of long-term toxicity to mammals, new insecticides that are biodegradable and highly selective must be developed. It is essential to research and develop environmentally friendly insecticides in order to reduce pesticide use and preserve agricultural yields. In terms of minimizing adverse effects on human health and the environment, natural products are an excellent alternative to synthetic pesticides [4].

2. Literature Survey

Over 2000 plant species possess some insecticidal diseases. So chemical pesticides should have some alternative which can be used across all the plants in homes and gardens. This paper refers to the formulation of the different types of pesticides and their forms. Spreaders, stickers, wetting agents, and foaming agents are some of the additives used to boost the active ingredient's efficiency. at room temperature normal We acquire organic pesticide to combat insect pests. Additionally, to manage numerous household and field insect pests. Synthetic pesticides are primarily used to control insects, but their high operating costs and environmental pollution have necessitated an alternative method [2]. Due to synthetic pesticide it also affect the fertility of soil but in the case of organic pesticide with the accurate active ingredients they become more efficient with the low cost as compared to synthetic pesticide. Steam distillation is a common conventional method for extracting the essential oil from catnip. This can be done by cutting up the flowers, stems, and leaves of the plant and cooking them in a pressure cooker for about an hour. At normal room temperature. Catnip oil is obtained which is highly effective for the insect pest. There are some pest contexts where conventional products have demonstrated equivalent control [4].
A mixture that is made up of two or more liquid stages is referred to as an "emulsion." Emulsions are used in food production, healthcare, chemical synthesis, and firefighting, among other industries in the chemical, energy, and environmental sectors. Due to asphaltene, a naturally occurring surfactant, when oil and water are mixed during the oil production process, water-in-oil emulsions naturally form. This study also summarizes previous experimental and modeling research and methodology, focusing on reliable laboratory equipment, simulation software, and modeling approaches for examining emulsion stability and size distribution. This analysis includes systematic parametric sensitivity testing to examine the various effects of important media, process, and thermodynamic features on the variables of interest. In order to properly build and operate the accompanying gear, this review paper gives helpful recommendations for characterising and modelling emulsions and their behaviours in diverse industrial sectors [7].
In this study, tea saponin extracts from Camellia oleifera seeds were extracted using AB-8 microporous resin and gradient elution with ethanol in a variety of purities. AB-8 microporous resin and a gradient ethanol elution were used to purify the tea saponins from Camellia oleifera seeds. An ethanol-sulfuric acid color reaction and thin-layer chromatography were used on the obtained eluates. Together, these findings demonstrate that tea saponins have a potent insecticidal effect on the pest E. obliqua larvae that infest tea plantations. This study revealed that tea saponins were 99 percent pure and had insecticidal properties. The saponins have the ability to kill the larvae and their intestines when they come into contact with their exterior surfaces [8].

The primary topic of this essay is brand-new insecticide varieties. The poisons in the soil, air, and water are rising as a result of the fact that many of the modern pesticides remain in the soil for years. Organic pesticides typically have a narrow spectrum of activity, are less expensive, toxic to workers or consumers, and are safer for the environment and for beneficial insects than conventional pesticides like organochlorine. They also provide better control than conventional pesticides like those that contain organochlorine. They are often stomach poisons that operate fast to stop eating, which makes them sometimes much more selective for nuisance insects that feed on plants and less hazardous to beneficial insects. Some plants also have specialised gland cells that create essential oils. Control of cockroaches and other household pests using peppermint oil [9].

According to the report, excessive use of pesticides with high toxicity and low biodegradability leads to environmental concerns. More attention has been paid to some of the most recent advancements in green pesticides. Essential oils have been touted as safe and affordable green pesticides with current commercial breakthroughs, as well as their potential and limitations. Plants can use the fragrant qualities of essential oils to attract or repel insects, among other things. The article discusses a variety of essential oils used as natural insecticides and their beneficial effects, including Lemon grass (Cymbopogon flexuosus), on the other hand, is well renowned for its ability to repel pests. Rose (Rosa damascena) is well known in the perfumery and aroma sectors. The reason why so few pest control treatments are based on plant essential oils has also been discovered. Regulatory barriers to commercialization may be brought on by the expense of toxicological and environmental analyses or the fact that the effectiveness of essential oils against pests and diseases is not as obvious or visible as that found with currently available therapies [10].

3. Materials and Methods

Neem Oil is a holistic pesticide due to the nature of its active components. This means that it is effective at all stages of the insect's life cycle. Azadirachtin, a key element in neem oil, functions as a natural insecticide in organic farming.

Neem Oil was bought from the local market and orange peels were bought from the juice centres. The peel was separated followed by drying for 5-6 days. By utilising mixer grinding, dried peels were reduced to a coarse powder.

Figure 1. Orange Peel Powder.

4. Methodology

4.1. Soxhlet Extraction

The orange peel powder was extracted with different solvents: acetone and methanol by using Soxhlet extractor. 50 gm powder was used with 300 ml of acetone and 30 gm powder was used with 300 ml methanol for the extraction at 70°C by Soxhlet extraction method for 1 hour [3].

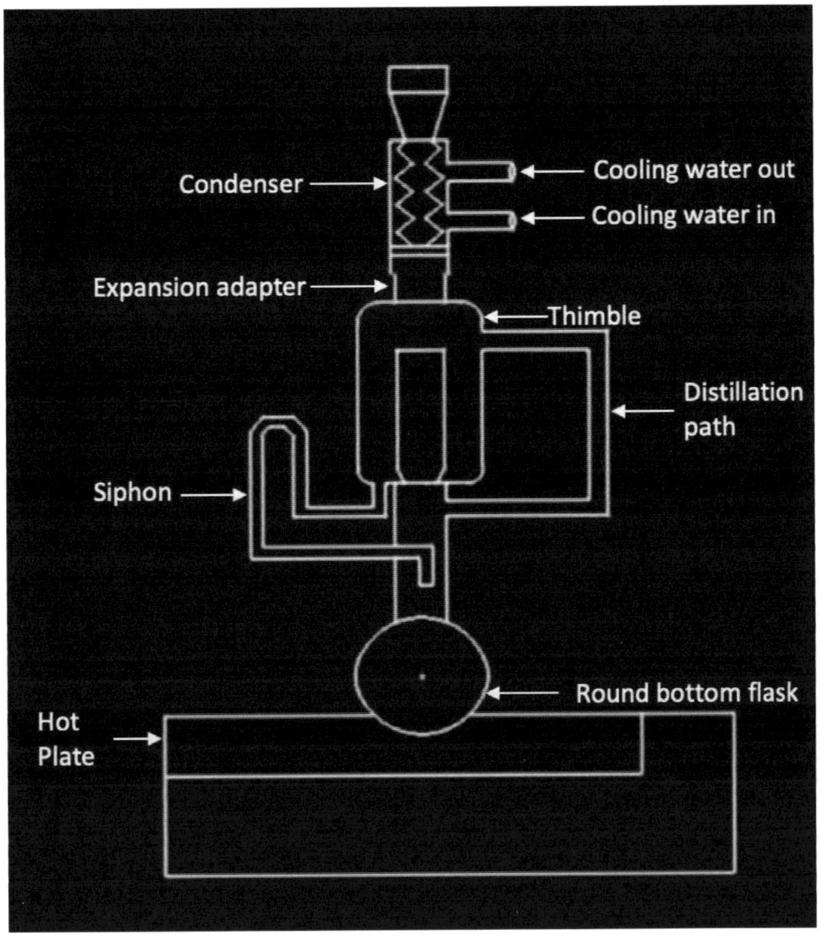

Figure 2. Soxhlet Apparatus

4.2. Distillation
As we need only the orange peel powder extract so we performed distillation for removing the solvents from the extract which we obtained from the Soxhlet Extraction process [3].

4.3. Phytochemical Analysis

4.3.1. Test for Saponins. In reality, saponins are one of the most well-known classes of naturally occurring emulsifiers, and their biological capabilities (anti-inflammatory, antibacterial, antioxidant, antiviral, hypo-cholesterolemic) and physicochemical characteristics (amphiphilic nature and high surface activity) extend their use to a variety of applications.
After boiling or combining two milliliters of orange peel extract with twenty milliliters of water, the mixture was filtered. 10 ml of this filtrate was vigorously shaken with 5 ml of distilled water to produce a stable, long-lasting froth. The appearance of foam is an indication that the extract contains saponins [3].

4.3.2. Test for Anthroquinones. After adding 0.5 grams of sulfuric acid (H2SO4) to the extract, it was heated for 10 milliliters and then filtered while it was still hot. Shaking the filtrate with 5 ml of chloroform was done. One milliliter of diluted ammonia was poured into the chloroform layer. The final solution's color variations were evaluated [3].

4.3.3. Test for Flavonoids. It has been demonstrated that some common food flavonoids can function as effective stabilisers of oil-in-water emulsions by adhering as water-insoluble particles to the oil droplets' surfaces, i.e., Pickering emulsions. A yellow colour formed after mixing 2 ml of orange peel extract with 1 ml of 2N sodium hydroxide suggests the presence of flavonoids [11].

4.3.4. Test for Tannin. The extract and 10 milliliters of water in a test tube were used to boil and filter the mixture. Three to four drops of ferric chloride at 1% were then added. After that, a blue-black or brownish-green color was observed [11].

4.3.5. Test for Terpenoids. 2 ml of chloroform was added to each 0.5 g of extract. Concentrated H2S04 (3 ml) was carefully added to form a layer. Terpenoids are indicated by the interface's reddish-brown appearance [11].

5. Results and Discussion

5.1. Extraction of orange peel extract by using various solvents
The yield of this study's soxhlet extraction of orange peel with various solvents varied from one attempt to the next. The yield of the extract varies with the solvent. We obtained a yield of 11 ml with acetone as the solvent and 10.7 ml with methanol.

From Literature we got the yield of acetone and methanol per 100 gm but we here performed for per 50 gm and acetone solvent showed the best results for all the analysis.

Table 1. Yield of orange peel extract by different solvent using Soxhlet extractor.

Solvent	Yeild per 50 gm
Acetone	11 ml
Methanol	10.7 ml

5.2. Phytochemical screening of orange peel extract
The results of the phytochemical screening of the various solvent extracts allowed for the identification of terpenoids, flavanoids, saponins, and tannins.
As shown in the table Anthroquinones was completely absent in orange peel in the performed method. While on other hand Methanol and Acetone showed the presence of Tannins, Terpenoids, Flavonoids and Saponins.

Table 2. Phytochemical constituents of Orange Peel Extracts by Soxhlet Extractor.

Extract	Anthroquinones	Tannins	Terpenoids	Flavonoids	Saponins
Methanol	-	+	+	+	+
Acetone	-	+	-	+	+

5.3. Proximate composition of orange peel powder
In Table No. 1, the approximate compositions of orange peel powder are listed. The shade-dried powdered sample's moisture content was determined to be 9.2%. Protein content on a dry weight basis was between 12% and 13%. (DW). This suggests that the orangepeel might possibly contain significant nutrients. The difference in the amount of carbohydrates was [3].

Table 3. Proximate analysis of orange peel powder

Sr.no	Proximate analysis	% content
1.	Moisture Content	9.2 ± 0.01
2.	Crude Fibre	14.17 ± 0.36
3.	Crude Protein	12.43 ± 0.20
4.	Ash Content	7.8 ± 0.01
5.	Carbohydrate Content	52.90 ± 0.43

5.4. Testing of solubility of extracts with neem oil

Testing outcomes showed that the extract obtained using methanol solvent is soluble in neem oil while the extract obtained using acetone solvent is insoluble.

Table 4. Solubility test of extracts with neem oil.

Sample	Results
Extract obtained by Acetone solvent	Insoluble
Extract obtained by Methanol solvent	Soluble

5.5. UV-Vis absorbance spectrum of orange peel extract

In contrast to a reference or blank sample, the amount of discrete wavelengths of UV or visible light that are absorbed by or transmitted through a sample can be determined using the analytical method of UV-Vis spectroscopy. The UV-Vis absorbance spectrum of orange peel exhibits a maximal 660 nm absorbance that is induced by a localised transition, as well as absorbances at 402 nm and 279 nm [14].

Table 5. Yield of orange peel extract by different solvent using Soxhlet extractor.

Wavelength (nm)	Absorbance (a.u)
279	3.39
402	2.737
660	3.154

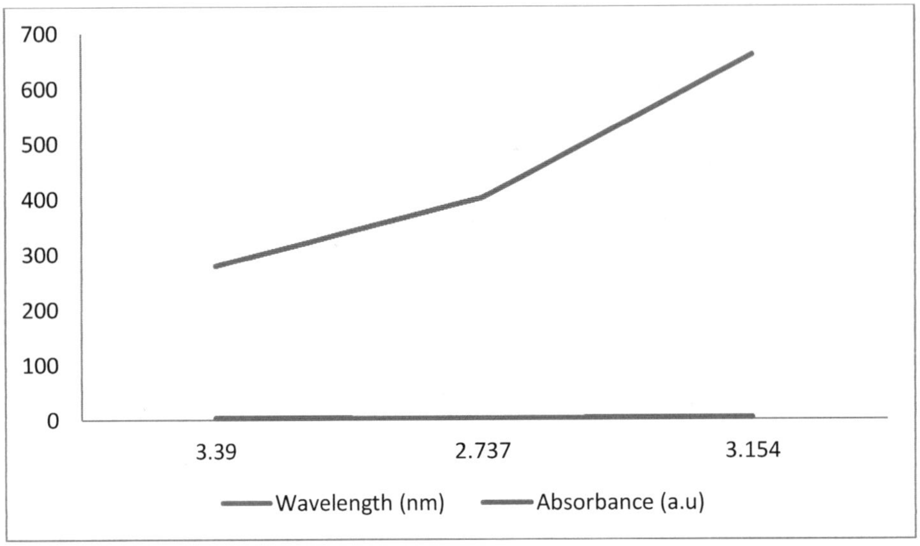

Figure 3. UV-Vis absorbance spectrum of orange peel

6. Conclusion and Future Enhancement

Synthetic pesticides have caused soil and air pollution, and some of these chemicals can take years or even decades to degrade. These insecticides are detrimental to human health as well as to animal, microbial, and plant life. Fortunately, there are lots of natural pesticides (called biopesticides) that are also efficient at controlling pests. People need to stop using dangerous pesticides and start utilising biopesticides, which decompose swiftly in the soil and sunlight. In this paper we got the best effective product as pesticide because neem oil is natural oil which works on insect pest and many companies combine it with chemical based emulsifier which is harmful in many aspects but here, we tried to replace the chemical based emulsifier with the orange peel extract which showed the best results.

With the help of different parameters we performed the experiments and tests and we got to know the presence of saponins and flavonoids which are one of the important properties of emulsifier for the emulsification of oil and water. We attempted to emulsify the orange peel with neem oil, and the extract obtained by using methanol solvent emulsified with the oil. Further we can study on different plant extract, gums which can work as emulsifier or stabilizer and after that the final product can be tested on the plants.

7. References

[1] Kalkura, Pramoda, Suhas Kashyap, and Ms Ramyashree. "Pest control management system using organic pesticides." *Global Transitions Proceedings* **2.2** (2021): 175-180.

[2] Libs, E., and E. Salim. "Formulation of essential oil pesticides technology and their application." *Agri Res Tech* **9.2** (2017): 555759.

[3] Gotmare, Sulekha, and Jaya Gade. "Orange peel: A potential source of phytochemical compounds." *Int. J. ChemTech* Res**11** (2018): 240-243.

[4] Koul, Opender, Suresh Walia, and G. S. Dhaliwal. "Essential oils as green pesticides: potential and constraints." *Biopesticides international* **4.1** (2008): 63-84.

[5] Owoeye, T. F., et al. "Phytochemical constituents and proximate analysis of dry pineapple peels." *IOP Conference Series: Earth and Environmental Science.* Vol. **993**. No. 1. IOP Publishing, 2022.

[6] Vinay, C. H., P. Goudanavar, and A. Acharya. "Development and characterization of pomegranate and orange fruit peel extract-based silver nanoparticles." *Journal of Manmohan Memorial Institute of Health Sciences* **4.1** (2018): 72-85.

[7] Goodarzi, Fatemeh, and Sohrab Zendehboudi. "A comprehensive review on emulsions and

emulsion stability in chemical and energy industries." *The Canadian Journal of Chemical Engineering* **97.1** (2019): 281-309.
[8] Cui, Chuanjian, et al. "Insecticidal activity and insecticidal mechanism of total saponins from Camellia oleifera." *Molecules* **24.24** (2019): 4518.
[9] Oguh, C. E., et al. "Natural pesticides (biopesticides) and uses in pest management-a critical review." *Asian Journal of Biotechnology and Genetic Engineering* **2.3** (2019): 1-18.
[10] Mohan, Manindra, et al. "Essential oils as green pesticides: for sustainable agriculture." *Research Journal of Pharmaceutical, Biological and Chemical Sciences* **2.4** (2011): 100-106.
[11] Ruvanthika, P. N., S. Manikandan, and S. Lalitha. "A comparative study on phytochemical screening of aerial parts of Nelumbo nucifera gaertn by gas chromatographic mass spectrometry." *International Journal of Pharmaceutical Sciences and Research* **8.5** (2017): 2258.
[12] Luo, Zijun, et al. "Particle-stabilizing effects of flavonoids at the oil– water interface." *Journal of agricultural and food chemistry* **59.6** (2011): 2636-2645.
[13] Schreiner, Tatiana B., et al. "Saponins as Natural Emulsifiers for Nanoemulsions." *Journal of Agricultural and Food Chemistry* (2022).
[14] Jayaprakash, N., et al. "One-step synthesis, characterisation, photocatalytic and bio-medical applications of ZnO nanoplates." *Materials Technology* **35.2** (2020): 112-124.
[15] Deore, Sharada Laxman, Bhushan Arun Baviskar, and Ashwini Suresh Rangari. "Rapid and high yield Extraction method for Saponins from Safed musli." *Pharmacognosy Journal* **7.4** (2015).
[16] Isman, Murray B. "Bioinsecticides based on plant essential oils: A short overview." *Zeitschrift für Naturforschung* C **75.7-8** (2020): 179-182.
[17] Mohaddes-Kamranshahi, Mahdiyeh, et al. "Evaluation of the saponin green extraction from Ziziphus spina-christi leaves using hydrothermal, microwave and Bain-Marie water bath heating methods." *Green Processing and Synthesis* **8.1** (2019): 62-67.
[18] Erukainure, Ochuko L., et al. "Orange peel extracts: Chemical characterization, antioxidant, antioxidative burst, and phytotoxic activities." *Journal of dietary supplements* **13.5** (2016): 585-594.
[19] Islam, Md Shafiqul, and Akib Morshed. "Study on Homemade Bio-Pesticides and Organic Pest Management in Organic Farming." *The International Journal of Engineering and Science* **2.7** (2013): 18-25.
[20] Oguh, C. E., et al. "Natural pesticides (biopesticides) and uses in pest management-a critical review." *Asian Journal of Biotechnology and Genetic Engineering* **2.3** (2019): 1-18.
[21] Raj, Aman, Ashwani Kumar, and Joanna Felicity Dames. "Tapping the role of microbial biosurfactants in pesticide remediation: an eco-friendly approach for environmental sustainability." *Frontiers in Microbiology* (2021): 3886.
[22] Arora, Mamta, and Parminder Kaur. "Phytochemical screening of orange peel and pulp." *International Journal of Research in Engineering and Technology* **2.12** (2013): 517-522.

8. Acknowledgement

This work was supported by Department of Chemical Engineering, Vishwakarma Institute of Technology, Pune. Authors are thankful to Prof. Dr. Tanushree Bhattacharjee for guiding and helping us throughout the project.

ANTITYROSINASE ACTIVITY OF PHYCOCYANIN AND CREAM FORMULATION FOR HYPERPIGMENTATION

Aishwarya Raju Muruganandam, Sivasankari Venkatasubramanian, Shail Adrian Jagmag, Veerabhuvaneshwari Veerichetty *

Department of Biotechnology, Kumaraguru College of Technology

Corresponding Author: Veerabhuvaneshwari Veerichetty *, veerabhuvaneshwari.v.bt@kct.ac.in, +91-9659023496, Department of Biotechnology, Kumaraguru College of Technology, Coimbatore

Abstract

Free radicals are involved in the catalytic reactions of tyrosinase to give dopaquinone in melanin biosynthesis. Screening of molecules with antioxidant activity from natural sources which inhibit tyrosinase has become important for cosmetic and medicinal topical products. Tyrosinase inhibitors are used in treating hyperpigmentation. Synthetic tyrosinase inhibitors possess side effects such as skin irritation, dermatitis, dryness, and inflammatory reaction. The objective of the study is to evaluate the tyrosinase inhibition potential of phycocyanin, which is a pigment extracted from the blue green algae Spirulina platensis and formulating a cream that can be used to treat hyperpigmentation. Phycocyanin possesses anticancer, antioxidant, antiviral and anti-inflammatory activities and exhibits the regulative ability of tyrosinase expression and thereby modulates melanogenesis. The spectral characterization of phycocyanin is carried out using UV-vis spectroscopy, FTIR and HPTLC. Phycocyanin showed tyrosinase inhibition potential with an IC50 of 30.88-39.87 µg/ml in SK-Mel-28 melanoma cells. In melanogenesis pathway, tyrosinase regulates the production of melanin by the skin cells and by incorporating phycocyanin in cream will reduce the melanin production and treat hyperpigmentation.

Keywords: Phycocyanin, Cream formulation, Tyrosinase inhibition assay, DOPA, Tyrosine, Hyperpigmentation, Melanogenesis

1. Introduction

Phycocyanin, a phycobiliprotein, is an important pigment found in Spirulina platensis that is utilized as dietary supplement in many countries. The commercialization of Spirulina for the valuable constituents it possesses such as several proteins and vitamins are currently underway in many places as it is inexpensive to obtain [1] [5] [8].

Tyrosinase is a vital enzyme which catalyzes the hydroxylation of L-tyrosine to 3,4-dihydroxyphenylalanine (DOPA) including the oxidation of DOPA to dopaquinone which polymerize spontaneously to form melanin in the first two consecutive steps of biosynthesis of melanin. The excess expression of tyrosinase promotes excessive melanin production, that paves to hyperpigmentation effects such as cervical poikiloderma, acanthosis nigricans melasma, age spots, and non-melanoma skin cancer. [4]

Insufficient or excessive production of melanin is the cause of several pigmentation issues. Pigmentation is a crucial defense mechanism of the human body and protects against oxidative stress and UV radiation. But abnormal hyperpigmentation of the skin is an aesthetic issue. Novel peptides with low molecular weight compounds are under investigation to regulate skin pigmentation. In-silico digestion of Phycocyanin from *S. platensis* allowed for construction of an internal library of peptides which were then analyzed for anti-melanogenic activity. [7] Phycocyanin is a good antioxidant and light harvesting protein which has been found to possess potential tyrosinase inhibition properties. [12]

IOP Conf. Series: Materials Science and Engineering 1291 (2023) 012039 doi:10.1088/1757-899X/1291/1/012039

The process of Monophenol ortho-hydroxylation and conversion to catechol and successive oxidation forming ortho-quinones is catalyzed by tyrosinase. Tyrosinase is the controller for browning hair and skin pigmentation. Alternatively, low tyrosinase can also cause anomalies like vitiligo or freckles. It also plays a huge role in cancer and Parkinson's disease which is a neurodegenerative disease. Since tyrosinase is a prominent target in categories like food, cosmetics, agriculture and medicine, the development and screening of effective tyrosinase inhibitors has great importance. [11]

Present tyrosinase inhibitors are hazardous and or sometimes ineffective, and there is enduring research for improved natural source inhibitors, which could possibly be free of negative side effects [3] Inhibition of the tyrosinase enzyme plays a significant role in the melanogenesis pathway helping in reduction of production of melanin in the skin, and thus regulating skin pigmentation. [10]

Some of the drawbacks of synthetic tyrosinase inhibitors are that they cause mild skin irritation and sensitization such as burning and stinging. They can also cause dermatitis, redness, and inflammatory reactions. Hence, natural tyrosinase inhibitors like phycocyanin cream are worth investigating to treat hyperpigmentation.

Creams are regarded to be one of the easiest and most convenient ways to use. People in any age group can use the cream since application is topical and on the surface of the skin only resulting in the formation of semi-solid emulsions. Creams are divided into two types: oil-in-water and water-in-oil. Oil-in-water (O/W) creams are made up of small drops of oil spread in a continuous water phase, and water-in-oil (W/O) creams are made up of tiny drops of water dispersed in a continuous oily phase. Oil-in-water lotions are more pleasant and cosmetically acceptable since they are less oily and simpler to wash off with water. Water-in-oil creams are a little more difficult to work with. phycocyanin was included in the cream formulation due to its tyrosinase inhibitory activity.

Fig. 1.1 represents the role of phycocyanin in melanogenesis. When the skin cells are over exposed to sunlight, UV radiation or chemotherapy, it attacks the normal melanocyte cells which results in abnormal melanocytes due to the production of increased ROS and tyrosinase. This favors hyperpigmentation disorders such as non-melanoma skin cancer, acanthosis nigricans melasma, cervical poikiloderma, lentigines and periorbital hyperpigmentation.

A tyrosinase inhibitor inhibits the production of tyrosinase in the melanogenesis pathway. Phycocyanin, a natural blue pigment obtained from Spirulina platensis, is used as a natural tyrosinase inhibitor for our study. Thereby reducing the melanin production by treating hyperpigmentation and its associated disorders. Hence, our study reveals that phycocyanin based cream is well suited to remove hyperpigmentation and help in further protection of UV and harmful sun radiations. [13]

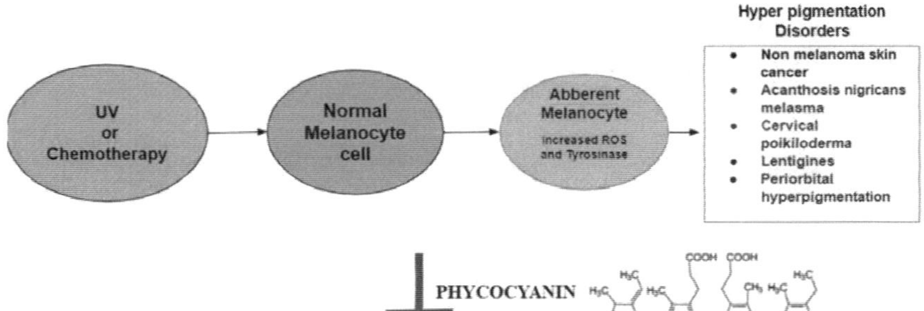

Fig. 1.1: Role of Phycocyanin in Melanogenesis

2.MATERIALS AND METHODS

2.1 CHARCTERISATION OF PHYCOCYANIN

The molecule structure of phycocyanin was measured by FTIR using spectra IR 4000- 1000 cm-1. FT-IR spectroscopy was performed to identify characteristic functional groups present in phycocyanin. Phycocyanin was subjected to UV visible spectroscopy characterisation to observe its peak absorption by using the UV visible spectrophotometer. Thin Layer Chromatography was done to validate the quality of Phycocyanin by optimising the solvents n-butanol, acetic acid and distilled water in the ratio of 10:6:7. Further it was characterised by HPTLC technique. It was performed in HPTLC sample applicator (Make: Aspire; Model: AE-05). Two bands were created each 10μl and 20μl was spotted in 10×10cm size precoated silica gel 60 aluminium sheet. The mobile phase solvent is n-butanol: Acetic acid: Distilled water in the ratio of 10:6:7. The sheet was then kept in the chamber for the mobile phase to react with the stationary phase and was then visualised in visible light, UV wavelength of short 265nm and long 364nm.

2.2 PHYCOCYANIN BASED CREAM FORMULATION

Table 2.2.1 Components and quantity of oil and aqueous phase in cream formulation

	COMPONENT	QUANTITY
Oil Phase	Stearic acid	5 g
	Coconut Oil	5 ml
Aqueous Phase	Glycerine	5 ml
	Cetostearyl Alcohol	0.5 g
	Sodium benzoate	0.1 g
	Phycocyanin	100 μl

The ingredients are chosen in such a way that they play a significant role in maintaining the homogeneity of the cream and also being skin friendly. Stearic acid is the base of the cream and gives consistency to the cream and a pearlescent appearance to the formulation. Sodium benzoate is used as a preservative so that the cream can be used for a long time. Coconut oil acts as a moisturizer that helps in maintaining the smoothness and hydration of the skin.

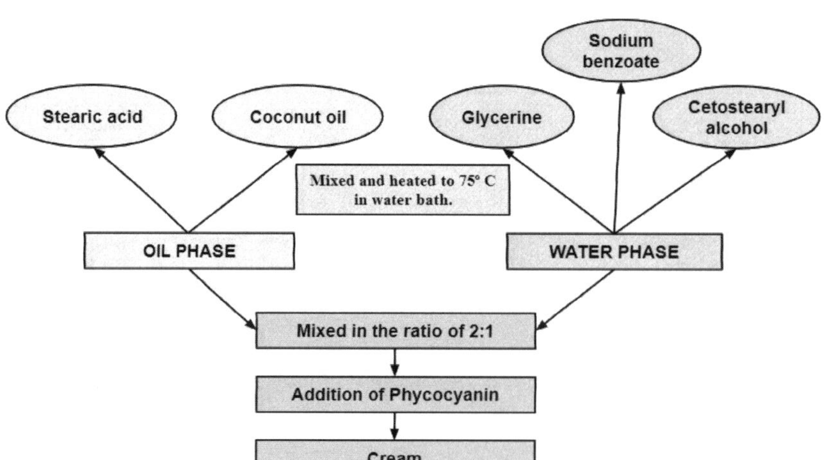

Fig. 2.2.1 Flow chart representing the process of cream formulation.

The oil in water emulsion was prepared for the cream formulation. The oil phase components stearic acid and coconut oil were first combined in the above-mentioned quantity and was then heated to 75°C in water bath. Simultaneously, the aqueous phase components Glycerine, Cetostearyl alcohol, Sodium benzoate were mixed and heated to 75°C in water bath. Then the aqueous phase was added slowly to the oil phase and combined well with continuous stirring. After cooling of the product, Phycocyanin was added and mixed well.

Table 2.2.2 Role of ingredients of oil and aqueous phase

S. NO	INGREDIENT	ROLE
1	**Stearic acid**	• Anionic surfactant • Lubricant • Governs consistency of cream • Pearlescent property
2	**Coconut oil**	• Emollient • Moisturizer
3	**Glycerine**	• Humectant • Surfactant • Averts evaporation of water from skin
4	**Cetostearyl alcohol**	• Emulsion stabilizer • Opacifying agent • Viscosity increasing agent
5	**Sodium Benzoate**	• Preservative
6	**Phycocyanin**	• Tyrosinase inhibitor

2.3 TYROSINASE INHIBITION ASSAY

SK-Mel-28 (Human Skin melanoma) Cells were maintained in Dulbecco's Modified Eagles Medium (DMEM-Hi media) - with 10% heat inactivated Fetal Bovine Serum (FBS) + 1% antibiotic cocktail containing Penicillin (100U/ml), Streptomycin (100µg/ml), and Amphotericin B (2.5µg/ml). The naturally present tyrosinase enzyme in SK-Mel-28cells is significantly higher than normal melanin cells. SK-Mel-28 cells were used for this \ assay TC flasks (25cm^2) of SK-Mel-28 cells were incubated at 37°C at 5% CO_2 environment with humidity in a cell culture incubator. L-DOPA of about 40 μL was combined with 80 μL of phosphate buffer with pH of 6.8 in the 96 well microtiter plate. The resulting mixture is kept in the incubator at 37°C for 10 minutes. 40 μL of varying concentrations Phycocyanin and 40 μL of SK-Mel-28 cells (250 U/mL, in PBS) were added to each of the wells on the plate. The absorbance characteristics of the mixtures were analysed using ELISA reader at 475 nm. The results were interpreted at 60-second intervals for a period of 120 minutes. The experiment was done with triplicate (n=3). IC50 value indicates the sample concentration or ascorbic acid concentration in need for inhibiting 50% of the enzyme activity. IC50 value was calculated using Graphpad prism with dose response curves.

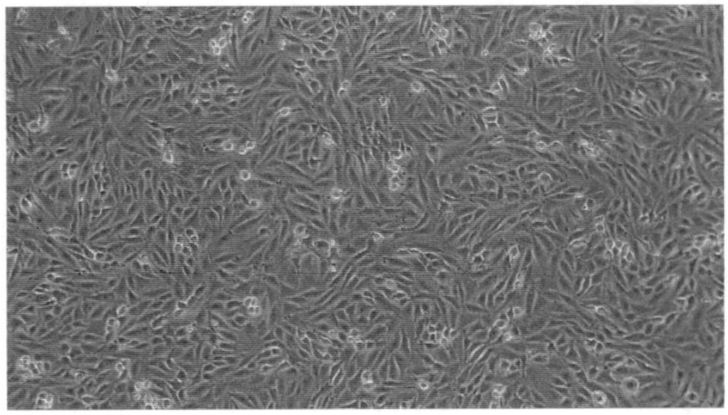

Fig. 2.3.1 Microscopic image of SKMEL cells at 10X (inverted phase contract)

3. RESULTS AND DISCUSSIONS

3.1 FTIR

Phycocyanin was characterised using Fourier Transform Infrared Spectroscopy from wavenumber range 500 to 4000 cm^{-1}. From the above figure of the infrared spectrum the evident peak at 3600.77519 cm^{-1} wavenumber denotes the O-H Stretch of free hydroxyl group. The peak at 3232.55814 cm^{-1} corresponds to C-H Stretch. Similarly peak at 1740.31008 cm^{-1} corresponds to C-O Stretch. The peaks at 1642.44106 and 1533.83539 cm^{-1} of the spectra corresponds to the amino N-H bend [2] [6][9].

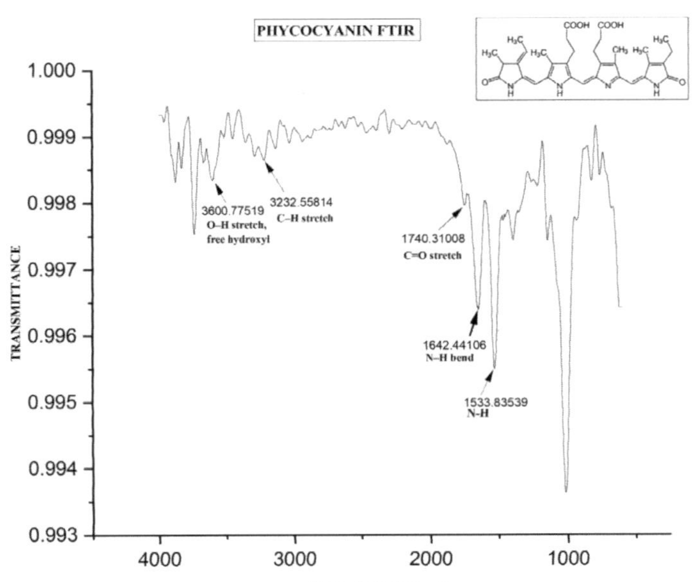

Fig. 3.1 FTIR of Phycocyanin

3.2 UV-Vis SPECTROSCOPY

UV-vis spectroscopy is a popular analytical method used for the characterization of phycocyanin purity and degradation. The long-wavelength band of native C-PC exhibits a distinct peak at 620 nm, with a larger shoulder at shorter wavelengths; the near UV-band is simply resolved into two bands of roughly similar intensity at 360 and 347 nm. The respective bands in denatured C-PC are considerably broadened. [2]

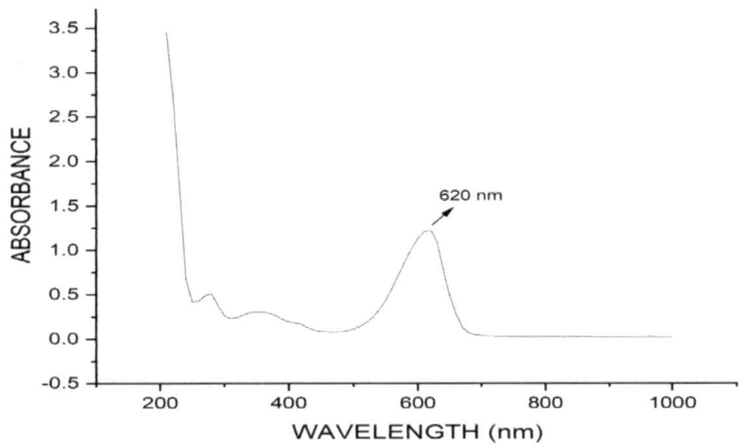

Fig. 3.2 UV-vis Spectra of Phycocyanin

3.3 Thin Layer Chromatography

The Slides were visualised in the visible and in Ultraviolet light at 254 and 365 nm and it confirmed the movement of the sample in the stationary phase. Figure 3.3.1(c) is captured under visible light, (a) is the image captured at 365 nm of UV light and (b) is at 254 nm. The three captures confirmed the movement of the sample.

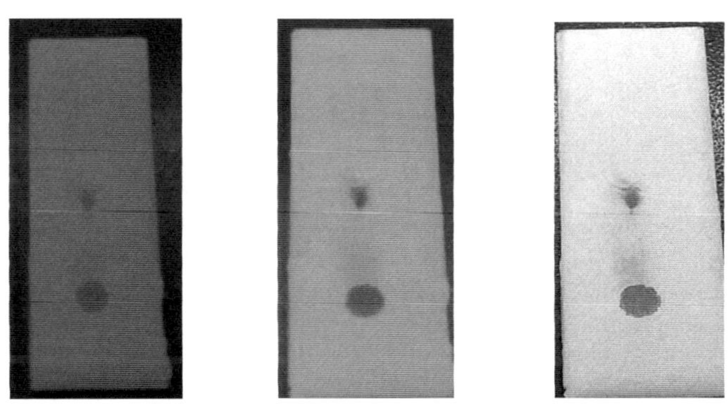

Fig. 3.3.1 (a) Fig. 3.3.1 (b) Fig. 3.3.1 (c)

TLC plate at (a) 365 nm, (b) 254 nm and (c) visible light

3.4 HPTLC

The HPTLC scan of the 2 lanes was obtained from the chromatogram photographed at low wavelength UV at 254 nm (Fig 4.4.1 (a)) and at long wavelength UV at 365 nm (Fig 4.4.1 (b)).

IOP Conf. Series: Materials Science and Engineering 1291 (2023) 012039 doi:10.1088/1757-899X/1291/1/012039

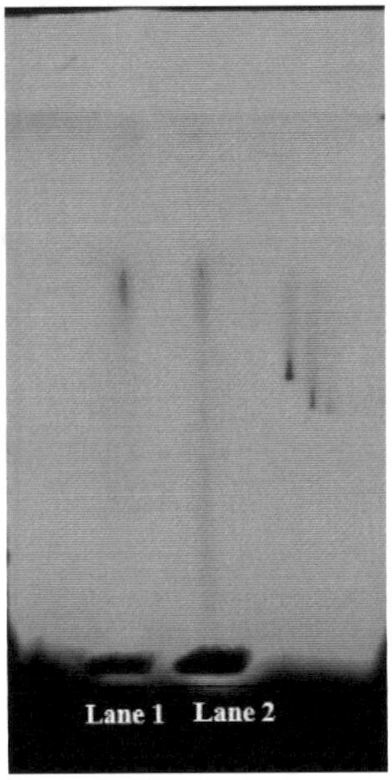

Fig 3.4.1 (a) Fig 3.4.1 (b)

HPTLC scan at (a) UV 254 nm and at (b) 365 nm

The lane 1 is result of the spot loaded with 10μ l of Phycocyanin and lane 2 is with 20μl.

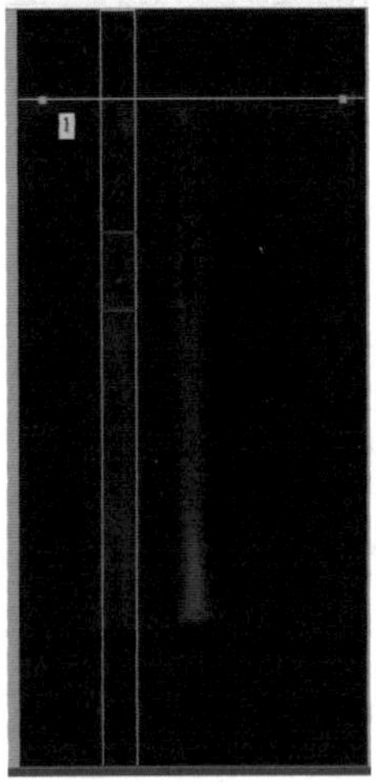

Fig 3.4.2 Phycocyanin lane 1 band 2 in UV 365 nm

The above figure is the UV long 365 nm scan of the sample and the band here indicates the Phycocyanin being illuminated by the UV light. Below is the chromatogram of the Lane 1 and Lane 2- band 2. Further the Rf value of 0.68 was obtained and it validated the compound in accordance with the literature review.

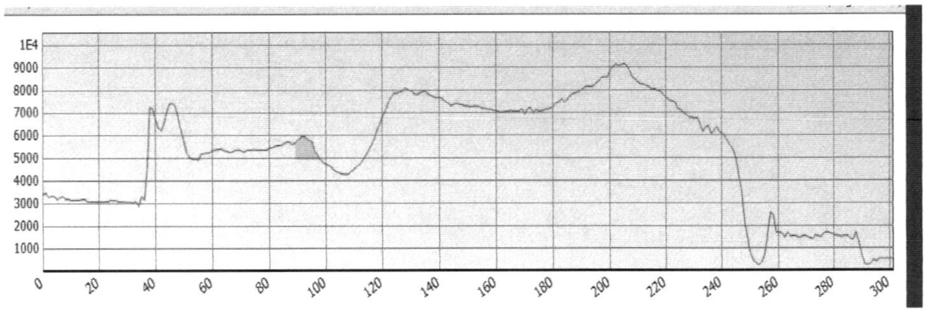

Fig 3.4.3 HPTLC chromatogram of Phycocyanin in Lane 1 band 2 at UV 365 nm

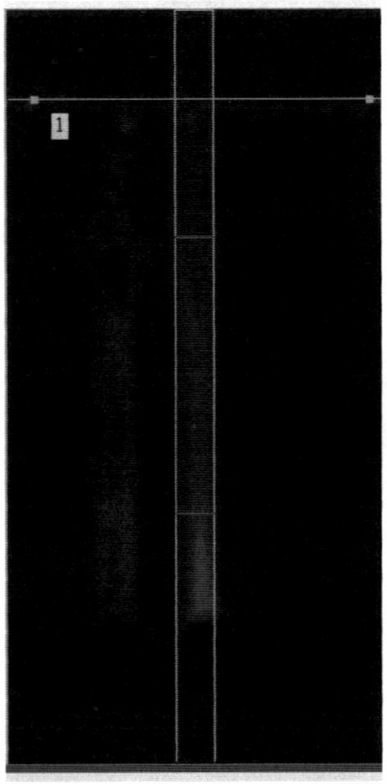

Fig 3.4.4 Phycocyanin lane 2 band 2 in UV 365 nm

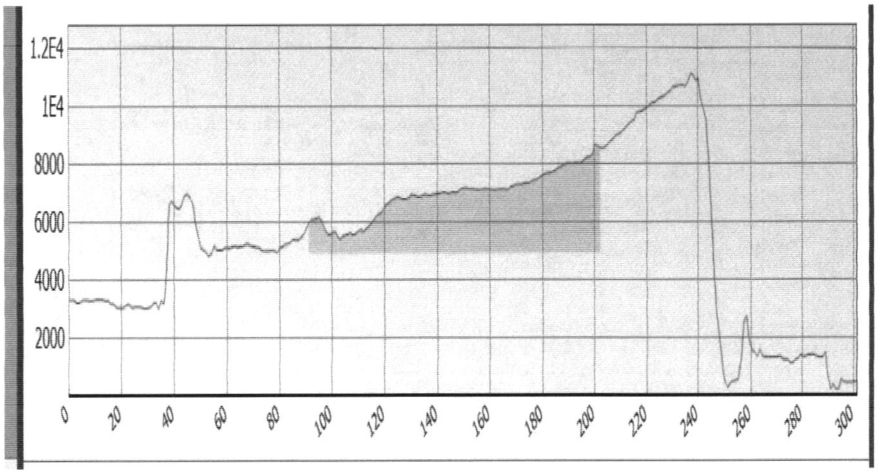

Fig 3.4.5 HPTLC chromatogram of Phycocyanin in Lane 2 band 2 at UV 365 nm

3.5 FTIR – CREAM WITH AND WITHOUT PHYCOCYANIN

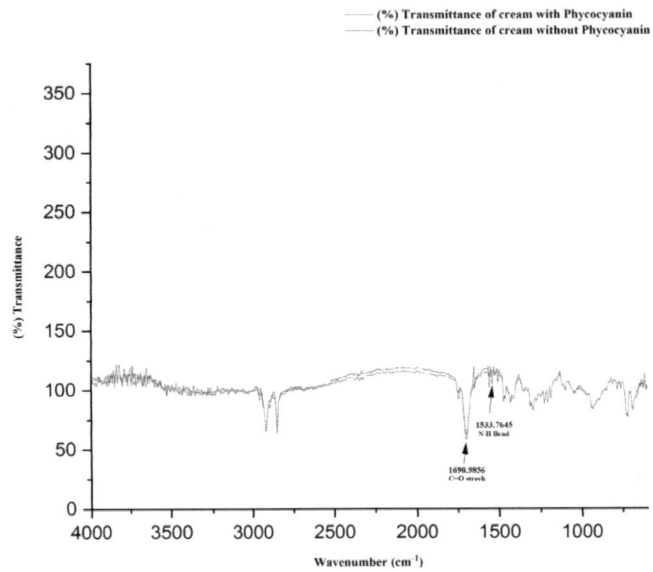

Fig. 3.5.1 FTIR spectra of cream

The formulated cream was analysed for its physicochemical properties by analysing using FTIR spectroscopy. It gave an insight into the functional groups indicating peak values and validated the presence and efficacy of phycocyanin in the formulated cream.

3.6 TYROSINASE INHIBITION ASSAY

3.6.1 TYROSINASE INHIBITION – COMPARATIVE RESULTS (60 MINS)

Tyrosinase inhibition assay was performed for phycocyanin and ascorbic acid at 60-minute time interval and the results were compared using GraphPad prism software.

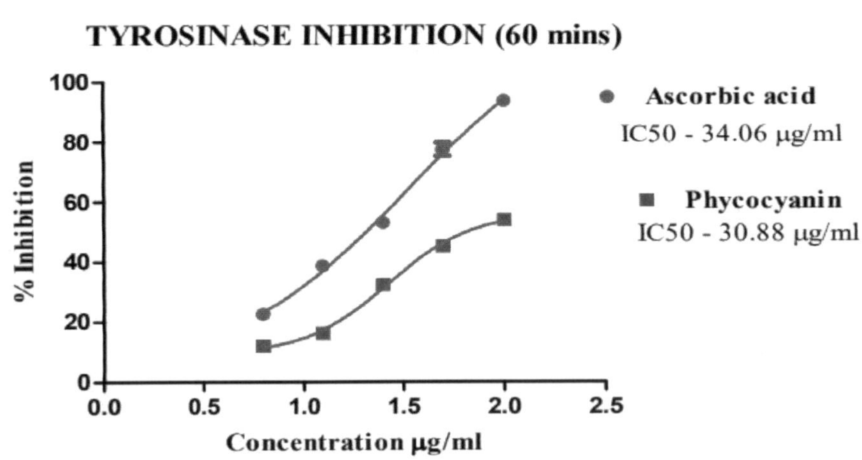

Fig 3.6.1 Tyrosinase inhibition of phycocyanin and ascorbic acid (60 mins)

Fig 3.6.1. From the results of tyrosinase inhibition assay, the concentration (µg/ml) and % inhibition was plotted for phycocyanin and ascorbic acid (control) for 60 minutes time interval. GraphPad prism software was used to find out the IC50 values for phycocyanin and ascorbic acid which was found to be 30.88 µg/ml and 34.06 µg/ml respectively. Avenanthramide from oats show concentration dependent inhibition of tyrosinase activity in α-MSH-induced SK-MEL-2 cells [14]. Flavan Kazinol U isolated from ethyl acetate extract of *B. kazinoki* showed tyrosinase inhibition in B16F10 mouse melanoma cells at the lower concentration of 20 µM [15].

3.6.2 TYROSINASE INHIBITION – COMPARATIVE RESULTS (120 MINS)

Tyrosinase inhibition assay was performed for phycocyanin and ascorbic acid at 120- minute time interval and the results were compared using GraphPad prism software.

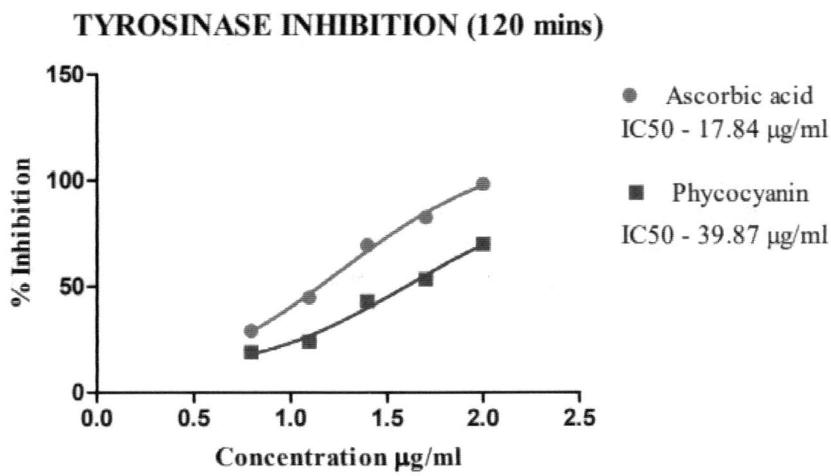

Fig 3.6.2 Tyrosinase inhibition of phycocyanin and ascorbic acid (120 mins)

From the results of tyrosinase inhibition assay, the concentration (µg/ml) and % inhibition was plotted for phycocyanin and ascorbic acid (control) for 120 minutes time interval. GraphPad prism software was used to find out the IC50 values for phycocyanin and ascorbic acid which was found to be 39.87 µg/ml and 17.84 µg/ml respectively.

3.7 CREAM FORMULATION

Phycocyanin incorporated cream was prepared in oil-in-water formulation, which resulted in a significant blue colour cream due to the presence of phycocyanin a naturally blue pigmented compound.

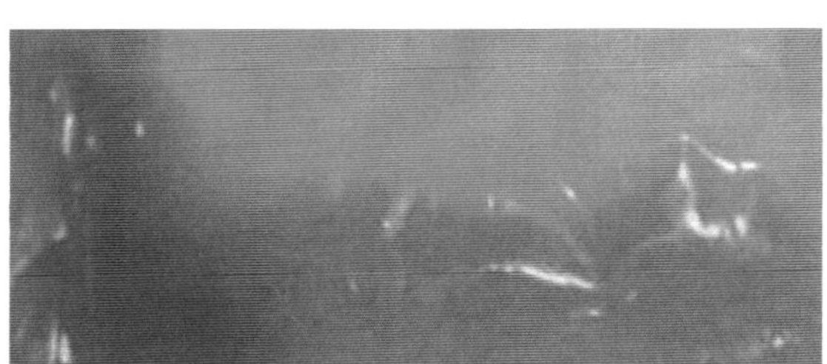

Fig 3.7.1 Phycocyanin incorporated cream

This cream is used in the treatment of hyperpigmentation by topically application to the affected areas in circular motion. Since phycocyanin is found to be an effective tyrosinase inhibitor, it affects the melanogenesis pathway by inhibiting tyrosinase and reduces the melanin production. Antifungal effect of phycocyanin topical cream was studied in mice infected with *Candida albicans in vivo* and wound healing potential of *Spirulina platensis* extracts on human dermal fibroblast cells *in vitro* [16, 17].

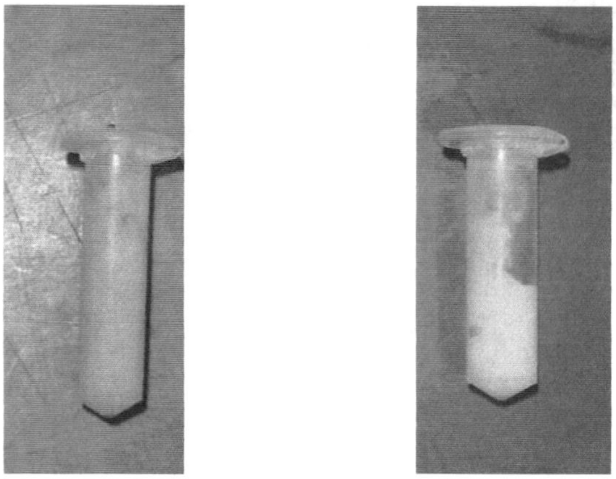

(a) (b)

Fig 3.7.2 (a) Cream with phycocyanin and (b) Cream without phycocyanin

3.7.1 PHYSICAL CHARACTERISATION OF CREAM

3.7.1.1 pH of the cream

The pH meter was used to examine the pH of the formulated cream. It was calibrated with Standard buffer. 0.5 g of the cream was dissolved in about 50 ml of distilled water and tested. The results revealed the pH to be 5.9 and indicate an appropriate match to the skin.

3.7.1.2 Homogeneity

The homogeneity of the cream was examined by visualising the consistency of the cream by touch and feel. It exhibited uniform distribution upon application to the skin.

3.7.1.3 After Application Feel

The cream was applied topically on the skin and spread in circular motions. It left no granules or other particles and applied and absorbed into the skin very smoothly. The emollience and slipperiness of the cream that was left after application was found to be good.

4. CONCLUSION

Phycocyanin showed concentration dependent tyrosinase inhibition in SK-Mel-28 cells against positive control ascorbic acid. Phycocyanin incorporated cream was formulated by taking appropriate ingredients in the oil and aqueous phase mixed in the ratio of 2:1 which resulted in a homogeneous mixture The cream had a significant blue colour due to the presence of phycocyanobilin in phycocyanin. This cream can be applied topically to the affected areas that are infected by over exposure to harmful sun and UV radiation. Thus, our study revealed that Phycocyanin, a naturally occurring blue colour pigment extracted from *Spirulina platensis* potentially inhibited the tyrosinase enzyme in the melanogenesis pathway. Tyrosinase inhibition will reduce the melanin production to a significant amount, and thereby could be used to treat hyperpigmentation and related disorders. Further *in vitro* dermal fibroblast and *in vivo* hyperpigmentation model studies will be performed to optimize the cream and concentration of phycocyanin in future. Thus, phycocyanin proves to be a non-hazardous natural tyrosinase inhibitor in addition to antioxidant [18], anti-inflammatory [19] and antiproliferative properties [20]. Phycocyanin has new scope to be utilized in cosmetic and dermatological fields of research with Generally regarded as safe category (GRAS).

REFERENCES:

[1] Avila Jerley, A., & Prabu, D. M. (2015). Purification, characterization, and antioxidant properties of C-Phycocyanin from Spirulina platensis. Journal of Agriculture, Plant Biotechnology and Bio Products, 2(1), 8-15.

[2] Izadi, M., Latifi, E. Comparison of the antibacterial properties of phycocyanin and its SNPs and their effects on rat blood cells and liver enzymes. Beni-Suef Univ J Basic Appl Sci 11, 56 (2022). https://doi.org/10.1186/s43088-022-00236-w

[3] Di Petrillo, A., González-Paramás, A. M., Era, B., Medda, R., Pintus, F., Santos-Buelga, C., & Fais, A. (2016). Tyrosinase inhibition and antioxidant properties of Asphodelus microcarpus extracts. BMC complementary and alternative medicine, 16(1), 1-9.

[4] D'Mello, S. A., Finlay, G. J., Baguley, B. C., & Askarian-Amiri, M. E. (2016). Signaling pathways in melanogenesis. International journal of molecular sciences, 17(7), 1144.

[5] Fernández-Rojas, B., Hernández-Juárez, J., & Pedraza-Chaverri, J. (2014). Nutraceutical properties of phycocyanin. Journal of functional foods, 11, 375-392.

[6] Jackson, M., & Mantsch, H. H. (1995). The use and misuse of FTIR spectroscopy in the determination of protein structure. Critical reviews in biochemistry and molecular biology, 30(2), 95-120.

[7] Kose, A., & Oncel, S. S. (2022). Design of melanogenesis regulatory peptides derived from phycocyanin of the microalgae Spirulina platensis. Peptides, 152, 170783.

[8] Kuddus, M., Singh, P., Thomas, G., & Al-Hazimi, A. (2013). Recent developments in production and biotechnological applications of C-phycocyanin. BioMed research international, 2013.

[9] Movasaghi, Z., Rehman, S., & ur Rehman, D. I. (2008). Fourier transform infrared (FTIR) spectroscopy of biological tissues. Applied Spectroscopy Reviews, 43(2), 134-179.

[10] Panzella, L., & Napolitano, A. (2019). Natural and bioinspired phenolic compounds as tyrosinase inhibitors for the treatment of skin hyperpigmentation: Recent advances. Cosmetics, 6(4), 57.

[11] Schurink, M., van Berkel, W. J., Wichers, H. J., & Boeriu, C. G. (2007). Novel peptides with tyrosinase inhibitory activity. Peptides, 28(3), 485-495.

[12] Zamani, N., Fazilati, M., Salavati, H., Izadi, M., & Koohi-Dehkordi, M. (2020). The topical cream produced from phycocyanin of Spirulina platensis accelerates wound healing in mice infected with Candida albicans. Applied Biochemistry and Microbiology, 56(5), 583-589

[13] Lee, S. Y., Baek, N., & Nam, T. G. (2016). Natural, semisynthetic and synthetic tyrosinase inhibitors. Journal of Enzyme Inhibition and Medicinal Chemistry, 31(1), 1-13.

[14] Park J-Y, Choi H-J, Park T, Lee M-J, Lim H-S, Yang W-S, Hwang C-W, Park D, Kim C-H. Inhibitory Effect of Avenanthramides (Avn) on Tyrosinase Activity and Melanogenesis in α-MSH-Activated SK-MEL-2 Cells: In Vitro and In Silico Analysis. International Journal of Molecular Sciences. 2021; 22(15):7814. https://doi.org/10.3390/ijms22157814.

[15] Lim, J., Nam, S., Jeong, J. H., Kim, M. J., Yang, Y., Lee, M. S., ... & Lim, J. S. (2019). Kazinol U inhibits melanogenesis through the inhibition of tyrosinase-related proteins via AMP kinase activation. British Journal of Pharmacology, 176(5), 737-750.

[16] Zamani, N., Fazilati, M., Salavati, H., Izadi, M., & Koohi-Dehkordi, M. (2020). The topical cream produced from phycocyanin of Spirulina platensis accelerates wound healing in mice infected with Candida albicans. Applied Biochemistry and Microbiology, 56(5), 583-589.

[17] Syarina PN, Karthivashan G, Abas F, Arulselvan P, Fakurazi S. Wound healing potential of Spirulina platensis extracts on human dermal fibroblast cells. EXCLI J. 2015 Mar 2;14: 385-93. doi: 10.17179/excli2014-697.

[18] Agrawal M, Bansal S, Chopra K. Evaluation of the in vitro and in vivo antioxidant potentials of food grade Phycocyanin. J Food Sci Technol. 2021 Nov;58(11):4382-4390. doi: 10.1007/s13197-020-04922-4.

[19] Bergandi L, Apprato G, Silvagno F. Antioxidant and Anti-Inflammatory Activity of Combined Phycocyanin and Palmitoylethanolamide in Human Lung and Prostate Epithelial Cells. Antioxidants. 2022; 11(2):201. https://doi.org/10.3390/antiox11020201

[20] Jiang L, Wang Y, Yin Q, Liu G, Liu H, Huang Y, Li B. Phycocyanin: A Potential Drug for Cancer Treatment. J Cancer. 2017 Sep 20;8(17):3416-3429. doi: 10.7150/jca.21058.

IOP Conf. Series: Materials Science and Engineering 1291 (2023) 012040 doi:10.1088/1757-899X/1291/1/012040

Effect of elevated temperature on rice husk ash based self-compacting alkali activated slag concrete mixes under different cooling regimes

Lekhan H N[a], Manjunath R[b], Neha S N[c] and M H Prashanth[d]

[a]PG student, Department of Civil Engineering, B.M.S College of engineering, Bangalore-560019, India
[b] Assistant Professor, Department of Civil Engineering, B.M.S College of engineering, Bangalore-560019, India
[c] Research Scholar, Department of Civil Engineering, National Institute of Technology Karnataka, Surathkal, Mangalore – 575025 (D.K), India
[d] Assistant Professor, Department of Civil Engineering, National Institute of Technology Karnataka, Surathkal, Mangalore – 575025 (D.K), India

[b] Corresponding Author: manjunathr.civ@bmsce.ac.in

Abstract: - Here an attempt has been made to develop self-compacting alkali activated slag concrete mix using Rice Husk Ash as Binder. Investigation on the mechanical properties and residual strength when subjected to temperature of 200,400,600,800°C retention period1hour. The Rice husk ash Self Compacting Alkali Activated Slag Concrete mixes was made with GGBS, Rice husk ash and Lime as the binder, with its content varying from 550 to 650 kg/m^{3}of fresh concrete. The W/B ratio of mixes varied in 0.7-0.88 range. The alkaline solutions had Na_2O percentages in the range 5-7 with constant activator modulus maintained at 1. Final mix considered had lower water content, higher percentage of RHA and higher compressive strength. Trails showed Processed Rice Husk Ash30%, lime20% and GGBS50% for binder content 600kg/m^3 and 6% Na_2O dosage showed greater compressive strength and lower water binder ratio compared to other trail mixes. Results showed the slump flow, L–Box, U-box and V-Funnel as greater than 650mm, 0.85, 20mm and 10s resp. Fire damage of RSAASC was measured by subjecting to elevated temperature between 200-800°C for a retention period of 1hour. Cooled to room using furnace cooling, air cooling and sudden cooling. Results indicated reduced compressive strength with increased temperature.

Keywords: *Rice husk ash, Elevated Temperature, Cooling regimes, Self-compacting concrete*

1. Introduction

Cement is the most important constituent of a concrete mix. Since 1800 Ordinary Portland Cement (OPC) is the most used binder material in concrete. The most widely used construction material in the world is concrete. Research estimates that each ton of Cement produced, about 0.84 ton of Carbon Dioxide is released into the environment. The total contribution of Cement to global Carbon Dioxide emissions is estimated to be between 5 and 8 percent Today around 4.2 billion tons of Cement are produced annually with an expected growth rate of 2.5% per annum. [1-3].

Geo-polymers are a new class of concrete mixes, with no Portland cement, and are produced through the reaction of alumino silicate-rich materials with an alkaline solution . Fly ash, Rice husk ash, GGBFS, Meta-kaolin is some of the materials being used for cement replacement. Again, some of these materials are pozzolanic in nature. GGBFS, material obtained from iron and steel industries, may be effectively used to provide high-strength concrete, below alkaline activation, under below ambient temperature conditions. GGBS consists of proper amounts of silicon, aluminum, and calcium oxides, among others, making it a suitable source material for generating Alkali Activated Slag Concrete (AASC) mixes [4-13].

In the earlier ranges, studies become extra focused at the behavior of alkali activated slag pastes and mortar mixes. In recent times, studies on AASC mixes have additionally gained greater attention Rice husk ash (RHA) will be found in rice-growing regions around the world, such as China, India, and the Middle East. RHA is a by-product of rice husk incineration. During the burning of rice husk, the majority of the evaporable components are lost slowly, and the silicates are the main left overs . Rice husks having a calorific value of about half that of coal, and assuming that the husks have about 8%–10% moisture content and no bran, the calorific value is calculated to be 15 MJ/kg. RHA is found to have pozzolanic properties as it is a silica-rich material. The major constituents were SiO_2 -90%, K_2O -4.60%, and P_2O_5 -2.43% [14-20]

Based on available literatures, there are several researches carried out on Alkali-Activated concrete mixes with different binders namely, GGBS, Metakaolin, Fly ash, RHA and many more. However, there are only few research carried out on Rice Husk Ash based alkali activated concrete mixes and there is currently no research on RHA based self-compacting alkali activated slag concrete mixes. Only few research has been carried out on elevated temperatures under different cooling regimes which could give an idea of a structure subjected to fire. Hence, the attempt has been made on effect of elevated temperature on RHA based self-compacting alkali activated slag concrete mixes under different cooling regimes. [21-35].

Rice husk ash based self-compacting alkali activated slag concrete mixes were developed with minimum to attain M30 grade concrete and compared with conventional OPC mix of similar grade. The main aim of the present study is to maximum utilization of Rice husk ash in self-compacting alkali activated slag concrete mixes. RSAASC mixes were prepared with GGBS, Rice husk ash and lime as a binder and alkali solution combination of sodium silicate and sodium hydroxide (NaOH) as an alkali activator. The fresh and hardened properties such as workability, compressive strength, split tensile strength, flexural strength, and modulus of elasticity of concrete mix were evaluated as per standard test procedures. RSAASC mixes were casted and exposed to different temperatures (200°C - 800°C) for a retention period of 1 hou Then they were cooled to room temperature with different cooling regimes like furnace cooling, air cooling and sudden cooling. After exposure to elevated temperature and cooling regimes, the residual compressive strength characteristics were studied.

The present study focused on development of Rice husk ash Self Compacting Alkali Activated Slag Concrete mixes (RSAASC) and to evaluate the mechanical properties of RSAASC mixes was main objective of this investigation. Further evaluation of residual strength of RSAASC mixes when subjected to elevated temperature under different cooling regimes and studies on microstructural characterization of RSAASC mixes after subjected to elevated temperature.

2. Experimental Program

2.1 Materials

2.1.1 Ground Granulated Blast Furnace Slag (GGBS)

GGBS is the most commonly used material in AAS mixes, due to its higher hydraulic activity when compared to other types [16]. Ground Granulated Blast-Furnace Slag (GGBS) conforming to IS 12089-1987 was procured from JSW Iron and Steel Plant, Bellary, India. GGBS as tested for Blaine's fineness, resulted about 390m^2/kg and specific gravity of 2.81. The Chemical composition (given by manufacturer) of the GGBS used is given in Table 1.

Table 1. Chemical Composition of GGBS.

Constituent (% by weight)	CaO	Al_2O_3	Fe_2O_3	SiO_2	MgO	Na_2O	K_2O	SO_3	LOI
Oxide content	37.34	14.42	1.11	37.73	8.71	0.16	0.07	0.88	0.04

2.1.2 *Rice Husk Ash (RHA)*

The material procured from Sri Shreesha Rice Industries, Tumkur. Rice Husk is the material burned under the temperature range between 600°C-800°C forms Rice Husk Ash [4,14]. The chemical composition of RHA consists of 90 percent amorphous silica, 5% carbon, and 2% K_2O make up the average composition of well-burned RHA [20,21].The material burned without constant temperature and may contain unburned rice husk called as Unprocessed Rice Husk Ash (URHA). The URHA is then taken to Grinding process called as Processed Rice Husk Ash (PRHA) as shown in **Fig 1**. Table 2 shows properties of PRHA and URHA.

Table 2. Properties of URHA and PRHA

Particulars	Values
Specific gravity of URHA	1.82
Fineness of URHA	210 m^2/kg
Specific gravity of PRHA	2.1
Fineness of PRHA	290 m^2/kg

URHA- Unprocessed Rice husk ash, PRHA- Processed Rice husk ash

Figure 1 Rice Husk Ash

2.1.3 *Aggregates*

In the present study, crushed stone sand also called manufactured sand is used as fine aggregate with maximum nominal size 4.75mm down size and natural aggregates of maximum nominal size of 12.5mm down size is used as coarse aggregates. The specific gravity, fineness modulus and water absorption of the fine aggregate were 2.65, 3.25 and 10%, respectively. The specific gravity and water absorption of the coarse aggregate were 2.67 and 0.5%, respectively. All the aggregates used herein were tested as per relevant Indian standard codes.

2.1.4 *Alkaline Solution*

Alkaline solution for alkali activation is a combination of sodium silicate solution (Na_2SiO_3) and sodium hydroxide (NaOH). The alkaline solution procured from Kamath's Rice Mill & Industries, Mangalore. Sodium hydroxide (NaOH) flakes having a purity of 97% and sodium silicate solution containing 14.7% Na_2O +32.8 SiO_2 + 52.5% H_2O, by mass and with a density 1570 kg/m^3 (given by manufacturer). In the present study, activator modulus i.e., ratio of SiO2/ Na_2O is kept constant as 1 but three different dosages of Na_2O such as 5%, 6% and 7% are used as alkaline solution.

2.2 Mixture proportioning, preparation and casting of mixtures

Alkali activated concrete mix design is based on trial-and-error basis. EFNRC guidelines were used to achieve self-compacting concrete. As we designed on trial-and-error basis, the water to binder ratio started with 0.4 and continued till the flow of fresh concrete achieves flow ability around 650-800mm. For reaction of alkalis present in the concrete, the Alkaline solution combined with Sodium silicate and Sodium hydroxide is used. The mix design starts with Na2O dosage and Modulus of sodium silicate. The binder content as per EFNARC guidelines is 500-700kg per meter cube. The binder used in this experiment are GGBS, Rice Husk Ash (URHA and PRHA) and Lime, the total binder content varying from 550-650 kg per meter cube. For all the mixes 1% of total binder content of super plasticizer is been used. The super plasticizer used in this study is Glenium ACE JP (30).

2.3 Arriving at Final mix

Initially started with maximum utilization of URHA by 75%, 50% and 25% replacement of binder content and in next mix we increased Na2O dosage by 5%, 6% and 7% of binder content. In addition to reduce setting time we introduced Lime in 10%,15% and 20% of binder content. In further trials, we reduced URHA by 10%, 15% and 20% of total binder. And finally, PRHA by 20%, 30% and 40% of total binder was used in the mix. As different combinations of trial mixes were used, lower water to binder ratio, flow ability and higher compressive strength were all the factors considered to finalize the mix. The final mix design consists of processed rice husk ash (PRHA) 30% of total binder, Lime 20% of total binder and GGBS 50% of total binder. The water to binder ratio is 0.7, flow ability is greater than 650mm and compressive strength for a period of 1,3 and 7day is 5.5MPa,14.15MPa and 20.5MPa respectively. The mix design consisting of unprocessed rice husk ash 15% and lime 20% comparably shown equal results of the mix consisting of PRHA 30% and Lime 20%. As we are looking for maximum utilization of Rice husk ash, the PRHA 30% and lime 20% and 50% GGBS of total binder was considered for further evaluation of mechanical properties and residential strength when subjected to elevated temperature. Finalized mix is shown in Table 3 for 1m^3 of concrete.

Table 3. Finalized Mix of RSAASC mix

Mix ID	Total Binder	GGBS	PRHA	Lime	Fine Agg.	Coarse Agg.	SS	NaOH	Water
RSAASC	Density(kg/m^3) 600		300	180	120	707.89	583.59	109.76	25.63
				182.38	2209.2				

PR/PRHA- Processed Rice Husk Ash, L- Lime, SS- Sodium Silicate

3. Results and discussions

3.1 Fresh Properties of RSAASC Mix

In the present investigation the levels of workability of the mixes were measured using the standard Abraham's slump cone and flow table. The slump cone with bottom diameter of 20cm, top diameter of 10cm and height of cone is 30cm. For all the trial mixes flow ability is checked to satisfy the EFNARC guidelines of flow 650-800mm. To achieve flow of the mixes, the water binder ratio is increased from 0.4 to 0.9 till the flow is greater than 650mm. The final fixed mix showed the slump flow values greater than 650mm showed in figure 2, with their L–Box ratios, U-box and V-Funnel values ranging between 0.85, 20mm and 10s respectively, satisfying the EFNARC guidelines.

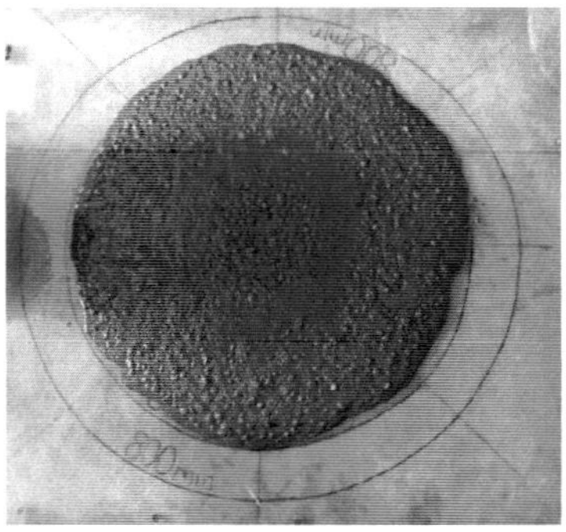

Figure 2. Slump flow and U-box test of RSAASC mix

3.2 Hardened Properties of RSAASC mix

3.2.1 Compressive strength of RSAASC and Control mix

Based on the trails, all self-compacting alkali activated slag concrete mixes with GGBS, URHA and PRHA replacing GGBS and introduction of lime in process of faster setting action herein tested as per IS 516:1999. From the trails, the mix with lower water binder ratio and higher compressive strength was finalized for further mechanical properties and residual strength when subjected to elevated temperature. From the trials conducted, it was observed that maximum utilization of Processed Rice husk ash (PRHA) shown compressive strength of about 30MPa, for a binder content of 600kg/m3 as shown in **table 4.** which consists of 30% of processed rice hush ash, 20% lime and 50% GGBS. Even though 15% Unprocessed rice hush showed 30MPa compressive strength, as we targeted maximum utilization of rice husk ash, hence 30% processed rice husk ash was fixed for further mechanical and residual strength. Control mix of concrete consisting of cement, fine and coarse aggregate is casted for M30 grade and the compressive strength at different age were shown in Table 4.

Table 4. Compressive strength, MPa

Duration, Days	3	7	14	28
RSAASC	14.15	20.05	25.36	30.30
Cement concrete	14.13	19.58	26.74	30.60

3.2.2 Residual Compressive strength of RSAASC and Control mix

Figure 3 shows an electric furnace. Temperature is set using the knob and temperature increased till it reached the target temperature. The specimen was subjected to exposure tests in an electrical furnace for temperatures of 200°C, 400°C, 600°C and 800°C with a retention period 1 hour. Three types of cooling regimes studied here are, furnace cooling, air cooling and sudden cooling. By furnace cooling, it is meant that, after completion of 1hour retention period at designated temperature, the furnace power supply is cut off and the door of the furnace shall not be opened till the specimen inside attain room temperature. The interior of the furnace temperature can be monitored digitally from outside display panel. It is a function of the temperature exposed to, it takes nearly one day for specimen subjected to 800°C, to reach room temperature. By air cooling, it is meant that, after retention period, that furnace door is opened and specimen are removed and allowed to cool in ambient/ room temperature. By Sudden cooling it is meant that, after completion of retention period at designated temperature, the door of the furnace shall be opened and the specimen shall be immediately quenched in water at room temperature.

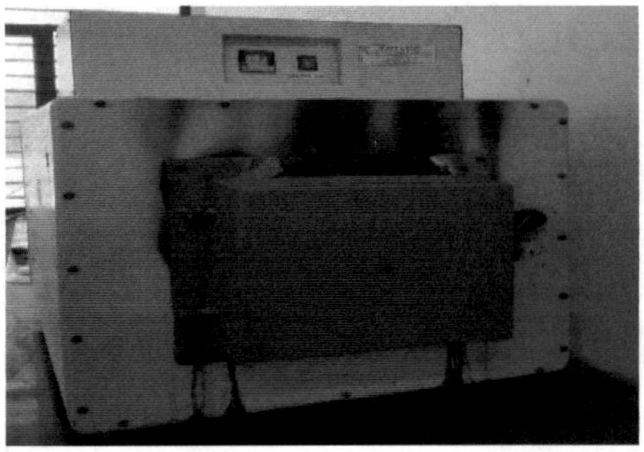

Figure 3. Electric Furnace

The specimens for a period of 28 days were casted and subjected to different elevated temperatures and followed by three different cooling regimes (Sudden cooling, air cooling and furnace cooling). Table 5 and Table 6 shows cube compressive strength at 28 days of RSAASC and Control mix which were subjected to different elevated temperature. Figure 4 shows variation of compressive strength when subjected to elevated temperature and three cooling regimes. From Table 5 and Table 6 it is observed that as temperature increases from 200°C to 800°C, the compressive strength decreases. As compared to control mix, RSAASC mix subjected elevated temperature relatively showed lower compressive strength. From both RSAASC and control mix, air cooling method is best cooling method which shows higher compressive strength compared to other cooling methods. Furnace cooling showed least compressive strength and sudden cooling showed medium compressive strength. For a temperature 800°C, the cube is being exposed to red hot and formation of wide range of cracks shown in figure 5.

Table 5. RSAASC mix cubes subjected to Elevated temperature at 28 days

28day residual compressive strength of RSAASC, MPa				
Temperature °C/ cooling methods	200 °C	400°C	600°C	800°C
Sudden cooling	18.25	12.64	8.33	5.68
Air cooling	20.48	15.83	11.58	6.29
Furnace cooling	13.95	9.76	7.49	4.17

Table 6. Control mix Cubes subjected to Elevated temperature at 28 days

28day residual compressive strength of Control mix, MPa				
Temperature °C / cooling methods	200 °C	400°C	600°C	800°C
Sudden cooling	19.67	15.76	13.98	10.69
Air cooling	23.05	17.53	15.13	12.23
Furnace cooling	18.64	15.09	11.76	7.89

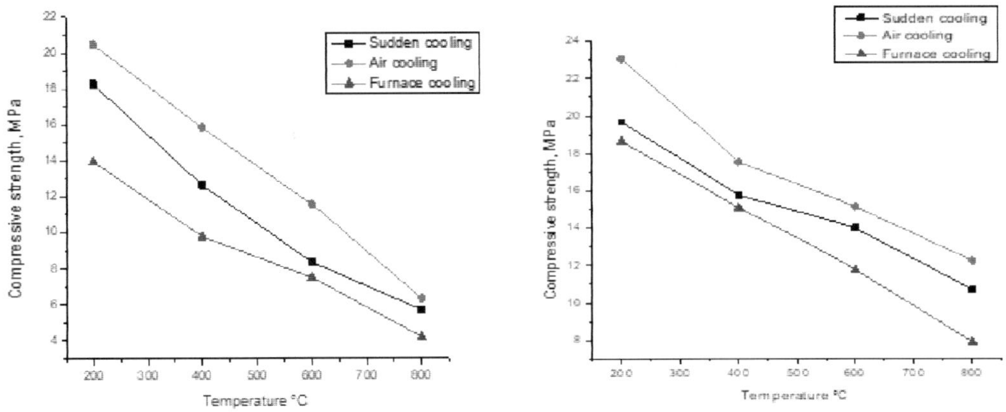

(a) (b)

Figure 4. Variation of 28day compressive strength subjected to elevated temperature (a) RSAASC mix (b) Control mix

From the Table 5 and Table 6, it is observed that maximum compressive strength 20.48 MPa and 23.05 MPa of RSAASC and control mix respectively exposed to air cooling method for temperature 200°C. Hence there is a reduction of 33% and 23% of compressive strength of RSAASC and control mix respectively. In a furnace cooling method, it is observed that the least compressive strength of 4.17MPa and 7.89MPa for a temperature of 800°C. About 86% and 73.5% reduction in compressive strength of RSAASC and control mix respectively. With different cooling regimes, air cooling showed higher compressive strength when compared with other two cooling regimes. Furnace cooling gave the least compressive strength while sudden cooling gave medium compressive strength of the cubes.

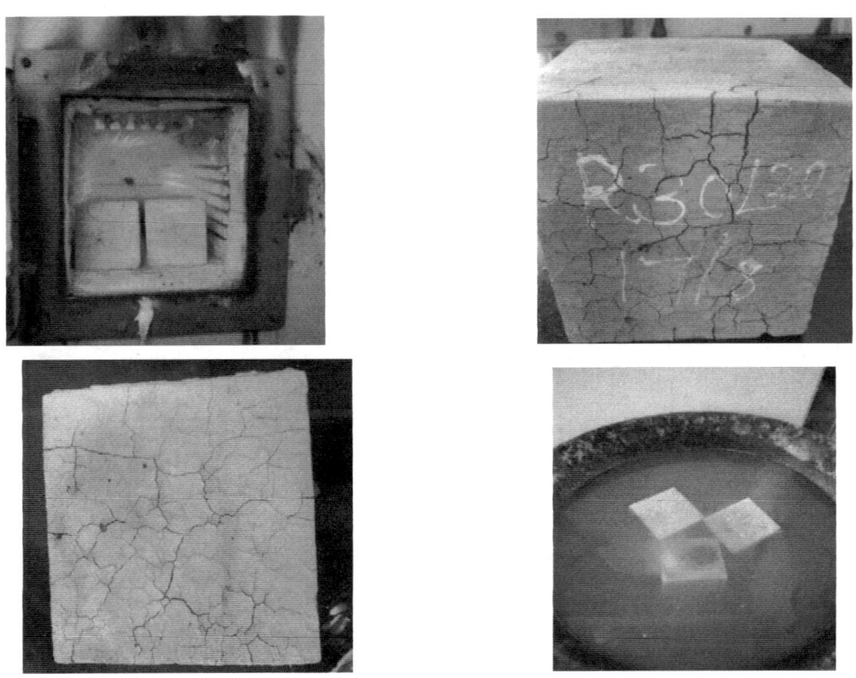

Figure 5. Specimens at 800°C and sudden cooling at 28 days

3.2.3 Split Tensile strength, Flexural strength and MOE of RSAASC and Control mix

Tensile strength test was conducted in laboratory by casting the cylinder specimen of size 100 mm dia. 200 mm height and tested using compression testing machine at 28 days for both ordinary concrete and RSAASC mixes. The results for split tensile strengths of RSAASC mix and control mix at 28 days is 1.56 MPa and 1.76MPa. Figure 6 shows split tensile test. Flexural experiment is carried in laboratory by casting beam specimen size 500mm x 100mm x 100mm using two-point loads in UTM at different duration for both conventional concrete and RSAASC mixes. The RSAASC mix consist of 30% PRHA, 20% lime and 50% GGBS with 6% Na2O dosage and modulus of sodium silicate as 1. The flexural strength of RSAASC mix and Control mix of 28 days is 2.33 MPa and 2.46MPa. Figure 7 flexural strength test. Modulus of elasticity (MOE) experiment carried out by casting cylinder of 150mm dia. and 300mm height and tested in universal testing machine. Figure 8 shows the experimental setup of MOE and testing. As per experimental results the RSAASC mix gave the result of 21GPa and Control mix showed result of 24GPa.

Figure 6. Split tensile test

Figure 7. Flexural Strength test

Figure 8. MOE Setup and Testing

3.2.4 Water absorption

Rice husk ash based self-compacting alkali activated concrete slag mix were subjected to water absorption test to know the pores inside the hardened concrete. It is observed that the water absorption of RSAASC mix showed an average water absorption of 3.25%. When comparing with control mix the water absorption of RSAASC mix showed higher water absorption value indicating that RSAASC mix sample is porous. The control mix showed water absorption of 2.5%.

3.2.5 Studies on microstructures of RSAASC mix

The RSAASC mix were subjected to elevated temperatures and treated with different cooling methods, and were tested for compressive strength. After testing for compression test, the samples were crushed to pieces. The crushed piece of compression tested samples were taken for microstructural studies. The smaller size pieces of respective temperature and cooling methods were subjected to SEM (Scanning electron microscopy) and EDX (Energy Dispersive X-Ray Analysis) analysis. Before subjecting to SEM and EDX, the samples were in oven dried condition.

(a) 1000x

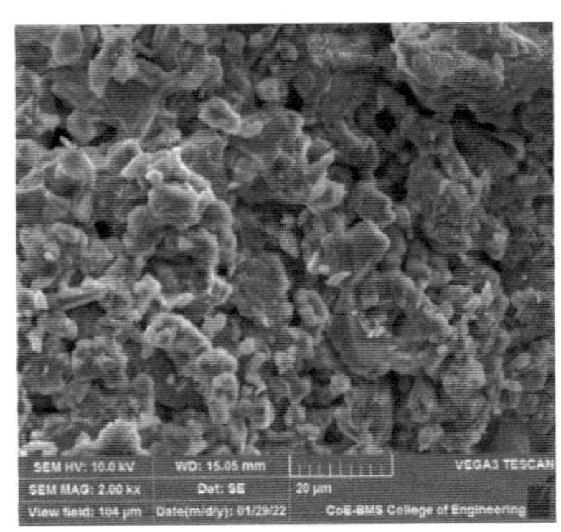

(b) 2000x

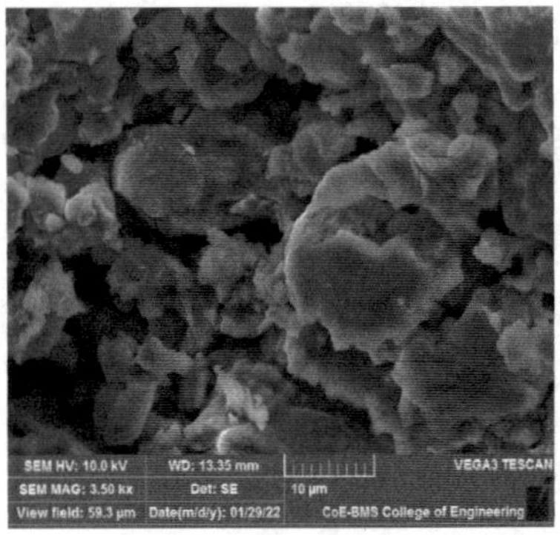

(c) 3500x

(d) 5000x

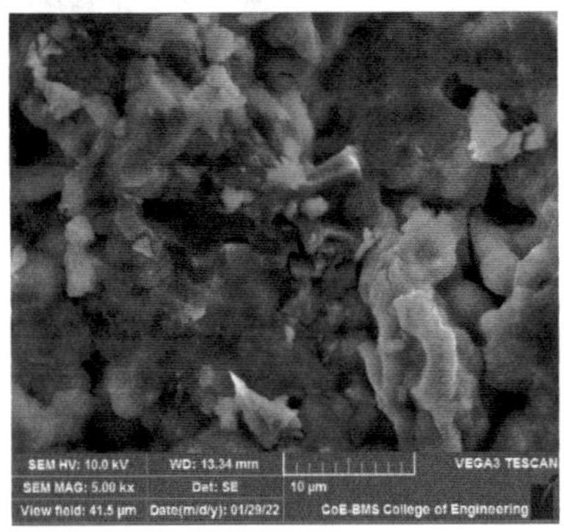

(e) 8000x

Figure 9. SEM analysis of RSAASC mix subjected to 400˚C and Air cooled (AC) sample with Different Magnification

Figure 9 shows SEM analysis with different magnification RSAASC mix subjected to 400˚C and air-cooled sample. The sample has been crushed to pieces and taken for SEM analysis in oven dried condition. SEM analysis shows that the slag grains and ash has been activated effectively. Densified structure of the sample can be seen in lower magnification. Smaller portion of the images are magnified further to check for activation. Magnified images show activation of slag grains and rice husk ash with negligible space occupied by unreacted ash and slag grains. From lower magnification significant number of pores and minor cracks can be seen which in turn lead to lower strength. At higher magnification the bonding of slag grains and ash material leading to strength development in the concrete. Formation of densified layer due to the activation of ash and slag grains, has probably led to disappearance of ITZ in and around the aggregates. As increase in magnification of the images, morphology of white patches shows the activation of slag grains and ash powder which are completely reacted by alkaline solution. At 400˚C and exposed to air cooled method, the images show lower number of pores and minor cracks in the concrete.

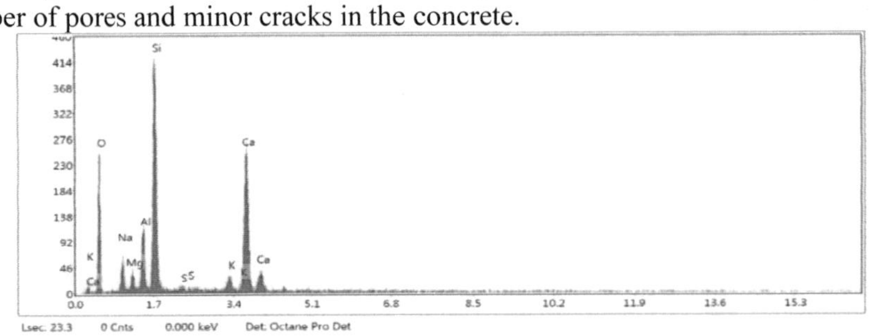

Figure 10. EDX analysis of RSAASC mix subjected to 400˚C and Air cooled (AC) sample

Figure 10 clearly shows EDX analysis of RSAASC mx subjected to 400°C and air-cooled sample. From the image it clearly indicating more prominent signatures of peaks for Si, Ca and O. As rice husk ash is composed of higher silica content which is why clear peak of silica element is high in EDX analysis. As per [16] investigation, EDX analysis shows peaks of Si and Ca in the graph are in same range for different binder content. As per this investigation, EDX analysis indicates the clear higher peak of Si element than Ca and O.

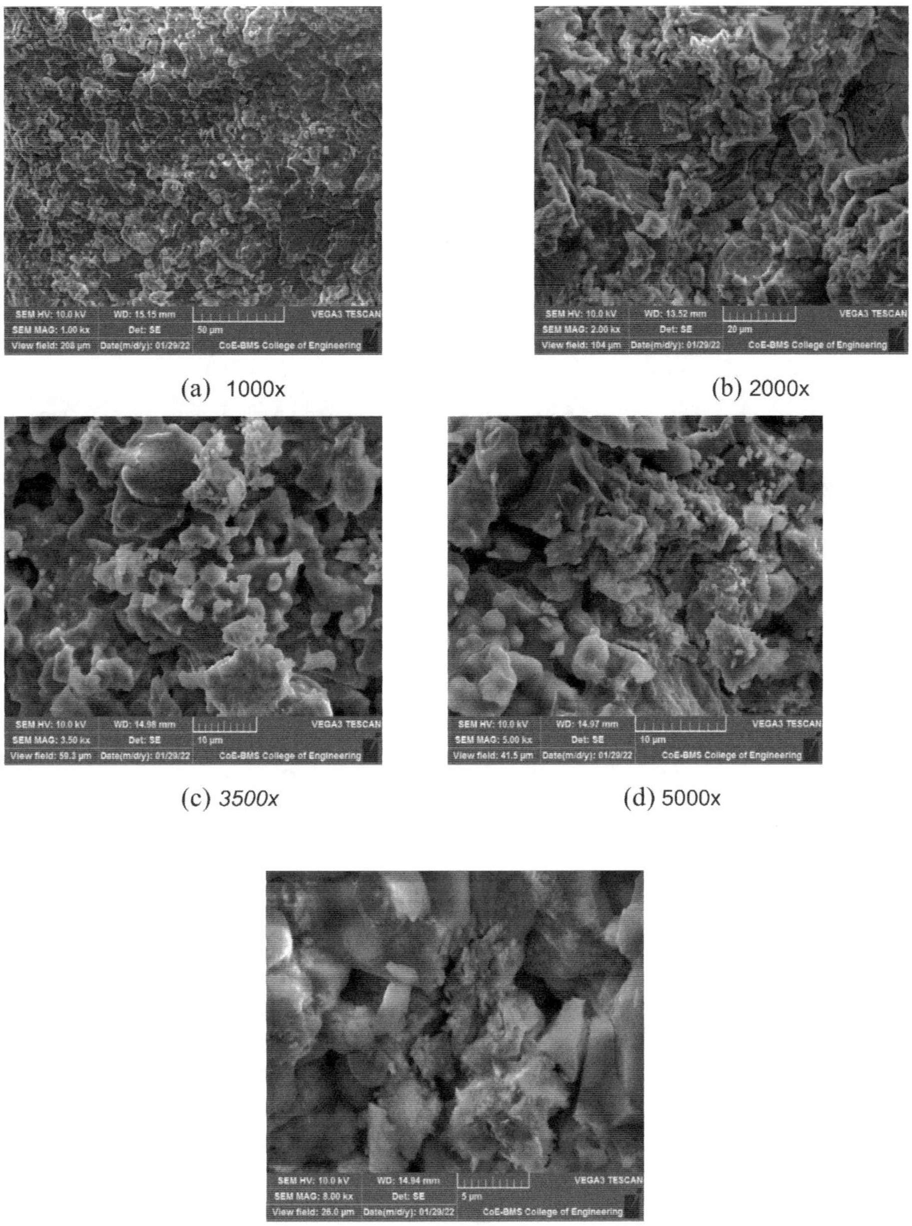

(a) 1000x

(b) 2000x

(c) *3500x*

(d) 5000x

(e) 8000x

Figure 11. SEM analysis of RSAASC mix subjected to 400°C and Furnace cooled (FC) sample with Different Magnification

Similarly, 400˚C and Furnace cooled (FC) shows number of pores and minor cracks in SEM analysis figure 11. As cooled in furnace, the sample cooled undergoes gradual decrease in temperature as it is kept inside muffle furnace and cooled to room temperature. SEM analysis shows white patches which indicates the activation of slag grains and ash material with alkaline solution complete reaction. EDX analysis shows elemental composition in **Figure 12** which clearly shows EDX analysis of RSAASC mix subjected to 400˚C and furnace cooled sample. From the image it clearly indicates the peaks for Si, Ca, Al, Mg, Na, K and O. As per this investigation, EDX analysis indicates the clear higher peak of Si element than Ca and O.

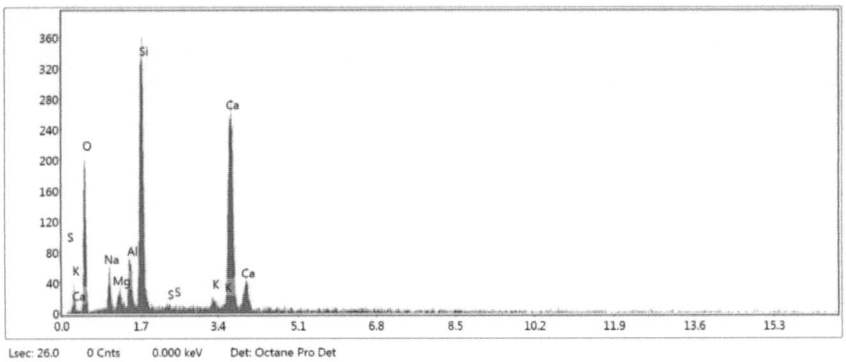

Figure 12. EDX analysis of RSAASC mix subjected to 400˚C and Furnace cooled (FC) sample

(a) 1000x

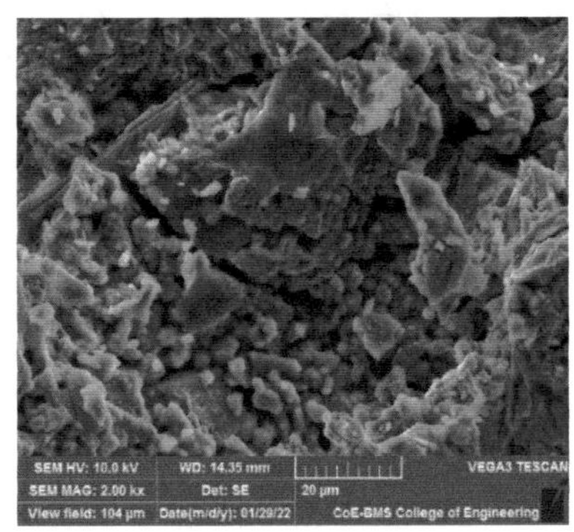

(b) 2000x

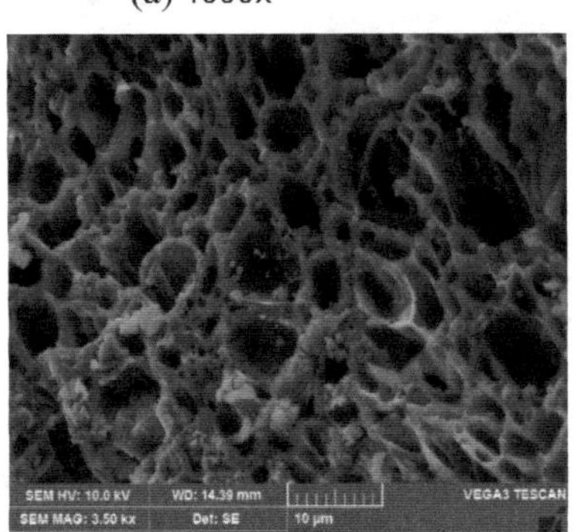

(c) 3500x

(d) 5000x

(e) 8000x

Figure 13. SEM analysis of RSAASC mix subjected to 600˚C and Sudden cooled (SC) sample with Different Magnification

Figure 13 shows SEM analysis with different magnification RSAASC mix subjected to 600˚C and sudden (water) cooled sample. The sample has been crushed to pieces and taken for SEM analysis in oven dried condition. SEM analysis shows that the slag grains and ash has been activated effectively with alkaline solution complete reaction. White patches in the images explains the reaction of alkaline solution. Densified structure of the sample can be seen in lower magnification. Smaller portion of the images are magnified further to check for activation. Magnified images shows activation of slag grains and rice husk ash with negligible space occupied by unreacted ash and slag grains. Magnification of 3500x shows significant amount of pores and 2000x, 5000x shows micro cracks. As increase in magnification of the images, morphology of white patches shows the activation of slag grains and ash powder which are completely reacted by alkaline solution.

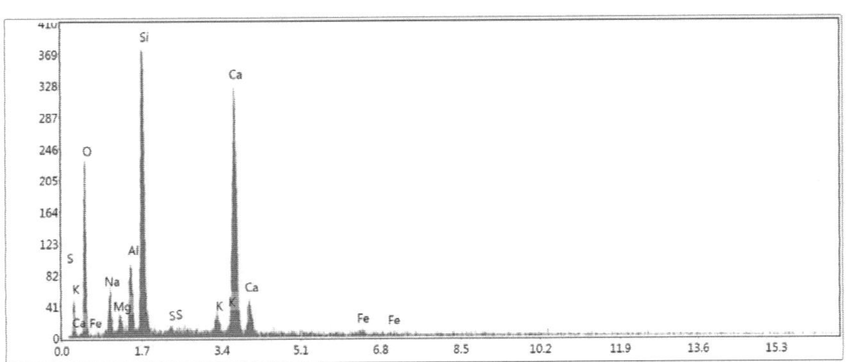

Figure 14. EDX analysis of RSAASC mix subjected to 600˚C and Sudden cooled (SC) sample

Figure 14 clearly shows EDX analysis of RSAASC mx subjected to 600˚C and sudden cooled sample. From the graph it is clearly indicating elements of Si, Ca, O, Na, Al, Mg, K and Fe. From the image it clearly indicating more prominent signatures of peaks for Si, Ca and O. As rice husk ash is composed of higher silica content which is why clear peak of silica element is high in EDX analysis.

IOP Conf. Series: Materials Science and Engineering 1291 (2023) 012040 doi:10.1088/1757-899X/1291/1/012040

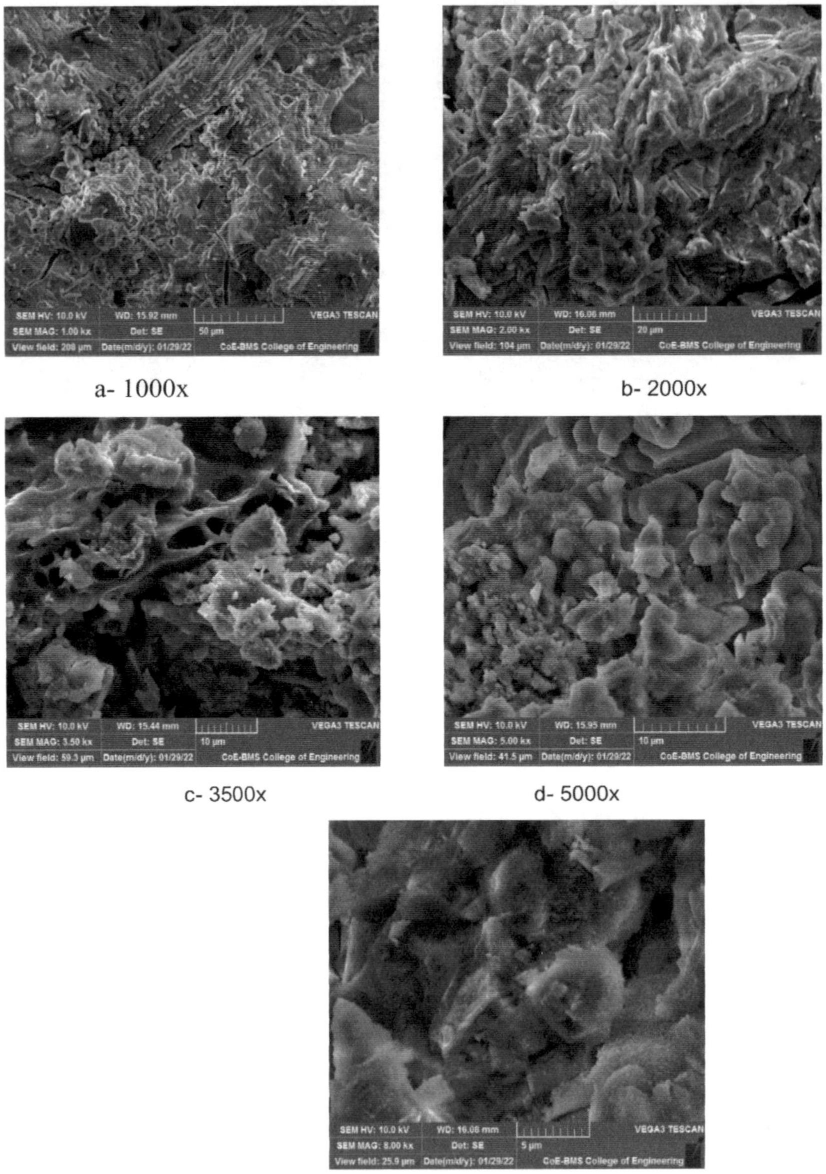

a- 1000x b- 2000x

c- 3500x d- 5000x

e- 8000x

Figure 15. SEM analysis of RSAASC mix subjected to 600°C and Furnace cooled (FC) sample with Different Magnification.

Figure 15 shows SEM analysis with different magnification RSAASC mix subjected to 600°C and furnace cooled sample. The sample has been crushed to pieces and taken for SEM analysis in oven dried condition. White patches in the images explains the complete reaction of alkaline solution. Magnified images shows activation of slag grains and rice husk ash with negligible space occupied by unreacted ash and slag grains. Magnification of 3500x shows significant amount of pores and 1000x, 2000x shows wider cracks in the images. As compared to sudden cooling, furnace cooling shows wider cracks. As increase in magnification of the images, morphology of white patches shows the activation of slag grains and ash powder which are completely reacted by alkaline solution.

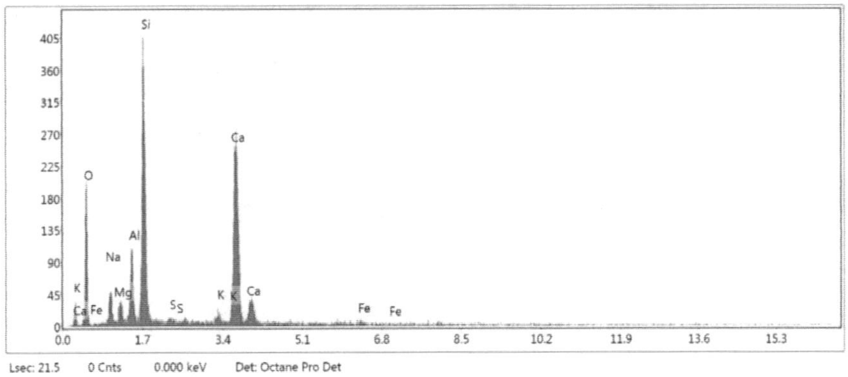

Figure 16. EDX analysis of RSAASC mix subjected to 600°C and Furnace cooled (FC) sample

Figure 16 clearly shows EDX analysis of RSAASC mx subjected to 600°C and furnace cooled sample. From the image it is clearly indicating elements of Si, Ca, O, Na, Al, Mg, K and Fe. From the image it clearly indicating more prominent signatures of peaks for Si, Ca and O. As rice husk ash is composed of higher silica content is the reason for concrete sample to show clear peak of silica element in EDX analysis.

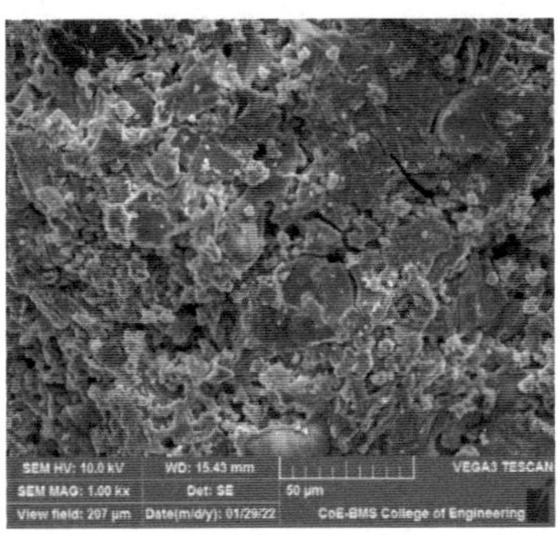

a- 1000x

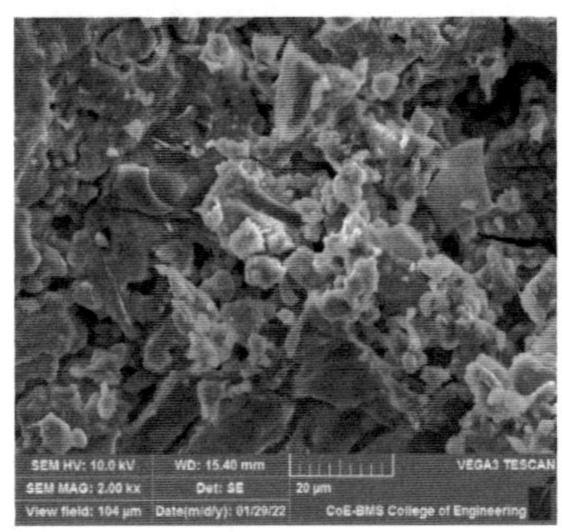

b- 2000x

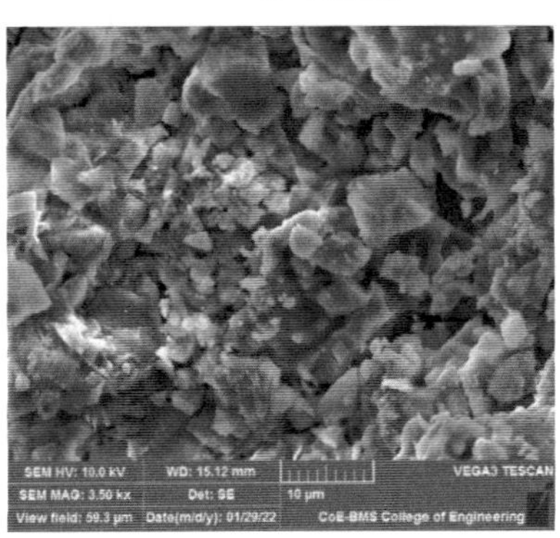

c- 3500x

d- 5000x

e- 8000x

Figure 17. SEM analysis of RSAASC mix subjected to 800°C and Air cooled (AC) sample with Different Magnification.

Figure 17 shows SEM analysis with different magnification RSAASC mix subjected to 800°C and Air cooled sample. The sample has been crushed to pieces and taken for SEM analysis in oven dried condition. White patches in the images explains the complete reaction of alkaline solution. Magnified images shows activation of slag grains and rice husk ash with negligible space occupied by unreacted ash and slag grains. As temperature increased to 800°C, wide range of cracks can be seen in 1000x and 2000x and even pores in the structure. As increase in magnification of the images, morphology of white patches shows the activation of slag grains and ash powder which are completely reacted by alkaline solution. At higher magnification of 8000x shows pores in the image and concludes the weak internal structure of the concrete.

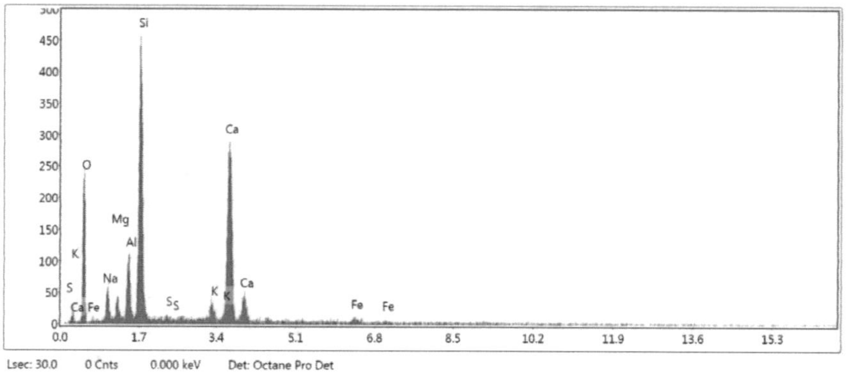

Figure 18. EDX analysis of RSAASC mix subjected to 800°C and Air cooled (AC) sample

Figure 18 clearly shows EDX analysis of RSAASC mx subjected to 800°C and Air-cooled sample. From the image it is clearly indicating elements of Si, Ca, O, Na, Al, Mg, K and Fe. From the image it clearly indicating more prominent signatures of peaks for Si, Ca and O. As rice husk ash is composed of higher silica content is the reason for concrete sample to show clear peak of silica element in EDX analysis.

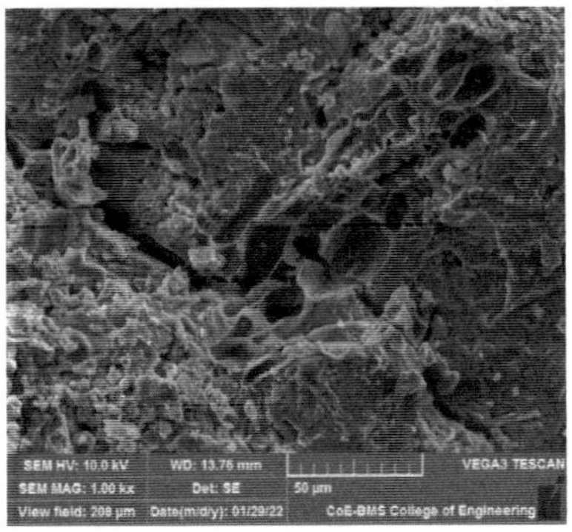

a- 1000x

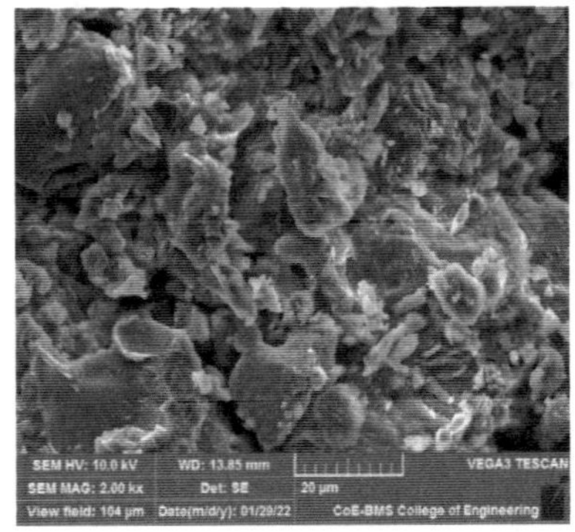

b- 2000x

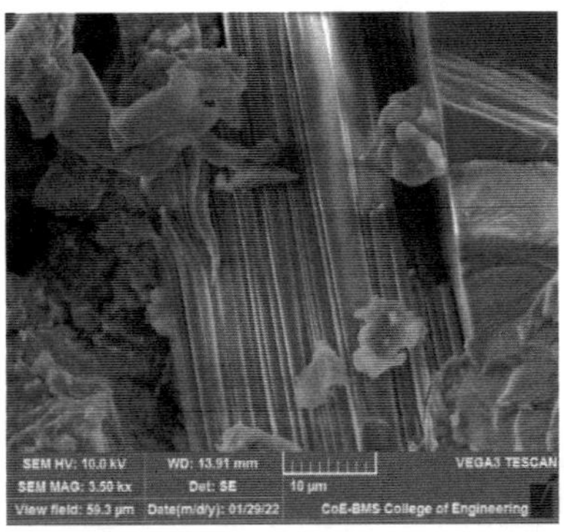

c- 3500x

d- 5000x

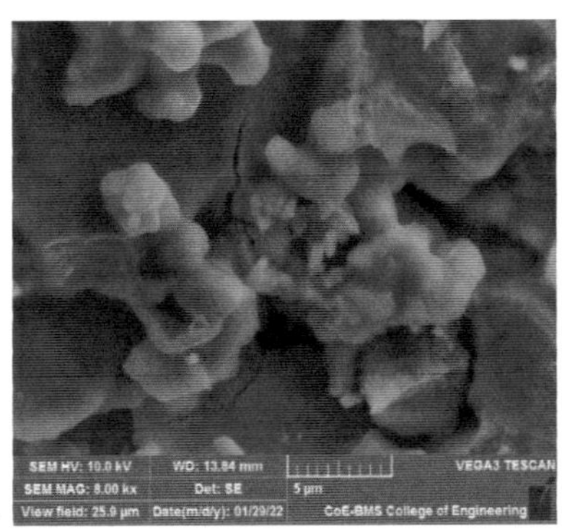

e- 8000x

Figure 19. SEM analysis of RSAASC mix subjected to 800°C and Furnace cooled (FC) sample with Different Magnification

Figure 19 shows SEM analysis with different magnification RSAASC mix subjected to 800°C and furnace cooled sample. The sample has been crushed to pieces and taken for SEM analysis in oven dried condition. White patches in the images explains the complete reaction of alkaline solution. Magnified images shows activation of slag grains and rice husk ash with negligible space occupied by unreacted ash and slag grains. As temperature increased to 800°C, wide range of cracks can be seen in 1000x and 2000x and even pores in the structure. As increase in magnification of the images, morphology of white patches shows the activation of slag grains and ash powder which are completely reacted by alkaline solution. At higher magnification of 8000x, image clearly shows the crack developed which leads to lower strength of the concrete. Comparing to air cooling method and furnace cooling method, air cooling is best cooling method with lower cracks and pores in the concrete.

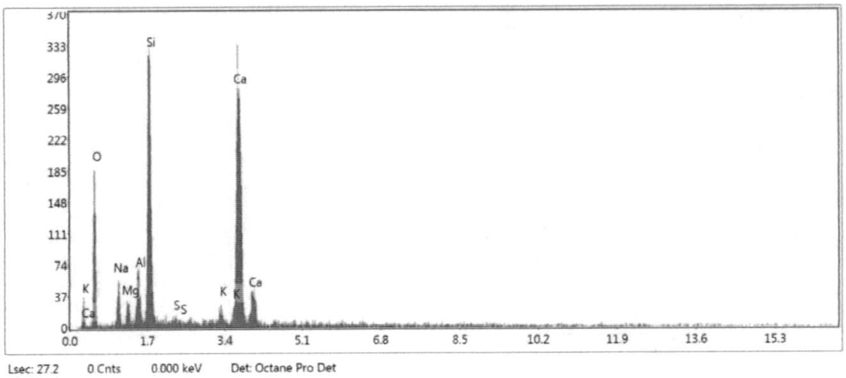

Figure 20. EDX analysis of RSAASC mix subjected to 800°C and Furnace cooled (FC) sample

Figure 20 clearly shows EDX analysis of RSAASC mx subjected to 800°C and furnace cooled sample. From the image it is clearly indicating elements of Si, Ca, O, Na, Al, Mg, K and Fe. From the image it clearly indicating more prominent signatures of peaks for Si, Ca and O. As rice husk ash is composed of higher silica content is the reason for concrete sample to show clear peak of silica element in EDX analysis.

4. Conclusions

1. A new form of self-compacting alkali-activated slag concrete mixes, there produced effectively using Rice husk ask as binder, flow greater than 650mm with L–Box ratios, U-box and V-Funnel values ranging between 0.85, 20mm and 10s respectively, satisfying the EFNARC guidelines.
2. Comparing to both unprocessed rice husk ash and Processed rice husk ash, processed rice husk improves reactivity and shows better result.
3. GGBS can be replaced by Processed rice husk ash (PRHA) by 30% which relatively gave equal compressive strength compared with cement based concrete.
4. Residual compressive strength of RSAASC mix showed lower compressive strength as temperature increased from 200°C to 800°C.
5. In a furnace cooling method, it is observed that the least compressive strength of 4.17MPa and 7.89MPa for a temperature of 800°C. About 86% and 73.5% reduction in compressive strength of RSAASC and control mix respectively.
6. With different cooling regimes, air cooling showed higher compressive strength when compared with other two cooling regimes. Furnace cooling gave the least compressive strength while sudden cooling gave medium compressive strength of the cubes.
7. Satisfactory performance in the split tensile strength, flexural strength and modulus od elasticity of 1.56 MPa, 2.21 MPa and 21GPa respectively.
8. From the microstructural studies, damage of internal structure can be seen due to rise in temperature from 400°C to 800°C of RSAASC mix leading to lower strength. Microstructural study concludes that as the temperature increases the number of pores and number of cracks increases.
9. Microstructure study also concludes that from three cooling regimes, air cooling regime showed micro cracks and pores while furnace cooling regime showed wide cracks and large number of pores in the sample at 400°C, 600°C and 800°C while sudden cooling at 600°C showed minor cracks and pores.
10. Microstructure study shows the activation of slag grains and ash material with complete reaction of alkaline solution and empty space occupied by unreacted rice husk ash and slag. It is clear evident that the formation of ITZ is very less.

References

[1] A. Ferhat Bingol and Rustem Gul (2009), "Residual bond strength between steel bars and concrete after elevated temperatures", Fire Safety Journal 44 854–859.
[2] A. Palomo, A. Fernandez Jimenez, C. Lopez Hombrados, J.L. Lleyda (2007), "Railway sleepers made of alkali activated fly ash concrete", Revista Ingeniería de Construction Vol. 22 22, 75–80.
[3] Abdelalim, A.M.K., Abdel-Aziz, G.E., El-Mohr, M.A.K. and Salama, G.A. (2009), "Effect of elevated fire temperature and cooling regime on the fire resistance of normal and self-compacting concretes", Engineering Research Journal 122, 63-81.
[4] Bhupinder Singh (2018), "Waste and Supplementary Cementitious Materials in Concrete", Wood head Publishing Series in Civil and Structural Engineering, 417-460.
[5] Chao-Lung Hwang and Trong-Phuoc Huynh (2015), "Effect of alkali-activator and rice husk ash content on strength development of fly ash and residual rice husk ash-based geopolymers", Construction and Building Materials 101, 1–9.
[6] D. Chetan and A. Aravindan (2020), "An experimental investigation on strength characteristics by partial replacement of rice husk ash and Robo sand in concrete", Construction and Building Materials 287, 39–49.
[7] Duc-Hien Le, Yeong-Nain Sheen and My Ngoc-Tra (2018), "Fresh and hardened properties of self-compacting concrete with sugarcane bagasse ash–slag blended cement", Construction and Building Material 185, 138–147.
[8] EFNARC, The European Guidelines for Self-Compacting Concrete: Specification, Production and Use, 2005.
[9] Gritsada Sua-iam, Natt Makul, Shanshan Cheng and Prakasit Sokrai (2019), "Workability and compressive strength development of self-consolidating concrete incorporating rice husk ash and foundry sand waste – A preliminary experimental study", Construction and Building Material 228, 116-813.
[10] IS 1727 (1967): Methods of test for pozzolanic materials [CED 2: Civil Engineering]
[11] J. Davidovits 1991, "Geopolymers", Journal of Thermal Analysis 37, 1633–1656
[12] Jagmeet Singh and SP Singh (2019), "Development of Alkali-activated Cementitious Material using Copper Slag", Construction and Building Materials 211, 73-79.
[13] K. Behfarnia, M and Hasanzadeh (2014), "Advanced Topics in Cement and Pozzolan", Technology, Isfahan University of Technology, Iran.
[14] Kishor Chandra Panda, Subhrasweta Behera and Shradha Jena (2020), "Effect of rice husk ash on mechanical properties of concrete containing crushed seashell as fine aggregate", Construction and Building Materials 266, 349–548.
[15] M.M. Hossain, M.R. Karim, M.M.A. Elahi, M.N. Islam and M.F.M. Zain (2020), "Long-term durability properties of alkali-activated binders containing slag, fly ash, palm oil fuel ash and rice husk ash", Construction and Building Materials 251, 119-094.
[16] Manjunath.R, Mathur C.Narasimhan (2018), "An Experimental investigation on Self-Compacting alkali-activated slag concrete mixes", journal of Building Engineering 17,1-12.
[17] Mustafa F. Hasan, Erhan Güneyisi and Yahya R. Atewi (2019), "Fresh and rheological properties of glass fiber reinforced self-compacting concrete with nanosilica and fly ash blended", Construction and Building Materials 211 349–362.
[18] N. Van Tuan (2011), "Hydration and microstructure of ultra-high performance concrete incorporating rice husk ash", Ceme. Concr. Res. 41 (11) (2011) 1104–1111
[19] S M Mousavi Alizade, A Rezaeian, Iraj R and Behzad T (2021), "Compressive stress-strain model and residual strength of self-compacting concrete containing recycled ceramic aggregate after exposure to fire", Journal of Building Engineering 38, 102-206.
[20] S. Varadharajana, Animesh Jaiswala and Shwetambara Verma (2020), "Assessment of mechanical properties and environmental benefits of using rice husk ash and marble dust in concrete", Structures 28 (2020) 389–406.
[21] Saloni, Parveen andThong M. Pham (2020), "Enhanced properties of high-silica rice husk ash-based geopolymer paste by incorporating basalt fibers", Construction and Building Materials 245, 118-422
[22] Shihada, S. (2011), "Effect of polypropylene fibers on concrete fire resistance", Journal of Civil Engineering and Management 17(2), 259-264.
[23] Subhash C. Yaragal and Narayan K. (2012), "Strength characteristics of concrete exposed to elevated temperatures and cooled under different regimes", J. Struct. Fire Eng., 3(4), 301-310.
[24] Subhash C. Yaragal and S. Ramanjaneyulu (2016), "Exposure to elevated temperatures and cooled under different regimes–a study on polypropylene concrete", Advances in Materials Research, Vol. 5, No.1, 21-34.
[25] Taylor H.F.W (1990), "Cement Chemistry. Academic Press", London, 475p.
[26] V. Jittin, A. Bahurudeen, S.D. Ajinkya (2020), "Utilisation of rice husk ash for cleaner production of different

construction products", Journal of Cleaner Production 263 121-578
[27] Y Fernando Silva, Rafael Andres R, P E Mattey and Silvio D (2016), "Properties of self-compacting concrete on fresh and hardened with residue of masonry and recycled concrete", Construction and Building Materials 124, 639–644.
[28]Zhihong Pan, Juanlan Zhou, Xin Jiang and Yidong Xu (2019), "Investigating the effects of steel slag powder on the properties of self-compacting concrete with recycled aggregates", Construction and Building Materials 200 570–577.
[29]Ghasan Fahim Huseien and Kwok Wei Shah (2020), "Durability and life cycle evaluation of self-compacting concrete containing fly ash as GBFS replacement with alkali activation", Construction and Building Materials 235 117-458.
[30]Sherin Khadeeja Rahman and Riyadh Al-Ameri (2021), "A newly developed self-compacting geopolymer concrete under ambient condition", Construction and Building Materials 267 121-822.
[31]Lingling Hu, Zhen He and Shipeng Zhang (2020), "Sustainable use of rice husk ash in cement-based materials: Environmental evaluation and performance improvement", Journal of Cleaner Production 264 121-744.
[32]A.S.M. A Awal, I.A. Shehu and Mohammad Ismail (2015), "Effect of cooling regime on the residual performance of high-volume palm oil fuel ash concrete exposed to high temperatures", Construction and Building Materials 98 875–883
[33]F Koksal, Elif Tugce Kocabeyoglu, Osman Gencel and Ahmet Benli (2021), "The effects of high temperature and cooling regimes on the mechanical and durability properties of basalt fiber reinforced mortars with silica fume", Cement and Concrete Composites 121 104-107
[34]Y Wang, Faqi Liu, Liangfu Xu and Hui Zhao (2019), "Effect of elevated temperatures and cooling methods on strength of concrete made with coarse and fine recycled concrete aggregates", Construction and Building Materials 210 540–547
[35] Kannan Rajkumar P.R, Durga Prasad Mathangi, Sudha C and Neelamegam M, "Experimental Investigation of Reactive Powder Concrete exposed to Elevated Temperatures", Construction and Building Materials 261 119-59.

Experimental Studies on Bagasse Ash-Based Self-Compacting Alkali Activated Slag Concrete

Nidhi K[a], Manjunath R[b*], Pruthvi Raj S[c] and Prashanth M.H [d]

[a]PG student, Department of Civil Engineering, B.M.S College of engineering, Bangalore-560019, India
[b] Assistant Professor, Department of Civil Engineering, B.M.S College of engineering, Bangalore-560019, India
[c] Assistant Professor, Department of Civil Engineering, National Institute of Technology Karnataka, Surathkal, Mangalore – 575025 (D.K), India
[d] Research Scholar, Department of Civil Engineering, National Institute of Technology Karnataka, Surathkal, Mangalore – 575025 (D.K), India

**Corresponding Author: manjunathr.civ@bmsce.ac.in*

Abstract. In the present study, an attempt was made to develop Self Compacting Alkali Activated Slag Concrete mixes incorporating Bagasse Ash as binder content. This study investigates the physical and mechanical properties of Bagasse Ash based Self Compacting Alkali Activated Slag Concrete (BSAASC) with varying proportions of Bagasse Ash. Lime and GGBS (Ground Granulated Blast Furnace Slag) are used along with Bagasse Ash as binder materials in the BSAASC. The total binder content was restricted to $600kg/m^3$ containing Bagasse Ash at 10%, 20% and 30% of total binder content. The solution-to-binder ratio was 0.6. the dosage of Na_2O was varied by 5%, 6% and 7%. The molarity of the alkaline solution is 1M. By using Minitab Statistical Software nine mixes were produced with 3 factors and 3 levels. The analysis was done by Desirability Function Approach (DFA) to check the effectiveness of the considered factors. The workability tests for properties of concrete samples are carried out according to EFNARC guidelines. Microstructural studies such as SEM, EDX and XRD analysis were also carried out, showing denser morphologies indicating effective activation of slag by the alkaline solution.

Keywords: Bagasse ash, Self-compacting concrete, Alkali activated slag concrete, Desirability Function Approach

1. Introduction

Since the 1800s, Ordinary Portland cement is widely used in the construction industry. According to the research, for every ton of cement produced 0.84 tons of carbon dioxide is released to the environment directly during decomposition and indirectly during production as energy consumed. Approximately 5 to 8 percent of Global Carbon Dioxide emission is contributed by the production of cement]. With an expected growth rate of 2.5% per annum, around 4.2 billion tons of Cement are produced annually. At the same time, latently hydraulic cementitious matrices such as Fly Ash, Silica Fumes, and Slag which are by-products from various industries have been used as admixtures to replace some of the cement content. However, the presence of cement is required for cementitious matrices reaction to occur and produce binding material in cement. **[1-10]**

The construction industry has taken considerable strides forward over the last two or three decades with regard to trials in the use of one or another cementitious material generally identified as pozzolans, for the compounding of various cement-based products. This has not only resulted in improving the compressive strength value attained thereby but also in qualities like the ability to set and harden under water. Among this coal fly ash, blast furnace slag, rice hush ash, bagasse ash, silica fume, or metakaolin are the most common ones. Due to economic and environmental concerns, different methods of making cement products are being considered. One method to achieve the goal of reducing carbon dioxide emissions and greenhouse gases is to formulate cement using a lower portion of calcinated material, thereby reducing carbon dioxide emissions per unit of product. Another approach is to use a lower percentage of cement and or gypsum compared to standard cement or gypsum to ensure an increased compressive strength and or flexural strength is yet attained thereby. This one is durable and suitable for all types of applications, and also benefits the environment. Additionally, a need exists for improved cement and gypsum products that permit the use of less expensive aggregates to reduce the cost of the cement product.

Alkali-Activation is a technique for starting the reactivity of cementitious matrices without using cement. The reaction of concrete including both cement and a cementitious matrix, such as Fly Ash, is the finest demonstration of the Alkali-Activation process. Two chemical processes are going on at the same time. The first is the classic cement reaction, in which the active components of cement, Tricalcium Silicate (C3S) and Dicalcium Silicate (C2S), react with water to generate Calcium Silicate (CSH) and Calcium Hydroxide (CH). In the reaction, Calcium Hydroxide from the first reaction reacts with the Silicon Dioxide (S) present in Fly Ash to produce more Calcium Silicate (CSH) **[11-22].**

2. Materials and Methods

2.1. Materials

Bagasse ash is the prime material used in the study. Sugarcane bagasse ash was procured from NSL Sugars Lmt., Chikkonahalli, Koppa, Maddur, Karnataka. Generally, every ton of sugarcane bagasse ash contains approximately 48% cellulose, 26% of hemicellulose, 26% lignin, and 26% of bagasse [4][5]. From the review of the literature, it can be found that up to 30% of replacement can be done [8][10]. The specific gravity and fineness of the Bagasse Ash used for the study are 1.84 and 220 m2/kg respectively and the chemical properties of Bagasse Ash [5][9] from the review of literature are in **Table 1.**

GGBS, a by-product of iron production mainly consists of melted calcium silicate and alumino silicate removed from the blast furnace on regular basis. The chemical composition of GGBS is shown in **Table 2** (as given by the manufacturer) [6][7]. The specific gravity, fineness, and bulk density obtained from the tests carried out in the laboratory are 2.81, 390 m2/kg, and 1120 kg/m3. Lime is the inorganic substance calcium hydroxide (also known as slaked lime) and has the chemical formula Ca(OH)2. When quicklime (calcium oxide) is combined or slaked with water, it produces a colourless crystal or white powder. Hydrated lime is also known as caustic lime, builders' lime, slack lime, and pickling lime. In this study, lime is used as a replacement for binder content along with GGBS varying in the range of 10, 15, and 20% of total binder content in the concrete mix. It is used in the mixes to have a rapid setting property in the presence of an alkaline solution in the BSAASC mix.The aggregates used in this study were tested according to Indian Standard codes. As a coarse aggregate, self-compacting concrete. The coarse aggregate had a specific gravity of 2.67 and a water absorption of 0.5 percent, respectively.

For this study, manufactured sand that is readily available is employed, and the sand is evaluated in the laboratory to determine its physical qualities. IS stands for "fineness." IS 383 -1970 fine aggregate modulus. The sand utilized in the experiment belongs to zone II. M sand is free of silt and clay particles, resulting in improved abrasion resistance, increased unit weight, and decreased permeability. Because it decreases sand mining from riverbeds, it is less detrimental to the environment. M Sand's precise grading and cubical shape provide strong strength and long-term durability to concrete. The specific gravity, water absorption, and fineness modulus of the used fine aggregate are 2.65, 10%, and 3.25 respectively.

Alkaline solution for Geopolymer is a combination of sodium silicate and sodium hydroxide. The alkaline solution was procured from Kamath's Rice Mill & Industries, Mangalore, in which sodium silicates contain 32.8% SiO2, 14.7% Na2O, and 52.5% of water by weight (as per the information provided by the distributor).

Table 1. Chemical properties of bagasse ash.

Sl. No.	Component	Percentage
1	Silica	62
2	Alumina	31.5
3	Ferric Oxide	1.79
4	Manganese Oxide	0.004
5	Calcium Oxide	0.48
6	Magnesium Oxide	0.39
7	Loss of Ignition	0.71

Table 2. Chemical properties of GGBS.

Sl. No.	Particulars	% Composition
1	Calcium oxide (CaO)	37.34
2	Aluminium oxide (Al_2O_3)	14.42
3	Iron oxide (Fe_2O_3)	1.11
4	Silica (SiO_2)	37.73
5	Magnesium oxide (MgO)	8.71

2.2. Optimum mix design of BSAASC mixes

In the present study, the Desirability Function Approach (DFA) method was adopted for the development of BSAASC mixes. DFA is generally used for optimizing the number of experiments to be carried out. Here 3 factors were considered on 3 levels. Percentage of Bagasse ash content, lime content percentage, and dosage of Na2O. an initial set of nine trial mixes were formulated using the Box Behnken method as shown in **Table 3** and the performance of these mixes was tested in the laboratory.

Table 3. Levels and factors of concrete mixes.

Factors	Level 1	Level 2	Level 3
Bagasse ash content (in %)	10	20	30
Na_2O dosage (in %)	4	5	6
Lime content (in %)	10	15	20

2.2.1. Mixture proportioning, preparation, and casting of mixtures

The details of the proportions of concrete mixtures are shown in **Table 4**. Test specimens were cast and tested, using the initial nine trial mixes, designated as BSAASC-1 to BSAASC-9. The mix design for BSAASC concrete is calculated following the volume batching of EFNARC guidelines. In this method, the total binder content is restricted to 600 kg/m3. The ratio of proportions of fine to coarse aggregates in all the mixes was maintained constant at 60: 40. The molarity of the alkaline solution was kept at 1M throughout. The water-to-binder ratio was 0.60 and was kept constant for all the mixes.

Table 4. Details of mix designations and mix proportions of concrete mixes.

Mix ID	GGBS	BA	Lime	FA	CA	Alkaline solution
BSAASC1	480	60	60	592.59	488.53	437.80
BSAASC2	360	180	60	573.25	472.59	437.80
BSAASC3	420	60	120	584.90	482.20	437.80
BSAASC4	300	180	120	565.56	466.25	437.80
BSAASC5	450	60	90	604.37	498.29	424.80
BSAASC6	330	180	90	585.03	482.30	424.80
BSAASC7	450	60	90	573.12	472.48	450.70
BSAASC8	330	180	90	553.78	456.54	450.70
BSAASC9	420	120	60	598.54	493.44	424.80

3. Results and discussions

3.1. Fresh properties of BSAASC mixes
In the present investigation, the levels of workability of the mixes were measured using the standard Abraham's slump cone and flow table. The slump cone with a bottom diameter of 20cm, a top diameter of 10cm, and a height of the cone is 30cm is used. For all the trial mixes flowability is checked to satisfy the EFNARC guidelines of flow 600-800mm. To achieve the flow of the mixes, the water binder ratio is increased from 0.4 to 0.9 till the flow is greater than 650mm. The BSAASC mixes showed slump flow values ranging from 650-700mm, with their L–Box ratios and V-Funnel values ranging between 0.80 and 08 s respectively, satisfying the EFNARC guidelines as shown in **Fig 1.**

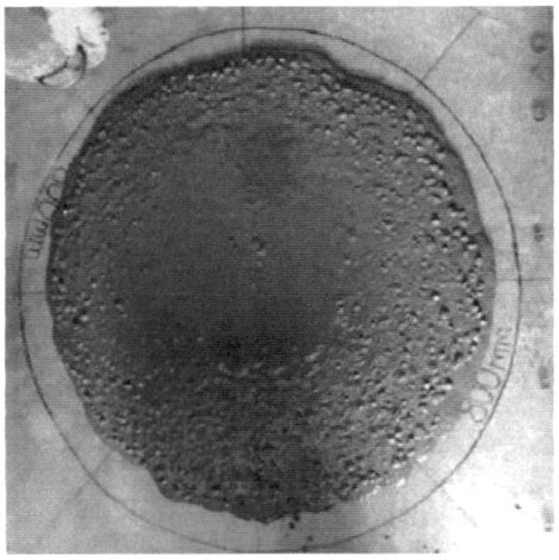

Figure 1. The slump flow of BSAASC mixes.

3.2. Tests on hardened SCAASC mixes

3.2.1. Compressive strengths of SCAASC mixes
All the BSAASC mixes tested here have shown the targeted strength values on testing as per IS 516:1999 as shown in **Table 5** and **Fig. 2** shows the variation in compressive strength. It was observed that compressive strength of about 40-50 MPa was obtained as shown in Fig. 2 for 600 Kg/m3 binder content. It is observed as the percentage of bagasse ash increases, there is a slight decrease in the compressive strength. The highest compressive strength value was observed for BSAASC-1 mix with 10% Bagasse

Ash, 10% Lime content, and 6% dosage of Na2O. And BSAASC-7 mix shows the least strength consisting of 10% Bagasse Ash, 15% lime, and 7% dosage of Na2O.

The results of 7-day and 28-day compressive strength obtained for various BSAASC mixes were statistically analyzed using Minitab Software. The main effect plot for compressive strength was generated as shown in **Fig. 3**. **Table 6** shows the compressive strength of 7 days and 28 days as analyzed by the DFA method.

$$Compressive\ strength\ 7D\ in\ Mpa = \begin{cases} 38.8 - 0.77\ BA + 0.51\ lime - 2.9\ Na_2O \\ -0.0324\ BA.lime + 0.199\ BA.Na_2O \\ +0.001\ lime.Na_2O \end{cases} \quad (1)$$

$$Compressive\ strength\ 28D\ in\ Mpa = \begin{cases} 38.8 - 0.77\ BA + 0.51\ lime - 2.9\ Na_2O \\ -0.0324\ BA.lime + 0.199\ BA.Na_2O \\ +0.001\ lime.Na_2O \end{cases} \quad (2)$$

Table 5. Compressive strength results

Mix no.	Compressive strength in MPa			
	3 Days	7 Days	14 Days	28 Days
OPC	14.13	19.58	26.74	30.6
BSAASC 1	25.44	32.42	37.38	50
BSAASC 2	24.18	35	43.54	42.68
BSAASC 3	26.86	34.04	36.93	38.7
BSAASC 4	23.62	30.15	32.46	39.76
BSAASC 5	24.86	32.38	36.46	42.25
BSAASC 6	24.18	26.46	33.73	39.15
BSAASC 7	22.56	28.97	32.64	38.62
BSAASC 8	26.42	31.02	37.67	43.56
BSAASC 9	20.35	24.58	32.48	43.79

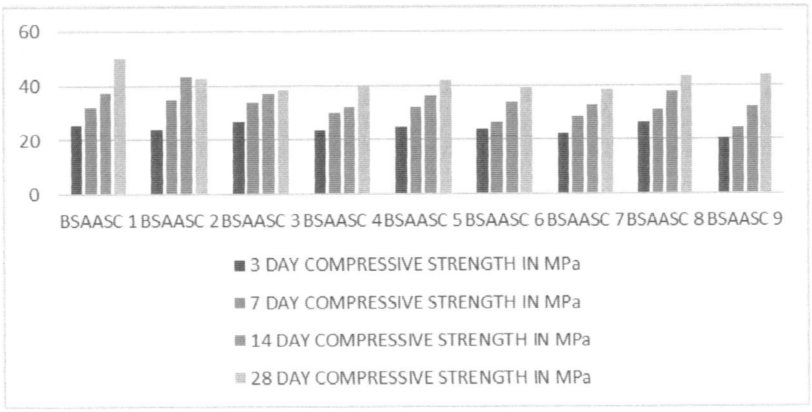

Figure 2. Variation of compressive strength for different BSAASC mixes.

Table 6. Compressive Strength of 7 days and 28 days as analysed by the DFA method

Mix no.	Compressive Strength in MPa	
	7 Days	28 Days
BSAASC 1	27.56	46.15
BSAASC 2	29.56	40.85
BSAASC 3	29.48	37.48
BSAASC 4	25.00	40.56
BSAASC 5	29.42	44.64
BSAASC 6	24.20	39.50
BSAASC 7	27.62	38.99
BSAASC 8	30.36	41.90
BSAASC 9	27.47	45.18

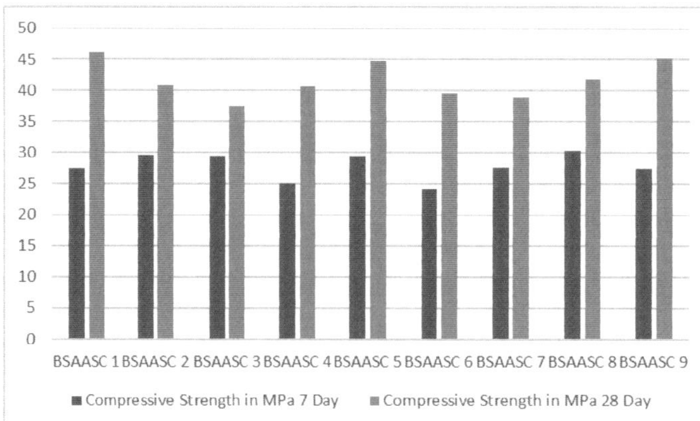

Figure 3. Variation of compressive strength for different BSAASC mixes after DFA analysis.

3.2.2. Split tensile strengths of BSAASC mixes

The split tensile strength tests were performed according to IS 5816 -1999 and the results are shown in **Table 7.** The values of split-tensile strengths of the various mixes, at different ages, are ranging from 1.7 MPa to 2.6 MPa, as shown in **Fig. 4.**

The result of the 28-day split tensile strength obtained for various BSAASC mixes was statistically analyzed using Minitab Software. The main effect plot for split tensile strength was generated as shown in **Fig. 5**. **Table 8** shows the compressive strength of 7 days and 28 days as analyzed by the DFA method.

$$Split\ tensile\ strength\ in\ Mpa = \begin{cases} 3.23 - 0.04\ BA - 0.058\ lime - 0.068\ Na_2O \\ +0.00155\ BA.lime + 0.002\ BA.Na_2O \\ +0.003\ lime.Na_2O \end{cases} \quad (3)$$

Table 7. Split Tensile strength results

Mix no.	Split tensile strength (in MPa)
BSAASC 1	2.23
BSAASC 2	2
BSAASC 3	2.04
BSAASC 4	2.12
BSAASC 5	1.89
BSAASC 6	1.74
BSAASC 7	2.08
BSAASC 8	2.01
BSAASC 9	2.22

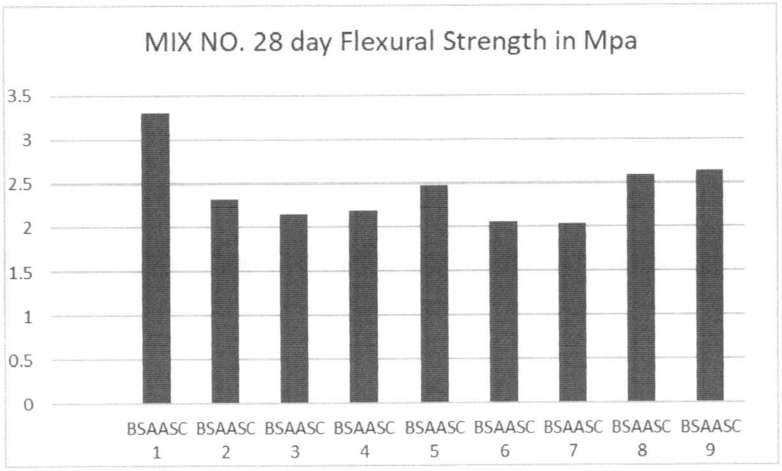

Figure 4. Variation of Split Tensile Strength for different BSAASC mixes.

Table 8. Split Tensile Strength as analysed by the DFA method.

Mix no.	Split tensile strength (in MPa)
BSAASC 1	2.3
BSAASC 2	2.04
BSAASC 3	2.05
BSAASC 4	2.11
BSAASC 5	2.18
BSAASC 6	2.04
BSAASC 7	2.17
BSAASC 8	2.17
BSAASC 9	2.17

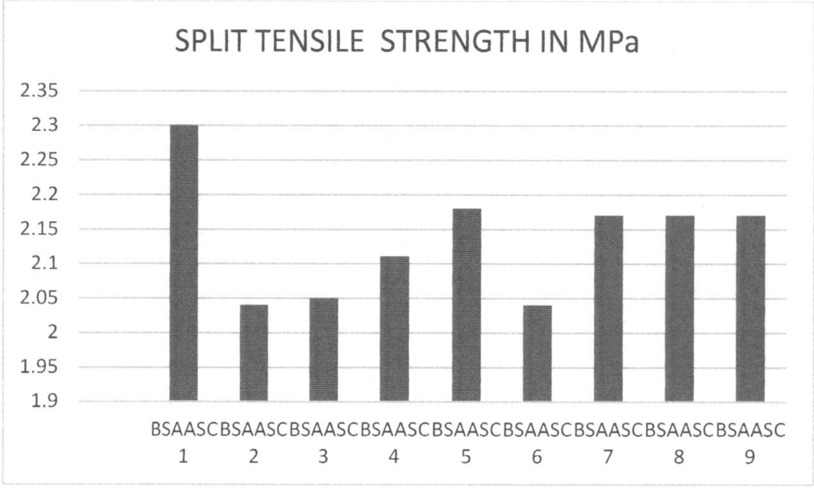

Figure 5. Variation of Split Tensile strength for different BSAASC mixes after DFA analysis.

3.2.3. Flexural strength of BSAASC mixes

The Flexural strength tests were performed according to IS 516 -1959 and the results of both normal strength and analysed strength are shown in **Table 9**. The values of normal flexural strengths of the various mixes over 28 days are ranging from 2 MPa to 2.7 MPa, as shown in **Fig. 6.**

$$Flexural\ strength\ in\ Mpa = \begin{cases} 9.77 - 0.2313\ BA - 0.262\ lime - 0.851\ Na_2O \\ +0.00505\ BA.lime + 0.0243\ BA.Na_2O \\ +0.02\ lime.Na_2O \end{cases} \quad (4)$$

Table 9. Flexural Strength for different BSAASC mixes on 28 days and DFA analysed Flexural Strength.

Mix no.	28 Days flexural strength (in MPa)	Analysed flexural strength (in MPa)
BSAASC 1	3.3	2.9
BSAASC 2	2.32	2.19
BSAASC 3	2.15	1.97
BSAASC 4	2.18	2.29
BSAASC 5	2.47	2.75
BSAASC 6	2.06	2.06
BSAASC 7	2.03	2.12
BSAASC 8	2.59	2.42
BSAASC 9	2.64	2.71

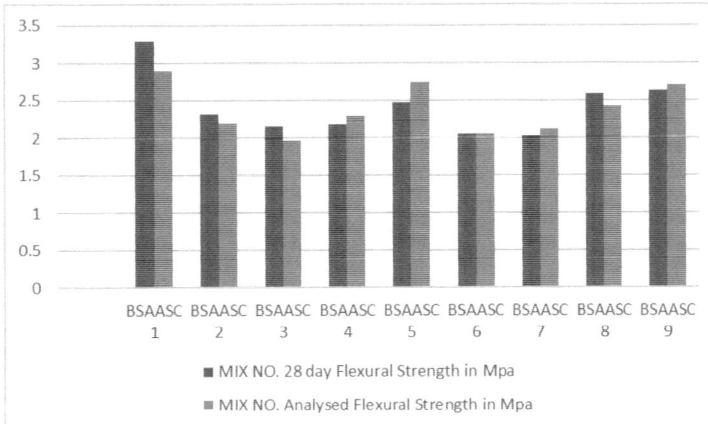

Figure 6. Variation of Flexural Strength for different BSAASC mixes on 28 days and DFA analysed Flexural Strength.

3.2.4. Performance of regression equations – tests on mixes in the verification Phase
The initially identified nine trial mixes (shown in Table 4) showed satisfactory properties for both fresh and hardened states. The predictive capabilities of these strengths were further tested by casting test specimens using six additional mixes (Verification phase) (denoted BSAASC-10 to BSAASC-15) and testing their strengths. All the details of the mixes and test results obtained are shown in **Table 10** and **11** respectively. The predicted strengths, as computed from regression equations (1), (2), (3) and (4) are also tabulated in Table 11 for all verification BSAASC mixes

Table 10. Verification BSAASC mixes.

Mix ID	GGBS	BA	Lime	FA	CA	Alkaline solution
BSAASC10	360	120	120	590.86	487.10	424.80
BSAASC11	420	120	60	567.30	467.68	450.70
BSAASC12	360	120	120	559.61	461.34	450.70
BSAASC13	390	120	90	579.08	477.39	437.80
BSAASC14	390	120	90	579.08	477.39	437.80
BSAASC15	390	120	90	579.08	477.39	437.80

Table 11. Results of verification BSAASC mixes.

Mix ID	Normalized Strengths (in MPa)						Analysed strength (in MPa)			
	compressive				Split tensile	Flexural	Compressive		Split tensile	Flexural
	3Days	7Days	14Days	28Days			7Days	28Days		
BSAASC10	19.68	23.52	30.82	40.27	2.05	2.27	26.14	38.96	2.05	2.10
BSAASC11	21.37	28.47	31.18	38.46	2.03	2.13	29.65	41.82	2.18	2.38
BSAASC12	22.34	27.44	31.56	38.43	1.92	2.16	28.34	39.08	2.15	2.17
BSAASC13	18.34	21.43	30.58	41.63	2.57	2.25	27.90	41.26	2.13	2.34
BSAASC14	18.34	21.43	30.58	41.63	2.57	2.25	27.90	41.26	2.13	2.34
BSAASC15	18.34	21.43	30.58	41.63	2.57	2.25	27.90	41.26	2.13	2.34

IOP Conf. Series: Materials Science and Engineering 1291 (2023) 012041 doi:10.1088/1757-899X/1291/1/012041

3.3. Studies on microstructures of BSAASC mixes

Here five BSAASC were chosen for analysing the microstructural details. Among these BSAASC-1, BSAASC-5, BSAASC-8 and BSAASC-9 were selected for high compressive strength and BSAASC-12 for low compressive strength when compared with other BSAASC mixes.

Crushed pieces of failed compression test samples (28 days) were taken for microstructural analysis. The smaller particles were subjected to Scanning Electron Microscope (SEM) and Energy Dispersive X-Ray Analysis (EDX) and finely powdered fractions passing through a 90μm sieve of the representative sample were taken for X-Ray Diffraction (XRD).

Fig. 7 to Fig. 11 shows the SEM, EDX and XRD analysis of the considered samples. They distinctly show the activation of the slag particles depending on the composition of the mixes. From the EDX images, it is clear that Si and Ca are the peak elements when compared to other elements.

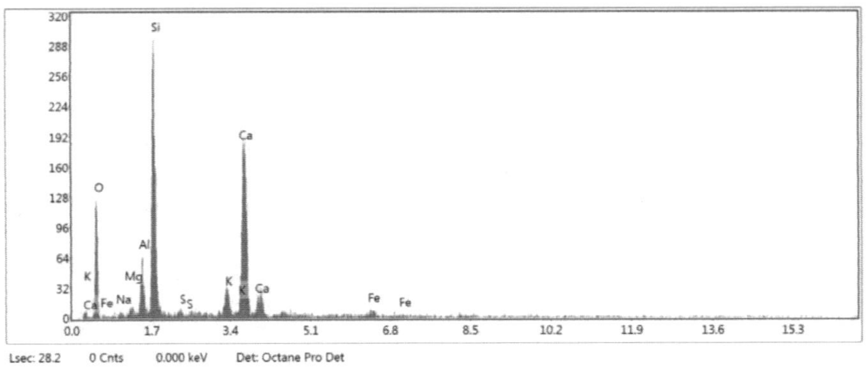

a.

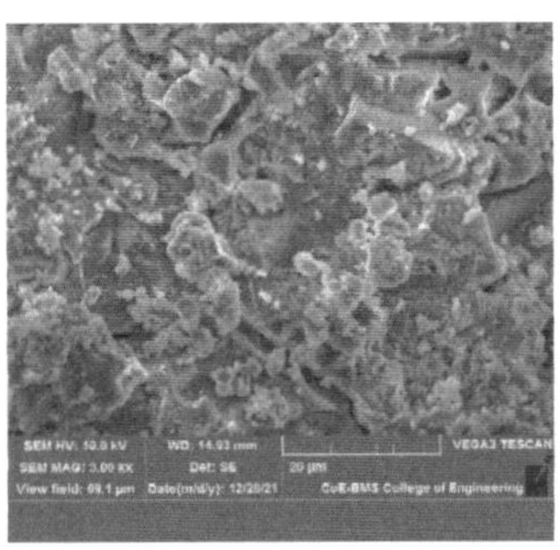

b.

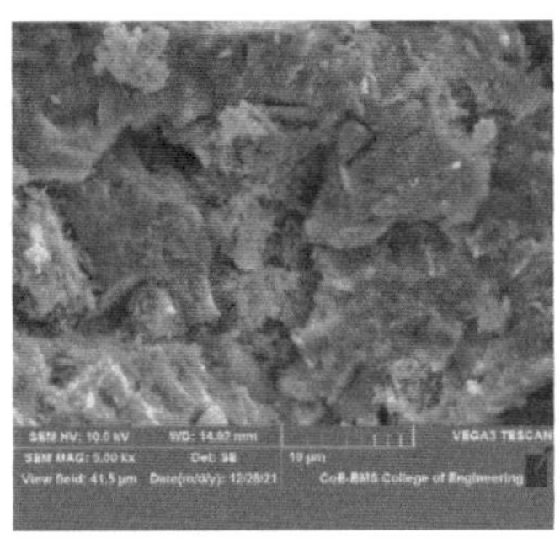

d.

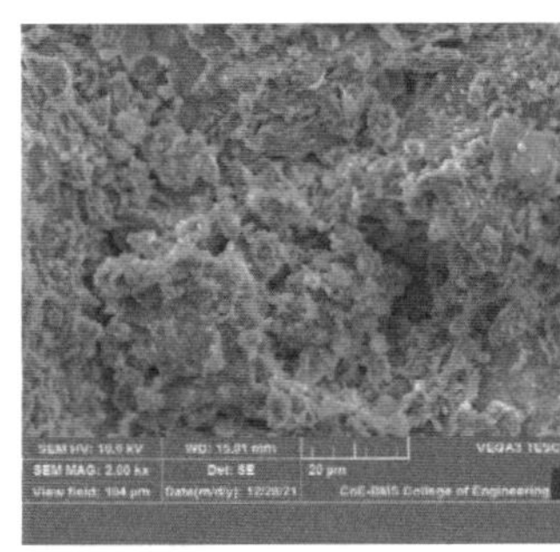

e.

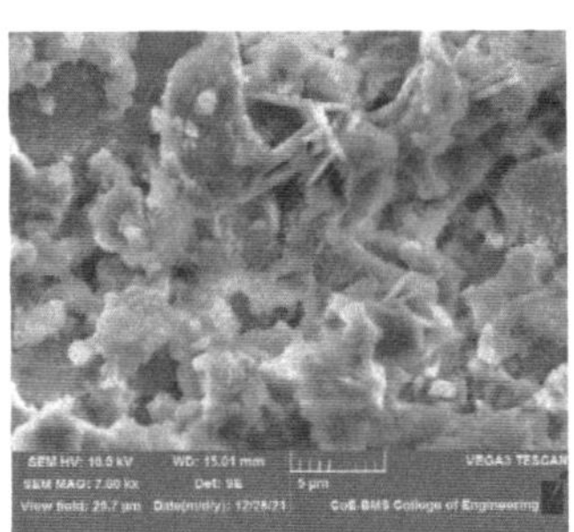

e.

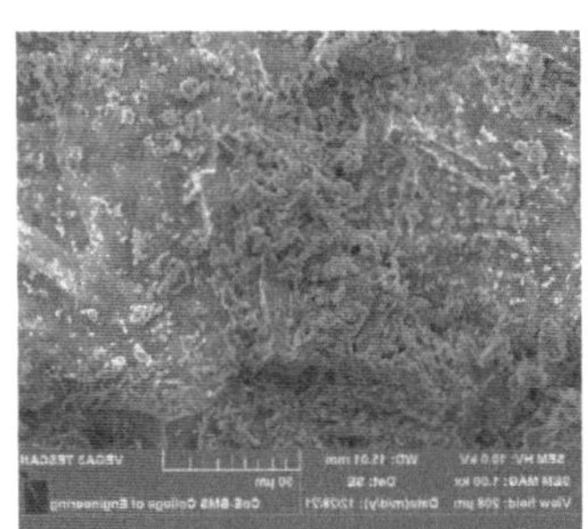

f.

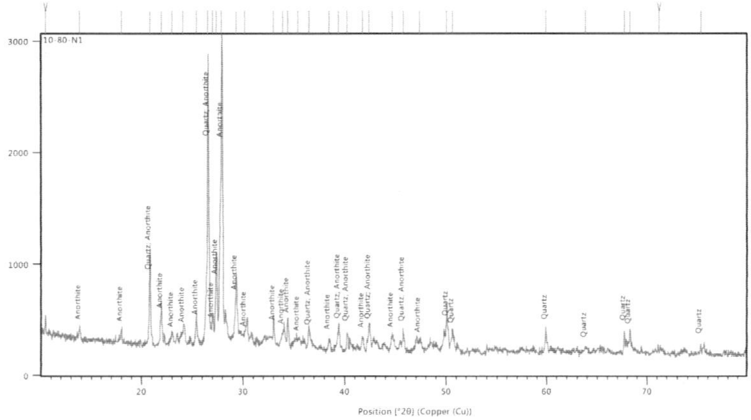

g.

Figure 7. shows ERD (a), SEM (b to f) and XRD (g) of BSAASC-1 Mix.

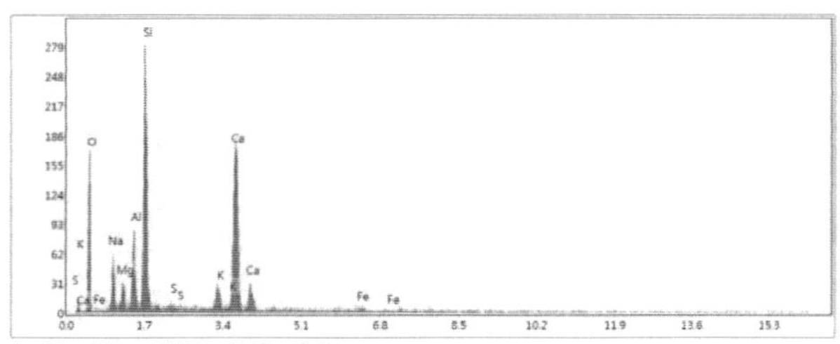

a.

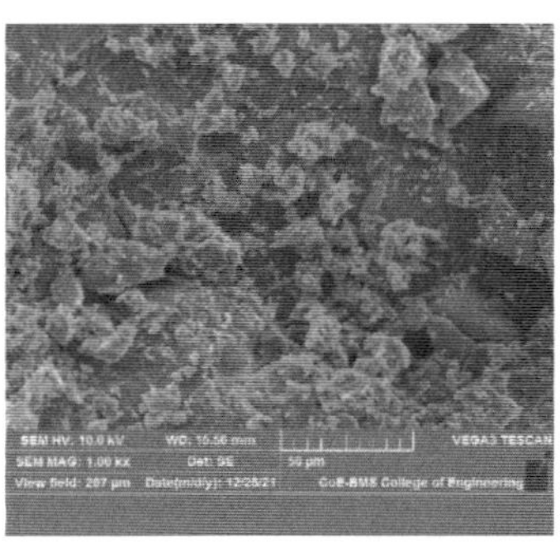

b.

c.

d.

e.

f.

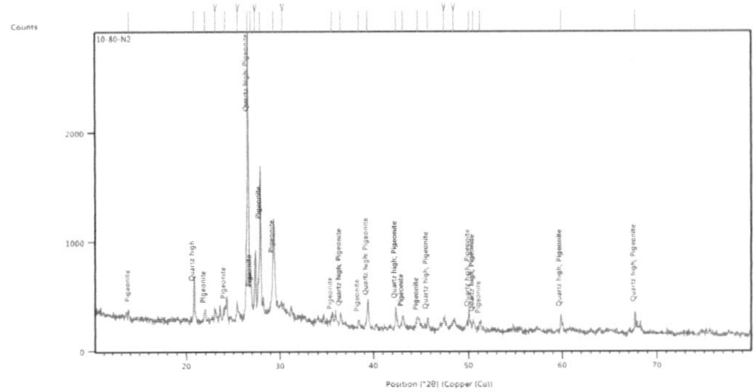

g.

Figure 8. shows ERD (a), SEM (b to f) and XRD (g) of BSAASC-5 Mix.

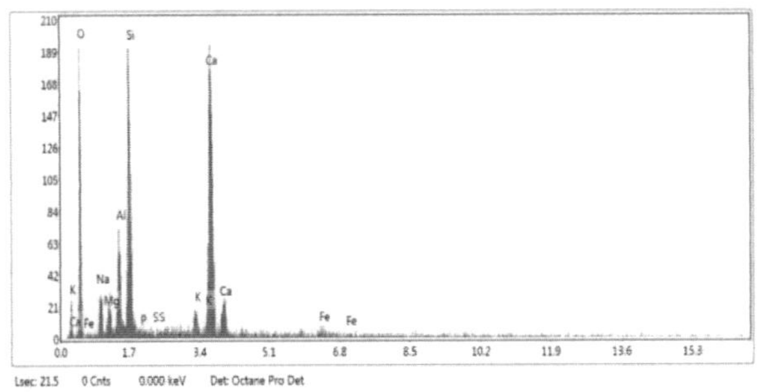

a.

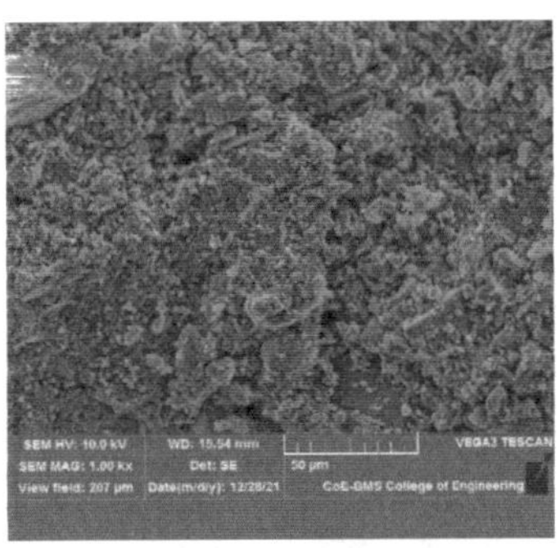

b.

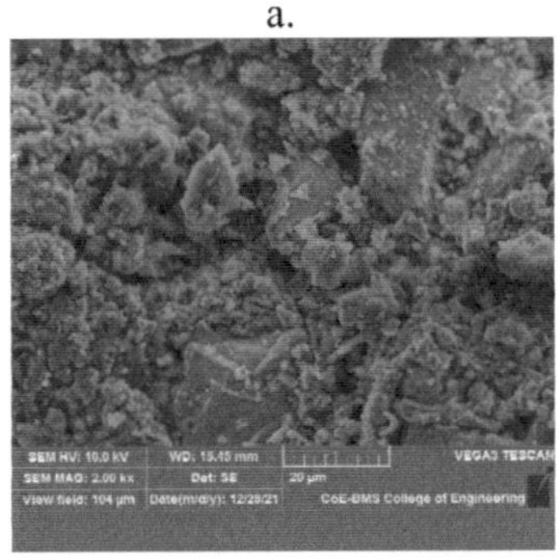

c.

d.

e.

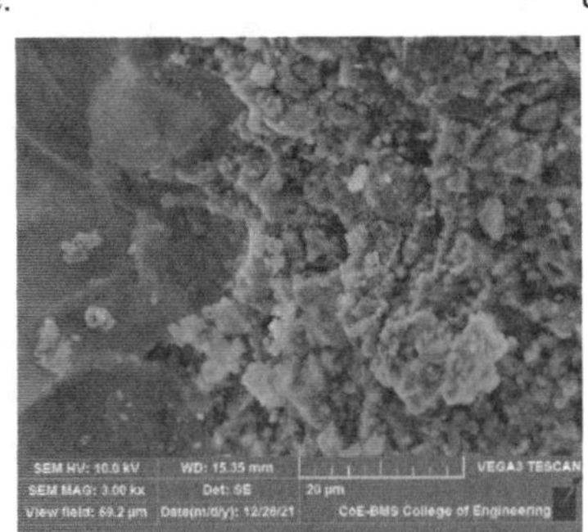

f.

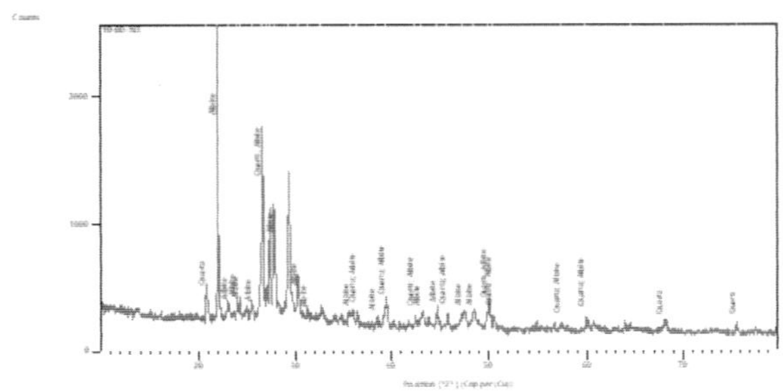

g.

Figure 9. shows ERD (a), SEM (b to f) and XRD (g) of BSAASC-8 Mix.

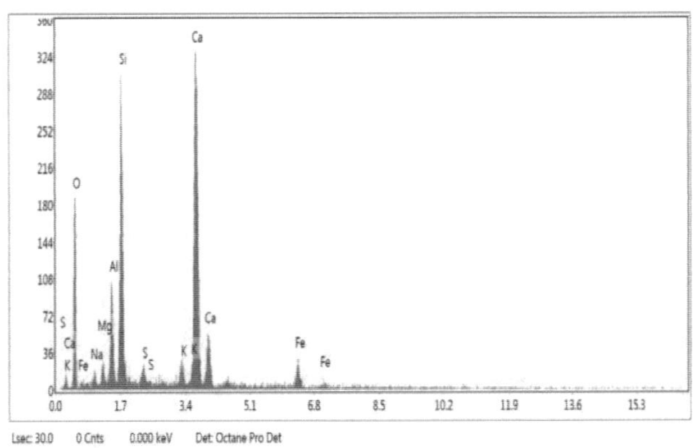

a.

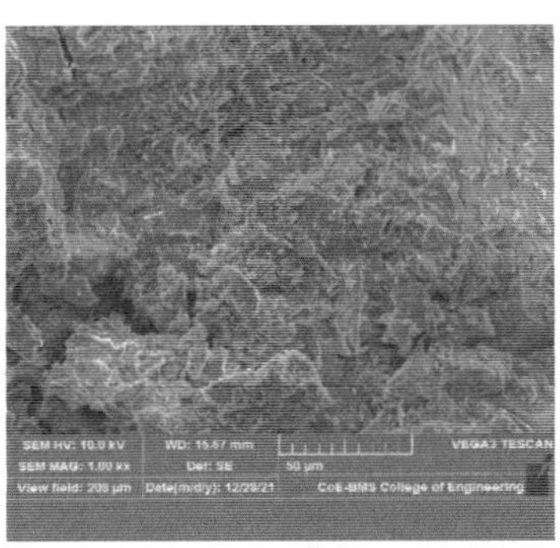

b.

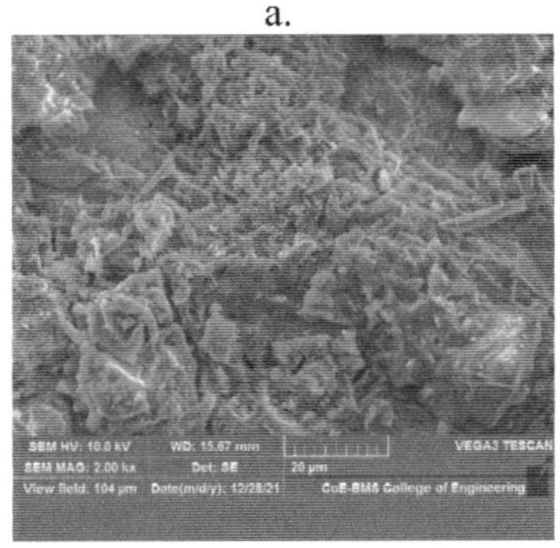

c.

d.

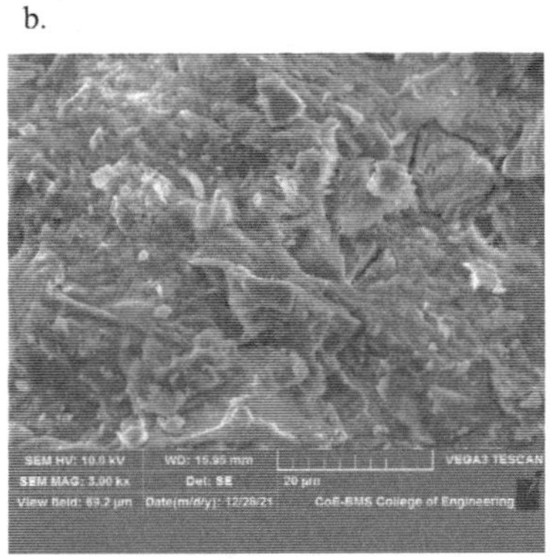

e.

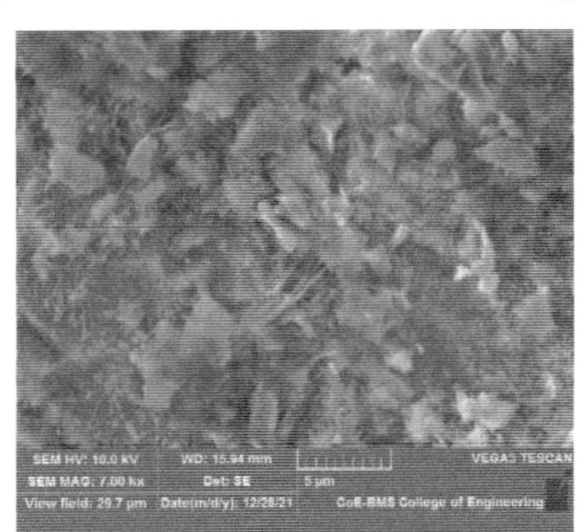

f.

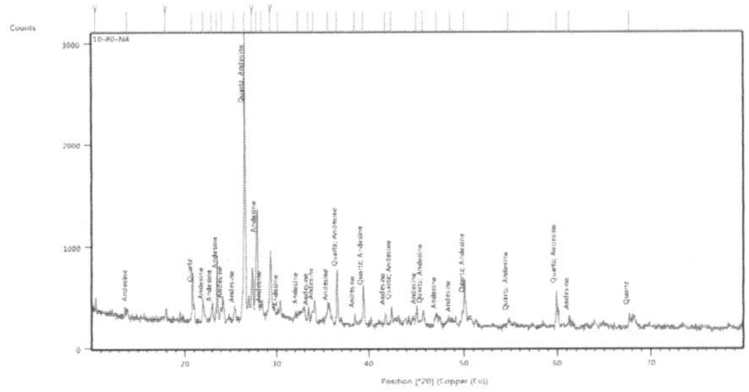

g.

Figure 10. Shows ERD (a), SEM (b to f) and XRD (g) of BSAASC-9 Mix.

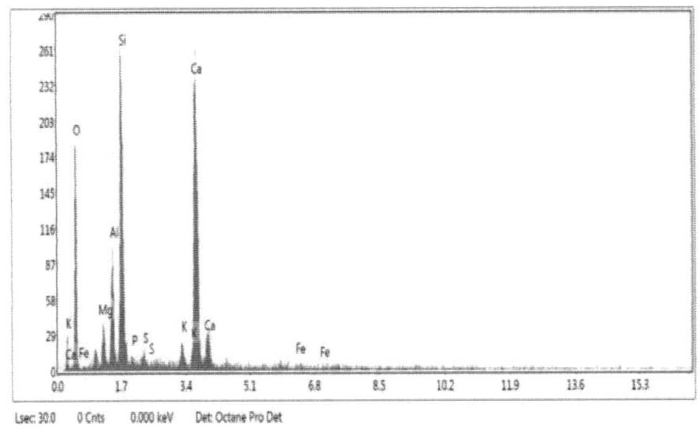

a.

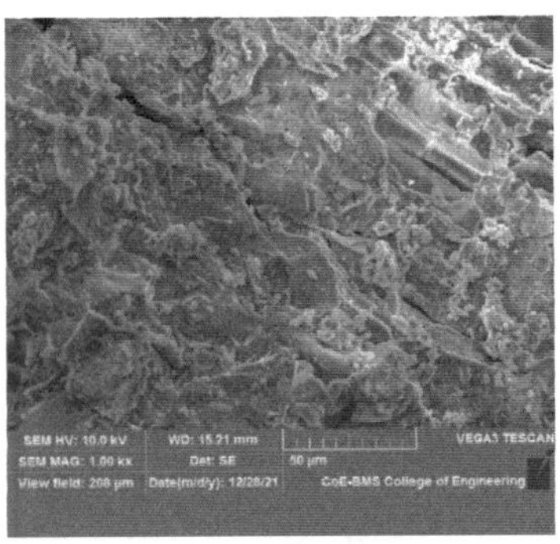

b.

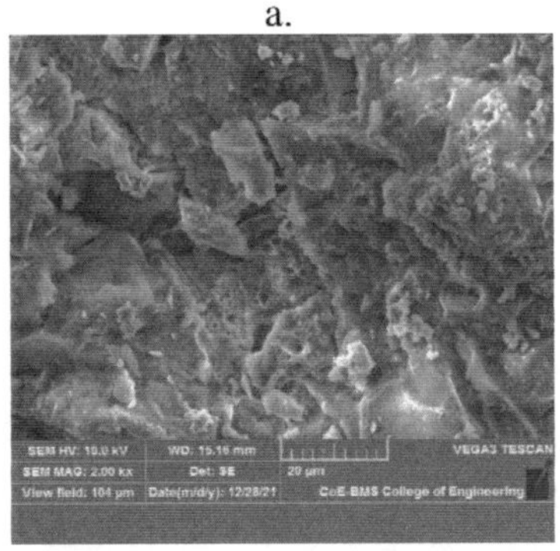

c.

d.

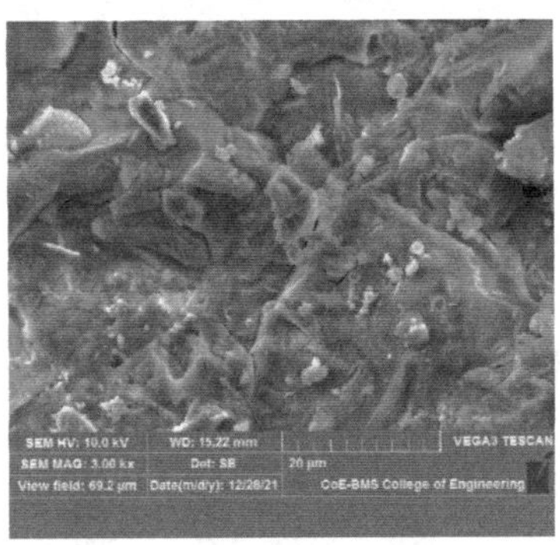

e.

f.

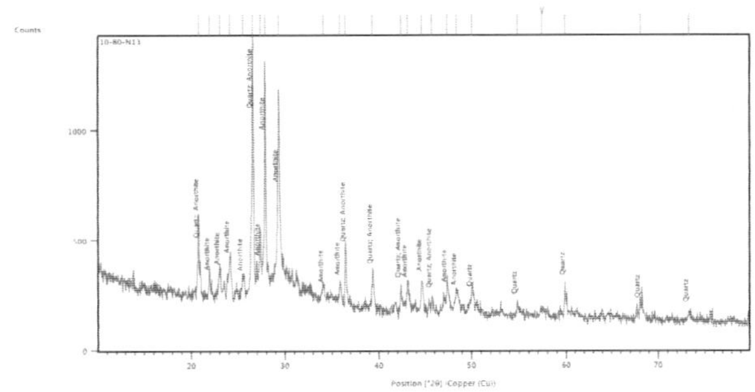

g.

Figure 11. Shows ERD (a), SEM (b to f) and XRD (g) of BSAASC-12 Mix.

4. Conclusions

- Bagasse Ash-based Self Compacting Alkali Activated Slag Concrete (BSAASC) mixes were effectively developed using Bagasse Ash at varying proportions. The mixes showed satisfying EFNARC guidelines, showing enhanced behaviour in terms of flowability and engineering properties.
- The compressive strength of 10% replacement of Bagasse Ash showed good strength when compared with 20% and 30% replacement of Bagasse Ash.
- BSAASC mixes developed here show compressive strength values ranging between 40 MPa to 50 MPa without affecting the flowability properties of BSAASC mixes. The split tensile strength and flexural strength also show satisfactory performance with values ranging from 1.7 MPa to 2.6 MPa and 2.0 MPa to 2.7 MPa respectively.
- The enhanced compressive strength may be due to the presence of lime and GGBS as binder content leading to the formation of a large amount of C-S-H and C-A-S-H gel. It is evident from the microstructural analysis.

5. References

[1] Yadav A L, Sairam V, Muruganandam L and Srinivasan, K. (2019). An Overview of the influences of mechanical and chemical processing on Sugarcane bagasse ash characterisation as a Supplementary Cementitious Material *Journal of Cleaner Production*. 118854–. doi:10.1016/j.jclepro.2019.118854

[2] Akbar Arslan, Farooq. Furqan, Shafique Muhammad, Aslam Fahid, Alyousaf Rayed and Abduljabbar Hisham. (2020). Sugarcane bagasse ash-based engineered geopolymer mortar incorporating propylene fibers. *Journal of Building Engineering.*101492–. doi:10.1016/j.jobe.2020.101492

[3] S K Rahman and R Al-Ameri. A newly developed self-compacting geopolymer concrete under ambient condition. *Construction and Building Materials* 267 (2021) 121-822 doi:10.1016/j.conbuildmat.2020.121822

[4] Zareei Seyed Alireza, Ameri Farshad and Bahrami Nasrollah. (2018). Microstructure, strength, and durability of eco-friendly concretes containing sugarcane bagasse ash. *Construction and Building Materials*. 184, 258–268. doi:10.1016/j.conbuildmat.2018.06.153

[5] Loganayagan S, Chandra Mohan N and Dhivyabharathi S. (2020). Sugarcane bagasse ash as alternate supplementary cementitious material in concrete. *Materials Today: Proceedings*. S2214785320318344–. doi:10.1016/j.matpr.2020.03.060

[6] Manjunath R and Narasimhan Mattur C. (2018). An experimental investigation on self-compacting alkali activated slag concrete mixes. *Journal of Building Engineering*. 17(), 1–12. doi:10.1016/j.jobe.2018.01.009

[7] Manjunath R, Narasimhan Mattur C, Umesh K M, Shivam Kumar and Bala Bharathi U K. (2019). Studies on development of high performance, self-compacting alkali activated slag concrete mixes using industrial wastes. *Construction and Building Materials.* 198(), 133–147. doi:10.1016/j.conbuildmat.2018.11.242

[8] Bahurudeen A, Kanraj Deepak, Gokul Dev V and Santhanam Manu. (2015). Performance evaluation of sugarcane bagasse ash blended cement in concrete. *Cement and Concrete Composites.* 59(), 77–88. doi:10.1016/j.cemconcomp.2015.03.004

[9] Bahurudeen A, Marckson A V, Kishore Arun, Santhanam Manu. (2014). Development of sugarcane bagasse ash-based Portland pozzolana cement and evaluation of compatibility with superplasticizers. *Construction and Building Materials.* 68(), 465–475. doi:10.1016/j.conbuildmat.2014.07.013

[10] Almeida Fernando C R, Sales Almir, Moretti Juliana P and Mendes Paulo C D. (2019). Use of sugarcane bagasse ash sand (SBAS) as corrosion retardant for reinforced Portland slag cement concrete. *Construction and Building Materials.* 226(), 72–82. doi:10.1016/j.conbuildmat.2019.07.217

[11] EFNARC, The European Guidelines for Self-compacting Concrete: Specification, Production and Use, 2005.

[12] Shutong Y, Jinjin X, Chaohui Z, Rui L, Qiubo Y and Shuguang S. Mechanical properties of alkali-activated slag concrete mixed by seawater and sea sand. *Construction and Building Materials*. 2019. 196. Pp. 395–410. DOI: 10.1016/j.conbuildmat.2018.11.113

[13] Guneet S and Uthej V. Assessing properties of alkali activated GGBS based self-compacting geopolymer concrete using nano-silica. *Case Studies in Construction Materials*. 2020. 12. e00352. DOI: 10.1016/j.cscm.2020.e00352

[14] Chaitanya S T and Gunneswara R. Effect of mix design parameters on mechanical and durability properties of alkali activated slag concrete. *Construction and Building Materials*. 2018. 193. Pp.173–188. DOI: 10.1016/j.conbuildmat.2018.10.189

[15] Subhash C Y, B Chethan and K C Jitin. Durability studies on ferrochrome slag as coarse aggregate in sustainable alkali activated slag/fly ash based concretes. *Sustainable Materials and Technologies*. 2019. DOI: 10.1016/j.susmat.2019.e00137

[16] Ali A A, Abd E M A E and Mohammed A E. Factors affecting the mechanical properties of alkali activated ground granulated blast furnace slag concrete. *Construction and Building Materials*. 2019. 197. Pp. 339–355. DOI: 10.1016/j.conbuildmat.2018.11.086

[17] Nabeel A F, Neaz M S and Muhammad N S H. Investigation of engineering properties of normal and high strength fly ash based geopolymer and alkali-activated slag concrete compared to ordinary Portland cement concrete. *Construction and Building Materials*. 2019. 196. Pp. 26–42. DOI: 10.1016/j.conbuildmat.2018.11.083

[18] Puertas F B, Gonzalez F I, Gonzalez T M M, Alonso M, Torres C G, Rojo F and Martinez A. Alkali-activated slag concrete: Fresh and hardened behaviour. *Cement and Concrete Composites*. 2018. 85. Pp. 22-31. DOI: 10.1016/j.cemconcomp.2017.10.003

[19] Md Manjur A E, Md Maruf H, Md Rezaul K, Muhammad F M Z and Christopher S. A review on alkali-activated binders: Materials composition and fresh properties of concrete. *Construction and Building Materials*. 2020. 260. Pp. 119788. DOI: 10.1016/j.conbuildmat.2020.119788

[20] Athira V S, V Charitha, G Athira and A Bahurudeen. Agro-waste ash based alkali-activated binder: Cleaner production of zero cement concrete for construction. *Journal of Cleaner Production*. 2021. 286. Pp.125429. DOI: 10.1016/j.jclepro.2020.125429

[21] Xijun S, Anol M and Dan Z. Sustainability assessment for Portland cement concrete pavement containing reclaimed asphalt pavement aggregates. *Journal of Cleaner Production*. 2018. DOI: 10.1016/j.jclepro.2018.05.004

[22] Nabil H, Hima K S, Mothi K M, Arjun H R, Santhosh G and Jorisa C. 2019. Alkali-activated concrete paver blocks made with recycled asphalt pavement (RAP) aggregates. *Case Studies in Construction Materials*. 2019. DOI: 10.1016/j.cscm.2019.e00322

IOP Conf. Series: Materials Science and Engineering 1291 (2023) 012042 doi:10.1088/1757-899X/1291/1/012042

Design and Implementation of Hybrid Energy Storage System Integrating Lithium-ion Battery and Wind Turbine

S.HARIPPRIYA, S.JAYANTHY, E.ESAKKI VIGNESWARAN
haripriya.2057002@srec.ac.in, jayanthy.s@srec.ac.in, esakkivignesh@srec.ac.in
Sri Ramakrishna Engineering College , NGGO Colony,
Vattamalaipalayam, Coimbatore, Tamil Nadu 641022

Abstract. A hybrid energy storage system (HESS) by integrating Lithium-Ion Battery and Wind Turbine System for Electric Vehicle is designed and implemented. An advanced model of lithium ion/wind turbine HESS model is developed to improve the battery's charging capacity. The behavior of this model is tested by using Machine Learning Algorithms to identify remaining battery capacity of the energy storage components. The Machine learning Algorithms such as Support Vector Machine (SVM), K-Nearest Neighbor (KNN), Random Forest (RF) Decision Tree (DT) and Linear Regression (LR) are implemented which helps in predicting the State of Charge (SoC) in the electric vehicle to estimate the battery's lifetime. The SoC of the EV without HESS is calculated and compared with the SoC of HESS implemented EV model. Sensors are used to measure the current and voltage parameters of the Lithium-Ion Battery setup (Internal Battery) and Storage Battery (External Battery). The measured data is sent to the LPC2148 ARM microcontroller Board. The data that is obtained from LPC2148 is given as input to the machine learning algorithms which are executed in the Computer. Performance metrics such as Mean Absolute Error (MAE), Root Mean Square (RMSE), R-Square Error (R2E) and Mean Square Error (MSE) are calculated and the SoC values are predicted. The Random Forest algorithm produces better results for the HESS model as the Error values produced are less than 2.5%. The battery capacity of the electric vehicle has been improved by using external battery that is charged by wind turbine. Thus, the driving range of the electric vehicle implemented with HESS is increased twice when compared to conventional electric vehicle [1].
Keyword - HESS (Hybrid Energy Storage System), ESS (Energy Storage System), EV (Electric Vehicle), ML (Machine Learning), LPC2148 ARM 7 microcontroller board, MSE (Mean Square Error), R2S (R2Score,) RMSE (Root Mean Square), MAE (Mean Absolute Error).

I. INTRODUCTION

In electric vehicles, the energy storage system used are high cost, limited in range, slow in recharge. The electric vehicle development on these days is providing solutions for improving the lifetime of these storage systems, driving range, sizing specifications. To improve EV's lifetime and efficiency that is embedded with ESS, a lithium-ion battery is integrated with wind turbine. The HESS model contains two batteries where one battery is the Lithium-Ion battery setup already present in the electric vehicle; the other battery is the storage battery which is charged from the wind turbine. EV's range is based on the SoC of battery [7].

SoC is the percentage of remaining charge available in the battery. SoC is calculated using the battery parameters current, voltage and temperature taken from both the batteries in HESS model. The LPC2148 controller is used to acquire the battery parameters from sensors. The LPC2148 controller is programmed to obtain and store the real time dataset from the sensors. The dataset is then uploaded to the Jupyter Notebook through an USB module connected to the controller. The machine learning algorithms such as Support Vector Machine, Decision Tree, K-Nearest Neighbor, Random Forest and Linear Regression are used to calculate the remaining charge. The machine learning metrices are used in calculating performance of the algorithms. These algorithms provide the aging by predicting the time until which the battery charge lasts.

II. BACKGROUND WORKS

An effective method to predict aging of the battery is proposed by Prakash et al (2021). Through real time estimation battery health can be predicted accurately. and State of Health (SoH) and Remaining Useful Life (RUL) are predicted by calculating SoC of Lithium-Ion battery. The Machine learning Algorithms are used to calculate the SoH of the battery. The results obtained from the neural network algorithm shows the error rate of ±5%. The LSTM algorithm works better in predicting RUL with ±10 cycles of accuracy [2].

Meshabi et al (2021) et al proposed an advanced electrothermal modelling of a HESS integrating lithium-ion batteries and supercapacitors. They have used particle swarm-Nelder-Mead (PSO-NM) optimization algorithm to evaluate the performance of the model. Electrothermal behaviors are analyzed in the hybrid system to estimate the performance. The experimental data is taken from the urban EV. The optimization algorithm produces high efficiency and the error obtained is less than 3% [3].

Lee et al (2021) proposed an energy management strategy for hybrid EV's namely Equivalent Consumption Management Strategy (ECMS) to reduce energy consumption. A reinforcement learning algorithm namely Deep Q-Networks (DQNs) algorithm is implemented to evaluate the control parameter for minimal energy consumption. The costate is estimated accurately using Deep Q-Networks (DQNs). The results perform better than the already existing adaptive technique for the control concept. Thus, the proposed study is feasible in optimizing the energy consumption [4].

Huilong et al (2021) proposed a HESS which combines lithium-ion batteries and supercapacitors. This method helps to overcome the drawbacks of ESS including short cycle life, low power density and high cost. Optimized HESS and better power management system can achieve better driving range and life cycle. In this paper Pareto optimization technique is used which provides both optimal design and efficient control. Battery Life is significantly increased with the current solution [5].

Min et al (2021) proposed an adaptive battery SoC estimation method for EV by using BMS technique. BMS manages charging and discharging rate of battery which provides the state of battery. SoC provides the battery state which need to be predicted accurately. In this paper to predict SoC a Kalman-filter (EKF) based method is used. Conventional Coulomb counting method and EKF is combined to predict SoC which shows less than 2% error in predicting SoC[6].

An energy management along with the optimization approach is proposed by Zhou et al (2020) for plug-in hybrid electric vehicle (PHEV) with a HESS containing a Ni-Co-Mn battery pack and Li-Ti-O battery pack. The performance of this system is tested using a test bench. Combination of both Energy management system and hybrid energy management system in PHEV has increased the efficiency up to 2.5% and the lifecycle is improved up to 203% [9].

III. METHODOLOGY

A. FLOW DIAGRAM

A flow of the proposed system is shown in Figure 1. The dataset is divided into testing and training dataset. The training data is the web data and the test data is the real-time data. Two different types of test data are fed into the algorithm. The first set of data is obtained from battery without integration of HESS model and the other data is evaluated from the battery in HESS model. The training data is given as input to ML algorithms. After training the dataset the testing is done which are then given to the model for predicting the battery's capacity.

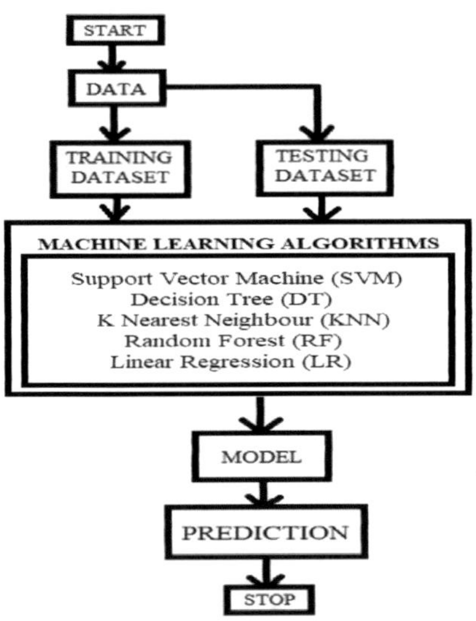

Fig 1 Flow Diagram of the proposed system

B. BLOCK DIAGRAM

Fig 2 shows the block diagram of the proposed system. The ARM 7 is the microcontroller used to get real-time data from sensors. Through kinetic energy the wind turbine is rotated. According to the speed of the wind turbine charge is generated and stored in the external storage battery. When the electrical motor in the EV runs in low speed the charge is taken from the external storage battery and in high-speed internal battery is used. The temperature of the battery is measured by LM395 sensor. The ACS712 sensor is used to measure the current discharge from the battery and the voltage of the battery is measured by voltage sensor. The data from LPC2148 Microcontroller is collected by CPC2102 USB to TTL module and stored in spreadsheet. Python IDLE is used for implementing the Machine learning algorithms to predict the aging of battery.

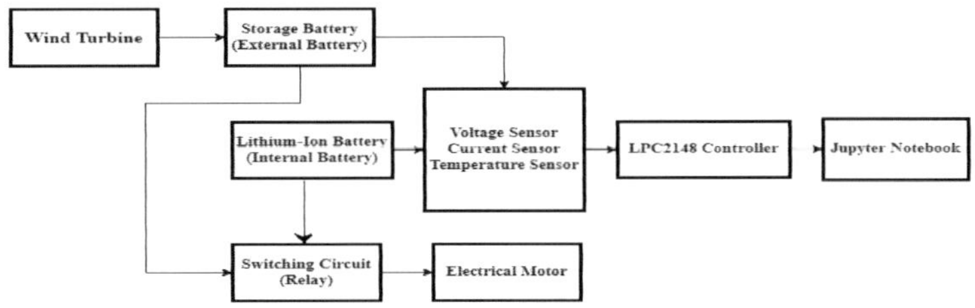

Fig 2 Block Diagram of the proposed system

C. SUPPORT VECTOR MACHINE (SVM)

Both classification and regression models use support vector machine. Decision boundaries are used in SVM where decision planes are predicted that are referred as hyperplane. The hyper plane separates two different set of points which has different class in the plane. The distance between the hyperplane and nearest points from either class is called is called as margin. The classification is done by figuring out the hyperplane that has the largest margin between the two classes. [7].

D. K-NEAREST NEIGHBOURS

The K-nearest neighbours works by classifying the data greatest number of common points according to the distance. The data point that needs to be predicted are considered from K numbered nearest points which belong to a specific class. The greatest number of points that belong to the points are the considered as the predicted values.

E. DECISION TREE (DT)

This algorithm works by creating a tree like structure. This tree structure is used to perform both classification and regression. The dataset is then divided into smaller nodes that are combined with the decision tree algorithm. The decision nodes and leaf nodes combine and form the tree structure. The nodes are the test data and the branch of the tree forms the result of the test data [8].

F. RANDOM FOREST (RF)

Random Forest algorithm helps in solving both regression and classification problems. Ensemble learning concept is used by RF algorithm. The RF algorithm takes the average value of numerous decision subsets to improve the precision of the dataset and provides the accurate prediction.

G. LINEAR REGRESSION (LR)

Linear Regression is based on supervised learning algorithm which performs regression for the given dataset. Independent variables are taken to identify the prediction values. Relationship between variables and forecasting are found out mostly using LR algorithm. For the given number of datasets LR algorithm plots the best fit line in the best possible way to fit the model.

IV. EXPERIMENTAL RESULTS AND ANALYSIS

A. EXPERIMENTAL SETUP

The experimental setup of the proposed system is illustrated in Figure 3. Sensors connected to the GPIO pins of LPC2148 MC collects the data from the current, voltage and temperature sensors. The data are collected from both internal and external battery. Data from LPC2148 is given to CPC2102 USB to TTL module which sends the data to an excel sheet displayed in PC.

Fig 3 Experimental setup of the Hybrid Energy Storage System

B. ANALYSIS 1: REAL-TIME DATASET 1

Fig 4 shows that the dataset contains real driving trips consisting of the temperature, elevation, speed, throttle, voltage, current, temperature, SoC, indoor temperature, heating power, etc. (Dataset obtained from kaggle) [10]

	Time	Velocity	Elevation	Throttle	Motor Torque	Longitudinal Acceleration	Regenerative Braking Signal	Battery Voltage	Battery Current	Battery Temperature	...	AirCon Power	Heater Signal	Heater Voltage	Heater Current	Ambie Temperatu
0	0.0	0.00	574.0	0.00	0.0	-0.03	0.0	391.40	-2.20	21.0	...	0.40	1	0	0	25
1	0.1	0.00	574.0	0.00	0.0	0.00	0.0	391.40	-2.21	21.0	...	0.40	1	0	0	25
2	0.2	0.00	574.0	0.00	0.0	-0.01	0.0	391.40	-2.26	21.0	...	0.40	1	0	0	25
3	0.3	0.00	574.0	0.00	0.0	-0.03	0.0	391.40	-2.30	21.0	...	0.40	1	0	0	25
4	0.4	0.00	574.0	0.00	0.0	-0.03	0.0	391.40	-2.30	21.0	...	0.40	1	0	0	25
...	...	...	...	...	...	...	...	...	...	...	...	...	...	...	...	...
10085	1008.5	20.71	565.0	25.06	-9.4	-0.28	0.0	387.91	-3.12	22.0	...	1.92	1	0	0	33
10086	1008.6	20.60	565.0	23.57	-8.9	-0.24	0.0	387.96	-2.37	22.0	...	1.92	1	0	0	33
10087	1008.7	20.44	565.0	22.55	-12.4	-0.31	0.0	388.01	-1.62	22.0	...	1.92	1	0	0	33
10088	1008.8	20.30	565.0	22.56	-15.2	-0.40	0.0	388.06	-0.92	22.0	...	1.92	1	0	0	33
10089	1008.9	20.14	565.0	22.43	-18.3	-0.39	0.0	388.10	-0.30	22.0	...	1.92	1	0	0	33

10090 rows × 28 columns

Fig 4 Set of Training Dataset

Test Data for Analysis 1:

Fig 5 shows the set of test data obtained from the obtained from real-time dataset that is collected without integrating HESS model (Dataset obtained from experimental setup). Temperature, Voltage and Current are the battery parameters obtained. The results contain predicted SoC values using regression algorithm.

	volt	current	temp
0	554	0	84
1	554	1	84
2	554	0	84
3	555	0	84
4	554	0	84
...	...	...	...
94	556	0	91
95	555	0	91
96	555	1	91
97	555	2	91
98	558	0	91

99 rows × 3 columns

Fig 5 Set of test data obtained from Electric Vehicle

Fig 6, shows the predicted State of charge values in support Vector machine algorithm, SoC predicted for particular voltage, current and temperature indicates the percentage of remaining charge in the battery.

```
[85.869985   86.41177718 85.869985   86.24068153 85.869985   85.869985
 86.09743955 86.24068153 87.26163244 86.24068153 85.869985   85.869985
 86.24068153 86.56155474 88.1554059  87.56513667 85.8975685  87.6438258
 85.72994141 87.56513667 85.72994141 85.72994141 85.39921442 85.72994141
 87.56513667 85.8975685  84.36049913 83.21810593 86.33356767 84.36049913
 86.20010296 83.53500394 83.21810593 83.53500394 83.21810593 83.53500394
 88.34585461 83.53500394 86.20010296 83.53500394 83.53500394 83.21810593
 86.49815308 83.21810593 81.66007003 84.75802143 81.66007003 83.75933096
 87.37282127 86.26819256 84.77929645 81.66007003 81.66007003 86.84687018
 84.60668276 85.71537185 85.11493071 85.29435178 82.01132208 84.19229077
 82.01132208 85.9760288  85.9760288  85.9760288  82.01132208 82.01132208
 80.56089302 82.01132208 83.03416845 85.32465404 82.01132208 82.01132208
 80.56089302 86.02560284 80.56089302 85.96440829 82.01132208 85.81146062
 81.77401205 81.77401205 80.27397272 81.88778001 81.88778001 80.27397272
 81.77401205 82.99794393 80.27397272 80.27397272 81.77401205 85.81146062
 84.5824935  81.77401205 84.5824935  80.80686343 80.80686343 82.11998944
 82.73463101 84.22479267 83.43781552]
```

Fig 6 Predicted SoC value for SVM algorithm

Fig 7 shows line graph for predicted SoC values for SVM algorithm are used to track variations of numerical values between actual and predicted SoC measured. Actual values indicate the SoC in training data and predicted values are the SoC obtained from testing dataset.

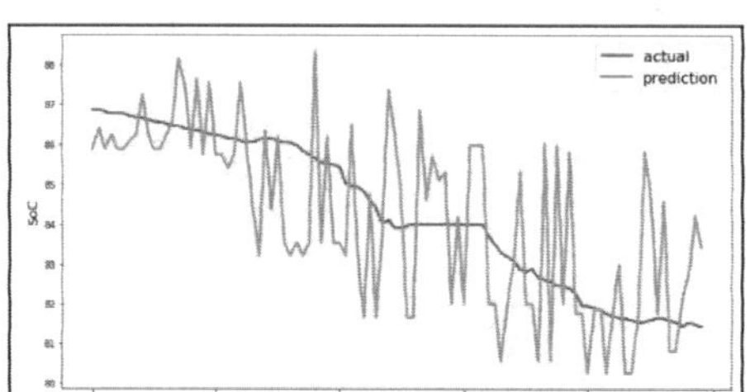

Fig 7 Line graph of SVM algorithm

Fig 8 shows the performance of SVM regression model which is evaluated by calculating the regression metrics such as Mean Absolute error, Root Mean Square Error, R-Square Value and Mean Square Error Mean. The Error values shows the closest prediction values with respect to their expected values.

```
Mean Absolute Error:
 1.4146130039678517

Mean Squared Error:
 3.0015779531478466

Root Mean Square Error:
 1.7325062635234099

R-Square:
 0.3612963645331496
```

Fig 8 Error values calculated for SVM algorithm

Fig 9, shows the predicted State of charge values and performance of K Nearest Neighbor algorithm, SoC predicted for particular voltage, current and temperature indicates the percentage of remaining charge in the battery. KNN regression model performance is evaluated by calculating the regression metrics. K values are calculated from the range of 1 to 20.

```
 [81.7 ]
 [81.7 ]
 [81.7 ]
 [86.11]
 [84.1 ]
 [81.7 ]
 [84.1 ]
 [81.8 ]
 [81.8 ]
 [81.8 ]
 [81.8 ]
 [84.1 ]
 [81.8 ]]
Mean Squared Error: 3.2490741900824336
RMSE value for k=  1 is: 1.8025188459715016
Mean Absolute Error: 1.308520113890596
R-Square Value: 0.3057840383829912
```

```
 [82.9355]
 [82.1645]
 [82.1645]
 [82.1645]
 [84.497 ]
 [84.497 ]
 [82.1645]
 [84.497 ]
 [82.1645]
 [82.1645]
 [82.1645]
 [83.3805]
 [84.497 ]
 [82.1645]]
Mean Squared Error: 2.0179012415419253
RMSE value for k=  20 is: 1.4205285078244383
Mean Absolute Error: 1.1943089431300098
R-Square Value: -0.08092046200958292
```

Fig 9 Predicted SoC value for KNN algorithm

Fig 10 shows line graph for predicted SoC values for KNN algorithm are used to track variations of numerical values between actual and predicted SoC measured.

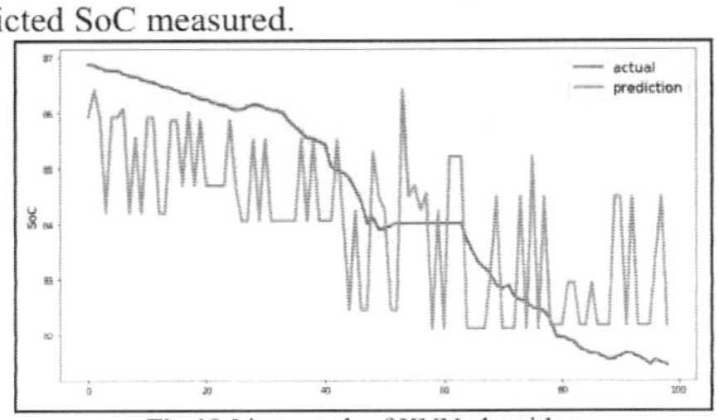

Fig 10 Line graph of KNN algorithm

Fig 11, shows the predicted State of charge values in Decision Tree algorithm, SoC predicted for particular voltage, current and temperature indicates the percentage of remaining charge in the battery.

```
[86.39 86.8  86.39 86.39 86.39 86.39 86.59 86.39 86.59 86.39 86.39 86.39
 86.39 86.39 86.59 86.59 86.3  86.59 86.3  86.59 86.3  86.3  86.3  86.3
 86.59 86.3  82.99 82.99 86.4  82.99 86.19 82.99 82.99 82.99 82.99 82.99
 86.4  82.99 86.19 82.99 82.99 82.99 86.19 82.99 81.7  84.7  81.7  81.7
 85.7  85.7  84.7  81.7  81.7  86.9  81.7  81.7  84.7  86.9  81.7  81.7
 81.7  86.2  86.2  86.2  81.7  81.7  81.7  81.7  83.6  86.11 81.7  81.7
 81.7  86.11 81.7  86.2  81.7  86.11 81.76 81.76 81.76 83.6  83.6  81.76
 81.76 83.6  81.76 81.76 81.76 86.11 86.2  81.76 86.2  81.76 81.76 81.76
 83.6  84.1  81.76]
```

Fig 11 Predicted SoC value for Decision Tree algorithm

Fig 12 shows line graph for predicted SoC values for Decision Tree algorithm are used to track variations of numerical values between actual and predicted SoC measured.

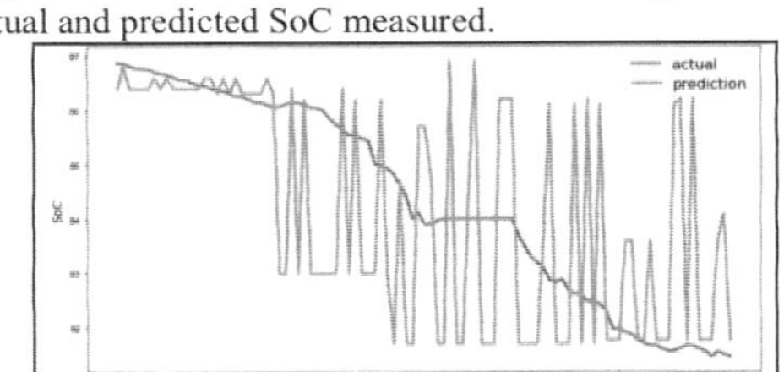

Fig 12 Line graph of Decision Tree algorithm

Fig 13 shows the error rate of the Decision Tree regression model is evaluated by calculating the regression metrics.

```
Mean Absolute Error:
 1.395710188079986

Mean Squared Error:
 3.5541389869098357

Root Mean Square Error:
 1.8852424212577639

R-Square Value:
 0.17663692712908186
```

Fig 13 Error values calculated for Decision Tree algorithm

Fig 14, shows the predicted State of charge values in Random Forest algorithm, SoC predicted for particular voltage, current and temperature indicates the percentage of remaining charge in the battery.

```
[86.058  86.8187 86.058  85.9117 86.058  86.058  86.3899 85.9117 86.4082
 85.9117 86.058  86.058  85.9117 85.9117 86.4578 86.4578 85.9683 86.438
 85.9683 86.4578 85.9683 85.9683 85.9683 85.9683 86.4578 85.9683 83.4094
 83.4094 86.2463 83.4094 86.2505 83.4094 83.4094 83.4094 83.4094 83.4094
 86.2463 83.4094 86.2505 83.4094 83.4094 83.4094 86.2505 83.4094 82.1226
 84.8695 82.1226 82.1226 85.704  85.7296 84.8695 82.1226 82.1226 86.2539
 82.6581 82.6581 84.8695 86.2819 81.7517 82.3087 81.7517 86.0098 86.0098
 86.0098 81.7517 81.7517 81.7517 81.7517 83.7464 86.0176 81.7517 81.7517
 81.7517 86.0176 81.7517 86.0098 81.7517 85.4442 81.7675 81.7675 81.7675
 83.7731 83.7731 81.7675 81.7675 83.7731 81.7675 81.7675 81.7675 85.4442
 85.4317 81.7675 85.4317 81.7675 81.7675 81.7675 83.801  84.116  81.7675]
```

Fig 14 Predicted SoC value for Random Forest algorithm

Fig 15 shows line graph for predicted SoC values for Random Forest algorithm are used to track variations of numerical values between actual and predicted SoC measured.

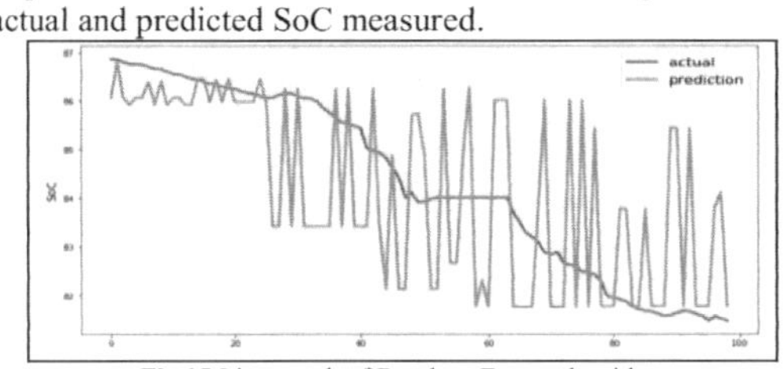

Fig 15 Line graph of Random Forest algorithm

Fig 16 shows the error rate of Random Forest regression model is evaluated by calculating the regression metrics.

```
Mean Absolute Error:
 1.3003849125466795

Mean Squared Error:
 2.7981428594561595

Root Mean Square Error:
 1.6727650341444131

R-Square:
 0.19074861343823546
```

Fig 16 Error values calculated for Random Forest algorithm

Fig 17 shows the predicted State of charge values in Linear Regression algorithm, SoC predicted for particular voltage, current and temperature indicates the percentage of remaining charge in the battery.

```
[[88.10797011]
 [91.52097746]
 [88.10797011]
 [88.15134879]
 [88.10797011]
 [88.10797011]
 [90.39037085]
 [88.15134879]
 [91.56435615]
 [88.15134879]
 [88.10797011]
 [88.10797011]
 [88.15134879]
 [88.19472748]
 [89.88081985]
 [89.92419854]
 [86.4678125 ]
 [93.33720589]
```

Fig 17 Predicted SoC value for Linear Regression algorithm

Fig 18 shows line graph for predicted SoC values for Linear Regression algorithm are used to track variations of numerical values between actual and predicted SoC measured.

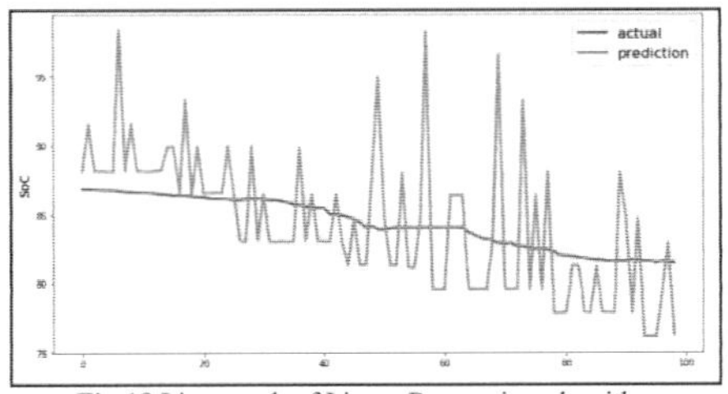

Fig 18 Line graph of Linear Regression algorithm

Fig 19 shows the error rate of the Linear Regression model is evaluated by calculating the regression metrics.

```
Mean Absolute Error:
 3.259415731339784

Mean Squared Error:
 17.362316460311

Root Mean Square Error:
 4.166811306060187

R-Square:
 0.3002320667459747
```

Fig 19 Error values calculated for Linear Regression algorithm

C. ANALYSIS 2: REAL-TIME DATASET 2

Analysis 2 consists of experimental results obtained from real-time dataset that is collected by integrating HESS model in the electric vehicle. Temperature, Voltage and Current are the battery parameters collected for both internal (Li-Ion battery setup) and external (Storage Battery) battery which is connected to Wind Turbine. The results contain predicted SoC values using regression algorithm.

Test Data for Analysis 2:

Fig 20 shows the set of test data obtained from the real-time dataset that is collected without integrating HESS model.

	volt	current	temp
0	502	0	66
1	502	1	66
2	502	0	66
3	502	1	66
4	504	0	66
...	...	...	...
94	515	3	66
95	515	0	66
96	515	0	66
97	515	0	66
98	515	0	66

99 rows × 3 columns

	voltage	current	temperature
0	249	0	94.4
1	249	0	94.4
2	249	1	94.4
3	249	0	94.4
4	249	2	94.4
...	...	...	...
94	253	0	97.2
95	253	2	97.2
96	253	1	97.2
97	253	0	97.2
98	253	3	97.2

99 rows × 3 columns

Fig 20 Set of test data obtained from Internal Battery and External Battery

Fig 21 shows the predicted State of charge values in support Vector machine algorithm, SoC predicted for particular voltage, current and temperature indicates the percentage of remaining charge in the battery.

```
[85.43976742 86.08855402 85.43976742 86.08855402 84.78409758 86.45099709
 84.78409758 84.7688602  83.10705194 84.7688602  83.10705194 86.8236557
 87.42734203 83.10705194 85.53044515 87.42734203 83.10705194 83.10705194
 85.41125029 84.38196852 86.39612293 84.38196852 86.39612293 84.38196852
 87.05439459 83.95202619 83.95202619 83.95202619 83.95202619 86.48923666
 83.95202619 83.95202619 86.39566227 83.51766729 83.51766729 83.51766729
 86.39566227 83.51766729 84.95998493 83.51766729 85.97276261 83.51766729
 87.11539256 82.47785227 87.84224082 82.47785227 82.47785227 82.47785227
 87.7713721  87.11539256 82.2660769  82.2660769  82.2660769  84.38520526
 82.2660769  84.38520526 82.2660769  82.2660769  87.19020824 82.2660769
 82.2660769  86.54363621 82.52541861 82.52541861 82.52541861 86.37783142
 82.52541861 84.43275592 82.52541861 86.15254759 82.52541861 82.52541861
 82.52541861 84.43275592 82.7948451  84.49945091 82.7948451  86.23824664
 82.7948451  84.49945091 82.7948451  84.49945091 82.7948451  86.77264589
 87.29410392 82.7948451  82.7948451  82.7948451  82.7948451  84.49945091
 82.7948451  82.7948451  82.7948451  82.7948451  87.29410392 82.7948451
 82.7948451  82.7948451  82.7948451 ]
```

```
[85.78664159 85.78664159 86.13133326 85.78664159 86.22747909 85.78664159
 86.22747909 85.78664159 85.78664159 86.13133326 85.78664159 85.53735481
 85.77113453 86.20577581 85.6622615  86.78387082 85.6622615  85.6622615
 85.6622615  86.70975974 85.6622615  86.70975974 85.6622615  87.24726242
 85.6622615  86.24864207 84.98186313 86.24864207 87.8569915  84.5062426
 86.44586649 84.5062426  86.26336252 84.5062426  84.5062426  84.5062426
 84.53237153 82.67566306 84.53237153 82.67566306 86.4246314  82.67566306
 87.27713602 82.67566306 87.87722216 82.67566306 82.34863042 82.34863042
 87.79051072 86.94932257 80.90859076 80.90859076 80.90859076 82.43841804
 80.90859076 82.43841804 80.90859076 80.90859076 86.94932257 80.90859076
 80.90859076 82.43841804 80.90859076 84.56269766 80.28085958 81.03480886
 80.28085958 81.03480886 80.28085958 85.92694407 84.47520175 80.28085958
 80.28085958 80.28085958 80.28085958 81.03480886 80.28085958 80.28085958
 81.03480886 80.28085958 81.03480886 80.28085958 83.32604791 81.55814927
 82.10774735 81.55814927 82.10774735 81.55814927 85.69705086 84.6274404
 81.55814927 85.35759557 84.6274404  81.55814927 81.55814927 83.32604791
 82.10774735 81.55814927 84.6274404 ]
```

Fig 21 Predicted SoC value of SVM algorithm for Internal and External Battery

Fig 22 shows line graph for predicted SoC values for SVM algorithm are used to track variations of numerical values between actual and predicted SoC measured.

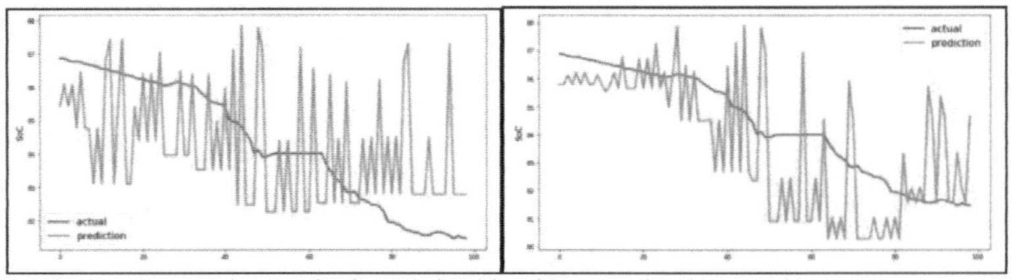

Fig 22 Line graph of SVM algorithm for Internal and External Battery

Fig 23 shows the error rate of the SVM regression model is evaluated by calculating the regression metrics.

```
Mean Absolute Error:
 1.7514381018007865

Mean Squared Error:
 4.35446759576148

Root Mean Square Error:
 2.0867361107148836

R-Square:
 -0.5126219679483746
```

```
Mean Absolute Error:
 1.588719671200316

Mean Squared Error:
 3.7182382774247436

Root Mean Square Error:
 1.9282733928114923

R-Square:
 0.3494689354254684
```

Fig 23 Error values calculated for SVM algorithm from Internal and External battery

Fig 24, 25 shows the predicted State of charge values and performance of K Nearest Neighbor algorithm, SoC predicted for particular voltage, current and temperature indicates the percentage of remaining charge in the battery.

```
 [81.7 ]
 [81.7 ]
 [81.7 ]
 [84.7 ]
 [81.7 ]
 [81.7 ]
 [81.7 ]
 [81.7 ]
 [86.19]
 [81.7 ]
 [81.7 ]
 [81.7 ]
 [81.7 ]]
Mean Squared Error: 6.766223115468663
RMSE value for k=  1 is: 2.6011964776749688
Mean Absolute Error: 2.0076512249757 2
R-Square Value: -0.4806738069494529 4
```

```
 [85.418 ]
 [83.019 ]
 [83.019 ]
 [83.019 ]
 [83.019 ]
 [84.1055]
 [83.019 ]
 [83.019 ]
 [83.019 ]
 [83.019 ]
 [85.418 ]
 [83.019 ]
 [83.019 ]
 [83.019 ]
 [83.019 ]]
Mean Squared Error: 3.701026974597994 5
RMSE value for k=  20 is: 1.923805336981368
Mean Absolute Error: 1.618299620404764 6
R-Square Value: -2.401053761154637 6
```

Fig 24 Predicted SoC value of KNN algorithm from Internal Battery

```
 [81.8 ]
 [81.74]
 [86.11]
 [84.1 ]
 [81.74]
 [84.06]
 [84.1 ]
 [81.74]
 [81.74]
 [82.4 ]
 [81.8 ]
 [81.74]
 [84.1 ]]
Mean Squared Error: 1.956581921206389
RMSE value for k=  1 is: 1.398778724890534 3
Mean Absolute Error: 0.995659567712972 5
R-Square Value: 0.586681399408771
```

```
 [82.2895]
 [82.1245]
 [82.2895]
 [82.1245]
 [84.497 ]
 [84.6065]
 [82.1245]
 [84.497 ]
 [84.6065]
 [82.1245]
 [82.1245]
 [83.7445]
 [82.2895]
 [82.1245]
 [84.6065]]
Mean Squared Error: 1.897466378904796 5
RMSE value for k=  20 is: 1.377485527657113 6
Mean Absolute Error: 1.143323169928518 5
R-Square Value: 0.206135530704608 1
```

Fig 25 Predicted SoC value of KNN algorithm for External battery

Fig 26 shows line graph for predicted SoC values for KNN algorithm are used to track variations of numerical values between actual and predicted SoC measured.

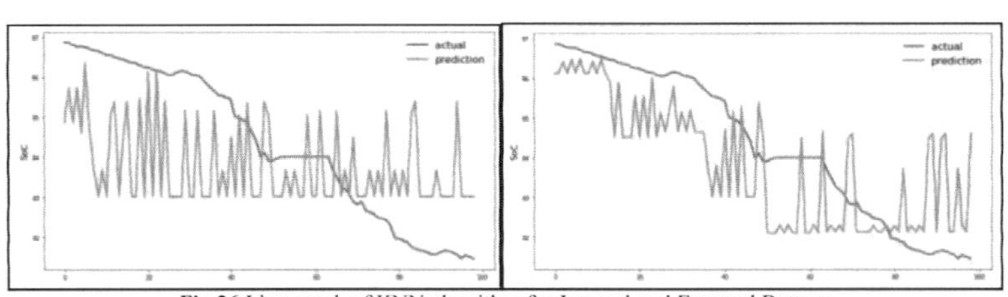

Fig 26 Line graph of KNN algorithm for Internal and External Battery

Fig 27 shows the predicted State of charge values in Decision Tree algorithm, SoC predicted for particular voltage, current and temperature indicates the percentage of remaining charge in the battery.

```
[82.99 86.64 82.99 86.64 82.99 86.6  82.99 86.1  82.99 86.1  82.99 86.4
 86.4  82.99 86.4  86.4  82.99 82.99 86.64 82.99 86.6  82.99 86.6  82.99
 86.4  82.99 82.99 82.99 82.99 86.4  82.99 82.99 86.4  82.99 82.99 82.99
 86.4  82.99 86.1  82.99 86.4  82.99 86.4  82.99 86.4  82.99 82.99 82.99
 86.4  86.4  82.99 82.99 82.99 86.1  82.99 86.1  82.99 82.99 86.4  82.99
 82.99 86.4  82.99 82.99 82.99 86.4  82.99 86.1  82.99 86.4  82.99 82.99
 82.99 86.1  82.99 86.1  82.99 86.4  82.99 86.1  82.99 86.1  82.99 86.4
 86.4  82.99 82.99 82.99 82.99 86.1  82.99 82.99 82.99 82.99 86.4  82.99
 82.99 82.99 82.99]
```

```
[86.39 86.39 86.5  86.39 86.79 86.39 86.79 86.39 86.39 86.5  86.39 86.79
 86.3  86.59 86.3  86.59 86.3  86.3  86.3  86.5  86.3  86.5  86.3  86.59
 86.3  86.2  82.94 86.2  86.59 82.94 86.2  82.94 86.59 82.94 82.94 82.94
 86.2  82.99 86.2  82.99 86.4  82.99 86.4  82.99 86.4  82.99 82.99 82.99
 86.4  85.7  81.74 81.74 81.74 81.74 81.74 81.74 81.74 81.74 85.7  81.74
 81.74 81.74 81.74 84.2  81.76 81.76 81.76 81.76 81.76 86.11 86.2  81.76
 81.76 81.76 81.76 81.76 81.76 81.76 81.76 81.76 81.76 81.76 84.2  81.76
 81.76 81.76 81.76 81.76 86.11 86.2  81.76 86.11 86.2  81.76 81.76 84.2
 81.76 81.76 86.2 ]
```

Fig 27 Predicted SoC value of Decision Tree algorithm for Internal and External Battery

Fig 28 shows line graph for predicted SoC values for Decision Tree algorithm are used to track variations of numerical values between actual and predicted SoC measured.

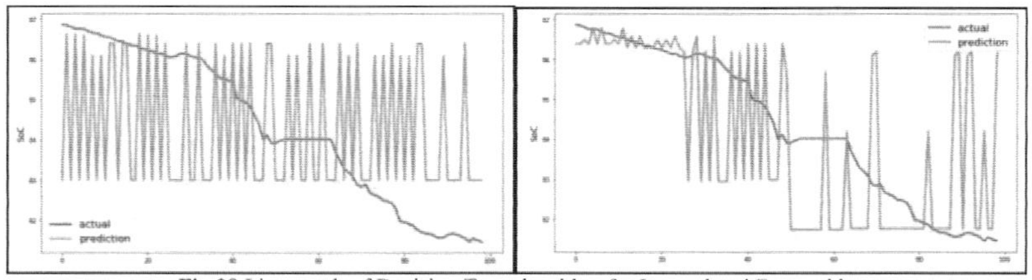

Fig 28 Line graph of Decision Tree algorithm for Internal and External battery

Fig 29 shows the performance of the DT regression model which is evaluated by calculating the regression metrics.

```
Mean Absolute Error:
 1.9075770346650471

Mean Squared Error:
 5.455875160217849

Root Mean Square Error:
 2.3357814881143844

R-Square Value:
 -1.0597673232534213
```

```
Mean Absolute Error:
 1.2834167529099287

Mean Squared Error:
 3.2423545282910706

Root Mean Square Error:
 1.8006539168566154

R-Square Value:
 0.29472722681575847
```

Fig 29 Error values calculated for Decision Tree algorithm from Internal and External Battery

Fig 30 shows the predicted State of charge values in Random Forest algorithm, SoC predicted for particular voltage, current and temperature indicates the percentage of remaining charge in the battery.

```
[83.3839    85.76666667 83.3839    85.76666667 83.3839    86.6353
 83.3839    85.46826667 82.9634    85.46826667 82.9634    86.2995
 86.2995    82.9634    86.2995    86.2995    82.9634    82.9634
 85.76666667 83.3839    86.6353    83.3839    86.6353    83.3839
 86.2995    82.9634    82.9634    82.9634    82.9634    86.2995
 82.9634    82.9634    86.3752    82.9634    82.9634    82.9634
 86.3752    82.9634    85.46826667 82.9634    86.2995    82.9634
 86.2995    82.9634    86.2995    82.9634    82.9634    82.9634
 86.2995    86.2995    82.9634    82.9634    82.9634    85.46826667
 82.9634    85.46826667 82.9634    82.9634    86.2995    82.9634
 82.9634    86.3752    82.9634    82.9634    82.9634    86.3752
 82.9634    85.46826667 82.9634    86.2995    82.9634    82.9634
 82.9634    85.46826667 82.9634    85.46826667 82.9634    86.3752
 82.9634    85.46826667 82.9634    85.46826667 82.9634    86.2995
 86.2995    82.9634    82.9634    82.9634    82.9634    85.46826667
 82.9634    82.9634    82.9634    82.9634    86.2995    82.9634
 82.9634    82.9634    82.9634    ]
```

```
[86.0677 86.0677 86.3483 86.0677 86.782  86.0677 86.782  86.0677 86.0677
 86.3483 86.0677 86.7784 86.089  86.438  85.9729 86.438  85.9729 85.9729
 85.9729 86.3557 85.9729 86.3557 85.9729 86.4621 85.9729 84.9175 83.8782
 84.9175 86.4633 83.8782 84.9175 83.8782 86.4392 83.8782 83.8782 83.8782
 84.2858 83.2125 84.2858 83.2125 86.3136 83.2125 86.3136 83.2125 86.3136
 83.2125 83.2125 83.2125 86.3136 85.7524 81.7466 81.7466 81.7466 81.7042
 81.7466 81.7042 81.7466 81.7466 85.7524 81.7466 81.7466 81.7042 81.7466
 84.2504 81.7675 81.7392 81.7675 81.7392 81.7675 85.6542 85.6348 81.7675
 81.7675 81.7675 81.7675 81.7392 81.7675 81.7675 81.7392 81.7675 81.7392
 81.7675 84.3141 81.7675 81.7392 81.7675 81.7392 81.7675 85.6542 85.6348
 81.7675 85.6542 85.6348 81.7675 81.7675 84.3141 81.7392 81.7675 85.6348]
```

Fig 30 Predicted SoC value of Random Forest algorithm for Internal and External Battery

Fig 31 shows line graph for predicted SoC values for Random Forest algorithm are used to track variations of numerical values between actual and predicted SoC measured.

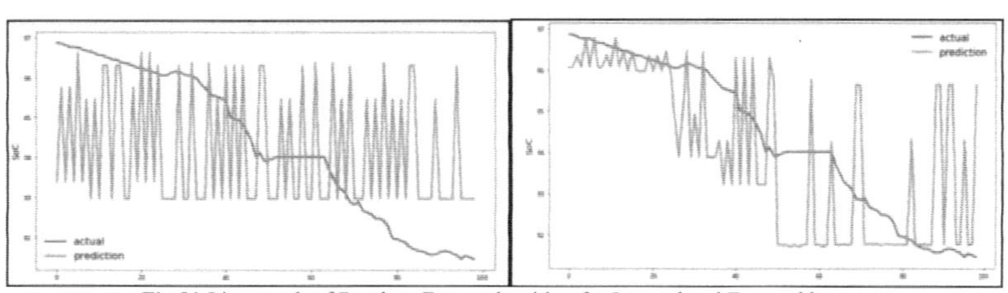

Fig 31 Line graph of Random Forest algorithm for Internal and External battery

Fig 32 shows the error rate of the Random Forest regression model is evaluated by calculating the regression metrics.

Mean Absolute Error: 1.8459793604570354	Mean Absolute Error: 1.2536975105783963
Mean Squared Error: 4.954465722840628	Mean Squared Error: 2.7072171916780374
Root Mean Square Error: 2.225862916452994	Root Mean Square Error: 1.6453623283878958
R-Square: -1.2234888768100172	R-Square: 0.2996660228252863

Fig 32 Error values calculated for Random Forest algorithm from Internal and External Battery

Fig 33 shows the predicted State of charge values in Linear Regression algorithm, SoC predicted for particular voltage, current and temperature indicates the percentage of remaining charge in the battery.

```
[[81.90593013]
 [84.19106106]
 [81.90593013]
 [84.19106106]
 [81.92953723]
 [86.4997991 ]
 [81.92953723]
 [84.26188236]
 [81.97675143]
 [84.26188236]
 [81.97675143]
 [93.40240609]
 [88.83214423]
 [81.97675143]
 [97.97266796]
 [88.83214423]
 [81.97675143]
 [81.97675143]
 [84.22647171]
 [81.94134078]
 [86.51160264]
 [81.94134078]
 [86.51160264]
 [81.94134078]
 [98.80853713]
 [81.95314433]
 [81.95314433]
```

```
[[88.37135889]
 [88.37135889]
 [90.57520361]
 [88.37135889]
 [92.77904833]
 [88.37135889]
 [92.77904833]
 [88.37135889]
 [88.37135889]
 [90.57520361]
 [88.37135889]
 [99.33109371]
 [86.10802541]
 [97.1646319 ]
 [86.14540831]
 [94.96078718]
 [86.14540831]
 [86.14540831]
 [86.14540831]
 [88.34925302]
 [86.14540831]
 [88.34925302]
 [86.14540831]
 [90.55309774]
 [86.14540831]
 [86.46314179]
 [84.25929707]
 [86.46314179]
 [88.7043694 ]
```

Fig 33 Predicted SoC value of Linear Regression algorithm for internal and external battery

Fig 34 shows line graph for predicted SoC values for Linear Regression algorithm are used to track variations of numerical values between actual and predicted SoC measured.

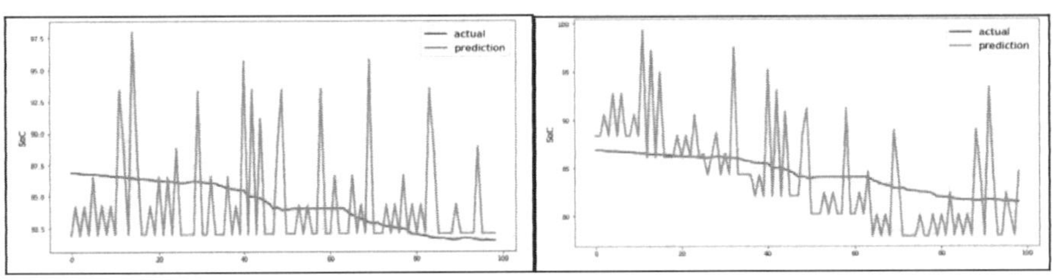

Fig 34 Line graph of Linear Regression algorithm for Internal and External battery

Fig 36 shows the error rate of the Linear Regression model is evaluated by calculating the regression metrics.

Mean Absolute Error: 3.1206647564015313	Mean Absolute Error: 3.3982528497713678
Mean Squared Error: 17.599814409451664	Mean Squared Error: 18.48002379804942
Root Mean Square Error: 4.1952132734167	Root Mean Square Error: 4.298839820003697
R-Square: -0.17637053029505378	R-Square: 0.3237399463402729

Fig 35 Error values calculated for Linear Regression algorithm from Internal and external Battery

IOP Conf. Series: Materials Science and Engineering 1291 (2023) 012042 doi:10.1088/1757-899X/1291/1/012042

D. ANALYSIS OF STATE OF CHARGE (SoC):

State of charge (SoC) is the amount of charge present in battery with respect to its capacity. 0% charge indicates the charge is empty and 100% implies full.

Table 1 table contains SoC estimated values in both analysis 1 and 2. It show that the report of estimated State Of Charge (SoC) in Random Forest Algorithm. The performance of random forest is better in predicting the SoC. The performance is analyzed using regression metrices.

	Battery Voltage (mV)	Battery Current (A)	Battery Temperature(°C)	State Of Charge (%)
Analysis-1	554	0	84	86.1
	554	1	84	86.82
	554	0	84	86.1
	555	0	84	86
	554	0	84	86
Analysis-2 (Internal Battery)	502	0	66	83.4
	502	1	66	85.8
	502	0	66	83.4
	502	1	66	85.5
	504	0	66	83
Analysis-2 (External Battery)	249	0	94.4	86.067
	249	0	94.4	86.067
	249	1	94.4	86.35
	249	0	94.4	86.067
	249	2	94.4	86.782

Table 1 Comparison of estimated State Of Charge (SoC)

E. ANALYSIS OF REGRESSION METRICES

The mathematical metrics which are used to identify the performance of the software model proposed is given below.

Mean Absolute Error:

MAE provides absolute difference between actual and predicted values.

$$MAE = \frac{1}{N}\sum_{i=1}^{N} 1\ |y_i - \hat{y}_i| \qquad (1)$$

Where y_i is the i'th expected value in the dataset, $\hat{y}_i$ is the i'th predicted value.

Mean Square Error:

MSE is the squared difference between actual and predicted value.

$$\text{MSE} = \frac{1}{n}\sum_{i=1}^{n}(\hat{y}_i - y_i)^\wedge 2 \qquad (2)$$

Root Mean Square Error:

$$RMSE = \sqrt{\frac{1}{n}\sum_{i=1}^{n}(y_i{\_}\hat{y}_i)^\wedge 2} \qquad (3)$$

Where *sqrt()* is the square root function.

R-Square Score:

R-Square Value is a metric that tells the performance of your model.

R 2 = 1 − sum squared regression / total sum of squares

$$SSR = \sum_{i=1}^{n}(y_i - \hat{y}_i)^\wedge 2 \qquad (4)$$

$$\text{SST} = \sum_{i=1}^{n}(y_i - \bar{y})^\wedge 2 \qquad (5)$$

Where y is each observed value y_i minus the average of observed values

Analysis Report for Regression Metrices

Table 6.2 show that the performance of the different algorithms for both data sets in terms of the above defined metrics. The analysis table shows that Random Forest algorithm produces less than 2.5% of Mean Absolute Error when tested using both the datasets. To conclude the performance of the algorithms, MAE is considered and shows less than 2.5% error for Random forest algorithm.

Algorithms	MAE	MSE	RMSE	R2S
Analysis 1: Real-time Dataset 1				
Support Vector Machine	1.414	3.001	1.732	0.361
K Nearest Neighbour	2.017	1.42	1.194	-0.08
Decision Tree	1.395	3.554	1.885	0.176
Random Forest	**1.3**	**2.798**	**1.672**	**0.19**
Linear Regression	3.259	17.362	4.166	0.3
Analysis 2: Real-time Dataset 2				
Internal Battery				
Support Vector Machine	1.845	4.954	2.225	-1.223
K Nearest Neighbour	3.701	1.923	1.618	-2.401
Decision Tree	1.907	5.455	2.335	-1.059
Random Forest	**1.751**	**4.354**	**2.086**	**-0.513**
Linear Regression	3.120	17.59	4.195	-0.176
External Battery				
Support Vector Machine	1.588	3.718	1.928	0.349
K Nearest Neighbour	2	1.377	1.143	0.206
Decision Tree	1.283	3.242	1.8	0.294
Random Forest	**1.253**	**2.707**	**1.645**	**0.299**
Linear Regression	3.398	18.48	4.298	0.323

Table 2 shows the comparison of regression metrices

V. CONCLUSION AND FUTURE WORK

A. CONCLUSION

In this paper the energy storage system is designed in order to predict the capacity of the Li-Ion battery. Which was developed and the ML algorithms used predicts the aging of the Battery. The Regression metrics such as Mean Absolute Error (MAE), Root Mean Square (RMSE), Mean Square Error (MSE), R-Square Error(R2E) are calculated and the SoC values are predicted. The Random Forest algorithm produces better results for the HESS model as the Mean Absolute Error values produced are less than 2.5%. This method helps the in estimating the driving range of the EV by predicting the remaining battery capacity. This technique of predicting the remaining charge is highly applicable for increasing the driving range during real time implementation.

B. FUTURE WORK

In further this work can be extended by implementing this technique using Deep Learning algorithms. The electric vehicle model can be created using simulation tool and the deep learning algorithms such as Convolutional Neural Networks (CNNs) , Artificial Neural Networks (ANNs) and Recurrent Neural Networks (RNNs), may produce more accurate results in estimating the remaining charge.

REFERENCES

1] S. Harippriya, E. Esakki Vigneswaran, S. Jayanthy, "Battery Management System to Estimate Battery Aging using Deep Learning and Machine Learning Algorithms", Journal of Physics: Conference Series, July, 2022

2] Prakash Venugopal, S. Siva Shankar , C. Phillip Jebakumar, Rishab Agarwal, Hassan Haes Alhelou, S. Sofana Reka, And Mohamad Esmail Hamedani Golshan, “Analysis of Optimal Machine Learning Approach for Battery Life Estimation of Li-Ion Cell”, IEEE Access, November, 2021.

3] T. Mesbahi, P. Bartholomeüs, N. Rizoug, R.Sadoun, F.Khenfri, P. Le Moigne, “Advanced Model of Hybrid Energy Storage System Integrating Lithium-ion Battery and Supercapacitor For Electric Vehicle Applications”, IEEE Transactions on Industrial Electronics, May, 2021.

4] Woong Lee, Haeseong Jeoung, Dohyun Park , Tacksu Kim, Heeyun Lee, And Namwook Kim, “A Real-Time Intelligent Energy Management Strategy for Hybrid Electric Vehicles Using Reinforcement Learning”, IEEE Access, May, 2021.

5] Huilong Yu, Francesco Castelli-Dezza, Federico Cheli, Xiaolin Tang, Xiaosong Hu, and Xianke Lin, “Dimensioning and Power Management of Hybrid Energy Storage Systems for Electric Vehicles with Multiple Optimization Criteria”, IEEE power electronics regular paper, May, 2021.

6] Min-Joon Kim, Sung-Hun Chae and Yeon-Kug Moon, “Adaptive Battery State-of-Charge Estimation Method for Electric Vehicle Battery Management”, International SoC Design Conference (ISOCC) February 2021.

7] Shiyao Zhou, Ziqiang Chen, Deyang Huang, Tiantian Lin, “Model Prediction and Rule Based Energy Management Strategy for a Plug-in Hybrid Electric Vehicle with Hybrid Energy Storage System “, IEEE Transactions on Power Electronics, October, 2020.

8] Yunfei Bai, Jianwei Li, Hongwen He, Ricardo Caneloi Dos Santos And Qingqing Yang, “optimal design of a hybrid energy storage system in a plug-in hybrid electric vehicle for battery lifetime improvement”, IEEE Access, August 2020.

9]Fen Zhao, Yinguo Li, Xinheng Wang, Ling Ba, And Tailin Liu, “Lithium-Ion Batteries State of Charge Prediction of Electric Vehicles Using RNNs-CNNs Neural Networks”, IEEE Access, May, 2020

10] Training Dataset Link: Link

Design and Comparison of three Unconventional Shaped DRAs for 60GHz Application

[1]Reyya Sohit Rao, [2]Rashmi Ranjan Sahu, [3]Shashi Shankar Mohanty, [4]Ribhu Abhusan Panda

[1]21ece008.reyyasohitrao@giet.edu,

[2]21ece015.rashmiranjansahu@giet.edu,

[3]21ece017.shasishankarmohanty@giet.edu

[4]ribhuabhusanpanda@giet.edu,

[1,2,3,4,] Department of Electronics and Communication Engineering, GIET University, Gunupur

Abstract. Shape modification and analysis of a conventional circular dielectric resonating antenna for 60 GHz application has been included in this paper. With The conformist shapes have been modified and a comparative study has been done among the unconventional designs based on scattering parameter, antenna gain and directivity. The circular shaped DRA has been modified as Biconvex, Plano convex and biconcave shaped DRA. With the dimension 20 mm × 20 mm × 1.6 mm, FR4 Epoxy material is used for the substrate. Alumina_92pct material with a dielectric constant of 9.2 has been used for the dielectric resonators. Frequency range from 55 GHz to 65 GHz has been considered and the resonant frequencies have been found as 58.7 GHz, 59.7 GHz and 55.5 GHz for Biconvex DRA, Plano Concave DRA and Biconcave DRA respectively. A comparative study has been done among all these three DRAs to find out the best DRA for 60 GHz WLAN application.

Keywords: DRA, alumina_92pct, S-Parameter, Antenna Gain, Directivity.

Introduction:

In recent past the development in the design related to dielectric resonating antennas has provided perturbed shaped DRAs for different applications. In the year 2021, dual mode cylindrical DRA has been designed. [1] A slot coupled annular DRA has been designed for both linear and circular polarisation [2]. The dielectric resonator has been reshaped for specific wireless applications [3-9]. In this paper three novel shaped DRAs have been designed based on arc to arc length. The arc to arc length has been taken into consideration after determination of the wavelength λ based on the designed frequency 60 GHz. The proposed antennas have been simulated by using HFSS software in the frequency range 55 GHz to 60 GHz.

Antenna Design:

Three unconventional shapes of dielectric resonator have been design on a substrate with length of 20 mm and width of 20 mm. The substrate has height of 1.6 mm. FR4 epoxy material which is having a dielectric constant 4.4.

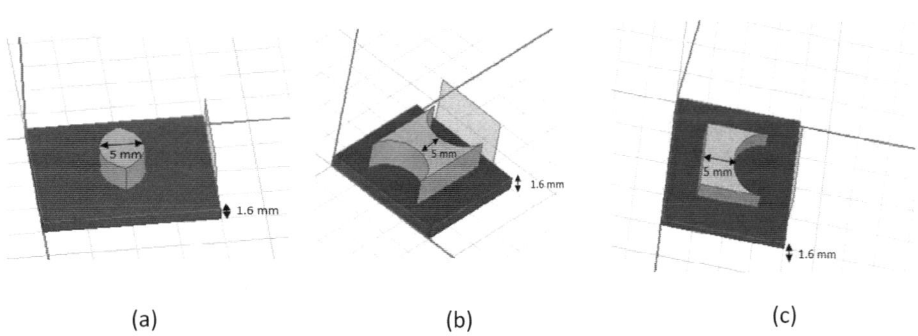

Figure 1. Design of Novel shaped DRA. (a) Bi Convex DRA. (b) Bi Concave DRA. (c) Plano Concave DRA

IOP Conf. Series: Materials Science and Engineering 1291 (2023) 012043 doi:10.1088/1757-899X/1291/1/012043

Result Analysis:

S-Parameter plays a very vital role for measuring the effectiveness of the DRA. Simulation has been done using the Ansys HFSS. Assessment of the S-Parameter is illustrated in figure 2. From the figure it is well observed that the Plano concave DRA has a sharp resonance at the frequency of 59.7 GHz with return loss of -19.2 dB. Compared to biconvex and biconcave DRA, Plano concave has better performance based on simulated parameters. The assessment parameters have been included in table 1. Comparison of antenna gain has been shown in figure 3. The antenna gain of the Plano concave DRA is around 4.03 dB which is more compared to other two designs. The antenna gain remains in the range of 3 dB to 4 dB within the frequency range 58 GHz to 60 GHz which makes the Plano concave DRA efficient one. Comparison of parameters has been included in Table 1.

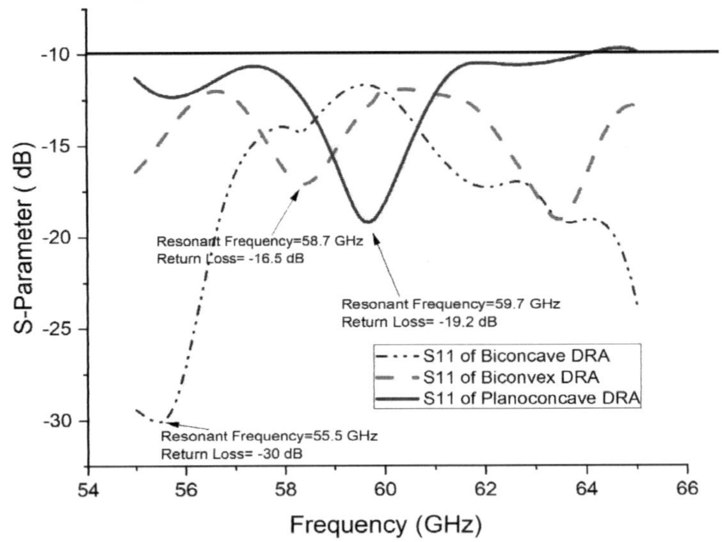

Figure 2. S-Parameters Comparison.

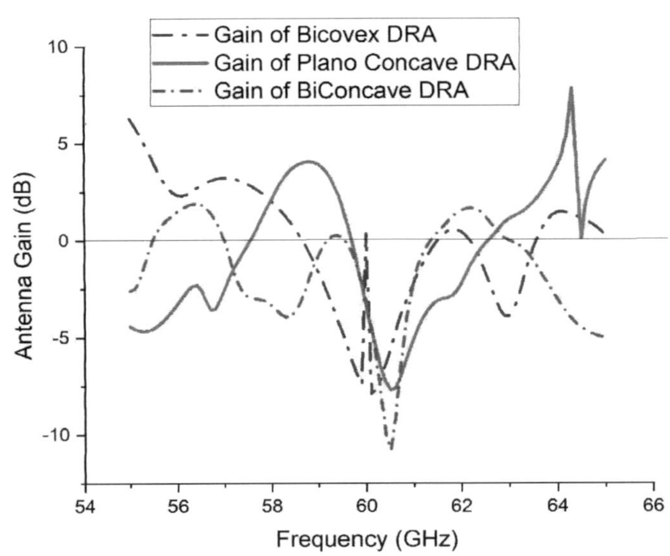

Figure 3. Antenna Gain with respect to frequency.

Table 1. Comparison of Parameters.

Parameters	Unconventional Shaped DRA		
	Biconvex	Biconcave	Plano Concave
Resonant Frequency (GHz)	58.7	55.5	59.7
Return Loss (dB)	-16.5	-30	-19.2
VSWR	1.83	1.56	1.24
Antenna Gain (dB)	1.92	1.88	4.03
Directivity (dB)	3.33	3.93	5.45
Area of the DRA (mm^2)	47.12	90	45
Area of the Surface (substrate)(mm^2)	150	150	150
Peak Radiation Efficiency (%)	32.25	32.7	60.4
Volume Current Density Distribution in Dielectric Resonator	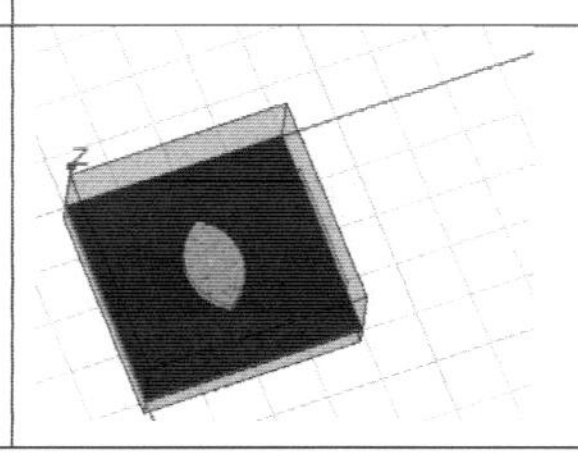	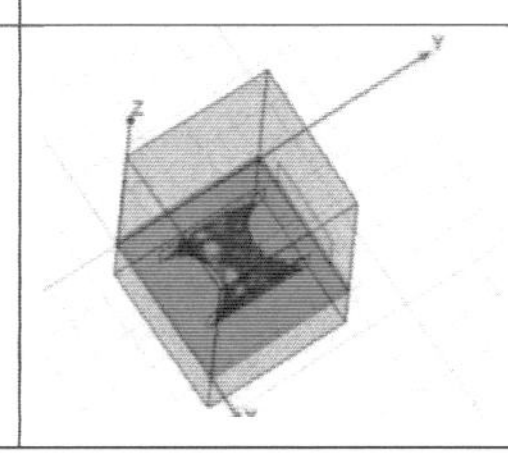	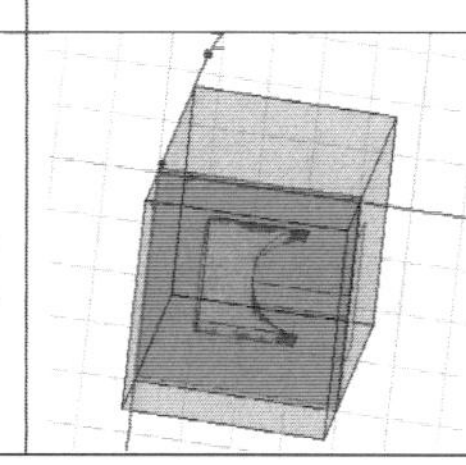

VSWR of Plano Concave DRA has been shown in figure 3. For the Plano concave patch the VSWR has been found to be 1.24. The desired VSWR value is 1 and the result is nearer to the desired value.

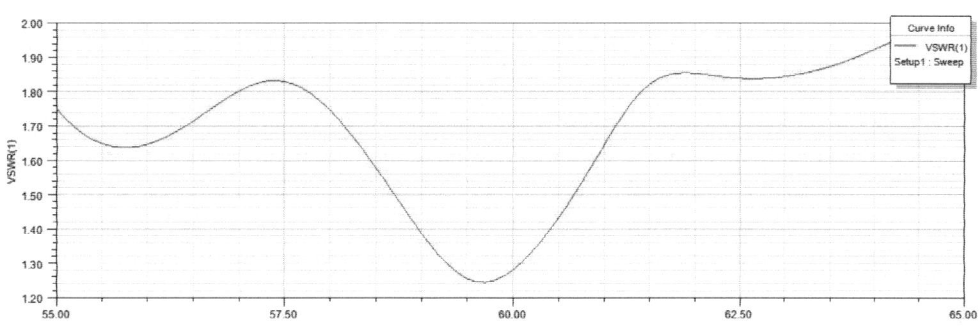

Figure 4. VSWR of Plano concave Patch.

The simulation results has indicated that Plano concave Dra has better antenna gain compared to other two unconventional shaped DRAs.The radiation efficiency of the proposed plano concave DRA is also above 60 %. The polar plots from the simulation have been shown in figure 4 to figure 6.

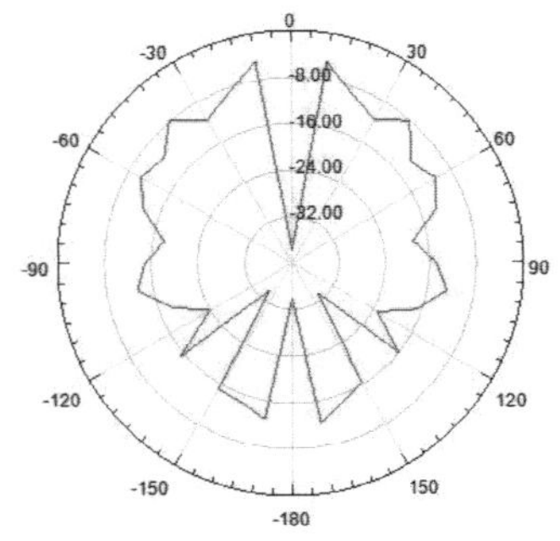

Figure 5. 2D Radiation Pattern at Phi=0^0.

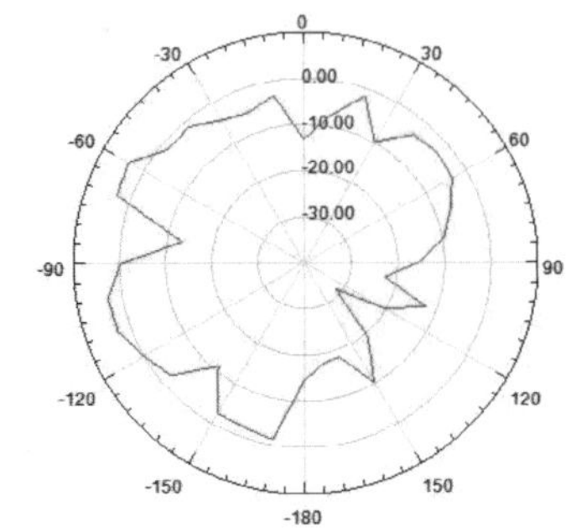

Figure 6. 2D Radiation Pattern at Phi =90.

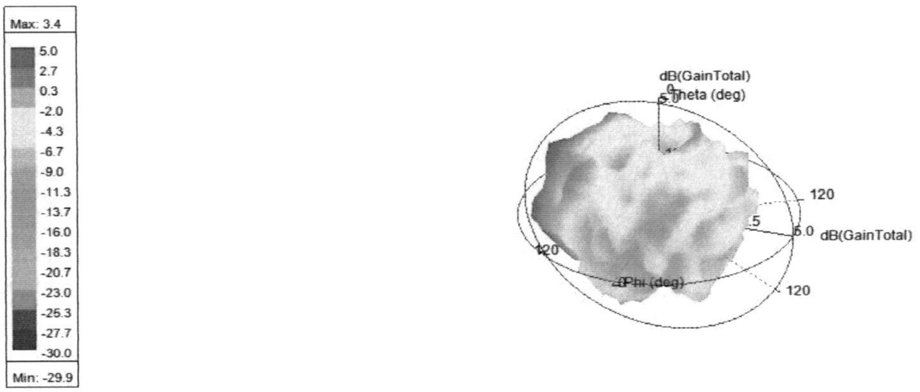

Figure 7. 3D Radiation Plot.

Conclusion:

The paper includes three unconventional shaped designs and the comparison has been done. From the comparison it has been found that the Plano Concave DRA has high antenna gain of 4.03 dB as well as a sharp resonant frequency at 59.7 GHz. The directivity of the proposed antenna is 5.45 dB so it can be efficiently used for 60 GHz WLAN, WiGig networks.

References:

[1] P. Gupta, D. Guha and C. Kumar, "Dual-Mode Cylindrical DRA: Simplified Design With Improved Radiation and Bandwidth," in *IEEE Antennas and Wireless Propagation Letters*, vol. 20, no. 12, pp. 2359-2362, Dec. 2021, doi: 10.1109/LAWP.2021.3110875.

[2] B. Ghosh, D. Bhattacharya, P. D. Sinha and D. H. Werner, "Design of Circular Waveguide Annular Slot-Coupled Two-Layer DRA for Linear and Circular Polarizations," in *IEEE Antennas and Wireless Propagation Letters*, vol. 19, no. 6, pp. 1012-1016, June 2020, doi: 10.1109/LAWP.2020.2986375.

[3] Panda, R.A., Patnaik, A. (2022). Design and Analysis of a Biconcave DRA by Using Machine Learning Algorithms for 5G Application. In: Udgata, S.K., Sethi, S., Gao, XZ. (eds) Intelligent Systems. Lecture Notes in Networks and Systems, vol 431. Springer, Singapore. https://doi.org/10.1007/978-981-19-0901-6_19

IOP Conf. Series: Materials Science and Engineering 1291 (2023) 012043 doi:10.1088/1757-899X/1291/1/012043

[4] Panda, R.A., Behera, P.P., Nayak, D., Patnaik, R.K. (2022). Design and Analysis of a Plano Concave DRA for 60 GHz Application. In: Suma, V., Fernando, X., Du, KL., Wang, H. (eds) Evolutionary Computing and Mobile Sustainable Networks. Lecture Notes on Data Engineering and Communications Technologies, vol 116. Springer, Singapore. https://doi.org/10.1007/978-981-16-9605-3_30

[5] Panda, R.A., Mishra, D. Mathematical analysis for resonant frequencies of a novel 3-element log periodic patch array. *Int J Syst Assur Eng Manag* (2022). https://doi.org/10.1007/s13198-022-01716-z

[6] Panda, R.A., Mishra, D. (2022). Mathematical Analysis and Optimization of a Remodeled Circular Patch for 5G Communication. In: Malik, P.K., Lu, J., Madhav, B.T.P., Kalkhambkar, G., Amit, S. (eds) Smart Antennas. EAI/Springer Innovations in Communication and Computing. Springer, Cham. https://doi.org/10.1007/978-3-030-76636-8_9

[7] Panda, R.A., Mishra, D. Efficient Design of Bi-circular Patch Antenna for 5G Communication with Mathematical Calculations for Resonant Frequencies. *Wireless Pers Commun* **112**, 717–727 (2020). https://doi.org/10.1007/s11277-020-07069-9

[8] P. R. Meher, B. R. Behera and S. K. Mishra, "Broadband Circularly Polarized Cylindrical Dielectric Resonator Antenna for Wideband Applications," *2019 TEQIP III Sponsored International Conference on Microwave Integrated Circuits, Photonics and Wireless Networks (IMICPW)*, 2019, pp. 524-528, doi: 10.1109/IMICPW.2019.8933171.

[9] P. R. Meher, B. R. Behera and S. K. Mishra, "Design of Different Shaped DRAs for 60 GHz Millimeter-wave Applications," *2018 IEEE Indian Conference on Antennas and Propogation (InCAP)*, 2018, pp. 1-4, doi: 10.1109/INCAP.2018.8770785.

IOP Conf. Series: Materials Science and Engineering 1291 (2023) 012044 doi:10.1088/1757-899X/1291/1/012044

Design of a single layer single band 5G FSS as shielding materials

Esther Florence[1,a], R Samuel Robinson[2,b*], R Vimal Samsingh[2,c] and Sakthi Abirami [3,d]

[1] Department of ECE, College of Engineering, Guindy, Chennai
[2] Sri Sivasubramaniya Nadar College of Engineering, Kalavakkam, Chennai,
[3] Department of ECE, Amrita school of engineering, Amrita Vishwa Vidyapeetham, Chennai

[a]esthervimal24@gmail.com, [b]samuel2010019@ssn.edu.in, [c]vimalsamsinghr@ssn.edu.in, [d]sakthiabirami1993@gmail.com

Abstract. This paper deals with a novel Frequency Selective Surface (FSS) Electromagnetic shield that operates in the low 5G vertical applications. This FSS is designed for the purpose ofserving as an electromagnetic shield to eliminate EMI. The FSS designed operates at a frequency of 5GHz with a bandwidth of 1.45GHz providing a shielding of more than 40dB. The proposed FSS uses a polyester sheet as substrate and uses silver for the metallization of the FSS. Since the proposed FSS is symmetric, it behaves similarly when a Transverse Electric (TE) or a Transverse Magnetic (TM) wave is incident. The structure was designed and simulated using CST software. The results show that this FSS blocks electromagnetic radiationat lower frequencies as intended.

1. Introduction

5G refers to a set of standards for the operation of mobile wireless. 5G is a new generation of cellular networking. It promises high data speeds and seamless connectivity wherever we are [2]. Compared toprevious cellular technologies like 2G,3G or 4G; 5G operates at a much higher frequency.

The National Communications Commission has recently finished gathering public input on the potential use of the frequencies between 4.8-4.9GHz for 5G vertical applications. 5G has many uses for several industries like cellular communication. The various uses for 5G technology are immense. 5G has the potential to enable technologies such as massive IoT, connected cars and many others which are yet to be determined. But there are some applications where transmissions in the 5G spectrum would not be desired. Stray EM waves in 5G frequencies interfere with the operation of existing antennas and sensors working in the same spectrum. It is known to interfere with ADAS and collision avoidance systems in cars [3]. There also exists a considerable overlap in the operating frequencies of 5G devices and the frequencies of operation of aviation-grade radios. There exists a riskof interference between terrestrial communication and crucial aviation communication links and interfere with airplane traffic and could cause mishaps. With the proliferation of 5G systems in metropolitan areas and areas which witness high air-traffic, flights in those regions are experiencing a higher amount of interference in their mission critical systems, like radio altimeter, aircraft radio, radars etc.[4]-[9]. The Federal Aviation Administration (FAA) had to set new operational directives to adapt to the changes [9]. 5G technology has also been postulated to affect human physiology. But research on the claim varies widely among the members of the scientific community [8]. The author Wersényi, György proposes several experiments that could shed light on the interaction of the human body with electromagnetic radiation but also shares their limitations. This field of study is a highly contested one with diverse and sometimes even biased views held by different members.

IOP Conf. Series: Materials Science and Engineering 1291 (2023) 012044 doi:10.1088/1757-899X/1291/1/012044

Still, there is a need for shielding of 5G frequency waves to protect critical equipment or biological tissues that might be affected by the transmission frequencies. Using a FSS based materialas a shield can decrease the electromagnetic radiation emissions in frequencies that are not present in band used for communication while enhancing the susceptibility from external radiated fields, thereby improving the overall performance of the device [7].

A FSS (Frequency Selective Surfaces) is an assembly of miniature elements arranged in a periodic manner in an array to achieve bandpass or band stop EM characteristics [1]. It is a burgeoningfield and has witnessed intense study since its inception. In this paper a novel FSS based shield is designed and the characteristics of the same is studied. The structure of the FSS is elaborated in Section 2 and the results are elucidated in Section 3.

2. Structure of FSS

The FSS was designed using CST Microwave Studio Suite. The FSS designed is a single layer FSS with polyester as the substrate and silver as the metallized element. Silver is a good conductor of electricity and has an electric conductivity value of 6.3012 x 10^7 S/m. The FSS has a thickness (H2) of 0.06mm. The substrate material 'polyester' has electrical permittivity of 3.3. The thickness of the polyester sheet used is H1 = 0.1mm. The advantage of polyester is that they are flexible and hence the designed FSS material can be conformed onto any surface.

Figure 1 shows the top view of the FSS unit cell designed. The grey regions are the silver conductive sheets and form the unit cell of the FSS.

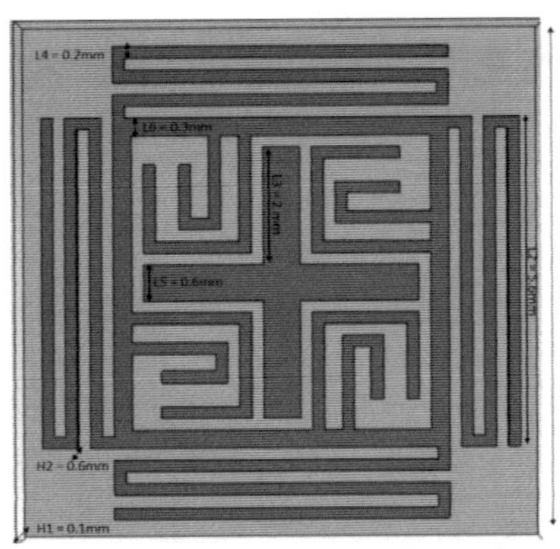

Fig. 1. Main FSS with dimensions

(a) (b) (c)

Fig 2. Evolution of the FSS (a) Evolution stage 1 (b) Evolution stage 2 (c) Evolution stage 3

IOP Conf. Series: Materials Science and Engineering 1291 (2023) 012044 doi:10.1088/1757-899X/1291/1/012044

The first step to design the FSS was to construct a simple cross structure of arm length L3 = 2mm and arm width L5 = 0.6mm and enclose it in square box of side length L2 = 5.6mm as shown in Figure 2(a). The result obtained by the simulation was not satisfactory as seen from Figure 3. The absorption characteristics did not show a minimum value at the frequency of interest. So inverted J shaped loading elements were added to the empty space between the cross and the square to fill it. To increase the attenuation in the transmission characteristics even more, meandered rectangular elements were added to the edges of the outer square; each of length L2 = 5.6mm and thickness L4 =0.2mm. The total length of the FSS unit cell became L1 = 8.6mm. All the measurements are mentioned in Table 1. The final FSS unit cell is shown in Figure 2(c).

Table 1. Table with dimensions

Parameter	Value (mm)	Parameter	Value (mm)
L1	8.6	L2	5.6
L3	2	L4	0.2
L5	0.6	L6	0.3
H1	0.1	H2	0.06

3. Simulation and Measured Results

To validate the proposed FSS, the unit cell is simulated using the electromagnetic (EM) simulation software CST Microwave Studio. The proposed unit cell is simulated with CST Software Suite with Floquet ports and boundary conditions to simulate an infinite array.
The CST solver uses Finite Integration Technique (FIT). FIT method was presented by Weiland in 1977. It discretizes the Maxwell-Heaviside equations which are in continuous form and represents them in dual-orthogonal grids [5,6]. Weiland introduced this novel technique as a Maxwell's equationsanalogue to ensure the physical consistency of computed fields. This method, an extension of the Finite Difference Time Domain (FDTD) approach and akin to the Finite Element Method (FEM), yields computationally efficient and memory-effective solutions.
The advantage of using the FIT method is that each further calculation of the time step requires one matrix-vector multiplication and as a result is advantageous as there are many matrix solvers capable of performing calculation at high speed [5,6]

3.1. Transmission Parameters
To achieve resonance at the frequency 5GHz, the FSS unit cell is designed as a simple cross in a square box. The frequency response of the FSS designed was inadequate. So, loading elements were used to give a staggered swastika shape with inverted J-shaped stubs. This increased the attenuation meanwhile resulting in shift of resonant frequency to 7GHz, which is greater than desired frequency range.
To lower the resonant frequency, meandered rectangular loading elements were used. The meandered rectangular elements were added to the outsides of the FSS to shift the operating frequency to the lower ranges. This resulted in the FSS meeting the design parameters. The transmission characteristicsas shown in Figure 3 showed increased absorption (minimum transmission) at the 5G frequency of 4.9GHz.

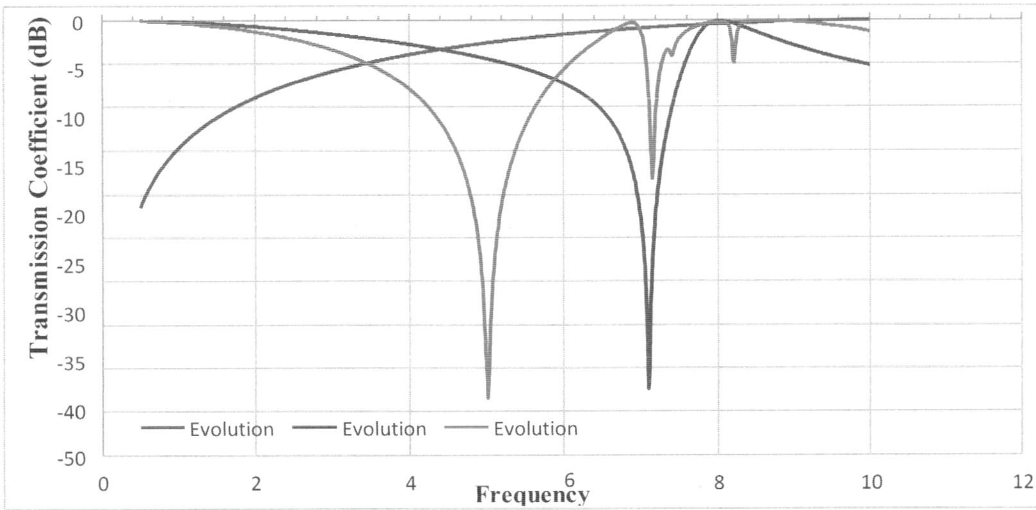

Fig 3. Transmission parameters of the different FSS evolutions designed

The simulated FSS has a transmission minimum of 45 dB at 4.98GHz. The bandwidth if the resulting structure is 1.45GHZ. The FSS has good polarization stability due to the symmetric shape of the structure. The polarization stability exhibits remarkable performance under normal incidence conditions, primarily attributable to the inherent symmetry present in the unit cell.

3.2. Surface Currents

To study the characteristics of the FSS with better clarity the surface current distribution of the FSS at the resonant frequency is plotted. The surface current distribution on the metallized regions of the FSS without the meandered rectangles (i.e.) the second stage of evolution of the design is shown in Fig 4a and 4b. The figures highlight the distributions at the resonant frequency in TE and TM modes at its resonant frequency of fr = 7.1 GHz.

When an incident EM wave falls on the FSS array, the energy of the wave in a particular frequency is dissipated as surface currents. This results in the attenuation of the incident wave.

(a) (b)

Fig 4. Surface current for wave incident on FSS without meandered stubs with angle of incidence (theta) equal to zero (a) TE mode (b) TM mode

The surface current distribution on the metal part of the FSS with the meandered rectangles in both TEand TM modes is also studied as shown in Figure 5 (a) and 5 (b). The surface current distribution on the metallized regions of the final FSS unit cell is shown in both TE and TM modes at frequency f = 4.9GHz. Studying the surface current distribution of the FSS for the Figure 5(a) and 5(b), it is observed that the surface currents are mainly concentrated on the outer meandered lines, thereby showing that the meandered lines are responsible for the maximum current distribution and hence maximum attenuation.

(a) (b)

Fig 5. Surface current for wave incident on FSS with meandered stubs with angle of incidence (theta) equal to zero (a) TE mode (b) TM

3.3. Angular Stability

The designed FSS was tested for angular stability to check position insensitivity to the incident wave and has been plotted in Fig 6. This is tested by choosing different incident angles of the EM wave and testing the transmission parameters of the FSS. A parametric sweep was performed on the FSS by the CST software from an incident angle of 0° to 60°. The angle of incidence was increased by 5° for every iteration on simulating, it has been realized that as the incident angle increases, the attenuation of the transmitted signal increased marginally without a change in the frequency of interest.

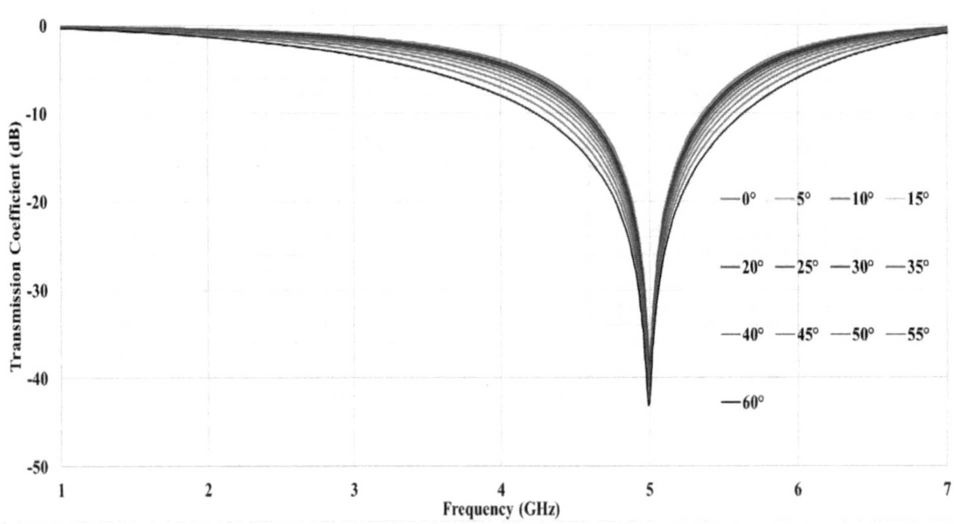

Fig 6. Graph depicting the transmission parameters for different angles of incidence

Improvement is also observed in the bandwidth. But most importantly the FSS remained stable during change in incident angle from 0° to 60°. This happens due to the reduction of resonator's loading factor. This can be attributed due to the symmetry of the structure designed.

4. Summary

In this paper, a simple FSS based spatial filter to shield low 5G frequencies with a simple cross geometry has been designed. This FSS' structure is a simple cross structure with inverted J shaped elements and meandered lines separating the individual elements. This FSS has been simulated in CST Design Studio and the results have been studied. A strong resonance peak has been obtained at a frequency of 4.98 GHz with a bandwidth of 1.45GHz. The shielding obtained is more than 40dB in the bandwidth of interest. The structure has also shown to be extremely tolerant to This structure is well suited for vertical applications of 5G in frequency shielding and can be easily fabricated by suitable methods like wet-etching, 3D printing etc.

5. References

[1] B.A.Munk, *Frequency Selective Surfaces: Theory and Design*. NewYork,NY, USA: Wiley, 2000.

[2] M. A. M. Albreem, "5G wireless communication systems: Vision and challenges," 2015 *International Conference on Computer, Communications, and Control Technology (I4CT), and Kuching, Malaysia,* 2015, pp. 493-497.

[3] H. Son and Y. Chong, "Analysis of the Interference Effects of 5G system on Automotive Collision Avoidance Radars," *2019 International Conference on Information Communication Technology Convergence (ICTC)*, Jeju, Korea (South), 2019, pp. 1463-1466.

[4] A. F. Coşkun, U. Saynak and B. Bati, "Practical Approaches on the Assessment Procedure pp. of Possible Interference Effects of 5G Base Stations on Radar Systems," *2020 28th Signal Processing and Communications Applications Conference (SIU),* Gaziantep, Turkey, 2020, 1-4.

[5] T. Weiland, "A discretization method for the solution of Maxwell's equations for six-component fields", *Electronics and Communication (AEÜ)*, vol. 31, p. 116, 1977.

[6] T. Weiland, "Time domain electromagnetic field computation with finite differencemethods", of *International Journal of Numerical Modelling*, vol. 9, pp. 295-319, 1996.

[7] J. F. Mologni, J. C. Ribas, A. C. Sodré and M. A. R. Alves, "Investigation on the deployment -5. FSS as electromagnetic shielding for 5G devices," *2017 SBMO/IEEE MTT-S International Microwave and Optoelectronics Conference (IMOC)*, Aguas de Lindoia, Brazil, 2017, pp. 1

[8] Wersényi, György. "Health issues using 5G frequencies from an engineering perspective: Current review" *Open Engineering*, vol. 12, no. 1, 2022, pp. 1060-1077.

[9] https://www.faa.gov/5g

Microstrip patch antenna design using mathematically modeled metamaterial cell

P R Satarkar[1], R B Lohani[2]
[1,2]Department of Electronics Telecommunication
Goa College of Engineering (Goa university/Govt. of Goa)
Farmagudi-Ponda-Goa-India, 403401
psatarkar10@gmail.com,

Abstract. Metamaterial is an artificial material made up of different types of structural designs on a dielectric substrate. In this paper, a broad investigation is being carried out by mathematically modeling and simulating a single negative metamaterial cell comprising a square split ring resonator with two rings. This metamaterial cell has negative permeability values for a particular frequency range. By enhancing properties like bandwidth, reflection loss, and gain in microstrip patch antennae, these metamaterial cells demonstrate unusually remarkable applications in this field. The resonance frequency obtained by simulating the metamaterial cell is compared with the calculated value from the circuit model. The resonance frequency expression is made simple by evaluating the effective permittivity of the substrate after loading the metallic structure on a dielectric substrate. The resonant frequency predicted analytically is found to be extremely close to the frequency obtained by modeling metamaterial structure. A parameter-retrieval method utilising the S parameters is used to determine the curves for the complex permeability of the metamaterial based on its unit element. In order to obtain the S parameters, modeling software is used. This modeled cell is used for designing a rectangular microstrip patch antenna. The parameters such as bandwidth, Gain, return loss, antenna, and metamaterial-based antenna are compared and recorded.

Keywords: Antenna, Metamaterials, Substrate, Permeability, Permittivity, Electromagnetics, Resonator.

1. Introduction

Although research on 'synthetic' materials for microwave applications began in 1948 [1]. Metamaterial's have only recently emerged as a breakthrough in the realm of electromagnetics. Metamaterials have piqued people's curiosity because of their unusual responses to electromagnetic radiation. These materials are created by structuring precise metallic inclusions onto a host dielectric substrate. The Structural design of metamaterial cells is responsible for the properties of the cell and not the chemical composition of their constituents [2-6]. The concept of metamaterials was given by Victor Veselago in 1968 who stated that the Poynting vector of a monochromatic plane wave is anti-parallel to its phase velocity in a medium having dielectric permittivity (ε) and magnetic permeability(μ), to be negative at the same time [7,8]. The materials were named by him as "Left-Handed Materials" (LHM). In 1999, J.B. Pendry proposed the artificial negative permeability medium by an arrangement of an array of small metallic structures called split-ring resonators (SRR). Rectangular rings having equal sides are called square split-ring resonators (SSRR). Through measurements and numerical simulations, Ellstein created analytical circuit models for single-layer and double-layer spiral resonators, proving that these models are superior than earlier circuit models. [17]. The authors noted in [19] that dielectric characteristics affect distributed capacitance associated with the rings of SRR, but inductance is unaffected. We've covered some of the key characteristics of mu negative metamaterials in this article. Mathematical modeling of mu negative metamaterial cells with two rings and structured on Rogers RT/Duroid 5880, Inductance and distributed capacitance of SSRR and hence the resonance frequency are calculated and compared with simulated value. Parameters of mu negative metamaterial cell are extracted using MATLAB script to obtain the curve of complex permeability. Different types of metamaterials are there and they are used for various applications [9-11]. Using this modeled cell Rectangular microstrip patch antenna (RMPA) has been designed. Its parameters are recorded and judged the advantages of loading SSRR cells on the surface of the patch.

2. Metamaterials

A metamaterial is a substance that has been artificially constructed and has exceptional electromagnetic properties that are not present in nature or are difficult to achieve. Metamaterials, a fast-expanding interdisciplinary field involving physics, electrical engineering, materials science, optics, and nanoscience, has arisen since the early 2000s. By modifying their internal physical structure, metamaterials can have their attributes precisely tuned. They so differ significantly from natural materials, whose characteristics are primarily governed by their chemical components and bonds.

2.1. Double negative or left-handed metamaterials.

The other name for this material is Left-Handed metamaterials. They are artificial materials with negative refractive indexes, permittivity, and permeability. Natural materials don't have these characteristics. The group and phase velocities of electromagnetic waves are opposite to one another because of the negative refractive index, and contrary energy flow is seen. Negative refraction can be achieved when both μ_r and ε_r are negative, [12-14]. Equations 1 and 2 are two of Maxwell's equations given to understand metamaterial.

$$\nabla \times E = -\mu_0 \mu_r \frac{\partial H}{\partial t} \tag{1}$$

$$\nabla \times H = -\varepsilon_0 \varepsilon_r \frac{\partial E}{\partial t} \tag{2}$$

Where ϵ_r and μ_r are relative permittivity and permeability respectively and Equation 3 is a wave equation derived from the above equation.

$$\nabla^2 \mathrm{E} = -\varepsilon_0 \mu_0 \varepsilon_r \mu_r \frac{\partial^2 E}{\partial t^2} \tag{3}$$

If ϵ_r and μ_r are considered real numbers, then one can observe that the wave equation does not change when signs of ϵ_r and μ_r are simultaneously changed. Due to the aforementioned facts, these materials are known as left-handed materials. The Drude–Lorentz model describes the material properties in classical electromagnetic. The effective permittivity and permeability are given in (4) and (5)

$$\varepsilon_{eff} = 1 - \frac{f_p^2}{f^2 - j\gamma f} \tag{4}$$

$$\mu_{eff} = 1 - \frac{f_{mp}^2 - f_0^2}{f^2 - f_0^2 - j\gamma f} \tag{5}$$

In (4) and (5), fp and fmp, respectively, stand in for the plasma frequency, the signal frequency, and the resonance frequency, f_0 as well as the damping factor γ, which is related to material losses. These equations can display a material's properties from the microwave to the optical range.

IOP Conf. Series: Materials Science and Engineering 1291 (2023) 012045 doi:10.1088/1757-899X/1291/1/012045

2.2 Design of mu Negative Metamaterial Cell on RT/Duroid 5880 Substrate.

The SSRR presented in figure 1 with some of the rings equal to two is designed on Rogers RT/ Duroid 5880 substrate with a permittivity value equal to 2.2 and thickness equal to 0.25 mm.

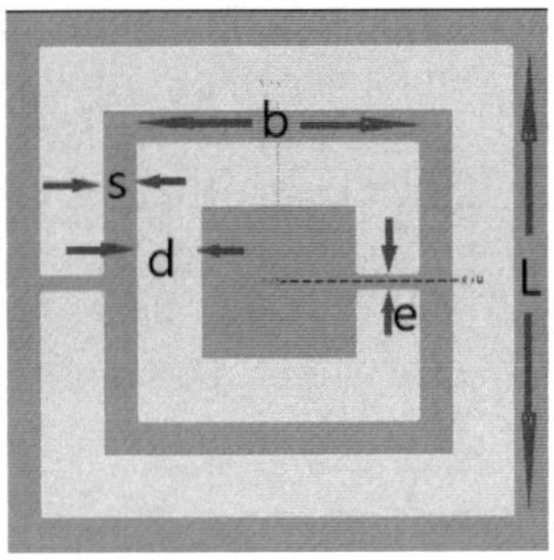

Figure 1. Square split-ring resonators (SSRR) metamaterial cell with two square split rings on top of the substrate.

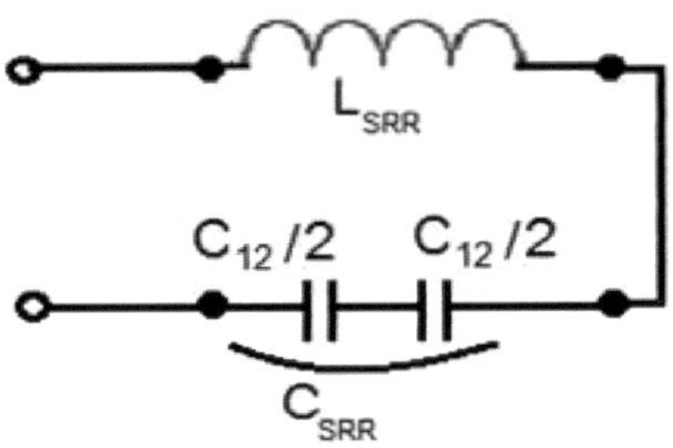

Figure 2. SSRR's equivalent circuit.

The dimensions of 1x1 mm SSRR cell are, L= 0.88 mm, b= 0.56 mm, Gap e=0.04mm, Strip width d = 0.1mm, and spacing between n and (n+1) th turn is S = 0.06 mm, The resonance frequency of SSRR is inversely proportional to the size of the cell, and it also depends on strip width, slit, spacing between (n+1) and nth ring. Figure 1 depicts the geometry of the SSRR, which consists of two rings with 180° of space between their slits. Figure 2 displays the SSRR equivalent circuit. When calculating the tank circuit inductance model Ls, it is assumed that the loop's current is constant. It is assumed that the tank circuit capacitance is equivalent to overall capacitances in each pair of adjacent loops in parallel. It is calculated the total strip length Ln of the nth ring of the SSRR and the total gap length Lg between the nth and (n+1) th turn of the SSRR are shown in equation 6 and 7. The following equations then display the total inductance and total dispersed capacitance between two subsequent rings of SSRR, of strip width d. [15-19]

$$Ln = 4\,[L - 2d\,(n-1) - 2s\,(n-1)] \tag{6}$$

$$L_g = 4[L - \left(n + \tfrac{1}{2}\right)(d+s) \quad , \quad n \geq 1 \tag{7}$$

$$C_{ssrr} = \frac{N-1}{2}\,[2L - (2N-1)(d+s)]C_{pul} \tag{8}$$

$$C_{pul} = \varepsilon_0 \varepsilon_{ref} \frac{K(k')}{K(k)} \tag{9}$$

C_{pul} is the capacitance per unit length of two parallel strips of width d apart by distance s. The dielectric substrate used in cell affects the distributed capacitance only and is given by equation (8).

$$\varepsilon_{ref} \cong \frac{1+\varepsilon_r}{2}$$

$$k' = \sqrt{1-k^2}$$

$$k = \frac{s}{s+2d}$$

$$\frac{K(k')}{K(k)} = \left[\frac{1}{\pi} In\left(2\frac{1+\sqrt{k\prime}}{1-\sqrt{k\prime}}\right)\right] \qquad \text{For } 0 \leq k \leq 0.7$$

IOP Conf. Series: Materials Science and Engineering 1291 (2023) 012045 doi:10.1088/1757-899X/1291/1/012045

Values of K(k) and K(K^{I}) also can be taken from figure 3. K(k) is called as the complete elliptical integral of the first kind, and K(k^{I}) is its complex value. It is defined for 0 < k < 1 by (10). Equation (11) presents the average length of the strip.

$$k(k) = \int_0^{\pi/2} \frac{d\theta}{\sqrt{1-k^2 sin^2\theta}} \quad (10)$$

$$L_{avg} = 4\,[L - (N-1)(s+d)] \quad (11)$$

The dielectric substrate used in SSRR does not affect the inductance. The inductance of SSRR is given by equation (12)

$$L_{SSRR} = \frac{\mu_0}{2}\,\frac{L_{avg}}{4}\,4.86\,[\,In\left(\frac{0.98}{\rho}\right) + 1.84\,\rho\,] \quad (12)$$

Where ρ is the fill ratio of SSRR and is stated as

$$\rho = \frac{(N-1)(s+d)}{L-(N-1)(s+d)}$$

$$f_r = \frac{1}{2\pi\sqrt{L_{ssrr}C_{ssrr}}} \quad (13)$$

The resonance frequency of the metamaterial cell is given by(13).

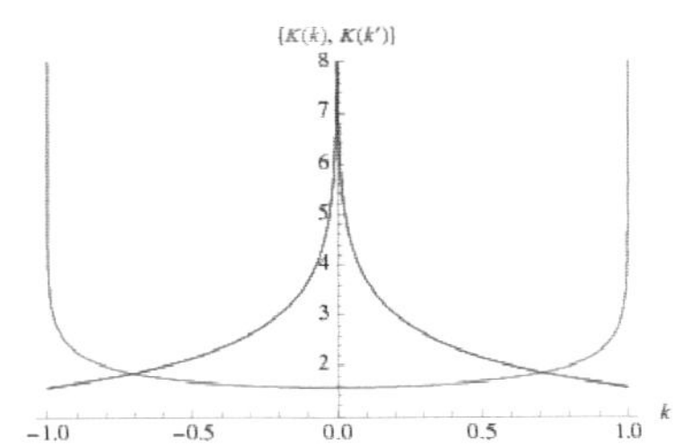

Figure 3. Graph of k v/s K(k) and K(k^{I}) [19]

3. Method to find various parameters of SSRR metamaterial cell.

Numerical simulations can reveal the SSRR's practical parameters. MATLAB code was used to create the artificial SSRR. This produces a metal surface, as seen in figure 1. The tank circuit capacitance, as shown in the diagram, is assumed to be the parallel equivalent of the capacitances in each pair of adjacent loops. negative μ near resonance frequency is produced by the SSRR. The magnetic field which is perpendicular to the SSRR was the only magnetic field that the metamaterial allowed to propagate backward waves. In order to confirm a cell's mu negativity, we have demonstrated the recovery of parameters from a metamaterial cell using MATLAB. Through a MATLAB script, the CST platform is invoked using an inherent feature of CST. [20-22]. The effective impedance of cell and refractive index has been shown in equation 14 and 15 respectively.

$$Zeff = \pm\sqrt{\frac{(1+S11)^2-(S21)^2}{(1-S11)^2-(S21)^2}} \quad (14)$$

$$\eta eff = \pm\frac{1}{kod}\left[\{[In(e^{inkod})] + 2m\pi\} - i[In(e^{inkod})]\right] \quad (15)$$

$$e^{inkod} = \frac{S21}{1-S11\left[\frac{(Z-1)}{(Z+1)}\right]}$$

where d is the substrate thickness, k_0 represents wave vector in vacuum, defined as $k_0=2\pi/\omega_o$, and m is an integer. We can find z_{eff} and n_{eff} from the given equation. The metamaterial are passive media with this consideration, That is a real impedance is positive and imaginary part of refractive index is negative. Further $\epsilon_{r,eff}$ and $\mu_{r,eff}$ can be obtained according to ,$\epsilon_{r,eff} = n_{eff}/z_{eff}$ and $\mu_{r,eff} = n_{eff}\, z_{eff}$. After simulation, It is observed that the metamaterial cell resonates at 28 GHz frequency. The graph of S11 and S21 v/s frequency is shown in figure 4. The resonating frequency thus observed which is 28 GHz is shown in the figure.

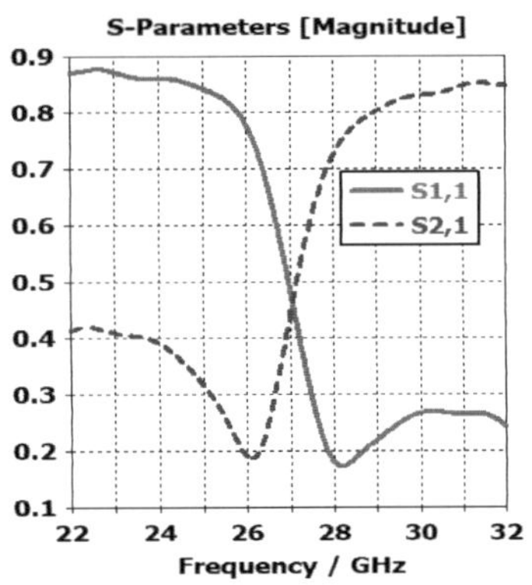

Figure 4. The graph of S11 and S21 v/s frequency of SSRR with Roger RT Duroid 5880 substrate.

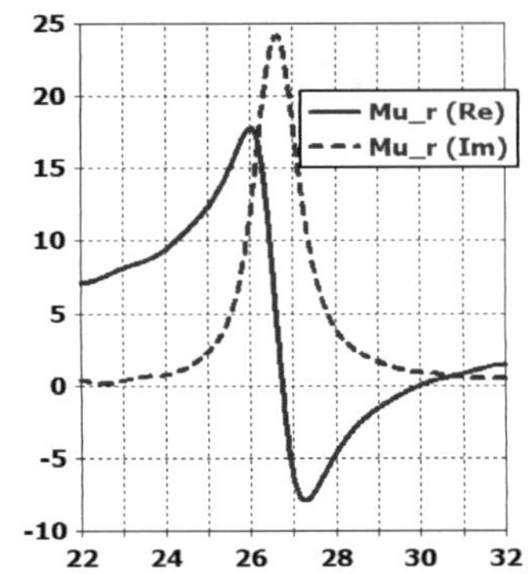

Figure 5.The graph of Permeability V/s frequency of SSRR with Roger RT Duroid 5880 substrate.

It is also observed from figure 5 that the permeability of the cell is negative at 28 GHz frequency. Also, inductance, capacitance, and resonance frequency of SSRR are shown in Table 1. The value of inductance and distributed capacitance is calculated using the equations shown above.

Table 1. Comparison of calculated and simulated resonance frequencies of SSRR.

Sr. No	Substrate material	Dielectric constant	No. of turns	Length of the side of SSRR (L)mm	Ls (nH)	C (pf)	Calculated resonance frequency(GHz	Measured resonance frequency (GHz)	Effe. Permeability at reson. freq
1.	Rogers RT/Duroid	2.2	02	0.88	3.38	0.01	27.2	28	- 5

4. Application of modeled metamaterial in designing rectangular shape microstrip patch antenna. (RMPA).

The design of antennas is done with SNG or DNG Metamaterial cells. Which are used to increase the performance of the system. These cells could significantly boost an antenna's gain and output power. These antennas can also enhance performance in terms of efficiency and bandwidth. For wireless communication, 4G and 5G applications, etc., a variety of metamaterial-based antennas can be used. [23-25].

4.1 RMPA Design on the substrate without metamaterial at 28 GHz resonating frequency.

A rectangular shape microstrip patch antenna (RMPA) at resonance frequency 28 GHz as used in 5G communication was designed on an RT Duroid substrate and tested using the FEKO software. Equation (16) gives the width of RMPA. In equations (17) and (18), the effective permittivity and delta L fringing length are correspondingly displayed.

$$w = \frac{c_0}{2f_r}\sqrt{\frac{2}{\varepsilon_r+1}} \tag{16}$$

$$\varepsilon_{reff} = \frac{\varepsilon_r+1}{2} + \frac{\varepsilon_r-1}{2}\left[1 + 12\,\frac{h}{w}\right]^{-1/2} \tag{17}$$

$$\Delta L = h(0.412)\frac{(\varepsilon_{reff}+0.3)\left(\frac{w}{h}+0.264\right)}{(\varepsilon_{reff}-0.258)\left(\frac{w}{h}+0.8\right)} \tag{18}$$

Effective length and actual length are given in eq 19 and 20.

$$L_{eff} = \frac{c_0}{2f_r\sqrt{\varepsilon_{reff}}} \tag{19}$$

$$\text{L} = L_{eff} - 2\Delta L \tag{20}$$

h=height of thesubstrate, w=width of the rectangular patch,L= Length of rectangular patch co = Electromagnetic wave speed [26].

4.2 conventional rectangular shape microstrip patch antenna (RMPA) design at 28 GHz resonating frequency.

Figure 6. shows the top view and design of conventional edge-fed RMPA on RT duroid 5880 substrates.

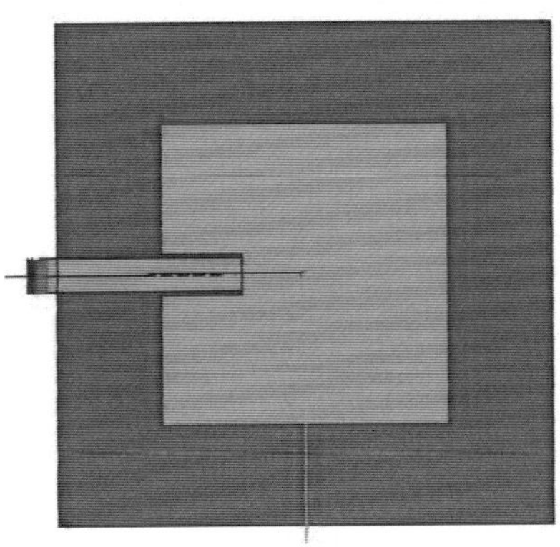

Figure 6. Conventional edge-fed RMPA (without metamaterial).

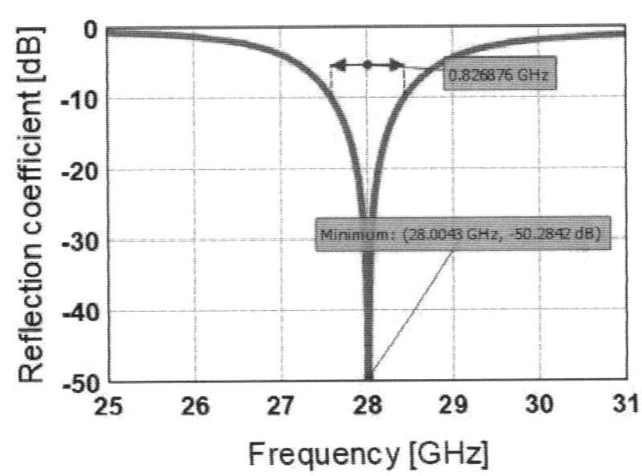

Figure **7.** Frequency v/s Reflection coefficient of conventional RMPA.

Figure 7 depicts the frequency v/s reflection coefficient graph of RMPA. These parameters determine the ratios of the amplitudes in the transmitted and reflected wave when a plane wave incident on the dielectric DNG medium. When a split-ring resonator is put onto the microstrip patch antenna, these factors improve its performance. The antenna radiates its greatest power at the resonance frequency because the return loss is relatively low (-50dB). At the resonance frequency, It is observed that the return loss is very low,Antenna radiates maximum power when return loss of antenna is very low. The BW, Gain, return loss, and VSWR are depicted in Table 4. Return loss Of the antenna depends on its impedance matching. Return loss is expected to be very low for the perfectly matched antenna. That is perfectly matched with source impedance.

Table 2 shows the design of conventional RMPA (without metamaterials). Resonance freq. = 28 GHz, Substrate material: RT Duroid 5880, with Loss tangent: 0.0009, ε_r= 2.2, height of substrate(h) = 0.25 mm.

Table 2. The design of conventional RMPA.

	Patch antenna calculated dimensions (mm)	Patch antenna optimized dimensions(mm)
Width of the patch (w)	4.24	4.11
Length of the patch(L)	3.48	3.36
Sub. width (W_s)	6.85	6.84
Sub.Length(L_s)	6.00	5.8
Feed length(a)	2.17	1.97
Feed width(b)	0.48	0.48

Figure 8 depicts a far-field radiating pattern in UN and VN planes at 28 GHz frequency. The antenna gain is 6.31 dBi at 0 degrees.

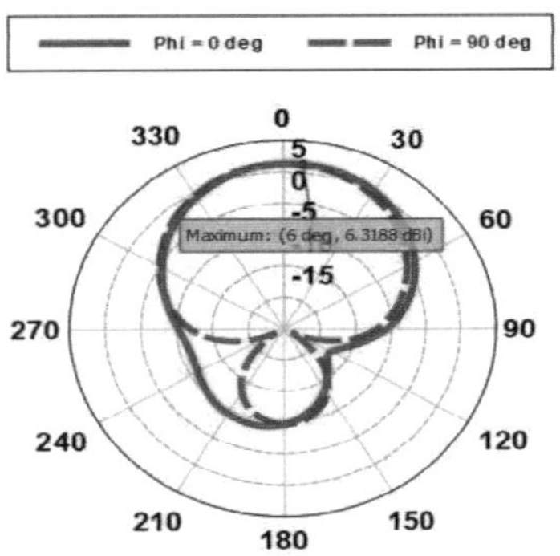

Figure 8. Far-field radiating pattern and total gain of RMPA at 28 GHz without metamaterial.

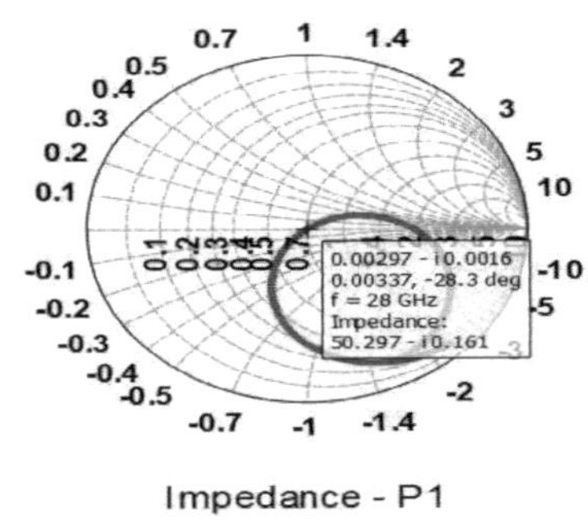

Figure 9. Impedance matching chart of RMPA at 28GHz.

Figure 9 presents the smith chart which depicts the impedance matching at 28 GHz of RMPA . The impedance is approximately 50 ohms at 28 GHz. VSWR is at 28 GHz is 1.02. Very important parameter of antenna is the gain of an antenna. Figure10 shows the far-field of RMPA and total surface current distribution at 28 GHz. The variation of current is shown in figure.

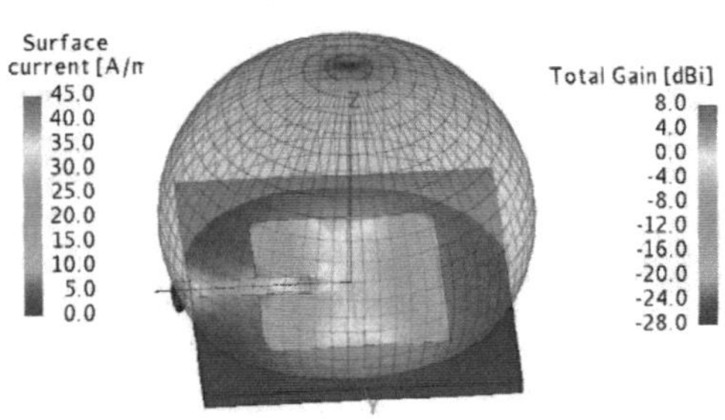

Figure10. 3D presentation of the far-field of RMPA and total surface current distribution at 28 GHz.

IOP Conf. Series: Materials Science and Engineering 1291 (2023) 012045 doi:10.1088/1757-899X/1291/1/012045

4.3 Metamaterial loaded RMPA design at 28 GHz resonating frequency.

Figure 11 and Figure 12 shows the top view RMPA and the bottom view of RMPA respectively. Figure 12 depicts the metamaterial cells on the bottom of the substrate. Frequency v/s reflection coefficient graph of metamaterial loaded RMPA is depicted in Figure 13. An antenna can emit its maximum power when very low return loss (-45 dB) at the resonance frequency.

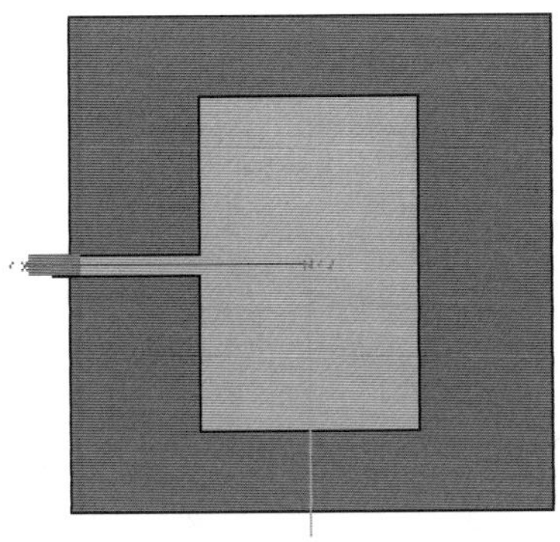

Figure 11. Top view of RMPA .

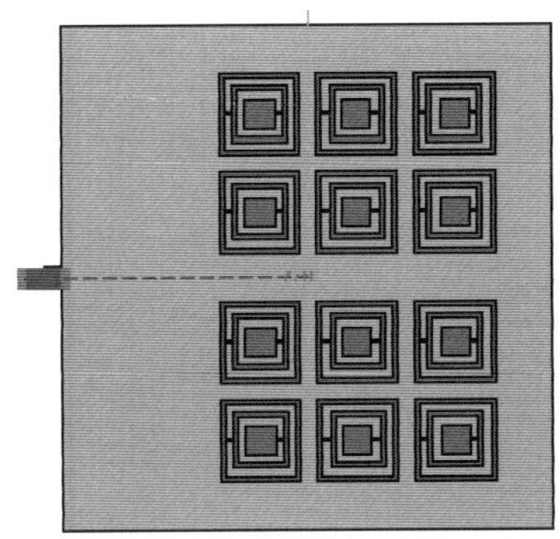

Figure 12. Bottom view of edge-fed RMPA with metamaterials.

The optimized design of metamaterial loaded RMPA shown in Table 3. The RT Duroid substrate is used to design RMPA. Frequency of resonance. = 28 GHz, Substrate: RT Duroid 5880, Loss tangent =0.0009 ε_r= 2.2, height of substrate(h) = 0.25 mm.

Table 3. Design of RMPA with metamaterial.

.	Optimized dimensions of RMPA (mm)
Width of the patch (w)	4.2
Length of the patch(L)	2.75
Sub. width (W_s)	6.2
Sub.Length(L_s)	6
Feed length(a)	1.63
Feed width(b)	0.28

The frequency v/s reflection coefficient graph of metamaterial loaded RMPA is depicted in figure 13, The reflection coefficient recorded from graph is -45 and bandwidth recorded is equal to 2.19 GHz.

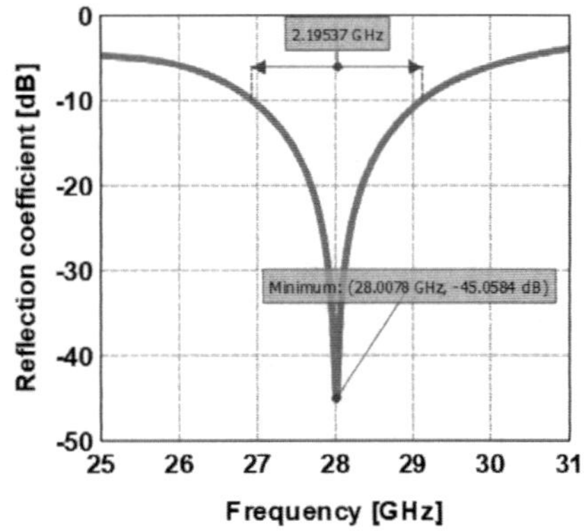

Figure 13. frequency v/s Reflection coefficient of metamaterial loaded RMPA.

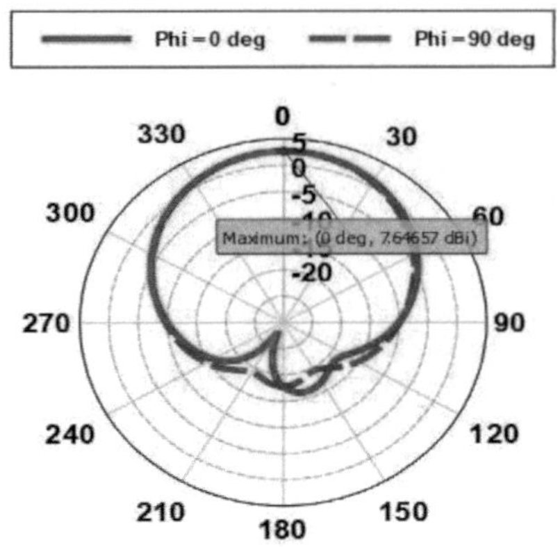

Figure 14. Far-field radiating pattern and total gain at 28 GHz of metamaterial-based RMPA.

Far-field radiating pattern of metamaterial loaded RMPA is depicted in figure 14 at 28 GHz resonance frequency. The antenna gain is 7.6 dBi at 0 degrees.

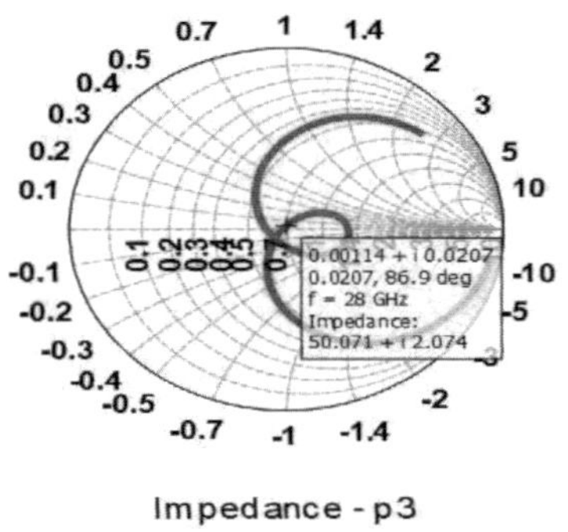

Figure 15. Smith chart showing impedance matching of metamaterial-based RMPA at 28 GHz.

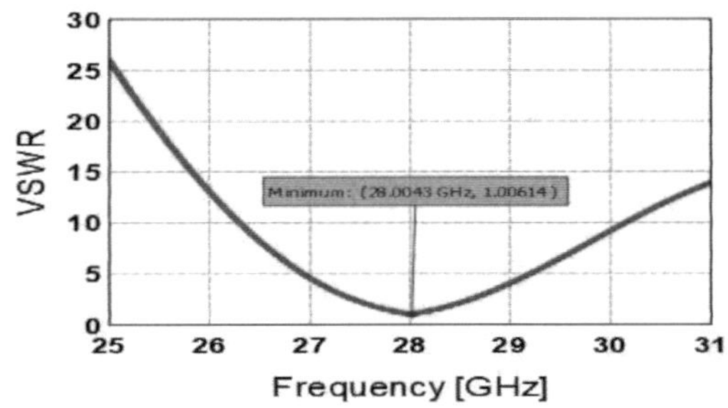

Figure 16.VSWR of RMPA with metamaterial.

Figure 15 shows the impedance matching of metamaterial loaded RMPA at 28 GHz. The impedance is almost 50 ohms at 28 GHz and the reactive impedance observed is close to zero, which matches the source impedance. Figure 16 presents VSWR, and it is equal to 1.00 which signifies perfect matching of source impedance of 50 ohms with an antenna.

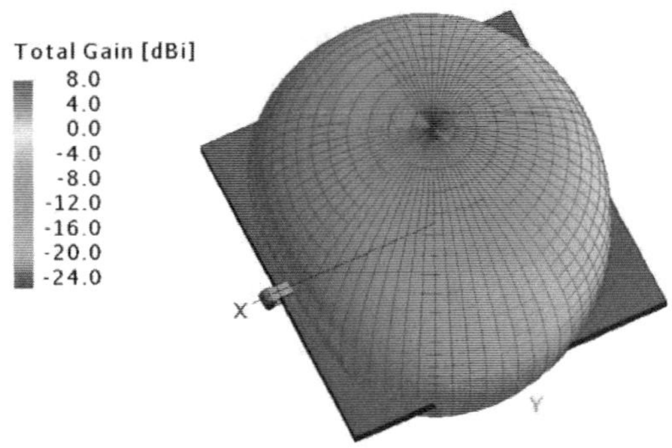

Figure 17. Far-field of RMPA with metamaterial at 28 GHz.

Figure 17 presents the far-field of RMPA with metamaterials at 28 GHz resonance frequency respectively Total gain has been depicted in the figure. The radiation performance in the antenna indicates the efficiency of the radiation measured in terms of percentage and is obtained by calculating the value of the ratio of the amount of power fed to the antenna to that of the power emitted (radiated) from an antenna. This radiation efficiency boosts the performance of the antenna useful for 5G wireless technology and its applications. The author's proposed antenna has higher radiation efficiency of 89.5% at 28 GHz.

In the one of the research paper metamaterial based antenna was designed with resonance frequency 28 GHz, The(-10 dB) band width obtained was equal to 1.7 Ghz [27]. In the second paper (-3 dB) bandwidth of metamaterial loaded patch antenna with resonance frequency 26 GHz was equal to 5.36 GHz [28]. In the third paper the patch antenna loaded with metamaterial resonates at 28.89 Ghz and the gain of the antenna was 7.7 dBi.[29]

5. Simulation results

The metamaterial cell comprising of SSRR is designed and simulated using the software. A metamaterial cell with only SSRR shows negative permeability to state it as a single or mu-negative metamaterial cell. The parameter comparison of RMPA and metamaterial-based RMPA with the same resonance frequency is listed in Table 4. Rectangular microstrip patch antenna for 5G wireless Network technology is effectively and efficiently designed, implemented, and compared for better performance at a 28 GHz resonance frequency. Conventional RMPA and RMPA loaded with metamaterial on the bottom or ground plane of substrate are simulated at a frequency of 28 GHz using FEKO software, When compared to conventional RMPA, it has been found that patch antenna bandwidth significantly outperforms RMPA in terms of both gain and bandwidth.

Table 4. Parameters of RMPA.

	RMPA without metamaterial	RMPA with metamaterial
Reflection coefficient	-50 dB	-45 dB
Bandwidth (10db)	0.826 GHz	2.19 GHz
Gain	6.31 dBi	7.64 dBi
VSWR	1.01	1.00

Conventional RMPA bandwidth and returns loss is 0.826 GHz and -50 dB respectively. Whereas metamaterial-based RMPA the values of bandwidth and return loss are 2.19 GHz &-45 dB respectively. The gain of RMPA without metamaterial is 6.31 dBi and with metamaterial cells is found to be equal 7.64 dBi. This results in a 1.36 GHz increase in bandwidth. Simulated findings have shown that the RMPA's bandwidth has increased. The VSWR of the conventional antenna is 1.01 and that of the metamaterial-equipped antenna is 1.00, indicating that the source impedance of 50 ohms and the antenna impedance are exactly matched.

6. Conclusion

A rectangular shape microstrip patch antenna is shown in this study both with and without metamaterial. By loading metamaterial on it, the bandwidth has increased significantly, according to design and simulation results. whereas a minor component has been added to the antenna's gain. The simulated outcome depicts the antenna and source being matched for impedance at 50 ohms at a frequency of 28 GHz. RMPA for 5G wireless Network technology is effectively and efficiently designed, implemented, and compared for better performance at a 28 GHz resonance frequency. Simulation and review of the proposed design are done using FEKO software. Table 4 highlights the substantial improvement in the proposed antenna design and shows better performance with an immense bandwidth for real-time smart device communication and other various applications of 5 G-enabled smart devices, improved return loss for good impedance matching, maximum power energy transfer with high radiation efficiency and good antenna gain for strong signal strength at long-range wireless communication.
In future metamaterial inspired microstrip patch antenna or MIMO antenna on flexible substrate for specific applications like 5G communication can be implemented. Phased array with metamaterial structure can be designed for electronically controlled beam steering function.

References

[1] Engheta, N. 2003. Metamaterials with Negative Permittivity and Permeability: Background, Salient Features, and New Trends. *International Microwave Symposium Digest,* 1: 187-190.

[2] Yongmin Liu, and Xiang Zhang."Metamaterials: Jan.2011 a new frontier of science and technology," *Chem. Soc. Rev.***, Vol 40**, pp. 2494-2507.

[3] J. D. Jackson, *Classical Electromagnetics,* John Wiley & Sons, New York, 3rd in, 1999.

[4] Nadar Eader Engheta Richard W.Ziolkowski. *Metamaterials Physics and Engineering Explorations* Edited by Published by John Wiley &Sons, Inc.

[5] Shridhar E. Mendhe, and Yogeshwar Prasad Kosta, June 2011 "Metamaterial properties and application," *International Journal of Information Technology and Knowledge Management,* **Vol. 4,** pp. 85-89.

[6] Pendry J.B, Holden A.J, Robbins D.J and Stewart, W.J. 1998. Low-Frequency Plasmons for Thin-Wire structure. *Journal of Physics: Condensed Matter,* 10(22):pp 4785–4809.

[7] Sihvola, A. 2007. Metamaterials in Electromagnetics. *Metamaterials: Springer*, 1: 2–11.

[8] P. R. Satarkar, R.B. Lohani, Metamaterial: July Sept 2020 Fundamental and engineering exploration. *Kala Sarovar ,UGC care Group 1 Journal,***Vol 23**,No 02 (V).ISSN : 0975-4520

[9] P. R. Satarkar, R.B. Lohani, Design and Characterization of Coaxial Feed Circular Patch Antenna on a Metamaterial Substrate, *Fourth World Conference on Smart Trends in Systems, Security and Sustainability (WorldS4)* 2020, IEEE

[10] J.D. Baena, J. Bonache, F. Martin, R.M. Sillero, F. Falcone, T. Lopetegi, M. Laso, J. Garcia-Garcia, I. Gil, M.F. Portillo, and M. Sorolla, (2005) Equivalent-circuit models for split-ring resonators and complementary split-ring resonators coupled to planar transmission lines, *IEEETrans Antennas Propag 53* ,1451–1461.

[11] Moataza A.Hindy, Ragab Elsagheer, and M.S. aseen, "Circular split-ring resonator (CSRR)

left-handed metamaterial (LHM) having simultaneous negative permeability and permittivity, " *International Journal of Hybrid Information Technology*, **Vol.10,** pp.171-178, 2017.

[12] Kaushal Gangwar, Paras, R.P.S. Gangwar, Jan 2014"Metamaterials: characteristics, process and application," *Advance in Electronic and Electrical Engineering,*" **Vol. 4**, No. 1, pp. 97-106,

[13] Yahong Liu, Xiaojing Guo, Shuai Gu, and xiaopeng Zhao, "Zero index metamaterial for designing high gain patch antenna," *International Journal of Antennas and Propagation,* Vol. **25,** pp. 1-12, Sept. 2013.

[14] S.S. Islam, M.R. Faruque and, T. Alam. "A new mu negative metamaterials, Jan. 2016" *In Proceedings of 2 nd international conference on Electrical Information and Communication Technology,* Bangladesh, ,pp.

[15] Toscano A and Vegni, Bilotti F, L. 2007. Design of Spiral and Multiple Split-Ring Resonators for the Realization of Miniaturized Metamaterial Samples. *IEEE Transactions on Antennas and Propagation*, 55(8): 2258-2267.

[16] Ellstein D, Wang B and Teo, K.H. 2012. Accurate models for Spiral Resonators. *9th European Radar Conference (EuRAD)*, Oct.-Nov. 2012, pp. 461-464.

[17] Rajni and Anupma Marwaha An accurate Approach of Mathematical Modeling of SRR and SR for Metamaterials*, Journal of Engineering science and technology Review* 9(6) (2016) 82-86.

[18] Bilotti F, Toscano A, Vegni L, Aydin K, Alici K.B and Ozbay, E. 2007. Equivalent-Circuit Models for the Design of Metamaterials Basedon Artificial Magnetic Inclusions. *IEEE Transactions on Microwave Theory and Techniques*, 55(12): 2865-2873.

[19] Eric Weisstein, Mathworld wolframe mathematical resources, the complete elliptic integral of the first kind,2021.

[20] MD Nazmul, Hassan.B.Monk, MATLAB to CST interfacing method (2000) Metamaterial critics & alternatives, INC publ.

[21] Numan A.B., & Sharawi M.S.,(Oct 2013) Extraction of Material for metamaterials using a Full Wave Simulator," IEEE Antenna and Propagation, Vol.55(5), pp.202– 211.

[22] Sarade S.S., Ruikar S.D., Bhaldar H.K., Design of Microstrip Patch Antenna for 5G Application, *Springer,* (2020).

[23] John Colaco, Rajesh Lohani, 2020 Design and Implementation of Microstrip Patch Antenna for 5G application. *(ICCES),*2020.

[24] A.A Deshmukh and K.P. Ray,2009"Compact broadband slotted rectangular microstrip antenna," *IEEE Antennas and Wireless Propagation,* Vol. 8, pp. 1410-1413.

[25] K.P. Ray, Girish Kumar, Broadband microstrip antenna, London, UK: Artech House,2003.

[26] Constantine A. Balanis, *Antenna Theory: Analysis and Design*, 3 rd ed. John Wiley & Sons, Inc., Publication, Hoboken, NJ, 2005.

[27] Bashar A. F. Esmail; Slawomir Koziel, Dec.2022 Design and Optimization of Metamaterial-based Dual-Band 28/38 GHz 5G MIMO Antenna With Modified Ground for Isolation and Bandwidth Improvement *IEEE Antennas and Wireless Propagation,* **Volume: 22 .**

[28] John Colaco and R.B. Lohani ,2021 ," Metamaterial Based Multiband Microstrip Patch Antenna for 5G Wireless Technology-enabled IoT Devices and its applications" *Journal of Physics: Conference Series*, **Vol 2070**

[29] Jalal Khan and Daniyal Ali Sehrai and Shakeel Ahmad ,2018, "Design and Performance Comparison of Metamaterial Based Antenna for 4 G / 5 G Mobile Devices*" International Journal of Electronics and Communication Engineering* **Vol:12**, No:6.

IOP Conf. Series: Materials Science and Engineering 1291 (2023) 012046 doi:10.1088/1757-899X/1291/1/012046

A Single Plane Low profile Patch Antenna π- Shaped Slots and Low Side Lobe Level for Directive Wireless Communications

Safa N. Nafea [1, 2]

[1] Computer Networks Engineering Department, College of Information Engineering, Al-Nahrain University, 10070, Baghdad, Iraq.

[2] Corresponding Author email:safanafea@gmail.com

Abstract. The patch antenna was used recently for ISM band (5.725 – 5.875) GHz, the patch antenna suffers of being an antenna with low gain as well as narrow operating bandwidth with high side lobe level (SSL) radiation pattern. A single plane with low profile and easy fabrication microstrip patch antenna has been proposed in this work was presented for directive communications and operates in Industrial, Scientific, and Medical (ISM) band 5.8 GHz with low SLL radiation pattern and gain of 5.20 dB for pint to point (directive). The gain of antenna has been increased while SLL decreased due to shifting the radiating patch and feeder of antenna with specific distance toward –ve side of the X- axis without any need for using techniques which would increase overall dimensions of antenna. The antenna has been printed on Arlon AD 250 substrate. The simulated results of antenna's gain, operating frequency, bandwidth, return loss, and radiation pattern are presented. CST Microwave studio software was used in this design simulation.

1. Introduction

The microstrip patch antenna has certain advantages such as being low cost antenna with easy fabrication requirements as well as being low weight antenna which has ability of operating with multiple operating frequencies [1]–[3]. On the other hand this type of antenna shows many disadvantages such us being low gain antenna which suffering from high side lobe level radiation pattern with narrow operating bandwidth as well as suffering from surface currents which reduce antenna's efficiency [1-3].

Many research activities have been conducted during last years to improve the operating bandwidth of microstrip patch antenna as well as achieving wideband operation [2-11] . On the other hand many researches had shown focus on enhancing gain of microstrip patch antenna and many techniques had been developed for this issue. Using grooved ground caused gain enhancement more than 10 dB for a patch antenna operating at 10.45 GHz in compare with a patch antenna with conventional ground plane [12]. Using array of patches is an ordinary technique to improve the gain of antenna, while using an array of 256 patches in arbitrary shapes caused to improve the gain of antenna as well as reducing SLL of radiation pattern [13]. In [14] using defected patch antenna led to achieve gain of 4.12 dB at 2.69 GHz ,while covering 4×1 patch array with three layers of high refractive index metamaterial with split ring resonator caused gain enhancement of 3.4 dB [15]. Moreover a triple operating frequencies using metamaterial loaded to patch antenna and caused gain of 3.1 dB at 5.8 GHz [16]. Using reflector layer with slot etched from ground of patch antenna caused to achieve gain of 7.35 dB [17]. In [18] using defected ground structure caused improvement around 1 dB in antenna's gain while adding a reflector layer under antenna's ground with a specific distance caused an enhancement of 2 dB in gain. Using multilayer antenna structure is very helpful technique, where the gain of conventional circular patch antenna was increased up to 10 dB at 5.8 GHz due to adding an FR4 superstrate with frequency selective surfaces above the patch [19]. Locating 155×155 mm^2 with 5×5 partially reflected surfaces on its bottom led to an enhancement in gain around 5.5 dB at 5.8 GHz [20]. Also gain of an elliptical

patch antenna had been increased from 8.5 dB to 14.1 dB at 5.8 GHz after adding 100×100 mm^2 superstrate with 7×7 partially reflected surfaces located on upper side of superstrate [21].
Using three superstrate layers of Roger RO3006 without any metallic objects also can increase gain of patch antenna and reduce SLL after studying height of each superstrate layer from antenna's ground which caused an increment in gain from 8.23 dB to 11.7 dB at 5.8 GHz [8], while multilayer antenna in [22] increased gain up to 12.7 dB due to adding three layers of Rogers RO6006 superstrate as well as adding a reflector under antenna's ground with optimizing height of each of four layers. Another effective technique to enhance the gain of patch antenna is etching slots from surface of radiating patch [10]. Moreover, the effects of thickness of copper of the patch on antenna's gain have been investigated to achieve better performance in general with highest gain at a certain frequency [11].
In this article a single plane low profile patch antenna with π – shaped slots operates at ISM Band at 5.8 GHz has been presented. The gain of patch antenna has been increased and SLL of radiation pattern have been eliminated based on investigating position of both microstrip line feeder and patch without need for using ordinary techniques; which described earlier in this section, to increase antenna's gain and reduce SLL.
The proposed antenna achieved gain of 5.20 dB which is higher than 4.30 dB and 4.12 dB achieved by antenna array proposed in [7], and defected patch structure proposed in [11], respectively. The proposed antenna has a radiation pattern with eliminated SLL and reduced size compared to antennas in [7-8], [16], and [19].

2. Antenna Design

The design of the proposed antenna is a single plane low profile patch antenna which is composed of semi-rectangular patch printed on Arlon AD 250 substrate with dimensions of (L×W×h) for length, width, and height, respectively and dielectric constant (ϵ_r) 2.5. The dimensions of the rectangular patch are (P_L×P_w). Two circles with different radius for each were located on the upper right and left edges of the rectangular patch with radius of (r_R) and (r_L) for the right and left hand circles, respectively. Both circles have an etched shape of three quarters of a circle with radius of 0.4 mm. Three rectangular slots were etched from rectangular patch to form the π- shaped . These slots are composed of two vertical slots with length and width of (L_{VS}) and (W_{VS}), respectively. The third slot is a horizontal slot with length and width of (L_{HS}) and (W_{HS}), respectively. The patch antenna has been fed using microstrip line feeder with length of (F_L) and width of (F_W). the thickness of annealed copper of the patch is (t).
The feeding microstrip line as well as the radiating patch have been shifted with distance (a) from the center of the substrate toward (-ve) X-axis to achieve minimum SLL in radiation patter as well as increasing gain of antenna. Top view for the single plane low profile antenna with π–shaped slots is shown in Fig.1 and list of antenna parameters' values are presented in Table 1.

Table 1. List of Feeding Patch Parameters' Values.

No.	Parameter	Value (mm)
1	L	38
2	W	60
3	h	0.762
4	P_L	16
5	P_w	31
6	r_R	1.45
7	r_L	0.9
8	L_{VS}	9
9	W_{VS}	0.5
10	L_{HS}	1.5
11	W_{HS}	6
12	F_L	16.38

13	F_W	2.22
14	a	9
15	t	0.035

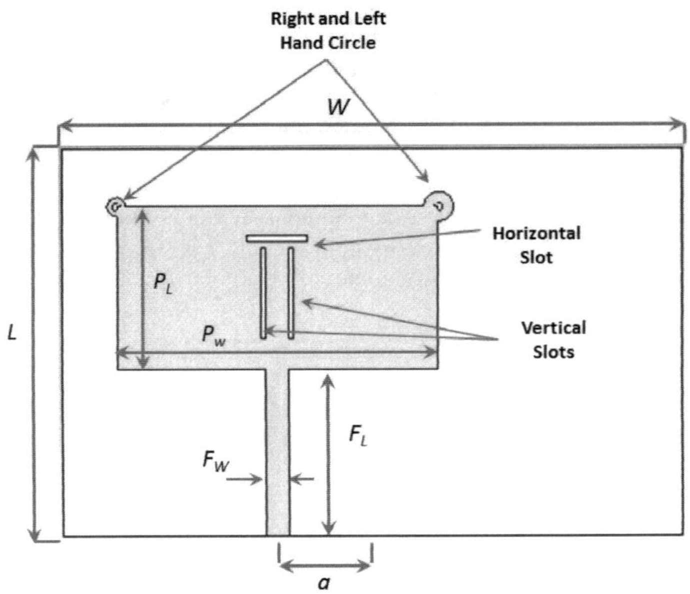

Figure 1. Top View For Proposed Antenna

3. RESULTS

This section will discuss in details steps of designing single plane and low profile microstrip patch antenna with low SLL radiation pattern. The following steps present parametric study antenna design with considering performance of each step in terms operating frequency, reflection coefficient (S_{11}), gain, bandwidth, and SLL.

1- The conventional rectangular patch with optimized dimensions of $P_L \times P_w$ resonated at 5.692 GHz with operating bandwidth of 164 MHz and reflection coefficient of -19.01 dB while gain of antenna at resonance frequency found to be 3.08 dB.

2- To improve performance and achieve an operating frequency within the ISM band 5.8 GHz; two circles have been added on the upper right and left hand edges of rectangular patch with radius of r_R and r_L, respectively, both circles have an etched shape of three quarters of a circle with radius of 0.4 mm. Adding these two circles caused shifting in operating frequency to 5.876 GHz. Moreover the reflection coefficient improved to be -30.55 dB while operating bandwidth decreased to 120.81 MHz with gain of 3.17 dB.

3- To achieve an operating frequency closed to ISM band (5.725-5.875) GHz than achieved in Step (2), a vertical rectangular slot have been etched from surface of patch which caused improvement in operating bandwidth up to 168.55 MHz and resonance frequency of 5.86 GHz with S_{11} of -22.42 dB and gain of 3.22 dB.

4- Etching second rectangular vertical slot; which is in parallel to the first slot with distance of 2.3 mm separating between them, improved the performance and caused resonance at 5.796 GHz with S_{11} of -26.50 dB and gain had been improved to 3.70 dB while the operating bandwidth decreased to 123.40 MHz.

5- Etching horizontal slot to form π-shaped slot; which is isolated from two vertical slots, led to enhance performance of antenna in terms of operating frequency which found to be 5.8 GHz, reflection coefficient which found to be -25.44 dB, operating bandwidth with increment up to 141.46 MHz, and gain enhancement to 4.09 dB.

6- As shown in previous steps an optimized performance had been achieved but SLL in antenna's radiation pattern still high> In this step SLL will be eliminated due to shifting microstrip line feeder and radiating patch with a distance (a) toward –ve X-axis of the substrate as shown below:

6 – A: Shifting microstrip feeder and radiating patch with distance of (a = 3mm) will cause to have a resonance at 5.796 GHz, S_{11} = -24.08 dB, operating bandwidth of 127.92 MHz, and gain increased up to 4.49 dB due increment in SLL of antenna's radiation pattern.

6 – B: Shifting microstrip feeder and radiating patch with distance of (a = 6mm) will cause to have a resonance at 5.796 GHz, S_{11} = -22.92 dB, operating bandwidth of 123.43 MHz, and gain increased up to 4.64 dB with SLL reduction.

6 – C: Shifting microstrip feeder and radiating patch with distance of (a = 9mm) will cause to have a resonance at 5.788 GHz, S_{11} = -25.50 dB, operating bandwidth of 150.73 MHz, and gain increased up to 5.20 dB with SLL reduction.

Tabulated data for steps of design of the proposed antenna presented in Table 2. Fig. 2 shows reflection coefficient for designing a single plane patch antenna with low SLL. The radiation patterns of the proposed antenna; before and after shifting microstrip feeder and radiating patch with distance of 9 mm, are presented in Fig. 3, where the red curves refer to the far field region and blue lines refer to the direction of the main lobe of radiation pattern while green circles refers to the lowest back and SLL.

Table 2. Results of Steps of Designing Proposed Antenna.

Step	S_{11} (dB)	VSWR	Frequency (GHz)	Gain (dB)	Bandwidth (MHz)
1	-19.01	1.252	5.692	3.08	164.00
2	-30.55	1.061	5.876	3.17	120.81
3	-22.42	1.164	5.860	3.22	168.55
4	-26.50	1.099	5.796	3.70	123.40
5	-25.44	1.113	5.800	4.09	141.46
6-A	-24.08	1.133	5.796	4.49	127.92
6-B	-22.92	1.154	5.796	4.64	123.43
6-C	-25.50	1.112	5.788	5.20	150.73

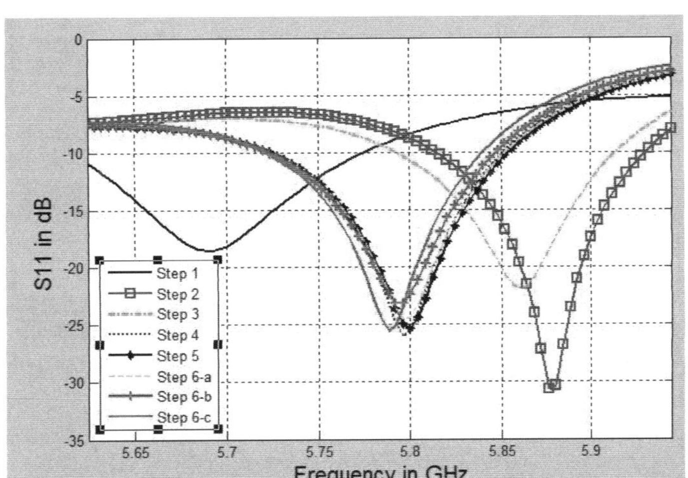

Figure 2. Reflection Coefficient for Steps (1 – 6-C)

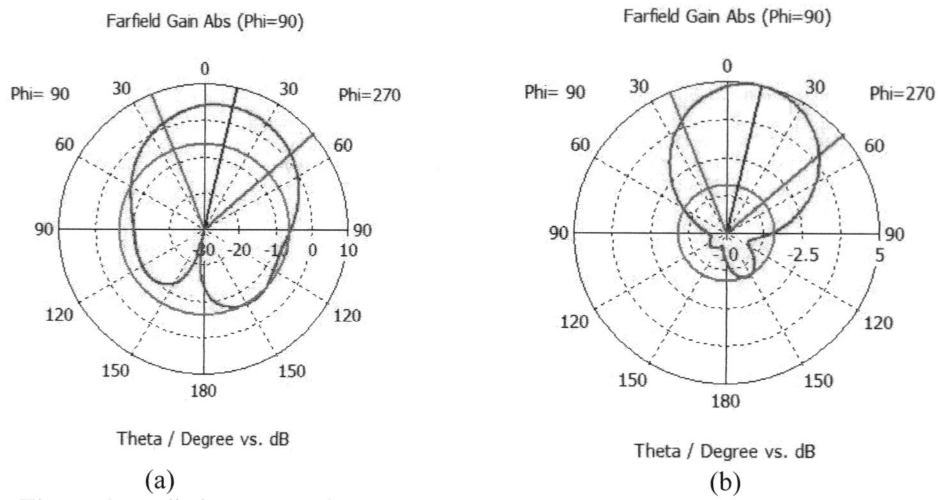

Figure 3. Radiation Pattern for Steps (a) 5 (a = 0 mm) and (b) 6-C (a = 9 mm).

4. DISCUSSION

The proposed single plane low profile patch antenna with π –shaped slots provided improvement in antenna's performance due to shifting microstrip line feeder as well as radiating patch toward the (-ve) X-axis. The proposed antenna achieved comparable gain compared to conventional rectangular and circular patch antenna's proposed by [14], [19]. The proposed antenna had achieved gain higher than 2×1 patch array proposed in [7]. The proposed antenna shows 0.6 dB of gain variation over ISM band (5.725 – 5.875) GHz as shown in Fig. 4. The proposed antenna radiation pattern has a Half Power Beam Width (HPBW) of 71 degree and front-to-back (F/B) ratio of 22.58 dB and proposed for directive point-to-point communications applications.

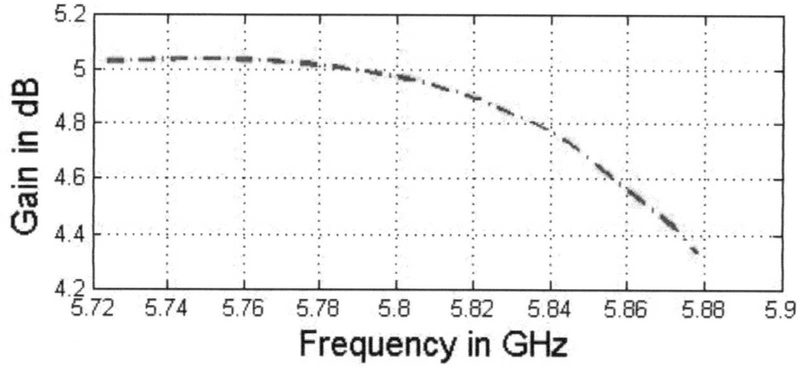

Figure 4. Gain over ISM Band (5.725 – 5.875) GHz for The Proposed Antenna.

Shifting the microstrip feeder and patch with a specific distance toward negative X-axis from center of substrate caused to eliminate the SLL of antenna's radiation pattern without need for using any technique which will lead to increase the overall dimensions of antenna. This led to achieve size reduction compared to antennas presented in previous works with reduction in SLL as shown in Table 3.

Table 3. Size Reduction Achieved by Proposed Antenna.

Ref. No.	Frequency (GHz)	Gain (dB)	Size Reduction Compared to The Proposed Antenna
[7]	5.80	4.30	98%
[8]	5.80	11.70	98%
[19]	5.80	9.85	99%
[22]	5.80	12.60	99%

5. CONCLUSION

The performance of microstrip patch antenna can be improved due to studying effects of position of microstrip feeder and radiating patch from the center of substrate on x-axis. The achieved enhancement in performance was in terms of increasing gain of antenna as well as covering the ISM band at 5.8 GHz with eliminating SLL in antenna's radiation pattern without using techniques which would increase the antenna's dimensions and led to present a single plane with low profile antenna compared to low SLL antennas in previous works.

REFERENCES

[1] J. Malik and M. V. Kartikeyan, "A stacked equilateral triangular patch antenna with sierpinski gasket fractal for WLAN applications," *Progress In Electromagnetics Research Letters*, vol. 22, no. December 2010, pp. 71–81, 2011, doi: 10.2528/PIERL10122304. DOI: 10.2528/PIERL10122304

[2] C. Mak, H. Wong, and K. Luk, "High-Gain and Wide-Band Single-Layer Patch Antenna for Wireless Communications," *IEEE TRANSACTIONS ON VEHICULAR TECHNOLOGY*, vol. 54, no. 1, pp. 33–40, 2005.

[3] J. W. Kim, T. H. Jung, H. K. Ryu, J. M. Woo, C. S. Eun, and D. K. Lee, "Compact multiband microstrip antenna using inverted-l-and t-shaped parasitic elements," *IEEE Antennas and Wireless Propagation Letters*, vol. 12, pp. 1299–1302, 2013, doi: 10.1109/LAWP.2013.2283796. DOI: 10.1109/LAWP.2013.2283796

[4] M. T. Islam, M. N. Shakib, and N. Misran, "Design analysis of high gain wideband L-probe fed microstrip patch antenna," *Progress in Electromagnetics Research*, vol. 95, pp. 397–407, 2009, doi: 10.2528/PIER09080204. DOI: 10.2528/PIER09080204

[5] F. Mohamadi Monavar and N. Komjani, "Bandwidth enhancement of microstrip patch antenna using Jerusalem cross-shaped frequency selective surfaces by invasive weed optimization approach," *Progress in Electromagnetics Research*, vol. 121, no. October, pp. 103–120, 2011, doi: 10.2528/pier11051305. DOI: 10.2528/pier11051305

[6] H. W. Lai, K. M. Mak, and K. F. Chan, "Novel aperture-coupled microstrip-line feed for circularly polarized patch antenna," *Progress in Electromagnetics Research*, vol. 144, no. October 2013, pp. 1–9, 2014, doi: 10.2528/PIER13101803. DOI: 10.2528/PIER13101803

[7] C. Arora, S. S. Pattnaik, and R. N. Baral, "SRR superstrate for gain and bandwidth enhancement of microstrip patch antenna array," *Progress In Electromagnetics Research B*, vol. 76, no. 1, pp. 73–85, 2017, doi: 10.2528/PIERB17041405. DOI: 10.2528/PIERB17041405

[8] S. N. Nafea, A. Ismail, and R. S. A. Raja Abdullah, "Low side lobe level multilayer antenna for wireless applications," *Progress in Electromagnetics Research Letters*, vol. 58, 2016, doi: 10.2528/PIERL15112202. DOI: 10.2528/PIERL15112202

[9] L. G. Ayalew and F. M. Asmare, "Design and optimization of pi-slotted dual-band rectangular microstrip patch antenna using surface response methodology for 5G applications," *Heliyon*, vol. 8, no. 12, p. e12030, 2022, doi: 10.1016/j.heliyon.2022.e12030. DOI: 10.1016/j.heliyon.2022.e12030

[10] H. Srivastava, A. Singh, A. Rajeev, and U. Tiwari, "Bandwidth and Gain Enhancement of Rectangular Microstrip Patch Antenna (RMPA) Using Slotted Array Technique," *Wireless Personal Communications*, vol. 114, no. 1, pp. 699–709, 2020, doi: 10.1007/s11277-020-07388-x. DOI: 10.1007/s11277-020-07388-x

[11] A. S. B. Mohammed *et al.*, "Mathematical model on the effects of conductor thickness on the centre frequency at 28 GHz for the performance of microstrip patch antenna using air substrate for 5G application," *Alexandria Engineering Journal*, vol. 60, no. 6, pp. 5265–5273, 2021, doi: 10.1016/j.aej.2021.04.050. DOI: 10.1016/j.aej.2021.04.050

[12] B. Lv, X. Wang, C. Zheng, J. Huangfu, C. Li, and L. Ran, "Radiation Enhancement for Standard Patch Antennas Using a Loosely Grooved Ground Plane," *IEEE Antennas and Wireless Propagation Letters*, vol. 11, no. 1, pp. 604–607, 2012, doi: 10.1109/LAWP.2012.2202364. DOI: 10.1109/LAWP.2012.2202364

[13] S. Ye *et al.*, "Design of arbitrarily shaped planar microstrip antenna arrays with improved efficiency," *International Journal of Antennas and Propagation*, vol. 2013, 2013, doi: 10.1155/2013/757061. DOI: 10.1155/2013/757061

[14] H. Elftouh, N. A. Touhami, and M. Aghoutane, "Miniaturized microstrip patch antenna with spiral defected microstrip structure," *Progress in Electromagnetics Research Letters*, vol. 53, no. March, pp. 37–44, 2015, doi: 10.2528/PIERL15031003. DOI: 10.2528/PIERL15031003

[15] and O. M. R. Hussein Attia, Omar F. Siddiqui, Na'el Suwan, "ANALYTICAL AND EXPERIMENTAL STUDY OF GAIN ENHANCEMENT IN ANTENNA ARRAYS COVERED WITH HIGH INDEX METAMATERIAL SUPERSTRATE," *MICROWAVE AND OPTICAL TECHNOLOGY LETTERS*, vol. 55, no. 1, pp. 215–218, 2013.

[16] A. K. Singh, M. P. Abegaonkar, and S. K. Koul, "Miniaturized Multiband Microstrip Patch Antenna Using Metamaterial Loading for Wireless Application," *Progress In Electromagnetics Research C*, vol. 83, no. March, pp. 71–82, 2018.

[17] H. A. Majid, M. K. Abd Rahim, M. R. Hamid, and M. F. Ismail, "Frequency reconfigurable microstrip patch-slot antenna with directional radiation pattern," *Progress in Electromagnetics Research*, vol. 144, no. October 2013, pp. 319–328, 2014, doi: 10.2528/PIER13102901. DOI: 10.2528/PIER13102901

[18] A. S. Mekki, M. N. Hamidon, A. Ismail, and A. R. H. Alhawari, "Gain Enhancement of a Microstrip Patch Antenna Using a Reflecting Layer," *International Journal of Antennas and Propagation*, vol. 2015, p. 7, 2015.

[19] K. Gupta, K. Jain, and P. Singh, "Analysis And Design Of Circular Microstrip Patch Antenna At 5.8 GHz," *International Journal of Computer Science and Information Technologies*, vol. 5, no. 3, pp. 3895–3898, 2014.

[20] A. R. Vaidya, R. K. Gupta, and S. K. Mishra, "Right-Hand / Left-Hand Circularly Polarized High-Gain Antennas Using Partially Re fl ective Surfaces," *IEEE Antennas and Wireless Propagation Letters*, vol. 13, pp. 431–434, 2014.

[21] X. Chen, Z. Luo, Z. Zheng, P. Feng, and K. Huang, "Effective reflective characteristics of superstrates and their effects on the resonant cavity antenna," *IEEE Transactions on Antennas and Propagation*, vol. 63, no. 4, pp. 1572–1580, 2015, doi: 10.1109/TAP.2015.2399508. DOI: 10.1109/TAP.2015.2399508

[22] S. N. Nafea, "Improving Performance of Patch Antenna For IEEE 802.16e Applications Using Multi-layer Antenna Structure with Reflector," *NTCCIT 2018 - Al Mansour International Conference on New Trends in Computing, Communication, and Information Technology*, pp. 18–22, 2019, doi: 10.1109/NTCCIT.2018.8681183. DOI: 10.1109/NTCCIT.2018.8681183

Study of low-cost, flexible graphite-based sensor for human finger movement

M Kaviguru, T Pavithra, A Sobana, S M Naveena, G Anitha , V Suresh*, S Gangadhara, T Venkadesh

Department of Electronics and Communication Engineering, National Engineering College, Kovilpatti-628503, India.

* *Corresponding Author: vsece@nec.edu.in*

Abstract. Paper-based sensor devices have grown prominently in recent decades owing to their flexibility, light weight, biodegradability, and low cost. Pencil traces on the paper are the protuberant component to sense the physical parameter variation, and the geometry on the paper substrate plays a vital role in sensing applications. This study used the commercially available pencil as an electrode on a paper substrate that senses the mechanical strain on human finger movements. Two different geometries of the paper substrate were considered, and the optimized geometry was identified. Two different grades (HB and 8B) pencils were used as the drawing instrument to draw the graphite on the paper. The repeated experimental results show a better version of the proposed sensor for measuring human finger movements. The grade HB pencil provides excellent results compared to the other gradepencils.

Keywords. Paper substrate, Graphite, Pencil on Paper, Variable Resistance, Wheatstone bridge circuit and Differential amplifier circuit.

1. Introduction

The production of flexible and wearable sensing devices has grown exponentially in recent decades due to material and structural science advancements. The nanostructure formation of thematerials plays a vital role in fabrication of flexible health monitoring devices, microfluidic devices, touchpads, transistors, displays, and energy-storing devices. Different substrates have been utilized to develop flexible sensors. The reported research outcomes show that polydimethylsiloxane (PDMS) is a substrate for flexible force sensors, textile material is used to detect human motion, and plants and plant-derived biomaterials may also use to develop the adjustable sensor [1-4]. The fabrication of flexible sensor is required the clean room facility, high cost material and also skilled person. However, the recent literature shows that the sensors can be made using the official material in the home. Hence, paper and pencil graphite is used to fabricate the sensor to measure the strain as human finger movement. Paper is a low-cost thin material prepared from cellulose and pulp fibers and a versatile material. There are different typesof papers available to fabricate the paper-based device. The type of paper has been selected basedon the sensing application [5].

The more excellent performance of wicking ability in Whatman brand chromatography paper is used to fabricate flexible electrochemical sensors. Due to the high protein-binding capabilities of the Nitrocellulose membrane, it has been used to develop the sensor based on gold nanoparticlebased assays. Office printing paper is also used to fabricate electronic, electrochemical, and wearable devices [6-8]. The paper has various properties, such as lightweight, biodegradability, biocompatibility, transport liquid, and flexibility. With these advantages, enormous paper-based devices have been developed such as paper-based pregnancy testers, paper- based glucose test kits, paper-based gas sensors, paper-based touch sensors, paper-based sound source devices, paper-based power sources, etc. Paper generally acts as a substrate in a paper-based device [9- 11]. Micro or Nano forms of a chemical component may deposit on the paper substrate, and these devices produce output signals under specific mechanical, physical, or chemical parameter changes [12].

Pencil graphite is highly reactive, sensitive, low cost, and easy to handle compared to other carbon-based materials. Pencil lead is graphite with clay particles and sometimes contains a little wax. The pencils have been categorized into grades (H and B) based on the graphite-to-clay ratio. The 'H' grade denotes the hardness due to the higher clay content. The other grade, 'B' is for blackness due to higher graphite content [13]. Pencil-trace drawn on printing papers is the simplest and easiest way of constructing graphite-based devices [14]. Pencil on Paper (PoP) based sensor fabrication technique is getting significant attention from researchers because of its simplicity, portability, environmental friendliness, and low cost. The PoP sensor can be fabricated minimally by supporting simple office tools (paper, pencil, and scissors) [15]. The fabrication process doesn't require a cleanroom facility and complicated process steps like the MEMS-based sensor fabrication techniques [16].

This paper is organized as follows. The graphite on the paper and its impacts are discussed in section 2.The proposed methodology for sensing the human finger movement and its output are presented in section 3. The signal conditioning circuit utilized in this work is discussed in section 4. The results & discussion is discussed in section 5 then followed by a conclusion in section 6.

2. Graphite on paper

The prime objective of this work is to develop a graphite-based sensor on paper to sense human finger movement. Pencil leads (HB, and 8B, Graded Pencil Usl- 967) were used for the electrode, and plain documents (A4, 80 GSM, India) were used as substrate materials. Two different rectangular geometries (50 X 5) mm and (30 X 5) mm have been considered to trace the pencil graphite on paper. Two different grade pencils such as HB (Hardness) and 8B (Blackness) are employed to achieve the additional amount of graphite on paper. MECO makes 45CF - 4 ½ Digit 20000 Count LCD Digital Multimeter were used to measure the resistance value. Figure 1. shows the 79.03 Kilo Ohm resistance output for the (50 X 5) mm rectangular geometry of twenty numbersfree hand traced HB pencil. Figure 2. shows the 49.97 Kilo Ohm resistance output for the (30 X 5) mm rectangular geometry of twenty numbers free hand traced HB pencil.

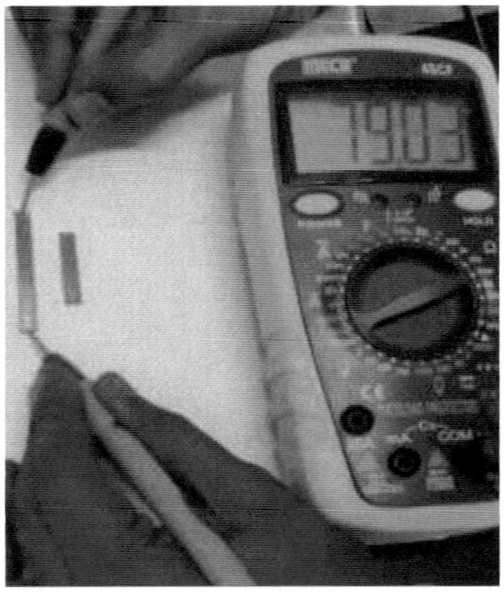

Figure 1. Resistance value of HB pencil traced (50 X 5) mm.

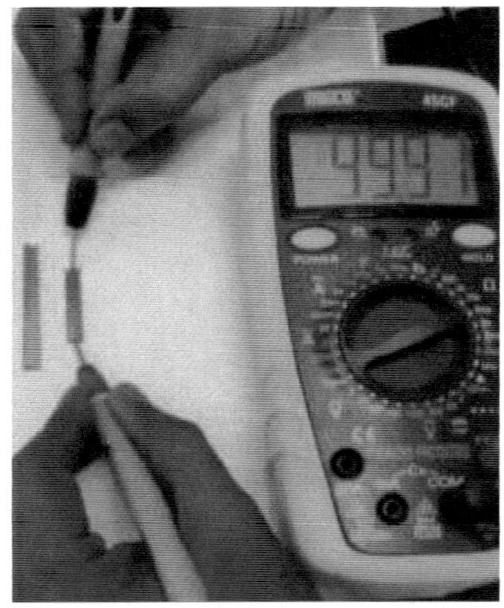

Figure 2. Resistance value of HB pencil traced (30 X 5) mm.

Figure 3. and 4. shows the 8.18 and 7.49 Kilo Ohm resistance of the (50 X5) mm and (30 X 5) mm rectangular geometry of twenty numbers free hand traced using an 8B grade pencil. It is observed that the geometry of the graphite region is minimized the resistance value is reduced. The hardness of the pencil grade also plays a vital role in resistance change, the higher hardness grade has a higherresistance value, and the higher blackness grade has a lower resistance value.

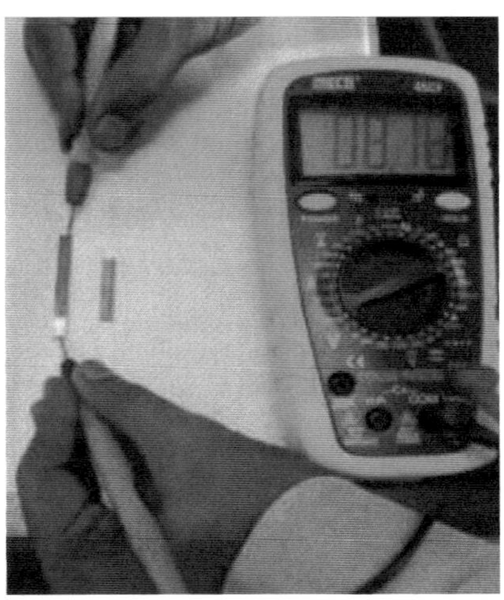

Figure 3. Resistance value of 8B pencil traced (50 X 5) mm.

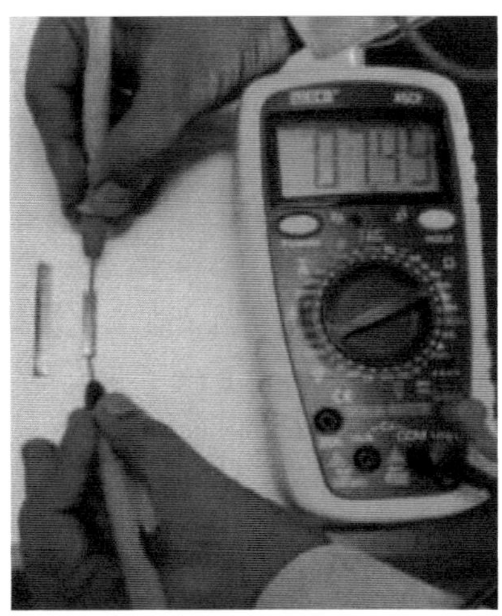

Figure 4. Resistance value of 8B pencil traced (30 X 5) mm.

IOP Conf. Series: Materials Science and Engineering 1291 (2023) 012047 doi:10.1088/1757-899X/1291/1/012047

The number of traces drawn on the paper also changes the resistance value. Table 1. and table 2. shows the resistance value for 10 and 20 trace drawn on the paper using HB and 8 B-grade pencils. The 10-number trace drawn on the paper has a higher resistance value, while the 20 traces drawn on the paper have a lower resistance value. And also, HB grade pencils for 20 traces have more resistance variation than the 8B grade pencil. More than 20 trace drawn on the paper has no significant changes in resistance value. For further study of this work, it is considered the 20 numbertrace drawn on the paper by using an HB grade pencil, to sense the finger movement of the Human.

Table 1. Resistance value for different traces of HB grade.

Pencil Grade	Length	Width	Pencil trace	Resistancekilo ohm
HB	30mm	5mm	10	110.50
HB	30mm	5mm	20	49.97
HB	30mm	5mm	10	290.00
HB	30mm	5mm	20	79.03

Table 2. Resistance value for different traces of 8B grade.

Pencil Grade	Length	Width	Pencil trace	Resistancekilo ohm
8B	30mm	5mm	10	18.96
8B	30mm	5mm	20	7.35
8B	30mm	5mm	10	87.21
8B	30mm	5mm	20	8.27

3. Sensor for finger movement

Figure 5. and Figure 6. shows the placement of twenty numbers free hand traced HB pencilsensor on a Human finger for a straight and bent position of (50 X 5) mm sensor.

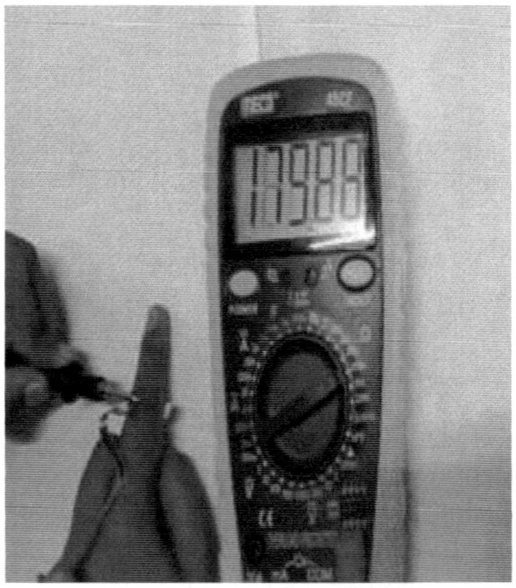

Figure 5. Resistance value of HB pencil – straight finger (50 X 5) mm.

Figure 6. Resistance value of HB pencil – bent finger (50 X 5) mm.

Figure 7. and Figure 8. show the placement of twenty numbers free hand traced 8B pencil sensor on a Human finger for a straight and bent position of (30 X 5) mm detector. Table 3. shows the resistance value of 20 traces drawn on the paper using an HB grade pencil for two rectangular geometry of (50 X 5) mm and (30 X 5) mm. It is observed that the rectangular geometry of (50 X5) mm has a resistance variation from 109.09 Kilo Ohm to 179.88 Kilo Ohm. The Wheatstone bridge circuit has been designed for measuring the resistance variation, and the circuit's output voltage occurs in the microcontroller for further processing.

Table 3. Resistance value for a straight and bent finger

Pencil Grade	**Length**	**Width**	**Straight Kilo Ohm**	**Bent Kilo Ohm**
HB	30 mm	5mm	103.60	67.10
HB	50 mm	5mm	179.88	109.09

IOP Conf. Series: Materials Science and Engineering 1291 (2023) 012047 doi:10.1088/1757-899X/1291/1/012047

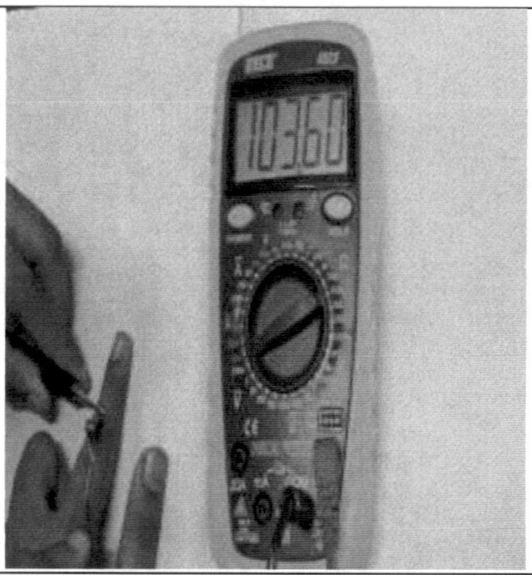

Figure 7. Resistance value of HB pencil – straight finger (30 X 5) mm.

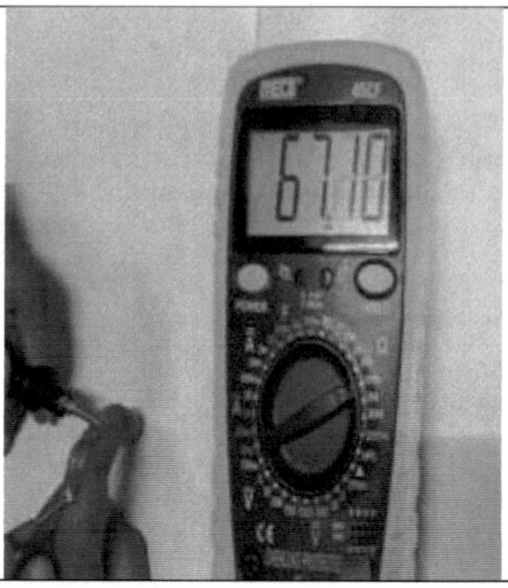

Figure 8. Resistance value of HB pencil – bent finger (30 X 5) mm.

4. Signal Conditioning Circuit

Wheatstone bridge circuit is considered for converting the change in resistance value (Rx) of sensorin to the voltage. The 110 Kilo Ohm resistance value is considered an R-value. The Rx value variesdepending on the finger position movement. The output voltage (Vbc) of the Wheatstone bridge isamplified with the support of an Instrumentation amplifier. The gain value of the amplifier is considered as '2'. Figure 9. shows the simulation study of the overall circuit considered for this work. Proteus 8 software is used for simulation study.

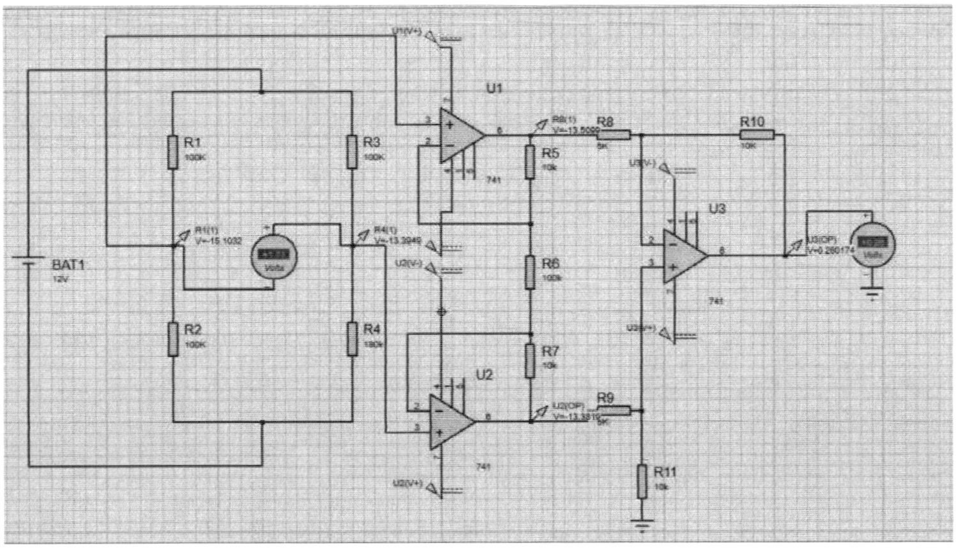

Figure 9. Circuit Simulation using Proteus 8 software.

5. Result and Discussion

The experimental output for four different positions of the human finger movement has been observed. The repeated experiment cycles for the exact position of the finger were noted 20 times,and the average value of the change in resistance is considered for the analysis. It is observed that the sensor has produced approximately 180 Kilo Ohm resistance value for the fully bent finger position, and the corresponding amplifier output voltage is 2.66 Voltage. The resistance output for the finger's small bent part is approximately 120 Kilo Ohm, and the corresponding amplifier output voltage is 0.434 Voltage. As the bent position of the finger increases, the changein the resistance value of the sensor also increases, and the output voltage increases. The theoreticaland experimental output for the different movements of the human finger is represented in table4.

Table 4. Resistance value for straight and bent finger

Finger position	**Change in Resistance – Rx (Kilo ohm)**	**Theoretical Vb – Vc - Volt**	**Experimental Vb – Vc -Volt**	**Theoretical V0 - Volt**	**Experimental Vo – Volt**
Straight	02	0.045	0.042	0.090	0.108
Slight Bent	10	0.217	0.224	0.434	0.451
Partial Bent	40	0.769	0.782	1.538	1.545
Fully Bent	70	1.206	1.219	2.412	2.425

The error percentage of theoretical and practical amplifier output for the finger movement of the Human is presented in Table 5. The formula utilized for error percentage calculation is represented in equation(5). The minimum error percentage of theoretical and practical amplifier output shows better performance of the proposed method. Figure 10. and 11. shows the experimental output for straight and bent movement of the human.

Error percentage = (Experimental (Vo) – Theoretical (Vo) / Theoretical (Vo)) * 100 (5)

Table 5. Error Percentage of Amplifier Output

Finger position	**Change in resistance – Rx(Kilo ohm)**	**TheoreticalVo – Volt**	**ExperimentalVo – Volt**	**Error percentage**
Straight	2	0.090	0.098	8.88
Slight Bent	10	0.434	0.451	3.91
Partial Bent	40	1.538	1.545	0.45
Fully Bent	70	2.412	2.425	0.54

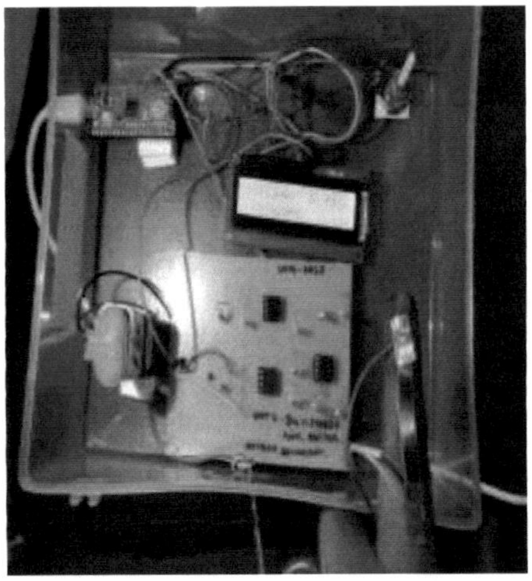

Figure 10. Experimental output for straight finger movement.

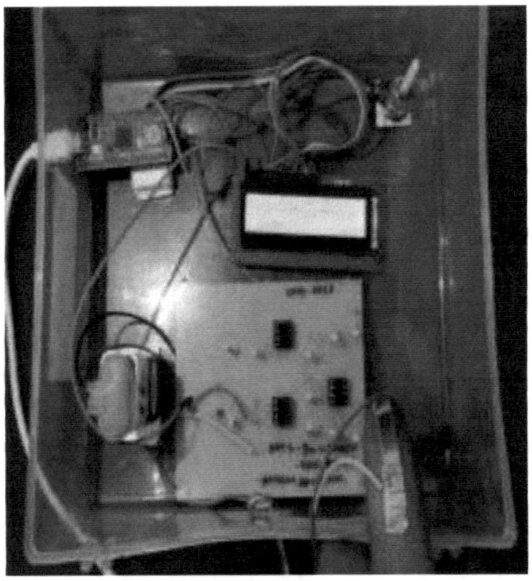

Figure 11. Experimental output for bentfinger movement.

6. Conclusion

The developed Pencil on Paper (PoP) based sensor is analyzed to sense the human finger movement. Two different sizes of the rectangular geometry of the paper substrate have been studied and observed that the (50 X 5) mm geometry produced better results. The performance of HB and 8B grade pencils has been analyzed, and it's concluded that the HB grade has higher resistivity because of the hardness. The number of traces drawn on the paper was also investigated, and it identified that the 20-number trace is optimized. The repeated experimental output shows that the better performance of the proposed graphite-based sensor is well-suited for sensing human finger movement. The minimum error percentage of theoretical and practical amplifier output agreements a better version of the proposed method. Due to the paper's low-cost, biodegradable, and environmentally friendly nature, this kind of device may be utilized to sense the Robot arm movement and its automation in the future.

References

[1] Choong CL, Shim MB, Lee BS et al 2014 *Highly stretchable resistive pressure sensors using a conductive elastomeric composite on a micropyramid array. Adv Mater,* vol 26, 3451–3458.

[2] Yang Z, Pang Y, Han XL et al 2018 *Graphene textile strain sensor with negative resistancevariationfor human motion detection. ACS Nano,* vol 12, 9134–9141.

[3] Wang L, Wang K, Lou Z et al 2018 *Plant-based modular building blocks for "green" electronicskins. Adv Funct Mater*, vol 28, 1–16. J.

[4] Yang *et al* 2022 Printable and Wearable Graphene-Based Strain Sensor With High Sensitivity forHuman Motion Monitoring *IEEE Sensors Journal*, vol. 22, no. 14, pp. 13937-13944

[5] Wang L, Wang K, Lou Z et al 2018 *Plant-based modular building blocks for "green" electronicskins. Adv Funct Mater*, vol 28, 1–16.

[6] Singh, A.T., Lantigua, D., Meka, A., Taing, S. Pandher, M., Camci-Unal, G. 2018 *Paper-BasedSensors: Emerging Themes and Applications. Sensors*, vol *18*, 2838.

[7] Ellerbee AK, Phillips ST, Siegel AC, Mirica KA, Martinez AW, Striehl P, Jain N, Prentiss M, Whitesides GM. 2009 *Quantifying colorimetric assays in paper-based microfluidic devicesbymeasuring the transmission of light through paper. Anal Chem.*, vol 81(20), 8447-52.

[8] Lisa M, Chouhan RS, Vinayaka AC, Manonmani HK, Thakur MS. 2009 *Gold nanoparticles based dipstick immunoassay for the rapid detection of dichlorodiphenyltrichloroethane: an organochlorine pesticide. Biosens Bioelectron.*, vol 25(1), 224-227.

[9] Santhiago M, Corrêa CC, Bernardes JS, Pereira MP, Oliveira LJM, Strauss M, Bufon CCB. 2017 *Flexible and Foldable Fully-Printed Carbon Black Conductive Nanostructures on Paper for High-Performance Electronic, Electrochemical, and Wearable Devices. ACS Appl Mater Interfaces.* vol 9 (28), 24365-72.

[10] Yao-Hung Tsai, Ting Yang, Ching-Fen Shen, Chao-Min Cheng, 2022 *Paper-based immunoassays for mobile healthcare: strategies, challenges, and future applications, Paper-based Analytical Devices for Chemical Analysis and Diagnostics,* Elsevier, Chapter 10, pp245-257,

[11] Tian H, Ren TL, Xie D, Wang YF, Zhou CJ, Feng TT, Fu D, Yang Y, Peng PG, Wang LG, LiuLT. 2011 *Graphene-on-paper sound source devices. ACS Nano.* vol 5(6), 4878-85.

[12] Isao Shitanda, Yukiya Morigayama, Risa Iwashita, Himeka Goto, Tatsuo Aikawa, Tsutomu Mikawa, Yoshinao Hoshi, Masayuki Itagaki, Hiroyuki Matsui, Shizuo Tokito, Seiya Tsujimura, (2021*). Paper-based lactate biofuel cell array with high power output, Journal ofPower Sources,* 489.

[13] Jeevanandam J, Barhoum A, Chan YS, Dufresne A, Danquah MK. (2018). *Review on nanoparticles and nanostructured materials: history, sources, toxicity and regulations.*

Beilstein J Nanotechnol. 9, 1050-74.

[14] Graphite Grading Scale Explained, (Accessed on 05.01.2022), https://pencils.com/pages/hb-graphite-grading-scale.

[15] Yan Liu, Shuting Mo, Siyao Shang, Peng Wang, Wei Zhao, Lin Li, 202 *Handwriting flexible electronics: Tools, materials and emerging applications*, *Journal of Science: Advanced Materials and Devices*, 5(4), 451-467.

[16] W. Mazurkiewicz, M. Podrażka, E. Jarosińska, K. Kappalakandy Valapil, M. Wiloch, M. Jönsson
-Niedziółka, E. Witkowska Nery (2020). *Paper-Based Electrochemical Sensors and How to Make Them (Work)*, *ChemElectroChem*, 7(14), 2939.

[17] René Harťanský , Martin Mierka , Mikuláš Bittera, Jozef Hallon, Ján Halgoš, Jaroslav Hricko, Robert Andok, Michal Rafaj, (2020). *Novel Method of Contactless Sensing of Mechanical Quantities* , *Measurement Science Review,* 20,(3), 150-156.

Wafer Defect Identification with Optimal Hyper-Parameter Tuning of Support Vector Machine using the Deep Feature of ResNet 101

Shishir Prasad Dash [1, a], J. Ramadevi [2, b], Rajat Amat [3, c],
Prabira Kumar Sethy [4, d] * Santi Kumari Behera [5, e] and, Sunil Mallick [6, c]

[1] Department of Electronics, Sambalpur University, Jyoti Vihar, Burla 768019, India

[2] Computer Science and Engineering Department, PVP SIDDHARTHA Institute of Technology, Vijayawada, India

[3] Department of ECE, SUIIT, Sambalpur University, Jyoti Vihar, Burla 768019, India

[4] Department of Electronics, Sambalpur University, Jyoti Vihar, Burla 768019, India

[5] Department of CSE, VSSUT Burla 768018, India

[6] Department of ECE, SUIIT, Sambalpur University, Jyoti Vihar, Burla 768019, India

[a]shishirprasaddash98@gmail.com, [b]K.ramakarthik@gmail.com, [c]amatrajat4@gmail.com,

[d]prabirsethy.05@gmail.com, [e]b.santibehera@gmail.com, [f]mallicksunil77@gmail.com

Keywords: Wafer, Defect identification, Resnet, Support vector machine, CNN.

Abstract: As semiconductor processing technologies continue to advance, semiconductor wafers are becoming more densely packed and intricate, resulting in a higher incidence of surface imperfections. Therefore, it is crucial to detect these defects early and accurately classify them to pinpoint the root causes of the defects in the manufacturing process, ultimately leading to improved yield. Therefore, defect detection is critical in the industrial production of monocrystalline silicon. This study employs deep learning techniques to propose a framework for detecting defects on silicon wafers, focusing on optimizing the hyperparameters of support vector machines (SVM). Three methods were utilized to fine-tune the SVM parameters: Bayesian optimization, grid search, and random search techniques. This study demonstrates how selecting optimal values for SVM parameters can lead to better classification. Additionally, real manufacturing data were utilized to evaluate the performance of the proposed SVM classifier, with a comparison to state-of-the-art techniques in the field. By using deep features from ResNet 101 and a support vector machine, this work achieves 74.5% accuracy in identifying wafer defects without employing any optimization technique. However, the performance of the model was further improved by utilizing the random search optimization technique, which yielded the best result among the three optimization techniques tested, with an accuracy of 88.1%.

Introduction

The manufacturing sector suffers enormous losses owing to semiconductor wafer defects. Additionally, it has a significant impact on product development. For wafer defect categorization, semiconductor engineers use a variety of methods such as manual data extraction of pertinent attributes or algorithms based on machine learning. However, these methods are unreliable and it is necessary to increase their classification power. To automatically identify wafer faults, this study proposes a convolutional neural network based on deep learning. The variety of potential faults makes it challenging to develop a reliable machine-vision system that can correctly identify and categorize numerous types of wafer defects. Determining the causes of flaws, which can be achieved by examining defect patterns, is essential for reducing the frequency of defects. The wafer test time can be reduced, and the yield can be increased using a productive defect detection method. Pre-processing and categorization techniques frequently lead to the loss of important sensor signal data that are necessary for identifying wafer defects. To manufacture semiconductors, wafer flaws must be identified and categorized accurately. They allow for the implementation of quality control and yield enhancement techniques, and provide useful insights into potential fault sources.

The DTL-MobileNetV2 model performed better in identifying and categorizing six types of monocrystalline silicon wafer defects, including cracks, double contrast, holes, microcracks, sawmarks, and stains, owing to the use of an existing deep-learning model to extract features from monocrystalline silicon wafer defects. When tested against the testing set, the DTL-MobileNetV2 model-based classification method for monocrystalline silicon wafer defects attained a remarkable accuracy of 98.99%. As a result, DTL is suitable for minimizing misclassification and increasing the total production capacity [1]. This demonstrates that DTL is a highly effective method for identifying various types of flaws in monocrystalline Si wafers. We used a data augmentation strategy to address the problem of class imbalance. The proposed model uses convolutional layers to extract significant characteristics rather than manually. Additionally, cutting-edge regularization methods, including batch normalization and spatial dropout, are used to improve the classification performance of the CNN-WDI model. Nine different types of wafer map defects were classified with an average accuracy of 96.2% using the CNN-WDI model [2]. a visual classification technique based on machine learning for visible surface flaws in semiconductor wafers. The proposed method uses deep learning-based convolutional neural networks to identify and classify four different types of surface flaws: center, local, random, and scrape. The method makes use of deep learning-based convolutional neural networks to recognize and classify four different types of surface flaws: center, local, random, and scraping flaws. The experimental findings showed that the proposed method, without any further improvement, had a high accuracy rate of 98–99%. Regarding its effectiveness for wafer-defect classification in experiments, the method performed better than previous machine learning approaches [3]. This study introduces DLADC, an ADC system that uses deep learning technologies and a deep convolutional neural network (CNN) architecture to identify and categorize surface defects on semiconductor wafers. Scanning electron microscopy (SEM) images are the input for the proposed system, which outputs the type and location of the identified fault. Additionally, knowledge about the size of the flaws can offer important insights into the cause of machine failure, which can help to enhance the manufacturing process and reduce defects. Therefore, the suggested system's subclassification of particle-type defects into different size groups is quite helpful for semiconductor makers. Furthermore, based on experimental findings using a genuine semiconductor defect dataset, an increased accuracy rate of 93.69% was attained [4]. constructed a machine learning system to look for surface flaws on semiconductor wafers. This technology combines a mask R-CNN-based AI algorithm with an optical scanning microscopy device. A dataset of microscopic images of wafers with different devices, including MEMS, silicon photonics, and superconductors, at different manufacturing phases, was used to train the system. Images of surface flaws were included in the training dataset. With an average precision of 92.6% and recall of 93.8, the experimental findings demonstrated that the proposed system achieved high flaw detection and classification accuracy. Results from testing on the most recent dataset of 192 photos with a resolution of 1600 × 1200 pixels and a 5x optical magnification revealed an accuracy of 86% and a detection time of 0.5 seconds per image [5]. The DEFF suggests that the ADC method makes use of an ensemble feature structure. The system has a decision network layer and featured network layer. The ensemble features are computed by concatenating the features learned by various pre-trained CNN models to represent wafer faults. Different types of wafer surface damage can be categorized automatically using the DEFF approach [6]. A CNN is a well-liked deep learning architecture, and its acceptance is credited with its efficiency. To perform wafer map retrieval tasks and classify defect patterns, this study suggests employing CNN and XGBoost. The authors assessed the performance of CNN and XGBoost against various machine-learning models, such as support vector machines, random decision forests, and adaptive boosting. The findings demonstrate that the proposed approach, which combines CNN and

XGBoost, outperforms competing models, with XGBoost being the most widely used machine-learning framework among data scientists. This is considered to be the first time such an accuracy has been recorded for this challenge, as the proposed solution employing CNN with extreme gradient boosting attained a high classification accuracy of 99.2% for the test dataset [7]. The suggested DL model was built on convolutional neural networks and utilized to identify wafer defect patterns. (CNNs). Convolutional and fully linked layers were employed with CNNs for feature extraction, classification, and pooling. Eight convolutional layers and three fully linked layers constitute the model. CNN is a reliable technique for locating flaws in semiconductor wafers. Even in the presence of random noise, it performs well for both single-defect detection and multiple-defect identification. In particular, the model exhibited 84 percent accuracy for mixed defect detection and 100 percent efficiency for single-defect pattern recognition [8]. The stacked denoising autoencoder (SdA), a deep-learning technique, was used in the proposed method for fault detection and classification (FDC). By locating global and invariant features in sensor signals, the SdA model can extract features and categorize them simultaneously. In comparison to 12 other models that used various feature extractors and classifiers, experiments on wafer samples obtained from a photolithography tool revealed that using a stacked denoising autoencoder (SdA) to create a fault detection and classification (FDC) model led to higher classification accuracy. In particular, when the measurement noise severity increased, the SdA model achieved up to 14% higher classification accuracy [9]. The YOLO architecture was assessed for wafer map defect identification and classification, the You Only Look Once (YOLO) architecture is assessed. On a dataset of 19200 wafer maps, the YOLOv3 and YOLOv4 variations achieved more than 94% classification accuracy in real time. assessment of the You Only Look Once (YOLO) architecture for the identification and localization of wafer defects. The categorization of wafer defects using the ResNet50 and DenseNet121 architectures was examined in the absence of localization capabilities. While ResNet50 and DenseNet121 only achieved 89% and 92% accuracy, respectively, the YOLOv3 and YOLOv4 variations of the YOLO architecture achieved over 94% classification accuracy in real time. Conclusion: Defects in semiconductor wafers can be identified and classified using object detection algorithms [10].

Dataset and Methodology

About Dataset

The dataset of wafer defects was obtained from the Kaggle website, comprising 54.02k images. The dataset was divided into two categories: balanced and imbalanced. Both varieties contained nine distinct types of wafer surface defects. "Experienced process engineers are hired to define wafer defect patterns and assign them unique labels such as Center, Donut, Local, Edge-Loc, Edge-Ring, Scratch, Random, Near-Full, and None. Additionally, these defects become more common owing to the increasing integration density of circuits and wafer design complexity. Each wafer defect occurs because of the specific abnormal behavior of certain fabrication processes. For instance, center defects may occur due to uniformity issues in chemical and mechanical planarization, Edge-Loc defects may occur because of thin film deposition, and Edge-Ring defects due to etching problems" [2]. In the imbalanced dataset, the "None" category has the highest number of images, with 36.4k, followed by "Edge_Ring" with 8,554 images, "Center" with 3,462 images, "Edge_Loc" with 2,417 images, "Scratch" with 500 images, "LOC" with 1,620 images, "Donut" with 409 images, "Random" with 609 images, and "Near_Full" with only 54 images.

In contrast, the balanced dataset had an equal number of images for each defect type, with 409 images per category. Each image in both the datasets had a resolution of 26 × 26 pixels. Figure 1 displays the samples of wafer surface defects.

IOP Conf. Series: Materials Science and Engineering 1291 (2023) 012048 doi:10.1088/1757-899X/1291/1/012048

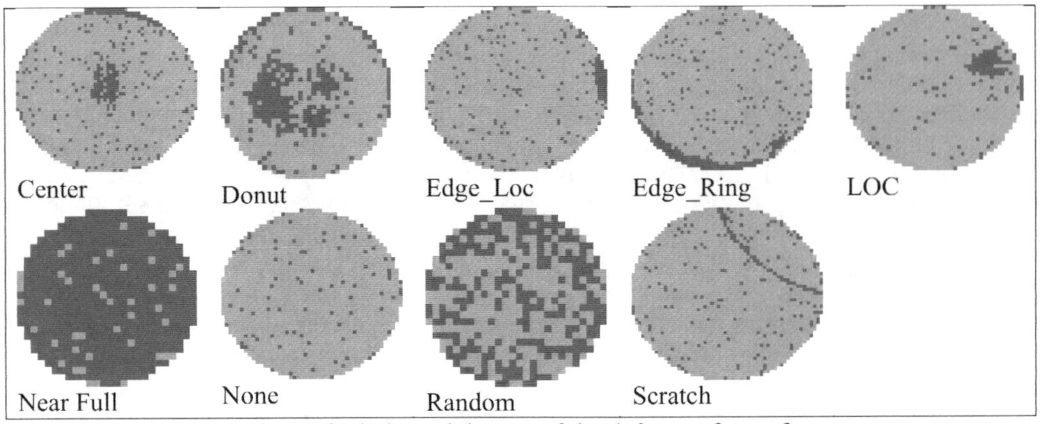

Fig. 1 Shows the balanced dataset of the defect wafer surface

Proposed Methodology

A strategy must be devised for constructing and sustaining the categorization algorithm in an actual production setting, where wafers go through multiple pieces of machinery before being assessed for their yield. To take advantage of early yield forecasting, the classification model must capture wafer measurements before their arrival at the end of the production line. Nonetheless, the manufacturing process can encounter alterations or deviations that affect the effectiveness of the classifier. Therefore, we suggest empowering an experienced process engineer to determine when to revise a model. This determination can be made with the aid of in situ monitors that are utilized in every processing equipment to monitor the key variables of the process. However, the SVM model must wait until several new wafers have been processed with the updated change and subsequently evaluated at the end of the line stage. The gathered measurements and yield values linked to these new m wafers were then employed as the new training set for updating the SVM model. The experimental section demonstrates how the adaptable SVM model enhances the yield classification when the model change is implemented following the desired process changes [11].

Support Vector Machine

Recently, SVM, developed by Vapnik in 1995, has emerged as a method that has gained significant attention owing to its impressive outcomes. The primary distinction between ANN and SVM lies in the principle of risk minimization. Whereas ANN employs empirical risk minimization to minimize the error in the training data, SVM implements the principle of Structural Risk Minimization by constructing an optimal separating hyperplane in the hidden feature space and utilizing quadratic programming to determine a unique solution. To achieve this, the SVM constructs an optimal separating hyperplane WXWX + b=0, thereby implementing the principle of Structural Risk Minimization.

To obtain the optimal hyperplane: {x ∈S(w,x)+b=0}, the vector w's norm must be minimized. Conversely, margin 1/||w|| between the two classes should be maximized. This is achieved by $\min_{i=1,2,..n} |(W, X) + b| = 1.$ (1)

The circled points in SVM are referred to as "support vectors." These vectors satisfy the condition Yi(W.Xi) + b = 0 and confine the margin. Therefore, any movement of these support vectors changes the normal vector W.

Nonlinear SVM utilizes a mapping function ϕ to map the training samples from the input space into a higher-dimensional feature space. In the nonlinear case, the data are first mapped to another Euclidean space H using a mapping function ϕ: Rd → H. Instead of dot products, a "kernel function" K is employed, such that K(Xi, Xj) = ϕ(Xi). ϕ(Xj). There are various kernel functions, including polynomial, radial basis function (RBF), and sigmoid kernel, which are represented as functions (2)

IOP Conf. Series: Materials Science and Engineering 1291 (2023) 012048 doi:10.1088/1757-899X/1291/1/012048

– (4), respectively. Properly setting the kernel parameters in these kernel functions is crucial for enhancing the classification accuracy.

Polynomial kernel: $K\ (X_i, X_j) = (1+ Xi.\ Xj)^{d}$, (2)

Radial basis function kernel: $K\ (X_i, X_j) = exp(-\gamma||\ K\ (X_i - X_j)\ ||)^2$ (3)

Sigmoid kernel: $K\ (X_i, X_j) = tanh\ (KXi\ .\ Xj - \delta)$ (4)

where γ is the kernel parameter. The kernel parameter γ is the most critical parameter, whereas the SVM technique is used to classify the samples [12].

Optimization Techniques

To achieve optimal results using deep learning algorithms, it is essential to tune their parameters appropriately. Selecting a powerful deep-learning algorithm and adjusting its parameters are critical for developing a high-accuracy classification model. However, manually performing parameter optimization can be extremely time-consuming, particularly when the learning algorithm has many parameters. One of the biggest challenges in setting up an SVM model is selecting an appropriate kernel function and determining the optimal parameter values. Poor classification outcomes often result from the selection of inappropriate parameter settings. Therefore, it is crucial to identify the most appropriate parameter settings for an SVM model in order to ensure the best possible classification results.

Hyperparameters in neural networks include the number of layers, learning rate, momentum, mini-batch size, and other factors that directly impact training performance. Therefore, optimizing the hyperparameters and selecting the most suitable set to achieve good training results, robustness, and strong generalization ability of the neural network is essential. However, optimizing hyperparameters is a complex combinatorial optimization problem that can be challenging. Furthermore, evaluating a set of hyperparameter configurations is time-consuming and costly. Therefore, it is crucial to carefully and efficiently optimize hyperparameters to achieve the best possible training outcomes [13].

This paper employed three methods to tune the SVM parameters: Bayesian optimization, grid search, and random search techniques. The aim of parameter tuning is to reduce testing time while maintaining test accuracy, and a feasible parameter setting must achieve comparable accuracy to the original model.

Grid Search Optimization Technique

The grid search method systematically tests a predefined subset of hyperparameter values by exhaustively evaluating all the possible combinations of hyperparameters within a specified range. Hyperparameters are typically defined using the minimum value, maximum value, and step size for each hyperparameter. The search can be conducted on linear, quadratic, or logarithmic scales. Performance metrics are then used to evaluate the performance of each combination of hyperparameters.

Grid search tunes the SVM hyper-parameters (such as C, γ, degree, etc.) using a performance metric based on the cross-validation (CV) technique. The primary objective is to find optimal hyperparameter values that lead to an accurate prediction of new data by the classifier while avoiding overfitting.

To determine the optimal values of C and γ using k-fold cross-validation, we divided the available data into k subsets (typically, k=10). One subset was used as the validation set and the model was trained on the remaining k-1 subsets. Next, we evaluated the performance of the SVM classifier on the validation set by using different values of C, γ, and other parameters. Next, we repeated this process for all k possible partitions of the data and calculated the average cross-validation error for each combination of hyperparameters. Finally, the combination that achieved the highest cross-validation accuracy (or lowest error) was selected and used to train an SVM on the entire dataset [14].

A. Random Search Optimization Technique

A grid search can result in unnecessary computations of the hyperparameters that have little impact on the performance of the model. A more efficient technique called random search can be used to overcome this problem. In a random search, hyperparameters are randomly sampled from the search space, and the configuration with the best performance is chosen. The intuition behind this approach is that a sufficiently large set of random samples is likely to include the global optimum or a close approximation with high probability. Moreover, random search is typically faster than grid search [13].

Bayesian Optimization Technique

The Bayesian optimization algorithm finds that the best parameter values are different for both grid search and random search. Bayesian optimization is an adaptive method for hyperparameter optimization. It uses information from previous experiments to predict the next set of parameters that might provide the greatest benefit. Bayesian optimization algorithms are different from grid search and random search in that they use all the information they already know when testing new parameter combinations. The Bayesian optimization algorithm starts by making a prior distribution based on what is known about the objective function. As new sampling points are tested, this information is used to update the prior distribution. This creates a posterior distribution that better represents the true objective function. The algorithm then tests the points in the area where the posterior distribution indicates that the global optimum is most likely to be found. This process of testing new points and updating the previous distribution occurs repeatedly until the algorithm determines the best set of hyperparameters. In Bayesian optimization, the algorithm balances exploration and exploitation so that it does not become stuck in a local optimal. Exploration refers to looking for new places to take samples from in the hyperparameter space, whereas exploitation refers to taking samples from places that have worked well in the past. By finding a middle ground between these two methods, Bayesian optimization can obtain the global optimum of the objective function with fewer samples than grid or random searches [13].

Resnet 101

The ResNet-101 network has 101 layers and is a deep convolutional neural network. Its design was based on the VGG-19 model, and it is one of the most complex structures proposed for the ImageNet competition. In a CNN, many layers are linked to one another and trained to perform different tasks. ResNet learns various feature levels through its layers. The convolutional layers of the model usually have 33 filters. One aspect that makes ResNet stand out is that each layer has the same number of filters for the same output feature map size. If the size of the output feature map is reduced by half, the number of filters doubles. This helps ensure that the temporal complexity of each layer remains the same. Direct downsampling is performed by combining two layers with a stride of two. ResNet ends with a global average pooling layer and a fully connected layer that is activated by SoftMax. ResNet's main characteristic is residual learning, which can be thought of as the subtraction of input features acquired from that layer. ResNet uses short-cut connections between each pair of 33 filters. The input of the kth layer was directly connected to the output of the (k+xth)th layer. This skips layers and helps prevent the problem of gradients that get smaller and smaller until they disappear. This is done by recycling the activations from the layer before the current layer until the layer after it has learned its weights. While the network was being trained, the weights made the next layer louder and the layer before it quieter. ResNet is easier to train than ordinary deep convolutional neural networks and solves the problem of accuracy loss [15].

Results and Discussion

In this experiment, two different methods were used to compare the performance of the SVM models. The first method involves using the SVM model without any optimization technique, whereas the second involves using the SVM model with optimization techniques to improve the classification accuracy. This experiment used three different optimization techniques: Bayesian, Grid Search, and Random Search. The proposed method was implemented on an HP Victus laptop with a 12th generation Intel Core i7 CPU, running on the Windows 11 operating system, including a built-in NVIDIA graphics processing unit, and employing MATLAB 2022a software. The training dataset comprised 70% of the data, the validation dataset comprised 20%, and the remaining 10% was reserved for testing the augmented dataset. The classification accuracy refers to the proportion of wafers that were classified correctly from the total number of wafers. In contrast, the false-negative rate is the percentage of bad wafers incorrectly classified as good.

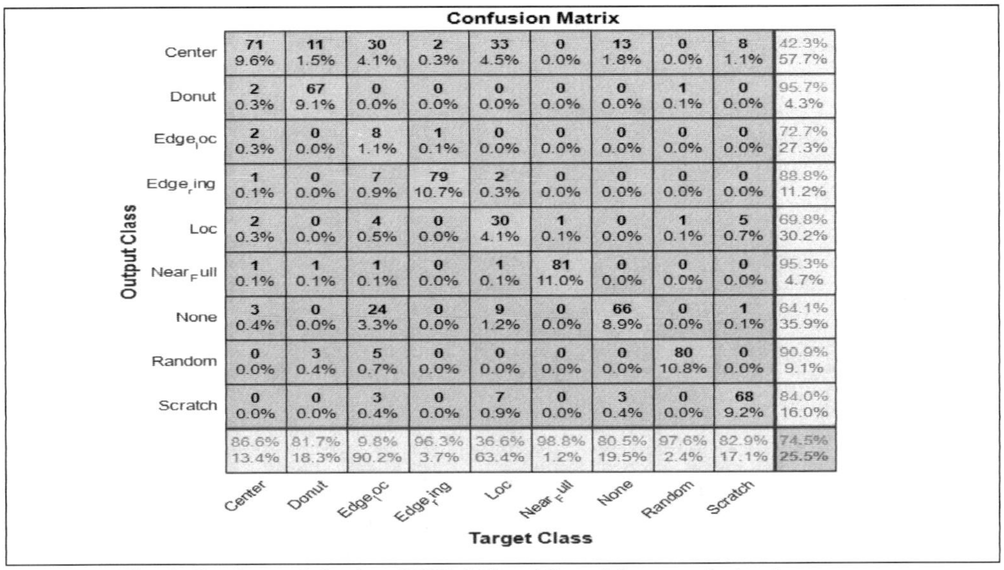

Fig. 2 Confusion Matrix of SVM Classifier without optimization

Figure 2 presents the accuracy scores of Resnet 101 using SVM hyperparameters for nine different types of wafer defect detection. As shown in the figure, the model achieved an accuracy score of 74.5%. However, because accuracy is critical in wafer defect detection in the manufacturing industry, it is necessary to improve it further. To achieve this, an optimization technique was employed with the SVM classifier to enhance the accuracy of wafer defect detection.

Bayesian optimization has emerged as a popular technique for addressing optimization problems in various domains where traditional numerical methods fall short. For example, one of its popular applications is hyperparameter tuning, which aims to minimize the validation error of a machine learning algorithm by tuning its hyperparameters. The process involves evaluating the objective function, which is the validation error, by training the machine learning model and assessing its performance on validation data. In this context, Bayesian optimization was applied in conjunction with an SVM classifier to improve the accuracy of wafer defect detection.

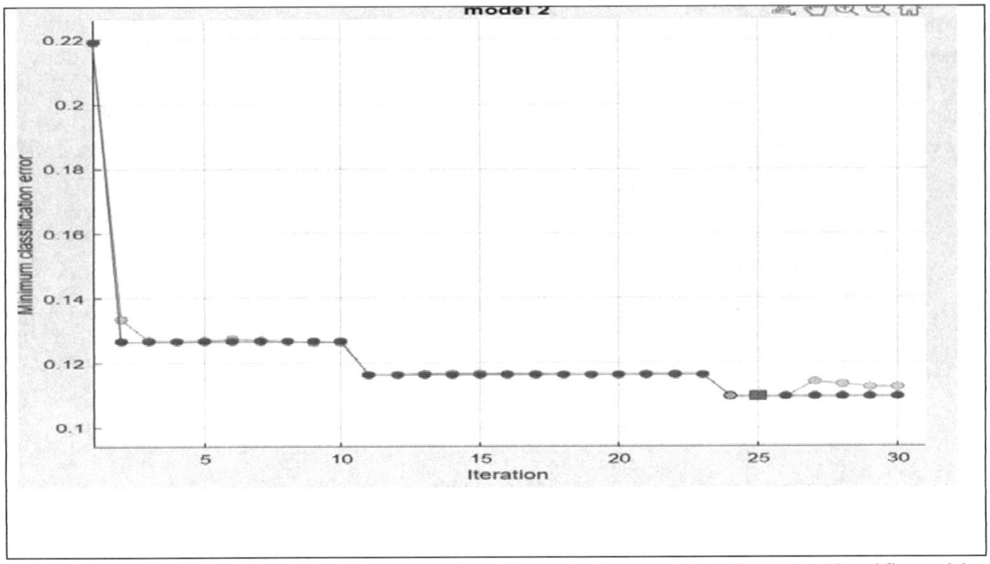

Fig. 3 Iteration Vs. Classifier Error graph and confusion matrix of SVM Classifier with Bayesian Optimization

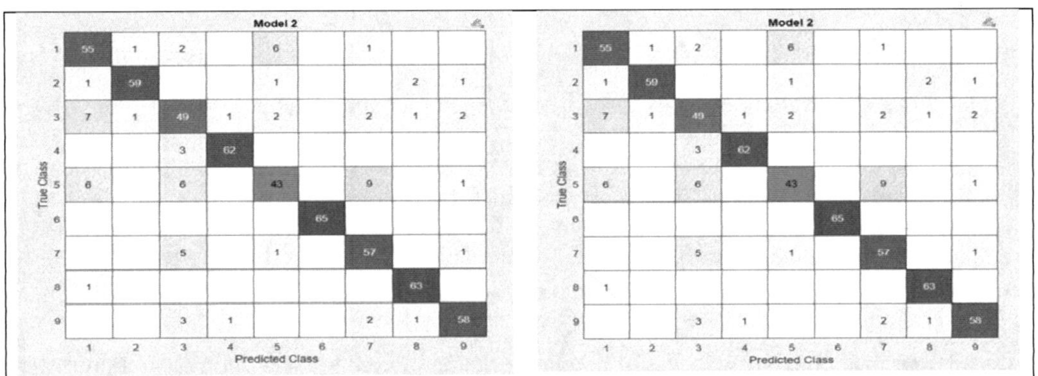

(a) (b)

Fig. 4 Performance of SVM with Bayesian optimization (a) confusion matrix in validation data (b) confusion matrix on test data

Bayesian optimization was utilized to tune the hyperparameters of the SVM method by employing a one-vs-one approach for multiclass classification. The hyperparameters chosen were the box constraint level and kernel scale, with values of 825.6872 and 12.9019, respectively. A Gaussian Kernel function is employed. To evaluate the performance of the proposed method, validation and evaluation were performed. After 30 iterations, the minimum classification error is 0.10946. The proposed method was validated using 20% of the dataset, and it achieved a validation accuracy of 87.8%. The testing accuracy was evaluated using 10% of the dataset, and an accuracy of 85% was achieved. Figure 3 presents the Iteration Vs. Classifier Error graph and confusion matrix of SVM Classifier with Bayesian Optimization The confusion matrix of this approach is shown in Figure 4.

A grid search is a systematic search based on a predefined subset of the hyperparameter space. It partitions the range of parameters to be optimized into a grid and evaluates all the points to obtain the optimal parameters. The grid search optimizes the SVM parameters using cross-validation as the performance metric. Our next strategy for enhancing accuracy is to utilize the grid search optimization technique with the SVM classifier.

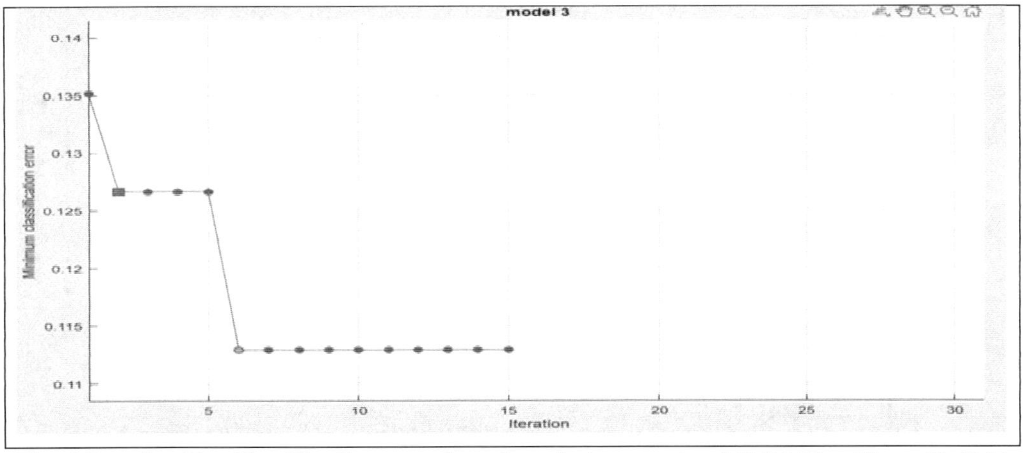

Fig. 5 Iteration Vs. Classifier Error graph and confusion matrix of SVM Classifier with Grid Search Optimization

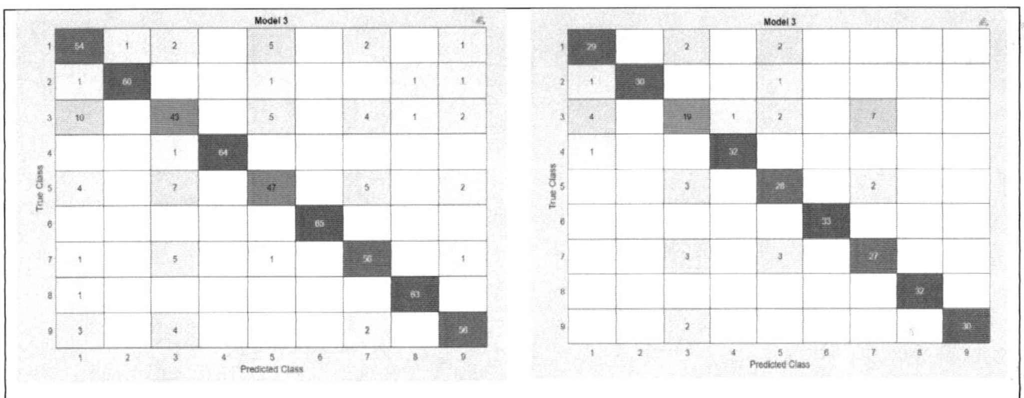

(a) (b)

Fig. 6 Performance of SVM with Grid Search optimization (a) confusion matrix in validation data (b) confusion matrix on test data

The Grid Search optimization technique was used to tune the hyperparameters of the SVM method for multiclass classification. The hyperparameter space was defined and divided into a grid, and all the points in the grid were evaluated to obtain the optimal parameters. The performance metric used for optimization was cross-validation. This approach used a linear kernel function with a box constraint level of 0.1. After 30 iterations, the minimum classification error is 0.11292. The proposed method was validated and evaluated using 20% and 10% of the datasets for validation and testing, respectively. As a result, the achieved validation accuracy was 87.3% and the testing accuracy was 88.4%. Figure 5 presents the Iteration Vs. The classifier Error graph and confusion matrix of the SVM Classifier with Grid Search Optimization The confusion matrix of the Grid Search optimization technique are presented in Figure 6.

The random search optimization technique selects hyperparameters by sampling them randomly from a defined search space and evaluating the model's performance using cross-validation. The advantage

IOP Conf. Series: Materials Science and Engineering 1291 (2023) 012048 doi:10.1088/1757-899X/1291/1/012048

of random search is that it can explore a wider range of hyperparameters than a grid search. Our next approach for improving the accuracy of wafer defect detection is to use a random search optimization technique with an SVM classifier.

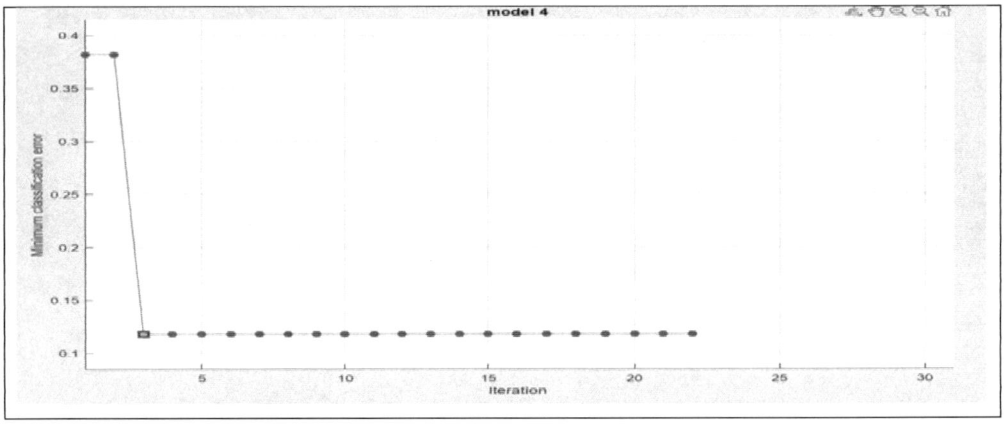

Fig. 7 Iteration Vs. Classifier Error graph and confusion matrix of SVM Classifier with Random Search Optimization

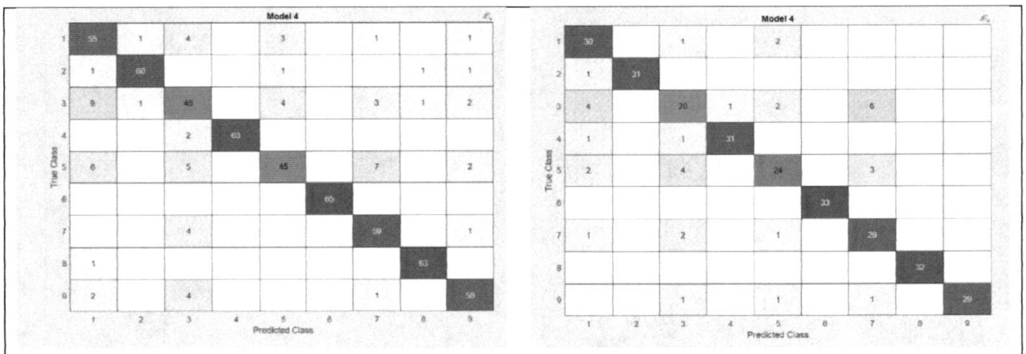

(a) (b)

Fig. 8 Performance of SVM with Random Search optimization (a) confusion matrix in validation data (b) confusion matrix on test data

Random Search hyperparameter tuning of the SVM method uses the one-vs-one approach for multi-class classification. The box constraint level for this approach is 0.062075, and the linear kernel function is used. Validation and testing were performed on the dataset to evaluate the performance of the proposed approach. After 30 iterations, the minimum classification error is 0.11804. The validation accuracy achieved was 88.1%, and the testing accuracy was 88.1% when 20% and 10% of the dataset were used for validation and testing, respectively. The confusion matrix for this approach is shown in Fig. 6.

IOP Conf. Series: Materials Science and Engineering 1291 (2023) 012048 doi:10.1088/1757-899X/1291/1/012048

Table 1. shows the comparison between three different optimization techniques.

Sl. No.	Optimization Technique	Validation Accuracy	Test Accuracy
1.	Bayesian	87.8%	85%
2.	Grid Search	87.3%	88.4%
3.	Random Search	88.1%	88.1%

Conclusion

In this paper, we propose a solution for classifying defects in monocrystalline silicon wafers using Resnet 101 deep features. We also explore the use of Bayesian optimization, Grid Search (GSGS), and Random Search (RSRS) techniques to fine-tune the hyperparameters of Support Vector Machines (SVMs) for accurate defect detection. This approach aims to enable semiconductor engineers to automatically classify wafers and diagnose defects early without relying on specialized or empirical knowledge. Based on the results presented in Table 1, the test accuracy achieved by Bayesian optimization is lower than the validation accuracy, indicating that the model may overfit the training data. Grid Search, on the other hand, has a larger difference between the test accuracy and validation accuracy compared to the Random Search optimization technique. Furthermore, Random Search showed no difference between test accuracy and validation accuracy. Therefore, Random Search is the preferred optimization technique for adjusting the hyper-parameters of the SVM using the deep features of Resnet 101 to detect defective wafers.

Funding

This research is not funded by any organization

Ethical Statement

In this research, there is no direct/ indirect involvement of any human beings or animals.

Disclosure of Potential Conflicts of Interest

The authors have no conflict of interest towards the publication of this research.

References

[1] A. Ganum, D. N. F. Iskandar, L. P. Chin, & A. H. Fauzi, Detection of Monocrystalline Silicon Wafer Defects Using Deep Transfer Learning, Journal of Telecommunications and Information Technology. (2022)

[2] M. Saqlain, Q. Abbas and J.Y. Lee, A deep convolutional neural network for wafer defect identification on an imbalanced dataset in semiconductor manufacturing processes, IEEE Transactions on Semiconductor Manufacturing. 33 (2020) 436-444.

[3] J.C. Chien, M.T. Wu and J.D. Lee, Inspection and classification of semiconductor wafer surface defects using CNN deep learning networks, Applied Sciences, 10 (2020) 5340.

[4] C.Phua and L.B.Theng, Dladc: Deep learning based semiconductor wafer surface defects recognition, IAENG International Journal of Computer Science, 49(2022) 20-30.

[5] D. Morits, M.R. Piton and T. Laakko, AI Machine Vision System for Wafer Defect Detection. In Industrial Artificial Intelligence Technologies and Applications, River Publishers, 2022, pp. 73-80.

[6] S. Misra, D. Kim, J. Kim, W. Shin and C. Kim, A voting-based ensemble feature network for semiconductor wafer defect classification. Scientific Reports, 12 (2022) 16254.

[7] Y. Yuan-Fu, A deep learning model for identification of defect patterns in semiconductor wafer map, In 2019 30th Annual SEMI Advanced Semiconductor Manufacturing Conference, 2019, pp. 1-6.

[8] B. Devika and N. George, N., Convolutional neural network for semiconductor wafer defect detection. In 10th International Conference on Computing, Communication and Networking Technologies, 2019, pp. 1-6.

[9] H. Lee, Y. Kim and C.O. Kim, A deep learning model for robust wafer fault monitoring with sensor measurement noise. IEEE Transactions on Semiconductor Manufacturing, 30 (2016) 23-31.

[10] P.P. Shinde, P.P. Pai and S.P. Adiga, Wafer defect localization and classification using deep learning techniques, IEEE Access, 10 (2022) 39969-39974.

[11] R. Baly and H. Hajj, Wafer classification using support vector machines, IEEE Transactions on Semiconductor Manufacturing, 25(2022) 373-383.

[12] C. Pao-Hua, W. Menq-Jiun and C. Kuang-Ku, Integrating support vector machine and genetic algorithm to implement dynamic wafer quality prediction system, Expert Systems with Applications 37(2022) 4413-4424.

[13] Q. Li, Z. Luo, H. Chen and C. Li, C., An overview of deeply optimized convolutional neural networks and research in surface defect classification of workpieces, IEEE Access,10 (2022) 26443-26462.

[14] I. Syarif, A. Prugel-Bennett and G. Wills, G., SVM parameter optimization using grid search and genetic algorithm to improve classification performance, TELKOMNIKA Telecommunication Computing Electronics and Control, 14(2016),1502-1509.

[15] P. Ghosal, L. Nandanwar, S. Kanchan, A. Bhadra, J. Chakraborty and D. Nandi, D., Brain tumor classification using ResNet-101 based squeeze and excitation deep neural network. In Second International Conference on Advanced Computational and Communication Paradigms, 2019, pp. 1-6.

Metamaterial Based Wearable Antenna with Low SAR

Mehaboob Mujawar [a*], Kaveri Basarakod [b], Aishwarya Kalal[c] and Bhagyashree Goudar [d]

Bearys Institute of Technology, Mangalore, Karnataka, India

[a]mehaboob311134@gmail.com, [b] kaverikavya494@gmail.com, [c] aishwaryark80@gmail.com, [d]bhagyashreegoudar123@gmail.com*

Abstract. In this paper, Wearable antenna based on metamaterial (MTM) with a low specific absorption rate (SAR) has been developed. The suggested antenna design has four metamaterial unit cells underneath the radiating element in the same plane as the ground, together with partial ground. Rectangular patches with I-shaped slots are used to create the MTM unit cells. The jeans fabric is positioned on top of the metamaterials. The MTM unit cells were significant in reducing SAR. Through comparative simulated analysis, such as reflection coefficient, VSWR, radiation patterns, and other antenna parameters, HFSS software presents the performance of the suggested antenna

Keywords: Metamaterial, Wearable, SAR, Gain, Return loss, Wi-Fi

Introduction

Artificial structures known as metamaterials are made of regular or irregular subwavelength macrocells. The ability to regulate the effective medium properties of metamaterials is one of their advantages. This is achieved by creating the macro units to create unique materials that do not exist in nature. The dimensions and bandwidth of the metamaterial antenna design are five times smaller. In contrast to conventional phased array antenna, it does not require active phase shifters or amplifiers. Wide angle scanning and superior beam performance are both offered. It features pointing and polarization that may be manipulated electronically. It uses incredibly little power. Metamaterial Antennas are compact, flat, and light in weight. Instead of using mechanical components, it steers using software [10]. Planar integration with other components is possible. Properties of metamaterials are obtained from their physical structures rather than their chemistry. They display characteristics not found in nature. They display qualities not seen in the components that make up their composition. A composite material that has been specifically designed to perform better. They display qualities that are not possible to achieve with regular material. The ratio between the capacitance of a capacitor utilizing that substance as its dielectric and a capacitor with a vacuum-based dielectric is known as relative permittivity. A composite medium with simultaneously negative permittivity and permeability had been postulated by D.R. Smith [4]. They present a composite medium that exhibits a frequency region in the microwave regime with simultaneously negative values of effective permeability and permittivity. This composite medium is based on a periodic array of interspaced conducting non-magnetic split ring resonators and continuous wires. Wearable refers to anything that can be worn on the body. The concept "wearable electronics" is relatively new, but it has started to gain traction among techies over the past ten years. A wearable electronics device is one that has intelligence, can process information, and can produce useful output. Products with wearable electronics that incorporate computing and electronic technologies into their functionality. The wearable antenna for the body area networks are used as off body/ on body communication, which find wide applications like Health care monitoring, military applications, personal Communications RFID tags etc. For the wearable technology the antenna should be flexible to mount on any part of the body. It is very important for the designer to design the wearable appropriately [5]. The main procedural steps to obtain a better performance wearable antenna include, Identification of the best design, checking for grip of the wearable, Ruggedness, developing wearables prototypes, Testing and verification, Selection of appropriate antennas and Flexibility of antenna. A wearable antenna uses two types of materials, namely conductive and substrate. The conductive materials for antenna depends on the thickness and conductivity parameters. Whereas, Substrate materials for antenna depends on dielectric constant and loss tangent parameters. Wearable material should ideally have Light weight [9]. The interactions between the lossy human body tissues and wearable antennas cause changes in the reflectivity and radiation properties of the antennas. Second, the type (skin, fat, muscle, etc.) and working frequency of human tissues affect their permittivity [6]. Finally, varying permittivity and conductivity levels primarily alter reflection coefficients, which in turn affect how much power is absorbed by the body. This results in a decrease in the antenna's radiated efficiency.

Literature survey

In this paper [1] the antenna is designed using polyester substrate. The use of copper material on the substrate is minimized to a great extent. The reduction in the size of the antenna results in the increase in resonant frequency. They have reduced the use of copper material and hence size of antenna is reduced. This antenna is designed for IEEE 802.11a wireless local area network applications. The substrate thickness is 1mm and dielectric constant is 1.39. copper tape with thickness of 0.1mm is used. The feedpoint location is 11mm in the x direction and 7mm in the y direction. Under unloaded condition, this antenna is resonating at 8.97 GHz and reflection coefficient is -19.5 dB. Under loaded condition, this antenna is resonating at 5.10 GHz and reflection coefficient is -24 dB. The suggested antenna's measured reflection coefficient is -30 dB. The suggested antenna has a 4 dBi gain. Using PVC pipe, the impact of bending on the performance of the antenna was also examined. The antenna's return loss was measured to be - 18 dB when bent in a 54.5mm radius. The antenna's return loss was measured to be - 21 dB when bent in a 44.5mm radius. Wearable microstrip patch antenna with interdigital capacitor and rectangular stub is presented in this research [2]. Inductance is produced by a rectangular stub, and capacitance by an interdigital capacitor. The ground plane is soldered properly to the polyester substrate's opposite side, which is where the antenna is constructed. The polyester substrate's utilised thickness is 3.14 mm, and its dielectric constant is 1.39. The copper tape has a 0.1 mm thickness. The feedpoint is situated at a distance of 7mm in the Y direction and 13mm in the X direction. At 2.45 GHz, the antenna's simulated reflection coefficient is -25 dB under no load. At 2.54 GHz, the antenna's observed reflection coefficient is -32 dB. The filters have been used to remove the unwanted bands. The proposed antenna offers a good gain. In this paper [3], they have designed aesthetic pattern antenna. They have used 3 petals and in the middle circular patch is placed, circular patch is used to connect 3 petals [8]. Spline curve method available in the CST software is used to design the petals. Depending on the requirements the ground structure can be changed. Partial ground has been used in order to increase the bandwidth. The substrate thickness has been increased by stacking one more substrate upon the another substrate. They have used two substrates since the available substrate is only 1mm thickness. By increasing the substrate thickness, the antenna bandwidth and gain has been improved. The SAR calculation performed with antenna placements at 0mm (touching the skin), 1mm and 3mm away from the body part or using phantom model.

Antenna design

Steps for Designing the Antenna in HFSS Software:
In step 1, create ground plane by using the rectangle which is the 2D and name it as ground. In step 2, After creating ground plane place a dielectric substrate, draw a box of same size and extend in the z-direction. In step 3, Click on the box substrate, jeans substrate will be used in our design. over this click

on create substrate dimensions are ok. But Z dimension has to be 1.6, click on ok, can be rotated to see the plane. In step 4, Over the substrate a patch has been placed with specified length and width, patch has to be placed on the top. field line impedance of 50 ohm is provided. In step 5, Once everything has been done. Then the design need to be validated. First save the design go to HFSS and click on validation check. In step 6, Go to the simulation and click on analysis and after that progress window can be seen, so once the analysis is completed, results can be seen and possible to find maximum gain. where it is occurring and at what frequency [7]. The proposed antenna design using HFSS software has been shown in figure 1 and 2. Figure 3 and 4 shows the fabricated proposed antenna design. The dimensions of the proposed antenna are listed in table.1 below.

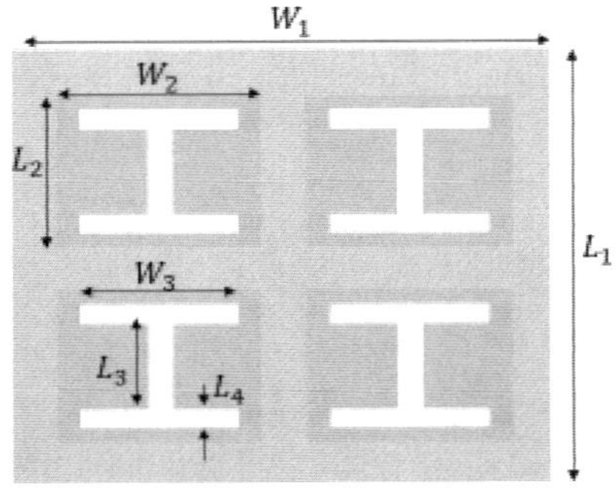

Fig.1: Bottom view of the proposed antenna

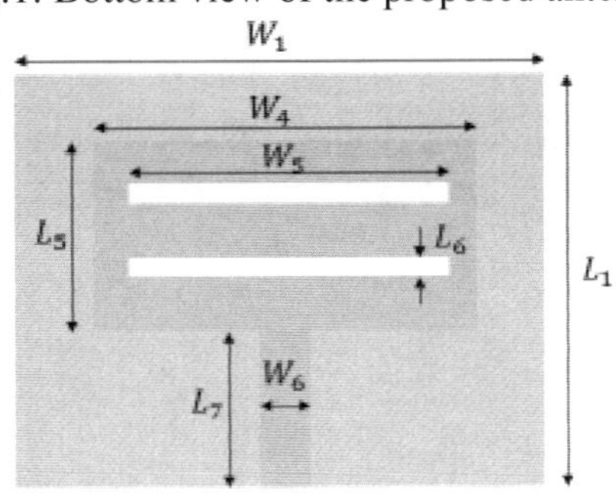

Fig.2: Top view of the proposed antenna

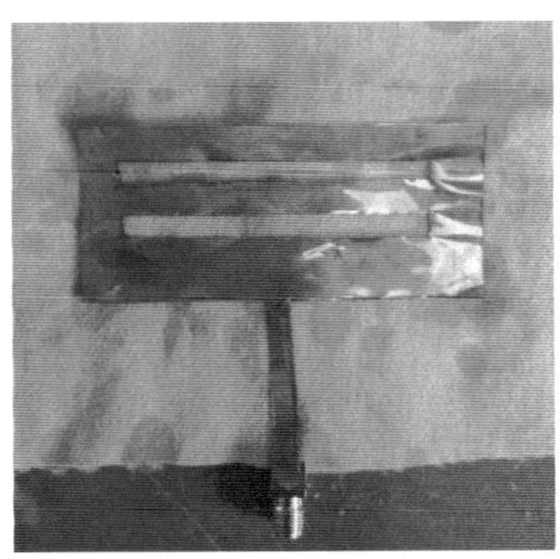

Fig.3: Top view of the fabricated proposed antenna

Table.1: Dimensions of the proposed antenna

Dimensions	Unit (mm)	Dimensions	Unit (mm)
W_1	28	L_1	27
W_2	11	L_2	9
W_3	10	L_3	5.4
W_4	25	L_4	1.2
W_5	24	L_5	14
W_6	1.2	L_6	0.6
-	-	L_7	9

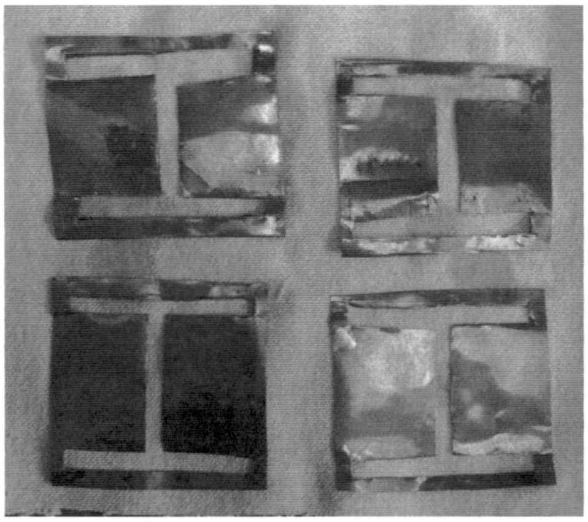
Fig.4: Bottom view of the fabricated proposed antenna

Results

As can be seen in fig.5, the reflection coefficient of the antenna for free space is -27.2 dB and the reflection coefficient of the antenna for on body is -26 dB. Figure 6, shows the VSWR characteristics of the proposed antenna. The VSWR of the antenna for free space is 1.15 and the VSWR of the antenna for on body is 1.09. The SAR value of the proposed antenna when placed on Phantom arm is 0.97 W/Kg. The suggested antenna design has four metamaterial unit cells underneath the radiating element in the same plane as the ground, together with partial ground. Rectangular patches with I-shaped slots are used to create the MTM unit cells. The jeans fabric is positioned on top of the metamaterials. The MTM unit cells were significant in reducing SAR. Through comparative simulated analysis, such as reflection coefficient, VSWR, radiation patterns.

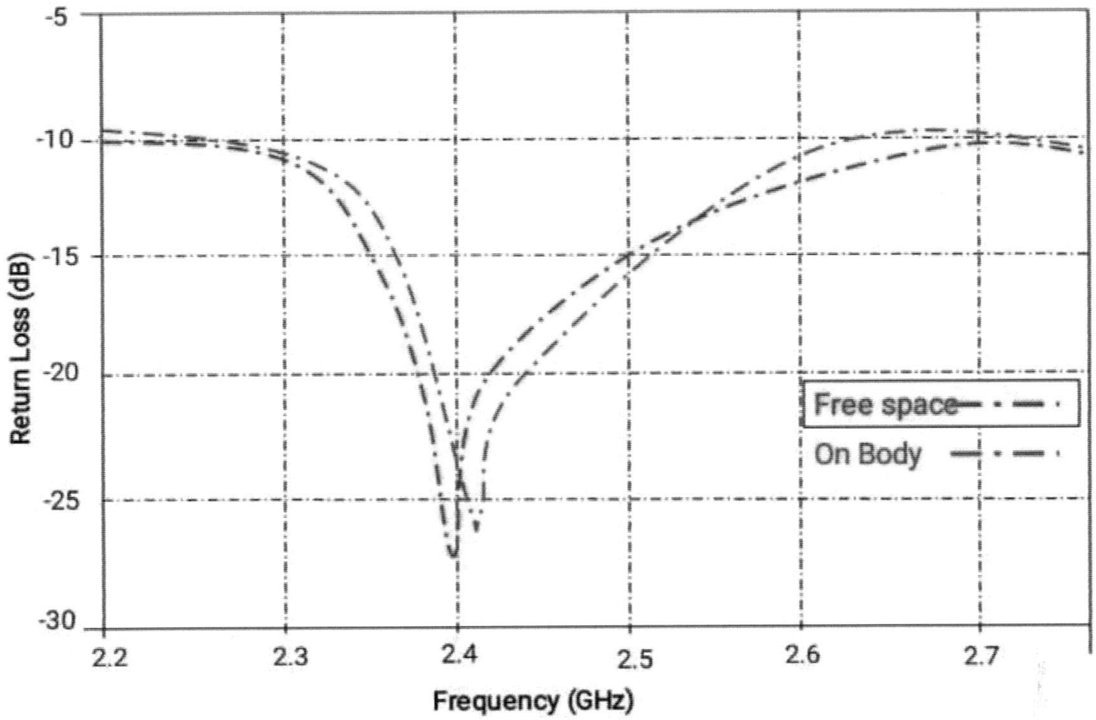

Fig.5: Reflection coefficient

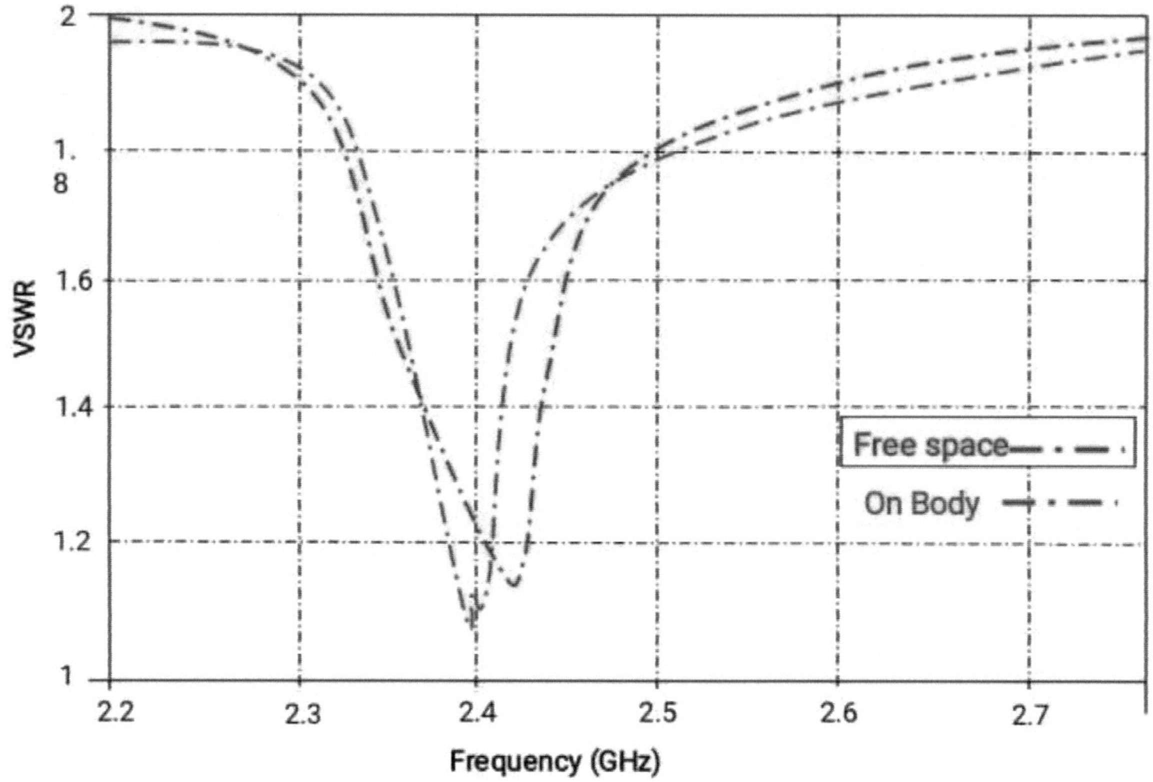

Fig.6: VSWR

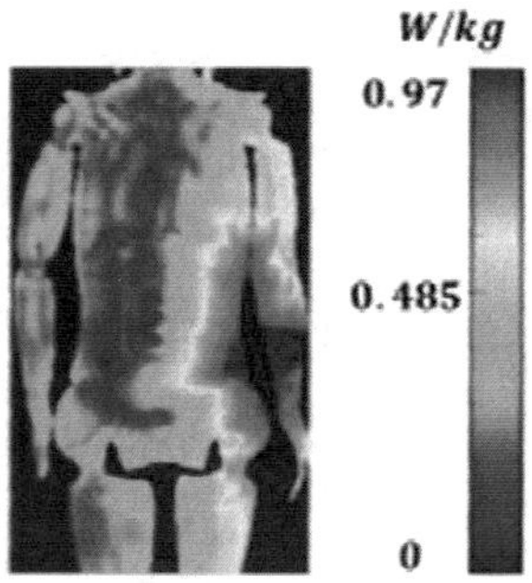

Fig.7: SAR analysis of the proposed antenna when placed on ARM

Conclusion

The simulation research demonstrated that the presence of the metamaterial unit cell improves the antenna performance and the reflection coefficient. According to FCC regulations, the SAR value has been decreased to below 1.6 W/Kg, making the developed antenna an excellent contender for wearable applications. The suggested antenna design has four metamaterial unit cells underneath the radiating element in the same plane as the ground, together with partial ground. Rectangular patches with I-shaped slots are used to create the MTM unit cells. The jeans fabric is positioned on top of the metamaterials. The MTM unit cells were significant in reducing SAR. Through comparative simulated analysis, such as reflection coefficient, VSWR, radiation patterns, and other antenna parameters.

References

[1]. Yan, S., L. A. Y. Poffelie, P. J. Soh, X. Zheng, and G. A. E. Vandenbosch, "On-body performance of wearable UWB textile antenna with full ground plane," 2016 10th European Conference on Antennas and Propagation (EuCAP), 1-4, Davos, 2016.

[2]. Ashyap AY, Zainal Abidin Z, Dahlan SH, Majid HA, Saleh G. Metamaterial inspired fabric antenna for wearable applications. International Journal of RF and Microwave Computer Aided Engineering. 2019;29(3):e21640. DOI: 10.1002/mmce.21640.

[3]. X. S. Fang and K. W. Leung, "Aesthetic transparent dielectric resonator antenna with omnidirectional radiation pattern," Proceedings of the 2012 IEEE International Symposium on Antennas and Propagation, Chicago, IL, USA, 2012, pp. 1-2, doi: 10.1109/APS.2012.6348955.

[4]. Mujawar, Mehaboob & Dommeti, Vijaya & Naz, Muhammad & Muduli, Arjuna. (2022). Design and performance comparison of arrays of circular, square and hexagonal meta-material structures for wearable applications. Journal of Magnetism and Magnetic Materials. 553. 169235. 10.1016/j.jmmm.2022.169235

[5]. Mujawar, Mehaboob. (2021). Compact Microstrip Patch Antenna Design with Three I-, Two L-, One E- and One F-Shaped Patch for Wireless Applications. 10.1201/9781003093558-7.

[6]. Mujawar, Mehaboob & Thangavel, Gunasekaran. (2022). Multiband Slot Microstrip Antenna for Wireless Applications. 10.1007/978-3-030-76636-8_3

[7]. Mujawar, Mehaboob. (2022). Arrow Shaped Dual-Band Wearable Antenna for ISM Applications. 10.1002/9781119792581.ch8.

[8]. Mujawar, Mehaboob. (2021). Antenna Array Design for Massive MIMO System in 5G Application. 10.1201/9781003175155-18

[9]. K. Li, C. Zhu, L. Li, Y. -M. Cai and C. -H. Liang, "Design of Electrically Small Metamaterial Antenna With ELC and EBG Loading," in IEEE Antennas and Wireless Propagation Letters, vol. 12, pp. 678-681, 2013, doi: 10.1109/LAWP.2013.2264099.

[10]. S. A. H. Saghanezhad and Z. Atlasbaf, "Miniaturized Dual-Band CPW-Fed Antennas Loaded With U-Shaped Metamaterials," in IEEE Antennas and Wireless Propagation Letters, vol. 14, pp. 658-661, 2015, doi: 10.1109/LAWP.2014.2376554.

AUTHOR INDEX

Institute of Physics
Dirac House, Temple Back
Bristol BS1 6BE UK

ISSN: 1757-8981
ISBN 978-1-7138-7970-1

9 781713 879701

CURRAN ASSOCIATES INC.
proceedings.com